Principles of Physiology for the Anaesthetist

Principles of Physiology for the Anaesthetist

Second edition

Ian Power
Professor
Anaesthesia, Critical Care and Pain Medicine
University of Edinburgh, The Royal Infirmary,
Edinburgh, UK

Peter Kam
Professor
Nuffield Professor of Anaesthetics
University of Sydney, Royal Prince Alfred Hospital,
Camperdown, NSW, Australia

HODDER ARNOLD
AN HACHETTE UK COMPANY

First published in Great Britain in 2001 by
Hodder Arnold, an imprint of Hodder Education, an Hachette UK
Company, 338 Euston Road, London NW1 3BH

http://www.hoddereducation.com

Distributed in the United States of America by
Oxford University Press Inc.,
198 Madison Avenue, New York, NY10016
Oxford is a registered trademark of Oxford University Press

Hachette UK's policy is to use papers that are natural, renewable and
recyclable products and made from wood grown in sustainable forests.
The logging and manufacturing processes are expected to conform to the
environmental regulations of the country of origin.

Whilst the advice and information in this book are believed to be true
and accurate at the date of going to press, neither the authors nor the
publisher can accept any legal responsibility or liability for any errors or
omissions that may be made. In particular, (but without limiting the gen-
erality of the preceding disclaimer) every effort has been made to check
drug dosages; however it is still possible that errors have been missed.
Furthermore, dosage schedules are constantly being revised and new
side-effects recognized. For these reasons the reader is strongly urged to
consult the drug companies' printed instructions before administering
any of the drugs recommended in this book.

British Library Cataloguing-in-Publication Data
A catalogue record for this book is available from the British Library

Library of Congress Cataloging-in-Publication Data
A catalog record for this book is available from the Library of Congress

ISBN   978-0-340-88799-8

2 3 4 5 6 7 8 9 10

Commissioning Editor: Sarah Burrows
Project Editor: Francesca Naish
Production Controller: Lindsay Smith
Cover Design: Helen Townson

10/12 Minion by Macmillan Publishing Solutions
(www.macmillansolutions.com)
Printed and bound in Malta

What do you think about this book? Or any other Hodder Arnold
title? Please visit our website: www.hoddereducation.com

Contents

Contributors

Professor Ian Power BSc (Hons) MD FRCA FFP-MANZCA FANZCA FRCS Ed FRCP Edin
Professor, Anaesthesia, Critical Care and Pain Medicine, University of Edinburgh, The Royal Infirmary, Edinburgh, UK

Professor Peter Kam MBBS, MD, FRCA, FANZCA, FCARCSI, FHKCA(Hon)
Nuffield Professor of Anaesthetics, University of Sydney, Royal Prince Alfred Hospital, Camperdown, NSW, Australia

Professor Michael J. Cousins MBBS, MD, FANZCA, FRCA, FFPMANZCA, FAChPM (RACP), DSc
Department of Anaesthesia and Pain Management, University of Sydney, Royal North Shore Hospital, St Leonards, NSW, Australia

Associate Professor Philip J. Siddall MBBS, PhD, FFPMANZCA
Department of Anaesthesia and Pain Management, University of Sydney, Royal North Shore Hospital, St Leonards, NSW, Australia

Preface to the first edition

The origin of this book lies in the many physiology teaching sessions we have held between us at various times in Australia, New Zealand, Hong Kong, Malaysia, Singapore and the United Kingdom for trainees attending preparatory courses for fellowship examinations in anaesthetics. One common lament of the participants, who were usually much too kind to criticize our modest teaching skills, was that there was a real need for a single text covering physiology for anaesthetists. Often the comparison was made with pharmacology, which, in contrast, was held to be well provided with textbooks for anaesthetists. We both gradually came to the same appreciation that there may well be some need for a succinct book presenting the understanding of physiological principles and the factual knowledge necessary for the practice of anaesthesia. Given the encouragement of our students and colleagues, we have attempted to produce a clear and concise presentation of physiology for anaesthetists. It is not our intention that this text should be used instead of standard physiology reference books; rather we seek to provide an overview of the subject in a single volume that will be useful for those preparing for postgraduate examinations. Whilst the title specifies anaesthetists, we hope that it might also contain some material of value for students in related areas.

We wanted to write most of the book and prepare the diagrams ourselves. However, we were enthusiastic to give prominence to the rapid advances made in the understanding of pain physiology and are delighted by the excellent chapter presented to us by our valued colleagues in Sydney, Professor Michael Cousins and Dr Philip Siddall, who are both recognized experts on this subject. We are especially grateful to the pharmaceutical companies Abbott, AstraZeneca, CSL and Roche in Australia for their generous financial support towards the artistic work involved in the preparation of the diagrams in this volume.

Our own enthusiasm for teaching physiology continues to be renewed by the interest and appreciation expressed by our many students in different countries, and we would like to dedicate this book to them.

Ian Power
Peter Kam

Preface to the second edition

We were delighted with the success of the first edition and hope that the second will be as well received. We have taken the opportunity to revise the entire text, pruning detail and repetition in some areas and updating the chapters. The cellular theory of coagulation and Stewart's physico-chemical theory of acid base disturbances have been included in the relevant chapters.

All chapters begin with learning objectives. A major feature of this second edition is the "Reflections" section at the end of each chapter that provides the essential points and a concise summary of the chapter. Another new addition is a summary of important equations at the end of the book. These new features should help the hard-pressed student at examination time.

New material should not be allowed to obscure basic concepts and their application in anaesthetic practice. The second edition of *Principles of Physiology for the Anaesthetist* remains a useful revision text and an invaluable reference for basic and applied physiology for anaesthetists and critical care physicians throughout their training and later in their practice.

Peter Kam
Ian Power

Abbreviations and measurements

ACh	acetylcholine	CRF	corticotrophin-releasing factor
ACTH	adrenocorticotrophin	CRGP	calcitonin gene-related peptide
ADC	antibody-dependent cytotoxic	CSF	cerebrospinal fluid; *also*, colony-stimulating factor
ADH	antidiuretic hormone (vasopressin)	CTZ	chemoreceptor trigger zone
ADP	adenosine diphosphate	CVLM	caudal ventrolateral medulla
AEP	auditory-evoked potential	DAG	diacylglycerol
AHG	anti-human globulin	DHEA	dehydroepiandrosterone
ALA	aminolaevulinic acid	2,3-DPG	2,3-diphosphoglycerate
AMH	A-mechanoheat (receptors)	DRG	dorsal root ganglion
AMP	adenosine monophosphate	ECF	extracellular fluid
AMPA	α-amino-3-hydroxy-5-methyl-4-isoxazolepropionic acid	ECG	electrocardiogram
ANF	atrial natriuretic factor	EDRF	endothelium-derived relaxing factor
AP-1	activator protein 1	EEG	electroencephalogram
ASIC	acid-sensing ion channel	ELISA	enzyme-linked immunosorbent assay
ATP	adenosine triphosphate	EPP	end-plate potential
AV	atrioventricular	EPSP	excitatory postsynaptic potential
BAEP	brainstem auditory-evoked potential	ER	endoplasmic reticulum
BCDF	B cell differentiation factor	ERV	expiratory reserve volume
BER	basic electrical rhythm	ESPV	end-systolic pressure–volume
BFU-E	burst-forming unit-erythroid	FAD	flavin adenine dinucleotide
BK	bradykinin	FFA	free fatty acid
BMR	basal metabolic rate	fMRI	functional magnetic resonance imaging
BNP	brain natriuretic peptide	FRC	functional residual capacity
cAMP	cyclic adenosine 3, 5-monophosphate	FSH	follicle-stimulating hormone
CBF	cerebral blood flow	GABA	gamma-aminobutyric acid
CCK	cholecystokinin	G-CSF	granulocyte-colony stimulating factor
CFU(GEMM)	colony-forming unit (granulocyte, erythrocyte, monocyte, megakaryocyte)	GFR	glomerular filtration rate
		GH	growth hormone
CGRP	calcitonin gene-related peptide	GIP	gastric inhibitory peptide
CMG	cytomegalovirus	GRP	gastrin-releasing peptide
CNS	central nervous system	GSSG	glutathione disulphide
CoA	coenzyme A	GSH	glutathione
COMT	catechol-*O*-methyl transferase	GTP	guanosine triphosphate
CPD	citrate–phosphate–dextrose	GVHR	graft-versus-host reaction
C-PMN	C-polymodal nociceptors	HAR	high-altitude resident
CPP	cerebral perfusion pressure	HCG	human chorionic gonadotrophin

HDL	high-density lipoprotein	NFP	net filtration pressure
HIV	human immunodeficiency virus	NGF	nerve growth factor
HLA	human leukocyte antigen	NK	natural killer
HMG-CoA	hydroxymethylglutaryl CoA	NK-1	neurokinin-1
HMP	hexose monophosphate	NMDA	N-methyl-D-aspartate
HMWK	high-molecular-weight kininogen	NO	nitric oxide
HPL	human placental lactogen	NSAID	non-steroidal anti-inflammatory
HR	heart rate		drug
HVGR	host-versus-graft reaction	NTS	nucleus tractus solitarius
5-HT	5-hydroxytryptamine (serotonin)	PAG	periaqueductal grey
ICAM	intercellular adhesion molecule	PAH	*para*-aminohippuric acid
ICF	intracellular fluid	PDGF	platelet-derived growth factor
ICG	indocyanine green	PET	positron emission tomography
ICP	intracranial pressure	PGE	prostaglandin E
IFN	interferon	PGI_2	prostacyclin
Ig	immunoglobulin	PGO	pontogeniculo-occipital
IGF	insulin-like growth factor	PHA	phytohaemagglutinin
IL	interleukin	PHSC	pluripotential haemopoietic stem
IP_3	inositol triphosphate		cell
IPSP	inhibitory postsynaptic potential	PIP_2	phosphatidy linositol bisphosphate
IVC	inferior vena cava	PK	potassium permeability
LAP	left atrial pressure	PKC	protein kinase C
LDL	low-density lipoprotein	PLC	phospholipase C
LGL	large granular lymphocytes	PNMT	phenylethanolamine-N-
LH	luteinizing hormone		methyltransferase
LHRH	luteinizing hormone-releasing	PNS	peripheral nervous system
	hormone	PRL	prolactin
LISS	low ionic strength saline	PTH	parathyroid hormone
LOS	lower oesophageal sphincter	PVA	pressure–volume area
LPH	lipotrophin	PVN	paraventricular nucleus
LTB_4	leukotriene B_4	PWM	pokeweed mitogen
LTP	long-term potentiation	RA	right auricle (atrium)
LVEDV	left ventricular end-diastolic volume	RANTES	regulated upon activation normal
LVESV	left ventricular end-systolic volume		T-cell expressed and secreted
MAC	membrane attack complex	RAST	radioallergosorbent test
MALT	mucosa-associated lymphoid tissue	RBC	red blood cell (erythrocyte)
MAP	mean arterial (blood) pressure	REM	rapid eye movement
MAO	monoamine oxidase	RER	respiratory exchange ratio
MCP	monocyte chemotactic peptide	RIA	radioimmunoassay
M-CSF	macrophage-colony stimulating	RQ	respiratory quotient
	factor	RV	right ventricle
MEPP	miniature end-plate potential	RV	residual volume
MH	malignant hyperpyrexia	RVLM	rostral ventrolateral medulla
MHC	major histocompatibility complex	SA	sinoatrial
MMC	migrating motor complex	SAGM	saline, adenine, glucose and
MOPG	3-methoxy-4-hydroxy phenyl glycol		mannitol
MSFP	mean systemic filling pressure	SBP	sulphobromophthalein
NAD	nicotinamide adenine dinucleotide	SCF	stem cell factor
NADP	nicotinamide adenine dinucleotide	SCUBA	self-contained underwater
	phosphate		breathing apparatus

SDA	specific dynamic action
SEP	somatosensory evoked potential
SID	strong ion difference
SON	supraoptic nucleus
SP	substance P
SPA	staphylococcal protein A
SVC	superior vena cava
TBG	thyroxine-binding globulin
TBPA	thyroxine-binding pre-albumin
TCR	T-cell receptor
TEA	tetra-ethylammonium
TEC	triethylcholine
TGF-β	transforming growth factor-β
TLC	total lung capacity
TNF	tumour necrosis factor
tPA	tissue plasminogen activator
TRF	thyrotrophin-releasing factor
TRH	thyrotrophin-releasing hormone
Trk	tyrosine kinase

TRP	transient receptor potential
TRPV	vanilloid receptor-related TRP
TSH	thyrotrophin
TSP	thrombospondin
TV	tidal volume
VDCC	voltage-dependent calcium channels
VIP	vasoactive intestinal peptide
VLDL	very low-density lipoprotein
VMA	vanillylmandelic acid
VMN	ventromedial nucleus
VEP	visual-evoked potential
vWF	von Willebrand factor
WDR	wide dynamic range

UNITS OF MEASUREMENT

7.5 mmHg	=	1 kPa (kilopascal)
	=	10.2 cmH$_2$O
1 kilocalorie	=	4.187 kJ

Physiology of excitable cells

LEARNING OBJECTIVES

After studying this chapter the reader should be able to:

1. Describe the nature of the cell membrane and the ionic basis of the resting membrane potential
2. Describe the nerve action potentials in nerves, cardiac cells and muscle cells
3. Describe the propagated action potential
4. Understand ion channels that are voltage gated: sodium, potassium and calcium channels
5. Describe neurotransmitters and receptors, including ion channels, G proteins and second messengers
6. Understand the structure and function of the specialized connection between nerves and striated muscle (the neuromuscular junction), with comprehension of the mechanism of neuromuscular transmission
7. Understand the different structures of skeletal, cardiac and smooth muscle
8. Describe the role of muscle spindles, Golgi tendon organs and spinal reflexes
9. Understand the classification and mechanisms of activation of sensory receptors, including those for pain

INTRODUCTION

An excitable cell responds to a stimulus by a rapid change in the electrical charge of the cell membrane. Nerve, muscle and sensory cells can be included in this classification, as the excitability of these tissues lies in the changing electrical properties of their membranes.

This chapter examines the physiology of some excitable cells. First, the structure of the cell membrane and the ionic basis for the resting membrane potential are examined, followed by the form of the action potential for both nerves and muscle. Next, voltage-sensitive membrane ion channels are described and their role in the action potential is discussed. In the next section, chemical communication between excitable cells is examined, by investigating the interaction of receptors, ion channels, G proteins and intracellular messengers. The physiology of neuromuscular transmission is then presented, followed by a discussion of skeletal, cardiac and smooth muscle cells, together with details of contractile proteins, the mechanical response to stimulation and excitation–contraction coupling. In the next section, organs involved in the regulation of muscle movement, muscle spindles and tendon organs are discussed, together with an examination of stretch reflexes. Finally, sensory receptors are classified, and their physiology is described.

THE MEMBRANE POTENTIAL

THE CELL MEMBRANE

Cell membranes (Fig. 1.1) are bilayers of phospholipids with the polar heads on the outside and the lipophilic fatty tails to the inside. This structure separates the cellular contents and the cytoplasm from the extracellular fluid. Integral proteins traverse the membrane, whilst peripheral proteins sit

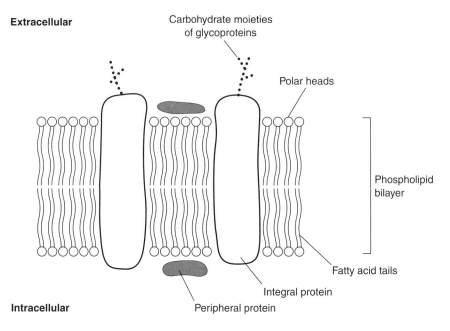

Figure 1.1　*Diagrammatic representation of the cell membrane.*

on the cytoplasmic and extracellular surfaces. Peripheral proteins on the inside and outside have different functions; glycoproteins have their sugar residues on the outside and transporting enzymes bind adenosine triphosphate (ATP) on the inside.

The cell membrane can be thought of as a layer of insulation covered on both sides by conducting material. This structure is traversed by protein channels that determine ionic permeability and the resultant electrical potential across the membrane.

THE RESTING MEMBRANE POTENTIAL

Cell membranes have an electrical charge across them with the inside being 70–80 mV negative with respect to the outside (Fig. 1.2).

This electrical potential depends on two main factors: the selective membrane permeability to ions, and the different ionic concentrations on the inside and outside of the cell. The former is the more important factor, as it can change. At rest, most membranes are permeable to potassium but not sodium ions. Membranes can be described as being semipermeable to ions; potassium passes across them with ease, but most others do not (Fig. 1.3).

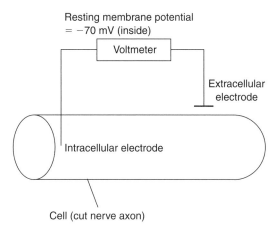

Figure 1.2　*The resting membrane potential.*

The potassium concentration is much higher inside (150 mM) than outside (5 mM) the cell. Potassium diffuses down this concentration gradient out of the cell, but this movement cannot continue indefinitely as it is opposed by electrical forces. As the membrane is impermeable to other cations, the movement out of positively charged potassium ions produces a negative charge on the inside of the cell. This charge tends to oppose and prevent further movement of potassium out of the

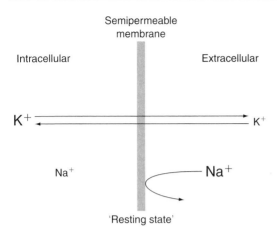

Figure 1.3 *Membrane permeability to sodium and potassium ions. (The chemical symbol size indicates the relative ionic concentrations of K^+, Na^+.)*

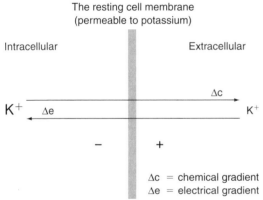

Figure 1.4 *Equilibration of potassium ions across the cell membrane – the electrochemical gradient.*

cell. Therefore, there are opposing chemical and electrical gradients acting on potassium ions (chemical – out; electrical – in). At the resting membrane potential of around $-70\,mV$, the electrical and chemical gradients acting on potassium ions balance. That is, the resting membrane potential is due to the equilibrium of the electrochemical gradients affecting potassium ions (Fig. 1.4).

The Na^+–K^+ pump which transports three sodium ions to the outside for every two potassium ions pumped to the inside causes a continual loss

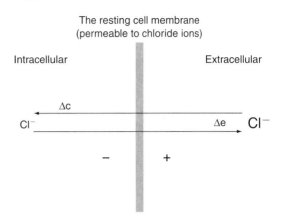

Figure 1.5 *Passage of chloride ions across the cell membrane.*

of positive ions from inside the membrane and therefore is electrogenic because it produces a net deficit of positive charges inside the cell. This causes an additional negative charge of about $-4\,mV$ inside the cell membrane.

It is important to realize that only a small number of potassium ions have to move to produce the resting potential. For a membrane potential of $-70\,mV$, the difference between the number of positive and negative charges inside the cell is only 0.000002 per cent.

Proteins are the main intracellular anions, but they also have little effect as they cannot traverse the membrane. The primary extracellular ion – chloride – can pass freely across the membrane (Fig. 1.5), but the resultant potential is similar to that set up by potassium. This can be demonstrated by considering the electrochemical forces on chloride ions.

Other cations – sodium included – contribute relatively little to the resting potential as the membrane is quite impermeable to them. If the membrane was permeable to sodium, then the potential would be completely reversed as the electrochemical gradient is the opposite to that of potassium (Fig. 1.6).

The Nernst equation

This calculates the potential difference that any ion would produce if the membrane was permeable to it (see below). The actual potential will

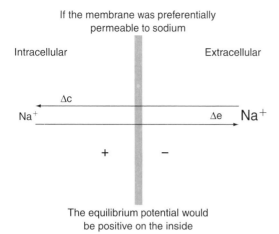

Figure 1.6   *Passage of sodium ions across the cell membrane.*

Table 1.1   *The Nernst potential of potassium, chloride and sodium ions*

Ion	Ionic concentration (mM)		≈Nernst potential (mV)
	Intracellular	Extracellular	
K^+	150	5	−90
Cl^-	10	125	−70
Na^+	15	150	+60

$$E = \frac{RT}{zF} \ln \frac{[ion]_o}{[ion]_i}$$

E = the equilibrium potential for a specific ion (inside of the cell with respect to the outside)

R = gas constant ($8.314\, J\, deg^{-1}\, mol^{-1}$)

T = absolute temperature (degrees Kelvin = 273 + degrees centigrade)

F = Faraday's constant (96 500 coulombs/ mol)

z = ionic valency (+1 for K^+, Na^+; −1 for Cl^-)

ln = logarithm to base e

o = ionic concentration outside the cell

i = ionic concentration inside the cell

The equation can be simplified:

$$E = 58 \log_{10} \frac{[ion]_o}{[ion]_i}\, mV$$

For example, for potassium:

$$E_K = 58 \log_{10} \frac{[5]}{[150]}\, mV\; (\text{approximate}$$
$$\text{concentrations})$$
$$= 58 \times -1.48$$
$$= -86\, mV$$

The Nernst equation

only be similar to the calculated Nernst potential if the membrane is permeable to that ion. At rest, the calculated Nernst potentials for potassium and chloride are similar to the real potential (Table 1.1), as these ions diffuse across the membrane with ease. This is not true for sodium as the resting membrane is relatively impermeable to this ion.

The Nernst equation was derived by proposing that the work required to move an ion across the membrane against firstly the concentration and secondly the electrical gradients will be equal at the resting potential, when the electrochemical forces balance.

Do other ions have any effect on the resting potential?

The immediate answer is, yes. If a nerve is bathed in different concentrations of potassium while the

membrane voltage is measured, it is found that the resting potential is lower (less negative) than the value calculated from the Nernst equation. The reason for this is that the membrane is not completely impermeable to sodium, and this reduces slightly the resting potential. The sodium permeability is much less than potassium, and the Nernst equation for the resting membrane can be reconstructed to account for this as shown here:

$$E_m = 58 \log_{10} \frac{[K^+]_o + 0.01[Na^+]_o}{[K^+]_i + 0.01[Na^+]_i}\, mV$$

Calculation of the resting membrane potential (E_m) using the Nernst equation

Membrane permeability

The membrane permeability is the central factor and the relative membrane permeability to different ions is important. To take account of this, the Nernst equation was expanded to the Goldman–Hodgkin–Katz form. P is the permeability to each ion; if this changes, then the membrane potential changes. This equation could be expanded to account for the effect of all ions and any changes in their membrane permeability (potassium, sodium and chloride are included in the equation):

$$
\text{Membrane permeability } (E_m) =
$$
$$
58 \log_{10} \frac{P_k[K^+]_o + P_{Na}[Na^+]_o + P_{Cl}[Cl^-]_i}{P_K[K^+]_i + P_{Na}[Na^+]_i + P_{Cl}[Cl^-]_o} \text{ mV}
$$

The Goldman–Hodgkin–Katz form of the Nernst equation

Function of the membrane potential

The presence of a membrane potential allows electrical communication between cells. For example, a motor nerve, when activated in the spinal cord, relays this information along the axon and releases transmitters by a progressive and propagated reversal of the membrane potential. Skeletal muscle contraction is produced by a propagated change in membrane potential spreading over the cell, precipitating the release of calcium from intracellular stores. Afferent nerves are activated by special sense organs and transmit this information by electrical axonal activation and release of neurotransmitters, altering the membrane potential and function of central nervous system cells. In the heart regular changes in the muscle cell ionic permeability and potential produce cardiac autorhythmicity. In all these examples alterations in the membrane potential lead to communication between cells.

ACTIVE TRANSPORT OF SODIUM OUT OF THE CELL

At the normal resting potential of -70 to $-80\,\text{mV}$, both the electrical and the chemical gradients of sodium tend to push it into the cell, despite the relative impermeability of the membrane to this

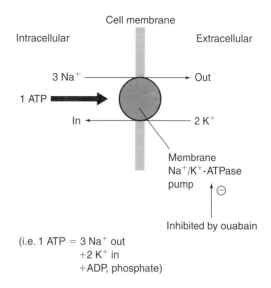

(i.e. 1 ATP = 3 Na$^+$ out
+2 K$^+$ in
+ADP, phosphate)

Figure 1.7   *The protein structure of the Na$^+$/K$^+$-ATP pump.*

ion. Some sodium does get in, and if this was not actively removed, the membrane potential would gradually diminish. This is confirmed by the observation that dinitrophenol, a metabolic inhibitor, stops active sodium extrusion, after which the membrane potential falls.

There is a membrane-bound protein pump that actively extrudes sodium in exchange for potassium ions and consumes ATP in the process. The protein consists of a large α and a smaller β subunit. The α unit contains the ATPase activity and is inhibited by ouabain. The pump straddles the membrane and is stimulated by sodium and potassium ions (Fig. 1.7). Sodium and potassium ions are not exchanged in equal ratio by the pump. Two potassium ions are taken up for every three sodium ions extruded. This means that the pump is electrogenic, contributing to the negative membrane potential by extruding more positive charges than it admits. The pump is energy-dependent, and one ATP molecule is split for every three sodium ions extruded.

The amino acid chain structure of the α-subunit of the mammalian sodium pump has been identified in considerable detail. It is embedded in and crosses the membrane repetitively (Fig. 1.8), and evidence of protein conformational changes upon binding of sodium and potassium ions may be indicative of the pumping mechanism.

CALCIUM

Calcium is used by cells as a physiological trigger, and intracellular concentrations are normally kept low: $0.1\,\mu M$ in axoplasm. For example, in muscle cells, calcium is sequestered by the sarcoplasmic reticulum and is only released during excitation–contraction coupling. As the extracellular calcium concentration is 5 mM, the chemical and electrical gradients both tend to push this ion across the membrane into cells. An increase in membrane permeability to calcium tends to depolarize cells.

There are two membrane mechanisms for removing calcium from the cell. The first is active at low internal calcium concentrations and is a membrane-bound pump activated by calmodulin; one ATP is split for each ion extruded. The second system involves the exchange of internal calcium for external sodium ions. The second mechanism is active at higher intracellular calcium concentrations, and is driven by sodium ions moving down their concentration gradient.

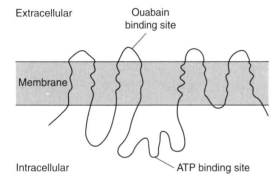

Figure 1.8   *The membrane orientation of the α subunit of the mammalian sodium pump.*

Electrical characteristics of the resting cell membrane

The resting nerve membrane has a transmembrane resistance and capacitance in parallel. This can be drawn as an electrical circuit diagram, including resistances for the external medium and the axoplasm (Fig. 1.9).

Low-intensity electrical stimuli applied to nerves produce electrotonic, or local, potentials,

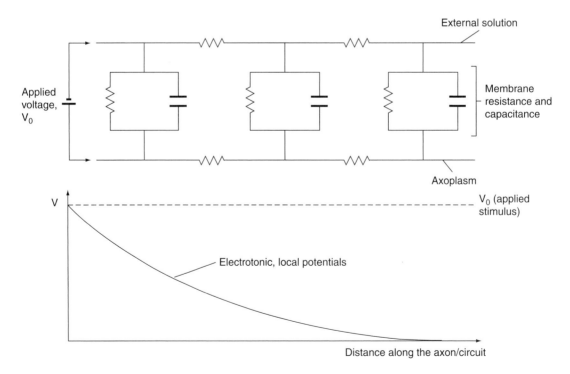

Figure 1.9   *Circuit diagram demonstrating the electrical characteristics of the resting cell membrane.*

which fall exponentially from the site of stimulus in accordance with the proposed electrical circuit model of the membrane.

THE FORM OF THE NERVE ACTION POTENTIAL

Intracellular recording from a nerve during application of stimulating voltages of increasing intensity reveals at first only local electrotonic potentials, but at a certain threshold stimulus, an action potential fires (Fig. 1.10).

During the action potential the membrane polarity reverses so that the inside has a positive charge for a short time. The stimulus threshold required is a depolarization of 10–15 mV. The electrotonic potentials depend only on the passive capacitance and resistance of the membrane, but the action potential is produced by biological changes in these basic electrical properties brought about by opening of protein channels for sodium.

The nomenclature of some aspects of the shape of the action potential can be confusing and dates from early extracellular recording techniques (Fig. 1.11).

The size of an action potential in a single nerve fibre is unaffected by stimulus intensity above the threshold. It is 'all or none'. Nerves transmit

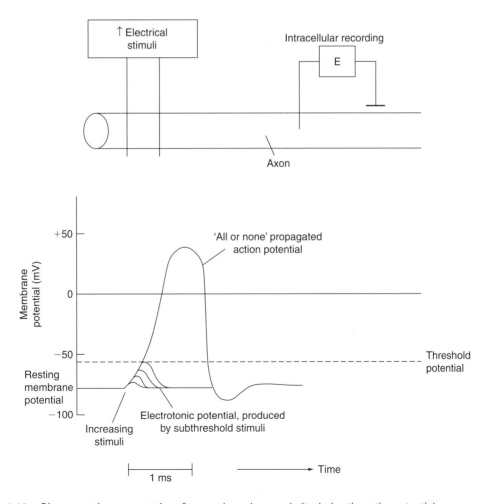

Figure 1.10 *Diagrammatic representation of reversed membrane polarity during the action potential.*

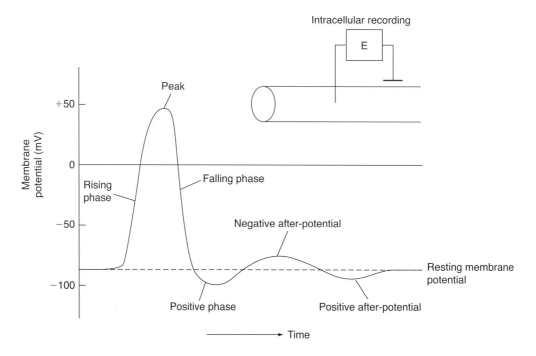

Figure 1.11   *Action potential nomenclature.*

information by the frequency, not the size, of action potentials.

THE IONIC BASIS OF THE NERVE ACTION POTENTIAL

Depolarization to the threshold potential briefly opens protein sodium channels in the membrane, allowing free movement of that ion into the cell. These are known as 'fast' sodium channels and they are blocked by tetrodotoxin and local anaesthetics. These specialized fast Na^+ channels are said to be voltage dependent because they are sensitive to the membrane potential and are activated when membrane depolarisation reaches a threshold of -55 mV. The voltage-dependent Na^+ channels have two different ionic gates that regulate the flow of Na^+ ions and can exist in three different states. At rest the activation gates are closed while the other – inactivation gate – is opened and thus the channel is closed. When the threshold potential is reached, the activation gate is opened and Na^+ ions flow unimpeded so that the channel is therefore opened. Rapidly after activation the

inactivation gate closes while the activation gate remains open and the channel is no longer permeable to Na^+ and the channels are open but inactivated. The channels are reactivated quickly but only when the membrane potential falls below -50 mV. The electrochemical forces affecting sodium mean that the membrane becomes positive on the inside. The number of sodium ions required to reverse the potential is relatively small, as was the case for potassium and the membrane at rest. During the action potential the membrane moves from the potassium to the sodium equilibrium potential and back again (Fig. 1.12). Termination of the action potential is caused by the activation of another set of voltage-sensitive channels permeable to potassium.

Another way of looking at this is to consider the effects of changes in sodium permeability on the Goldman–Hodgkin–Katz equation. Threshold depolarization increases the permeability of sodium, P_{Na}, so that it dominates the equation and the membrane potential moves towards the sodium equilibrium potential. A late increase in potassium permeability, P_K, restores the negative resting membrane potential.

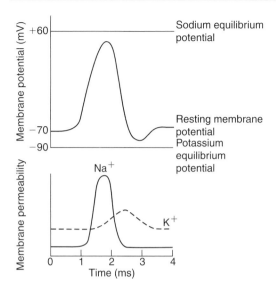

Figure 1.12 *Graph to show changes in the potassium and sodium equilibrium during the action potential.*

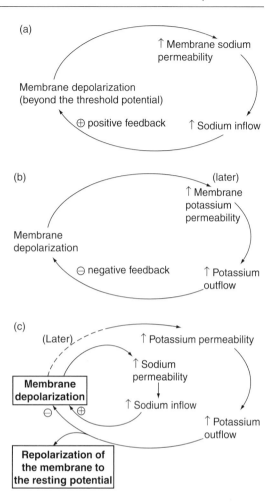

Figure 1.13 *Diagrammatic representation of 'positive and negative feedback' during the action potential. (a) Sodium permeability; (b) potassium permeability; (c) sodium and potassium permeability.*

The action potential is 'all or none' because, once the threshold has been attained, there is a 'positive feedback' between membrane depolarization and sodium permeability (Fig. 1.13(a)). If unopposed, this would produce permanent reversal of the potential, but depolarization also produces a delayed 'negative feedback' increase in potassium permeability (Fig. 1.13(b)), returning the membrane to the resting state. In addition, sodium channels can only open briefly before they must close again. The action potential is therefore produced by consecutive changes in sodium and potassium permeability (Fig. 1.13(c)).

In life, nerve action potentials are evoked by the central nervous system or sense organs rather than by laboratory electrodes, but the electrical process is the same. The sodium and delayed potassium channels opened by depolarization can be described as being 'voltage-gated' as they are controlled by changes in the membrane potential.

Calcium and nerve excitability

Low plasma calcium concentrations increase nerve (and muscle) excitability. This may appear contrary to the effects of low extracellular calcium on the Nernst potential, where the effect would be to hyperpolarize and reduce membrane excitability. The effects of low calcium are due to local charges on the membrane. There are fixed negative charges on membrane surfaces from polar phospholipids and proteins which are normally balanced on the outside by positive calcium ions. These fixed charges usually have little effect on the membrane potential as they are present on the inside and outside surfaces. A reduction in plasma calcium produces an imbalance and allows the fixed

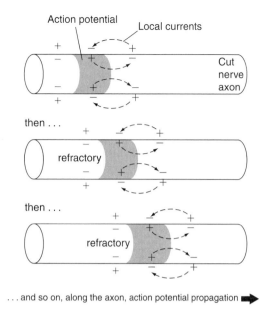

Figure 1.14   *Diagrammatic representation of action potential propagation along a nerve.*

negative charges on the outside of the cell to depolarize the membrane potential towards the threshold potential and increase cellular excitability and irritability.

THE PROPAGATED ACTION POTENTIAL

The action potential moves along the axon in a wave of changing membrane permeability and potential. This propagation of the action potential is produced by an interplay between the ionic events described above and the passive electrotonic nature of the membrane.

An action potential present at one point in a nerve axon sets up local, electrotonic, currents in adjacent resting membrane. These local currents depolarize the adjacent membrane towards the threshold, and a new action potential is then fired off. This process is continuous all along the axon and the action potential propagates from one end of the nerve to the other (Fig. 1.14).

For most neurons the action potential is initiated at the axon hillock (an initial segment of the axon as it is formed from the cell body) which

contains a high density of fast voltage dependent Na^+ channels and therefore a cascade of ionic events generated by reaching the depolarization threshold is easier to trigger. Although this local depolarization gives rise to a passive spread of current up and down the axon, the propagation of the action potential is always unidirectional because the regions behind the front end of the propagation are in various stages of refractoriness. The local current is therefore only effective at generating depolarizing in the regions ahead of the action potential by bringing resting Na^+ channels to threshold and generating subsequent action potentials.

The refractory period

During the 1 ms nerve action potential it is impossible to evoke another in the absolute refractory period when sodium channels are open. This is followed by a relative refractory period of 10–15 ms when another action potential can be elicited, but only with greater than normal stimuli. These refractory periods mean that action potentials are not only propagated in one direction, but also that the frequency of nerve impulse conduction is limited.

Saltatory conduction

In some large-diameter nerves the process of action potential conduction is not continuous along the length of the fibre. Instead, action potentials jump along from point to point in a 'saltatory' manner (Fig. 1.15(a)). This occurs in myelinated nerves where a fatty layer composed of overlapping Schwann cells covers the axon apart from at the regularly spaced nodes of Ranvier. The nodes of Ranvier are characterized by a high concentration of voltage-dependent Na^+ channels which mediate the action potentials. The myelin layer greatly increases the resistance and reduces the capacitance of the membrane, and this means that action potentials can fire off only at the nodes. Local, electronic, currents are responsible for depolarizing the next node along the axon to the threshold potential.

Again, the action potential only propagates in one direction as each node is refractory for a time after stimulation. Because of saltatory conduction, action potential propagation is more rapid in large nerve fibres when they are myelinated (Fig. 1.15(b)).

Another benefit of myelination is the conservation of energy because of the reduced flux of ions and hence reduced expenditure of energy required to restore ionic concentrations.

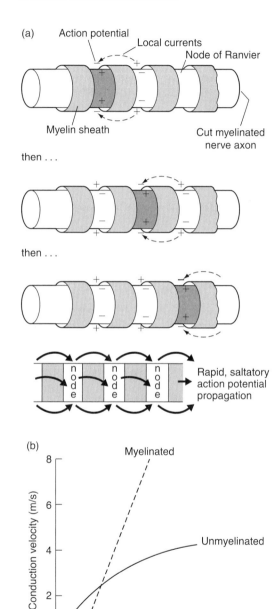

(a)

Figure 1.15 *Saltatory conduction in a large-diameter nerve. (a) Diagrammatic representation; (b) graph showing effect of myelination on speed of propagation.*

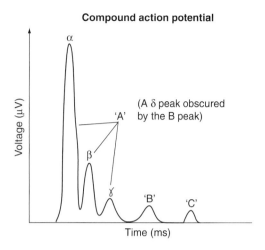

Compound action potential

(A δ peak obscured by the B peak)

Figure 1.16 *Graphical and tabular representation of compound action potential.*

Fibre	Function	Diameter (μm)	Conduction velocity (m/s)
A			
α	Skeletal motor, joint position	10–20	60–120
β	Touch, pressure	5–10	40–70
γ	Muscle spindle motor	3–6	15–30
δ	Pain, temperature touch	2–5	10–30
B	Preganglionic autonomic	1–3	3–15
C	Pain	0.5–1	0.5–2

Compound action potentials

Peripheral nerves contain a mixture of fibres, and these have been classified according to function, diameter and conduction velocity. Monophasic extracellular recording (obtained by placing one of the electrodes on an area of crushed nerve) reveals a compound action potential composed of A, B and C peaks (Fig. 1.16). The compound action potential is not 'all or none'; the various components have different threshold intensities because they represent simultaneous activity in fibres of different diameter and conduction velocity. The compound potential consists of a series of waves because the action potentials of fibres with higher

Table 1.2 *Alternative classification for sensory fibres in the nerves of mammalian muscles*

Group	Sensory ending	Diameter (μm)	Conduction velocity (m/s)
Ia	Muscle spindle, primary ending	12–20	72–120
Ib	Golgi tendon organ	12–20	72–120
II	Muscle spindle, secondary ending	4–12	24–72
III	Pressure/pain receptors	1–4	6–24
IV	Pain	0.5–1	0.5–2

conduction velocities reach the recording electrodes first. The A fibres include all the peripheral myelinated fibres and can be subdivided into α, β, γ and δ components in order of decreasing size and conduction velocity. The B group are the small myelinated preganglionic autonomic fibres in visceral nerves. The C group comprises all small-diameter unmyelinated motor and sensory fibres.

An alternative classification exists for sensory fibres in the nerves of mammalian muscles (Table 1.2).

THE IONIC BASIS OF THE CARDIAC ACTION POTENTIAL

Nerve action potentials can be explained by changes in membrane permeability to sodium and potassium ions, but in the heart calcium is also important. The reason for the autorhythmicity of the sinoatrial node is explained by cyclical changes in membrane ionic permeability.

Ventricular muscle

Ventricular muscle action potentials differ from those in nerves as they are longer in duration and have a distinct plateau phase during which depolarization is maintained (Fig. 1.17). The resting membrane potential of cardiac muscle is about −85 to −95 mV, and the action potential is 105 mV. The membranes are depolarized for 0.2 s in the atria and for 0.3 s in the ventricles.

- *Phase 0.* The ventricular muscle cell is depolarized by a rise in sodium permeability. Fast sodium channels open for only a few 10 000th of a second and a fast sodium current (i_{Na})

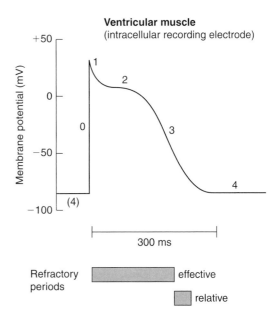

Figure 1.17 *Intracellular electrode recording of membrane potential in ventricular muscle.*

produces rapid depolarization. These are similar to those in nerves and are sensitive to tetrodotoxin. Potassium conductance decreases.

- *Phase 1.* Partial repolarization produced by a rapid decrease in sodium permeability.
- *Phase 2.* Plateau. Cardiac muscle has unique slow calcium channels. The cell permeability to calcium rises, maintaining depolarization and promotes cardiac muscle contraction. Sodium conductance continues to decline slowly. Another cause for the plateau of the cardiac action potential is a decrease in the cardiac potassium permeability which prevents return to the resting membrane potential.
- *Phase 3.* Repolarization. Potassium, sodium and calcium permeability return towards normal.

When the slow calcium channels close after 0.2–0.3 s, the potassium permeability increases rapidly and thus returns the membrane potential to its resting level.

- *Phase 4.* The resting potential. The membrane potential is mainly governed by potassium.

These phases are not separate events. Depolarization to the threshold affects consecutively sodium, calcium and potassium channels and produces the depolarization, the plateau and eventual repolarization of the ventricular cells.

The ventricular muscle action potential lasts for 250 ms. Of this, the absolute refractory period accounts for the first 200 ms and the relative refractory period for the other 50 ms.

The sinoatrial node

The sinoatrial node has no resting state; rather there is a pacemaker potential that generates cardiac autorhythmicity. The maximum diastolic potential of a sinus node cells is -45 to -55 mV and -50 to 65 mV in the AV node cells. Phases 1 and 2 are absent in the sinoatrial node as there is no depolarization plateau (Fig. 1.18).

- *Phase 4.* From the maximum diastolic potential, spontaneous diastolic depolarization slowly shifts the membrane towards the action potential threshold potential (approximately -40 mV). The pacemaker potential is produced by a fall in potassium permeability (i_K), a hyperpolarization activated mixed sodium-potassium 'funny' current (i_f) and an increase in a slow inward current. The slow inward current consists of a voltage-gated increase in calcium permeability via transient calcium channels and activity of the electrogenic sodium–calcium exchange system, driven by inward movement of calcium ions. This pacemaker activity brings the cell to the threshold potential.
- *Phase 0.* Depolarization is produced by opening of long-lasting voltage-gated calcium channels (i_{CaL}) and inward movement of positive ions. There is no sodium current involved in the sinoatrial node potential. The SA node potential reaches a peak at about 20 mV.
- *Phase 3.* Repolarization occurs as a result of a reduction in depolarizing currents (i_f and i_{CaL} are inactivated by positive potentials) and an

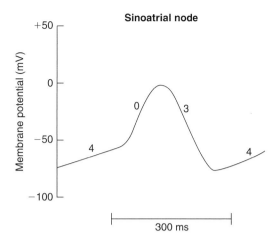

Figure 1.18 *Membrane potential in the sinoatrial node.*

increase in repolarizing currents (i_K becomes activated by positive potentials) causing a late increase in potassium permeability and outward flow of ions.

Again, the action potential is caused by a process of changing membrane permeabilities. A cycle of reduced potassium, increased calcium and then increased potassium permeability produces the sinoatrial node autorhythmicity.

Parasympathetic nerve stimulation increases potassium permeability of the sinoatrial node, hyperpolarizes the cell, and inhibits spontaneous cardiac activity. Sympathetic nerve activity has the opposite effect by opening calcium channels.

THE ACTION POTENTIAL IN MUSCLE

In striated muscle the action potential propagates over the cell surface in a similar manner to nerves. Smooth muscles may show spontaneous activity, and the principal inward current during depolarization is via calcium channels.

VOLTAGE-GATED ION CHANNELS

The basis of the action potential is that depolarization opens sodium, potassium or calcium channels that are 'gated' by the membrane voltage.

The 'voltage-clamp' and the 'patch-clamp' techniques have been used to investigate the function of ionic channels. The essential of the 'voltage-clamp' is that the positive feedback between depolarization and increased sodium permeability can be eliminated by passing current through the membrane; the current required to maintain a constant membrane potential reflects the ionic flow through channels. In 'patch-clamp' experiments, a small area of membrane is voltage-clamped so that individual channels can be observed.

THE SODIUM CHANNEL

A physiological model has been constructed to account for the known features of the sodium channel (Fig. 1.19); opened by membrane depolarization, blocked by tetrodotoxin and inactivated by a particle which can be removed by internal application of the enzyme pronase.

In the resting state, the positively charged sensor is attracted by the negative charge on the inside of the membrane. Depolarization to the threshold potential swings the sensor towards the outside of the membrane, and the activation gate opens. After opening, the channel is inactivated before it can open again.

The channel is selective and filters ions passing through it. The relative ionic permeability is lithium (1.1), sodium (1), potassium (1/12) and rubidium (1/40). During an action potential each sodium channel opens for 0.7 ms. Estimates of the membrane density of sodium channels suggest that there are 500–1000 channels per μm^2.

A considerable amount is known about the chemical composition of the sodium channel which is an internal membrane protein made up

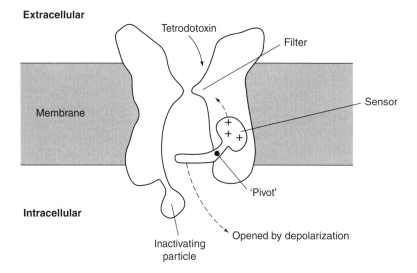

Figure 1.19   *Structural representation of the sodium channel.*

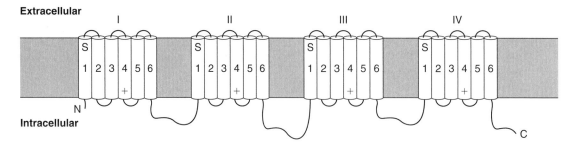

Figure 1.20   *Protein structure of the sodium channel.*

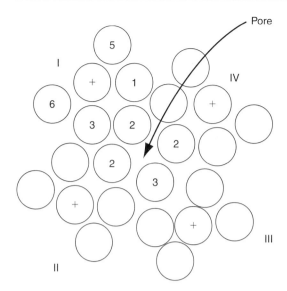

Figure 1.21 *The four subunits of the sodium channel are arranged around a central pore, which is closed during the resting state.*

of four domains, each containing six α helices (S1 to S6) (Fig. 1.20). The S4 helix has a positive charge, and S1 to S3 are negative. The four domains of this long protein chain enclose a pore through the membrane, which is lined by the negatively charged helices (Fig. 1.21).

In the resting state, the channel is closed. When the membrane is depolarized, the positively charged S4 segments move outwards, the whole structure changes conformation, and the central pore opens. Subthreshold depolarizations cannot fire off an action potential as they are insufficient to open the sodium channel (Fig. 1.22).

Tetrodotoxin and saxitoxin block the mouth of sodium channels from the outside. Sodium channel function is reduced by a high hydrogen concentration. Ester and amide local anaesthetics block the channel after diffusing through the membrane. Quaternary ammonium ions may block the channel at the same site, but they are not

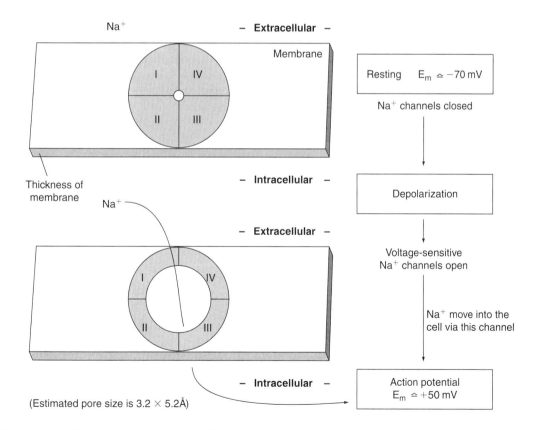

Figure 1.22 *Diagrammatic representation of the sodium channel, showing the voltage changes that occur to open the central pore or channel.*

lipid-soluble, and only work when applied internally. Some peptide toxins from scorpions and sea anemones prevent inactivation of sodium channels, producing hyperexcitation and pain.

POTASSIUM CHANNELS

A number of different types of potassium channel have been identified but 'delayed rectifier' potassium channels are involved in the nerve action potential. The membrane density has been calculated as 36 channels per μm^2 (much less than sodium channels). Unlike sodium, there is no evidence of an inactivating mechanism for potassium channels. The channels are 3–3.3 Å in diameter, and only potassium, thallium, rubidium and ammonium can pass. (Sodium ions are smaller than potassium, but they normally cross the membrane with a hydration shell which makes them too big for delayed rectifier channels.)

Delayed rectifier potassium channels are blocked by caesium, barium, 4-aminopyridine, strychnine and quinidine. Tetraethyl ammonium and other quaternary ammonium ions also block these channels, but only when applied from the inside of the membrane.

CALCIUM CHANNELS

Divalent cations (manganese, nickel, cobalt and cadmium), verapamil and dihydropyridines and *Conus* shell toxin block calcium channels. T (transient currents), N (inactivated at very negative potentials, neuronal) and L (long-lasting currents) types of calcium channel differ in their sensitivity to blockers and threshold potentials (Table 1.3).

Table 1.3 *Calcium channels: blocker sensitivity and threshold potentials*

	T	N	L
Threshold potential (mV)	−70	−10	−10
Cadmium block	+	+++	+++
Conus toxin block	+	+++	+++
Dihydropyridine sensitivity	−	−	+++

The molecular sequence of L calcium channels has been identified and is remarkably similar to that of sodium channels.

NEUROTRANSMITTERS AND RECEPTORS: ION CHANNELS, G PROTEINS AND SECOND MESSENGERS

OVERVIEW

Many chemical neurotransmitters have been identified, but all of them modify membrane ionic permeability and cellular excitability by one of three basic mechanisms:

- In the first type, the effect is direct as the ionic channel is an integral part of a large receptor protein, and the cellular response to the transmitter is rapid (Fig. 1.23(a)). Examples are the nicotinic acetylcholine, gamma-aminobutyric acid (GABA) and glycine receptors.
- In the second type, the receptor is linked to the ion channel via a membrane intermediary, a G protein (binds guanosine triphosphate) (Fig. 1.23(b)). For example, muscarinic acetylcholine receptors in heart muscle control potassium channels via G proteins.
- In the third type, the receptor is linked to a G protein, which stimulates production of an intracellular second messenger (Fig. 1.23(c)). For example, β-receptors are linked to cellular adenylate cyclase and cyclic adenosine monophosphate (cAMP) via a stimulatory G protein.

Neurotransmitters

Acetylcholine
Catecholamines: norepinephrine, epinephrine, dopamine
Serotonin
Histamine
Amino acids: gamma-aminobutyric acid, glycine, glutamate
Purines: adenosine, ATP
Neuropeptides: met-enkephalin, leu-enkephalin, dynorphin, substance P, calcitonin gene-related peptide

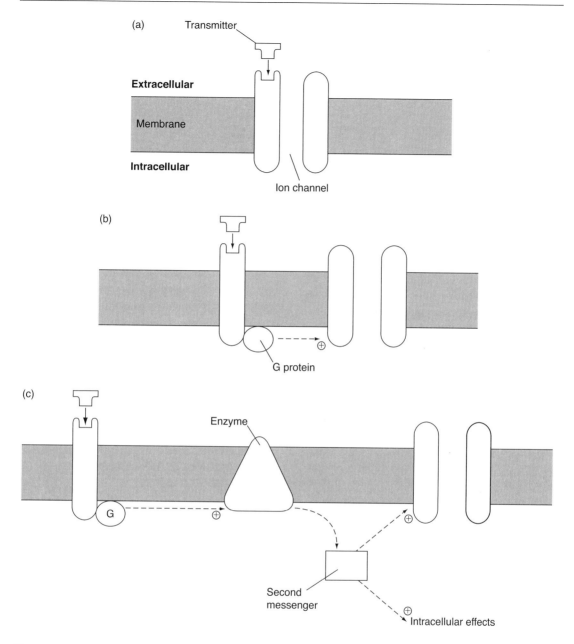

Figure 1.23   *The mechanisms of neurotransmitter action. (a) Direct interaction with a membrane ion channel; (b) indirect interaction with an ion channel via a G protein; and (c) indirect interaction with an ion channel via a G protein and intracellular second messenger.*

In general, potassium and chloride channels are opened by inhibitory transmitters. Excitatory transmitters open sodium and calcium channels.

G PROTEINS

G proteins are nucleotide regulatory proteins that bind guanosine triphosphate (GTP) and modulate membrane ion channels or cellular enzymes. Some are stimulatory (Gs) and others inhibitory (Gi).

The large heterotrimeric G proteins couple cell surface receptors to catalytic units that catalyse the formation of intracellular second messengers or couple the receptors directly to ion channels. They are therefore intermediates between the plasma membrane receptors and the intracellular processes they control. G proteins consist of α, β and γ chains and bind guanosine diphosphate (GDP). Upon receptor stimulation, the GDP is exchanged for GTP and the α unit dissociates from the $\beta\gamma$ chains. The active α-GTP complex then affects ion channels or second messenger systems. G proteins are inactivated when the α unit hydrolyses GTP to GDP and rejoins the $\beta\gamma$ complex.

The heterotrimeric GTP-binding proteins are classified on the basis of their α subunits. Gs are activated by receptors for epinephrine, norepinephrine, histamine, glucagons and others to activate adenyl cyclase and increase cyclic AMP. G_{olf} is involved in signalling in olfaction via an increase in cyclic AMP. Gt_1 and Gt_2 are important in modulating vision via a decrease in cAMP in the rods and cones. Gi is an inhibitory G protein receptor for norepinephrine, prostaglandins, opiates and many peptides that decreases cAMP. Gq protein is activated by receptors for acetylcholine via an increase in inositol triphosphate, diacylglycerol and Ca^{++}.

Small G proteins are involved in many intracellular functions such as the regulation of vesicle movement between the endoplasmic reticulum, Golgi apparatus and the cell membrane, regulating growth by transmitting signals from the cell membrane to the nucleus.

SECOND MESSENGERS

Second messengers are released inside cells by transmitters affecting enzymes on the inner surface of the membrane. The most important are the adenylate cyclase and the phosphatidylinositol systems.

Adenylate cyclase converts adenosine triphosphate (ATP) to produce the second messenger, cAMP. This system is modulated by neurotransmitters and hormones working via stimulatory and inhibitory Gs and Gi proteins.

Membrane phosphodiesterase converts phosphatidylinositol to yield two second messengers: inositol triphosphate (IP_3) and diacylglycerol (DAG). This conversion is stimulated by a G protein (Gp). The water-soluble IP_3 enters the cytoplasm and opens calcium channels in the membrane and the endoplasmic reticulum. DAG stimulates a membrane protein kinase and controls calcium, potassium and chloride channels.

ACETYLCHOLINE

There are two types of acetylcholine receptor: nicotinic and muscarinic (Table 1.4). In peripheral nerves, nicotinic receptors are present at the neuromuscular junction and in autonomic ganglia. Muscarinic receptors are present at parasympathetic postganglionic nerve endings (Fig. 1.24).

Table 1.4 *Acetylcholine receptors: agonists and antagonists*

Type	Agonist	Antagonist
Nicotinic	Nicotine	Curare
Muscarinic	Muscarine	Atropine

Nicotinic acetylcholine receptors

The nicotinic receptor is a large membrane protein incorporating an integral ion channel ($GABA_A$ and glycine receptors are similar). Acetylcholine attaches to the nicotinic receptor on the outside of the membrane and opens the ion channel.

As the receptor and the ion channel are part of the same structure, the membrane response to acetylcholine is rapid; motor nerve endings are a good example of this. The nicotinic ion channel is permeable to both sodium and potassium ions; it may be as large as 6.5 Å in diameter. When the channel opens, the membrane approaches a 'reversal potential' between the potassium and sodium equilibrium potentials. The reversal potential is held constant until acetylcholine is metabolized by acetylcholinesterase, and the channel closes.

Each nicotinic receptor protein comprises two α and one β, γ and δ subunits, and acetylcholine binds to the α chains (Fig. 1.25). It may be that

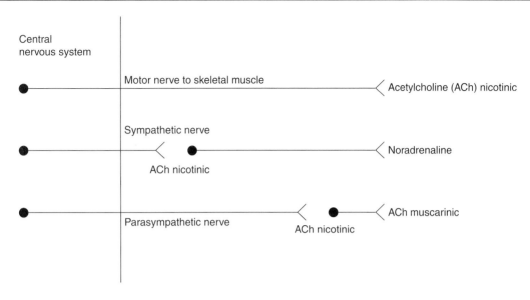

Figure 1.24 *Acetylcholine receptors in the nervous system.*

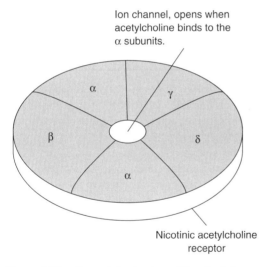

Figure 1.25 *The protein structure of the nicotinic acetylcholine receptor.*

acetylcholine must bind to each α chain before the receptor is activated – a process prevented by curare. Post-synaptic nicotinic receptors are blocked irreversibly by the snake toxin, α-bungarotoxin, and reversibly by tubocurarine.

Muscarinic acetylcholine receptors

Muscarinic receptors act indirectly on ion channels or second messengers via G proteins, and can

be differentiated by their response to antagonists. M_1 receptors are found in sympathetic ganglia, and have high affinity for the antagonist pirenzipine. M_2 receptors are found in the heart, and have low affinity for pirenzipine.

The molecular structure of M_1 receptors has been identified as a sequence of seven α helices which cross the cell membrane repeatedly. This structure is similar to that of β receptors.

M_1 receptors may work via G proteins, the phosphatidylinositol second messenger system and intracellular calcium to open membrane ion channels. M_2 receptors are also linked to G proteins, but act more directly on ion channels. In the heart, M_2 receptors depress autorhythmicity by opening potassium channels and hyperpolarizing the cell.

CATECHOLAMINES

Ahlquist classified adrenergic receptors by comparing the tissue effects of isoprenaline with those of norepinephrine and epinephrine (noradrenaline and adrenaline). He suggested that in tissues where isoprenaline is less active, α receptors are present. If isoprenaline is more effective, β receptors are present. α and β receptors have been further classified by their response to agonists and antagonists (Table 1.5).

Table 1.5 *Catecholamines: agonists and antagonists*

Type	Agonist	Antagonist
α_1	Phenylephrine	Prazosin
α_2	Clonidine	Yohimbin
β_1	Dobutamine	Practolol
β_2	Salbutamol	Butoxamine

- *α-receptors.* These work by the phosphatidylinositol system and increase the intracellular calcium concentration. α_2-receptors inhibit adenylate cyclase via the Gi protein.
- *β-receptors.* These act via the stimulatory protein Gs to stimulate adenylate cyclase and increase intracellular cAMP.

AMINO ACID TRANSMITTERS

In the central nervous system, gamma-aminobutyric acid (GABA) and glycine are inhibitory transmitters, and glutamate is excitatory.

Gamma–aminobutyric acid (GABA)

Two types of GABA receptors have been identified. $GABA_A$ receptors inhibit central nervous system cells by opening chloride channels. Benzodiazepines and barbiturates affect $GABA_A$ receptors and increase their inhibitory effect. $GABA_A$ receptors comprise two pairs of proteins forming an $\alpha_2\beta_2$ complex, and benzodiazepines and GABA bind to the α and β chains, respectively. $GABA_B$ receptors may be linked to the adenylate cyclase system.

Glycine

Glycine mediates postsynaptic inhibition in the spinal cord by opening chloride channels. Strychnine is an antagonist and produces convulsions.

Glutamate

Glutamate is the primary mediator of fast synaptic transmission in the spinal cord. The best

defined glutamate receptors open calcium channels, are blocked by magnesium and are activated by *N*-methyl-D-aspartate (NMDA).

Purines

Adenosine triphosphate (ATP) is released from synaptic vesicles with acetylcholine. In smooth muscle, ATP produces excitation by opening sodium and calcium channels. The stimulants caffeine and theophylline may block adenosine receptors in the central nervous system.

NEUROPEPTIDE TRANSMITTERS

Single chains of amino acids are produced in nerve cells and released as neurotransmitters. Examples are the enkephalins and substance P (Fig. 1.26).

OPIOIDS

There are at least three types of opiate receptor with different binding characteristics: μ – morphine, δ – enkephalins, and κ – dynorphin. Opioids produce hyperpolarization and depression of central nervous system cells by opening potassium channels, a mechanism common to α_2 receptors (this may explain the analgesic effect of the α_2 agonists clonidine and dexmedetomidine). Opioids inhibit gut motility by increasing potassium (μ) or reducing calcium (κ) permeability and reducing smooth muscle excitability.

NEUROTRANSMITTER RELEASE IN SYMPATHETIC GANGLIA

In sympathetic ganglia preganglionic fibres synapse with postganglionic effector cells, releasing a number of excitatory and inhibitory transmitters. In all ganglion cells, the acetylcholine released produces fast EPSPs (excitatory postsynaptic potentials) via nicotinic receptors. However, in some postganglionic cells acetylcholine produces either slow

met-enkephalin	Tyr – Gly – Gly – Phe – Met
leu-enkephalin	Tyr – Gly – Gly – Phe – Leu
substance P	Arg – Pro – Lys – Pro – Gln – Gln – Phe – Phe – Gly – Leu – Met – NH$_2$

Figure 1.26 *The amino acid sequence of the neuropeptide transmitters.*

EPSPs or slow IPSPs (inhibitory postsynaptic potentials) via muscarinic receptors. In addition, a peptide transmitter similar to luteinizing releasing hormone (LHRH) is released by some preganglionic fibres to produce a late IPSP in postganglionic cells. Activity in postganglionic sympathetic fibres therefore depends on the balance of the various excitatory and inhibitory transmitters acting upon them.

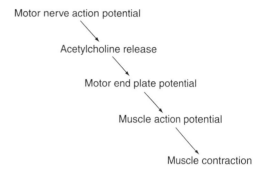

Figure 1.27 *An overview of neuromuscular transmission.*

NEUROMUSCULAR TRANSMISSION

STRUCTURE OF THE NEUROMUSCULAR JUNCTION

Skeletal muscle fibres are long cylindrical multinucleate cells innervated by a branch of a motor neuron. The point of contact is the motor end plate where an area of specialized muscle membrane forms a series of folds. The unmyelinated terminals of the motor nerve lie in gutters on the muscle end plate with a synaptic cleft of 500 Å between them. In the nerve terminal there are many synaptic vesicles containing acetylcholine, while plentiful nicotinic receptors lie opposite on the crests of the motor end plate folds. The enzyme acetylcholinesterase is present in the junctional cleft.

AN OVERVIEW OF NEUROMUSCULAR TRANSMISSION

The motor nerve action potential depolarizes the nerve terminal and releases acetylcholine from the synaptic vesicles into the junctional cleft. Acetylcholine excites the postjunctional nicotinic receptors, depolarizes the end plate, and generates

a propagated action potential in the surrounding muscle membrane (Fig. 1.27). Muscle shortening then follows by the process of excitation–contraction coupling. Acetylcholine is soon metabolized by acetylcholinesterase, and the end plate returns to the resting state. Neuromuscular transmission is an amplification process whereby a nerve action potential produces a much larger muscle action potential. Acetylcholine therefore has a central role in the process of neuromuscular transmission (Fig. 1.28).

Acetylcholine synthesis

Acetylcholine is synthesized in nerve axoplasm from choline and acetylcoenzyme A. The reaction is catalysed by choline-*O*-acetyltransferase, which is produced in the nerve cell body and transported to the axon. Choline-*O*-acetyltransferase activity is increased by steroid administration. Acetylcoenzyme A is produced in the mitochondria of axon terminals from pyruvate and the enzyme pyruvate dehydrogenase. Choline is obtained from the diet and liver synthesis, and is taken up by membrane

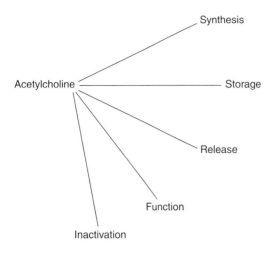

Figure 1.28   *The central role of acetylcholine in neuromuscular transmission.*

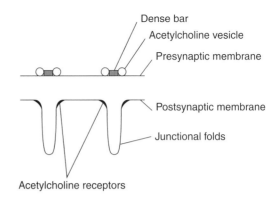

Figure 1.29   *Diagrammatic representation of the synapse.*

carrier mechanisms which are blocked by several quaternary ammonium compounds, including hemicholinium and triethylcholine (TEC). Nerve stimulation augments acetylcholine synthesis by increasing the intracellular sodium concentration.

Acetylcholine storage

Acetylcholine is stored mainly in vesicles in the nerve terminal. The active transport of acetylcholine into the vesicles involves an ATPase, and can be inhibited by vesamicol, tetraphenylboron and quinacrine. Some 80 per cent of the acetylcholine in the nerve can be released by action potentials, and this represents the amount in vesicles. The other 20 per cent cannot be released and forms a stationary store. Another surplus store can be detected only when intracellular cholinesterase is inhibited by physostigmine. The stationary and surplus stores consist of acetylcholine dissolved in the cytoplasm.

The presynaptic membrane has dense bars opposite the nicotinic receptors on the crests of the postsynaptic folds (Fig. 1.29). Some acetylcholine vesicles line up on either side of the dense bars to form active zones. The vesicles in these active zones are immediately available for release; the others form a reserve of acetylcholine.

Acetylcholine release

SPONTANEOUS ACETYLCHOLINE RELEASE

Miniature end-plate potentials (MEPPs) of 0.5–1 mV can be recorded from an intracellular electrode in the motor end plate in the absence of any nerve activity. MEPPs are spontaneous, random, and of constant amplitude, and each is the result of the quantal release of the acetylcholine contents of one vesicle. The vesicles fuse with the nerve membrane and empty into the junctional cleft by 'vesicular exocytosis'. The dense bars of the active zones may be the docking sites for vesicular exocytosis. The quantal theory of acetylcholine release states that one vesicle empties, releases 1500 acetylcholine molecules, and produces one MEPP:

> 1 Quantum = acetylcholine molecules
> in one vesicle (1500)
> → MEPP (0.5 − 1 mV)
>
> 1 Quantum − 1 vesicle − 1 MEPP

The quantal theory of acetylcholine release

The vesicles contain acetylcholine, ATP, vesiculin, cholesterol, phospholipids and calcium. After discharge, the vesicles and some of the products of their contents – including choline – are recycled by the nerve terminal.

MEPPs are abolished by curare and increased in frequency and amplitude by acetylcholinesterase inhibitors. MEPP frequency is directly related to extracellular calcium concentration, and inversely so to magnesium. Theophylline, catecholamines and cardiac glycosides increase MEPP frequency. The venoms of the black widow (α-latrotoxin) and the Australian red-back spiders increase MEPP frequency greatly by enhancing vesicle discharge and empty the nerve of acetylcholine. Botulinum toxin, tetanus toxin, β-bungarotoxin, Australian Tiger Snake venom (notexin), adenosine and GABA inhibit vesicle exocytosis and decrease MEPP frequency.

MOTOR NERVE ACTION–POTENTIAL–EVOKED ACETYLCHOLINE RELEASE

The 'vesicular exocytosis hypothesis' proposes that a motor nerve action potential evokes the exocytosis of 200 vesicles, containing 300 000 molecules of acetylcholine which produce the muscle end-plate potential (EPP). This represents a five-fold safety factor in the amount of transmitter released at the neuromuscular junction:

Nerve impulse $\rightarrow$ 200 vesicles $\rightarrow$ EPP

The 'vesicular exocytosis hypothesis'

Extracellular calcium is essential for this process; if it is absent then evoked release of acetylcholine is abolished. The nerve action potential stops at the last node of Ranvier and does not propagate into the terminal in the neuromuscular junction. Instead, depolarization from the node by local, electrotonic, circuits opens calcium channels in the prejunctional membrane. The prejunctional membrane potential is then restored to the resting state by a delayed opening of potassium channels.

Release of acetylcholine involves calcium-dependent fusion of the vesicle with the prejunctional membrane, permitting exocytosis of the contents. The mechanism by which calcium promotes exocytosis is unclear, but it may involve binding to the protein calmodulin forming the calcium–calmodulin complex, which activates various enzymes and affects vesicle structural proteins, including synapsin.

Evoked acetylcholine release is enhanced by tetra-ethylammonium (TEA), aminopyridines, catechol (and phenol), guanidine and germine. TEA, the aminopyridines and catechol block the potassium channels which normally repolarize the nerve terminal. This prolongs calcium channel opening and increases acetylcholine release at the neuromuscular junction. Guanidine and germine also delay the inactivation of calcium channels by blocking sodium channels in the prejunctional membrane.

Botulinum toxin

The anaerobic bacterium *Clostridium botulinum* produces an exotoxin which inhibits acetylcholine release from cholinergic nerves. Consequences include gastrointestinal and urinary dysfunction, blurred vision, and paralysis which spares limb but affects respiratory muscles. The aminopyridines may be used to treat the paralysis. *Clostridium* toxin (tetanus toxin) also prevents acetylcholine release, as well as producing generalized muscle spasms by removing spinal cord inhibition.

Acetylcholine function

Acetylcholine diffuses across the junctional cleft and attaches to the α subunits of nicotinic receptors on the crests of the folded motor end plate. It has been estimated that there are 10 receptors available for every acetylcholine molecule released. Upon excitation, the integral pore in the nicotinic receptor opens and the membrane becomes permeable to the cations sodium, potassium, calcium, magnesium and ammonium. Cation flow depolarizes the membrane, but the result is a localized EPP rather than a propagated action potential. The EPP approaches the 'reversal potential' of -15 mV, at which the various cation electrochemical potentials are in balance. EPPs are distinct and different from propagated action potentials in nerve or muscle (Table 1.6). The localized EPP generates, by local current flow, muscle action potentials which propagate over the surrounding membrane and lead to excitation–contraction coupling (Fig. 1.30). In denervated muscle fibres, acetylcholine receptors spread outside the end plate to cover the muscle membrane.

Table 1.6   *Acetylcholine effects*

Cause	EPP Acetylcholine	Action potential/ current flow
Change in permeability	↑ Na$^+$, K$^+$, Mg^{++}, Ca^{++}, NH$^+_4$	↑ Na$^+$ ... then ↑ K$^+$
Change in potential	To the 'reversal potential'	'All or none'
Propagated?	No	Yes
Refractory period?	No	Yes

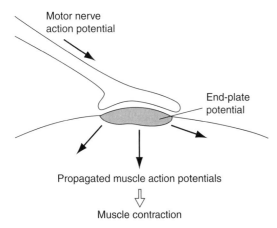

Figure 1.30   *Diagrammatic representation of the end plate potential (EPP) and the propagated muscle action potential.*

Acetylcholine inactivation

Each acetylcholine molecule can stimulate only one nicotinic receptor and open the ion channel for 1 ms before it is hydrolysed by acetylcholinesterase to choline and acetate. The acetylcholinesterase molecules are in the extracellular material (basal lamina) of the synaptic cleft and the junctional folds. Each acetylcholinesterase molecule has six enzyme sites, each with anionic and esteratic components that attract and hydrolyse acetylcholine.

Prejunctional acetylcholine receptors

Nicotinic receptors are also present on the nerve terminal, where they act in a positive feedback fashion to increase acetylcholine mobilization and release from vesicles. During high-frequency stimulation, released acetylcholine stimulates these prejunctional nicotinic receptors which increase transmitter release according to demand. Unlike those on the motor end plate, prejunctional nicotinic receptors are not blocked by α-bungarotoxin. There is some pharmacological evidence for other inhibitory and excitatory presynaptic muscarinic and nicotinic receptors, but their physiological significance has not been established.

Neuromuscular relaxants

Non-depolarizing muscle relaxants bind reversibly to, but do not activate, nicotinic receptors. They prevent neuromuscular transmission by competing with acetylcholine for nicotinic receptors on the motor end plate. All the non-depolarizing relaxants in clinical use also block prejunctional nicotinic receptors, and this may explain the fade observed during tetanic or 'train of four' neuromuscular stimulation with these agents (α-bungarotoxin produces muscle relaxation, but no fade). Depolarizing muscle relaxants maintain the motor end plate in a depolarized state (until they are metabolized) so that a zone of inexcitable membrane with increased potassium permeability develops, through which muscle action potentials cannot propagate.

MUSCLE

SKELETAL MUSCLE

Anatomy

Skeletal muscle fibres are 10–100 μm in diameter, as long as the muscle itself, and have multiple nuclei on the periphery of the cell. Each fibre comprises many myofibrils surrounded by cytoplasm containing mitochondria, the internal membranes of the sarcoplasmic reticulum and the T tubules, and glycogen. The striated pattern observed under light microscopy is formed by the regular organization of interdigitating thick myosin and thin

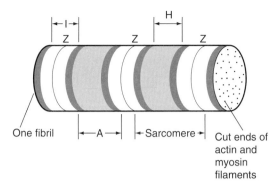

Figure 1.31 *The structure of skeletal muscle.*

actin filaments in the myofibrils (Fig. 1.31). The sarcomere is the basic unit of muscle contraction, and is enclosed by adjacent Z lines.

Innervation

Skeletal muscles are innervated by α motor fibres with a separate γ supply to muscle spindles. Each motor neuron innervates a number of fibres, comprising a motor unit. Most fibres are 'twitch', have one nerve terminal, and respond rapidly but briefly to stimulation. Other, 'tonic' fibres (extraocular muscles, the larynx and the middle ear) have many nerve terminals and contract in a slow, sustained manner.

The contractile proteins of the sarcomere

The I band of the sarcomere consists of actin filaments only. Myosin forms the A band. Cross-bridges of myosin heads form with the interdigitating actin filaments. At the M line the thick myosin filaments are connected together. The actin filaments are anchored at the Z line or disc. Each sarcomere therefore consists of one A band and two I band halves (Fig. 1.32). (N.B. The nomenclature for the bands in Figs 1.31 and 1.32 is from obsolete initial descriptions: Isotropisch, Anisotropisch, Zwischenscheibe, Hensen's disc and Mittelmembran.)

MYOSIN

The myosin molecule is a large complex protein (mol. wt. 520 000 Da) with a long tail and two

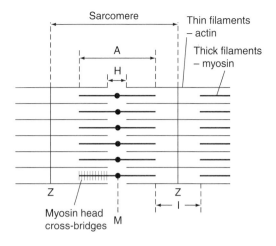

Figure 1.32 *The structure of the sarcomere.*

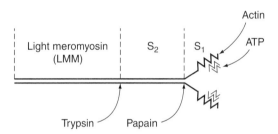

Figure 1.33 *The structure of myosin.*

globular heads which each bind actin and ATP (Fig. 1.33). The protein chains of the tail form α helices and are wound around each other. The heads comprise one heavy and two short protein chains, and there may be a flexible hinge in the S2 segment.

The myosin molecules fuse to form the thick filaments, with the long tails orientated towards the M line. The S1 heads and the S2 sections protrude out from the thick filament in a radial fashion (six actin filaments are arranged in a hexagonal fashion around each myosin filament).

ACTIN

The thin filaments are made up of two chains of F-actin formed from the polymerization of globular G-actin (mol. wt. 42 000 Da), which are twisted together like a double-stranded cord. Actin potentiates the ATPase activity of myosin.

Table 1.7 *Chemical interactions of actin and myosin*

Condition	In solution	In muscle
Actin and myosin: no ATP	Actomyosin formed; viscous solution	Rigor: muscle stiff, inextensible
Actin + myosin: + ATP – being split	Superprecipitation; plug forms	Contraction
Actin + myosin: + ATP – not being split (e.g. Ca^{++} absent)	Actomyosin dissociates; viscosity falls	Relaxation

CHEMICAL INTERACTIONS OF MYOSIN AND ACTIN

When mixed in solution, actin and myosin extracted from muscle form actomyosin and can produce reactions similar to those occurring in rigor mortis, contraction and relaxation, depending on the experimental conditions (Table 1.7).

TROPOMYOSIN

Tropomyosin (mol. wt. 66 000 Da) consists of two α-helical chains, and lies in the groove between the two actin polymers. Tropomyosin prevents the interaction of myosin with actin – an effect modulated by troponin.

TROPONIN

The globular protein troponin (mol. wt. 70 000 Da) is present with tropomyosin at regular intervals on the thin filaments with one molecule of each for every seven actin monomers. There are three subunits: troponin-T binds tropomyosin, troponin-I inhibits actomyosin ATPase, and troponin-C binds calcium. In the presence of calcium, the configuration of the troponin–tropomyosin complex alters, and myosin can interact with actin.

The sliding filament theory

Muscle contraction involves the thick and thin filaments sliding along each other. This sliding

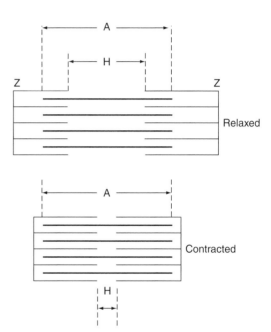

Figure 1.34 *The sliding filament theory.*

motion is produced by the myosin head cross-bridges pulling the actin fibres towards the centre of the sarcomere (Fig. 1.34).

Muscle shortening is produced by each cross-bridge undergoing cycles of attachment, pulling and detachment from actin. ATP hydrolysis provides the energy for each cycle. The ATPase myosin heads are independent force generators, and will undergo such cycles when actin is accessible, as regulated by calcium, troponin and tropomyosin. The physical process of cross-bridge activity is unclear; the myosin heads may swivel

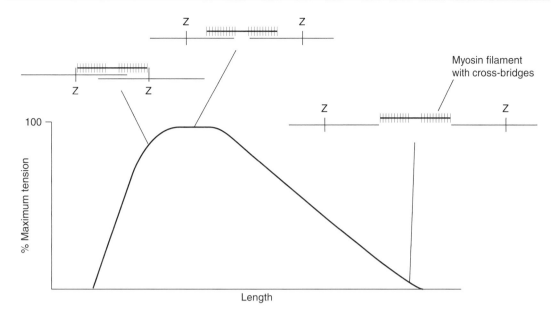

Figure 1.35   *The relationship between overlap of actin and myosin filaments and muscle-developed tension.*

on the tails, and the long chain may flex at the S2 segment.

The length–tension relationship of skeletal muscle

The sliding filament theory is confirmed by the observation that developed tension during an isometric contraction depends on the initial muscle length. The overlap of actin and myosin filaments in the sarcomere determines the number of cross-bridges developing tension, and this is reduced by excessive shortening or stretching (Fig. 1.35). If a muscle is stretched too much or not enough, it becomes inefficient. In life, most skeletal muscles are arranged near their optimal length.

The energy source for muscle contraction

ATP is the only energy source that the contractile proteins can use, but muscles only store enough for eight twitches. During muscle contraction ATP is rapidly regenerated from ADP by utilization

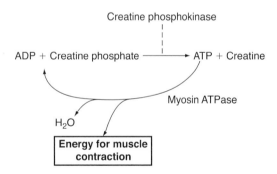

Figure 1.36   *The ADP–ATP cycle in muscle.*

of the more plentiful creatine phosphate stores (Fig. 1.36).

The net result is the hydrolysis of creatine phosphate, not ATP. Indeed, ATP concentrations in muscle only fall when creatine phosphate is depleted by violent exercise. Glycogen stores are metabolized aerobically by glycolysis, the tricarboxylic acid cycle and the electron transfer chain to produce ATP, which replenishes creatine phosphate by reversing the creatine phosphokinase reaction. Anaerobic metabolism produces lactic acid and muscle pain.

Excitation–contraction coupling

The propagated muscle action potential produces a much slower mechanical response from the muscle (Fig. 1.37). The muscle action potential, via the T tubule system, depolarizes the membrane and induces the release of calcium from the sarcoplasmic reticulum into the cytoplasm. The free calcium ion concentration in the cells increases from $0.1\,\mu M$ when resting to $10\,\mu M$ during activity. Calcium interacts with troponin-C, tropomyosin rolls deeper into the groove between the two actin strands exposing them to myosin, cross-bridges form, the inhibition of actomyosin ATPase by troponin-I is removed, and contraction proceeds. The muscle relaxes when the sarcoplasmic reticulum sequesters calcium and the troponin–tropomyosin-mediated inhibition returns.

The sarcoplasmic reticulum and the T tubules

The sarcoplasmic reticulum is a network of vesicular elements running longitudinally around the myofibrils. The sarcoplasmic reticulum sequesters calcium by a calcium- and magnesium-dependent ATPase pump. At regular places on the myofibril (A–I junction in most muscles, Z line in frog muscle), T tubules, invaginations of the muscle membrane, form triad structures with two lateral sacs of the sarcoplasmic reticulum. The T tubules and adjacent sarcoplasmic reticulum form electron-dense feet, although their lumina are not connected. The muscle action potential propagates down the T system, opening sarcoplasmic reticulum calcium-release channels, enabling contraction.

The mechanism by which T tubule action potentials influence the sarcoplasmic reticulum is unclear, but the movement of a charged particle within the membrane is implicated. Ryanodine, a plant alkaloid, locks open calcium-release channels in the sarcoplasmic reticulum by binding to specific receptors. The electron-dense feet of the T system and the ryanodine receptors may be part of the same protein molecule incorporating the calcium-release channels. A genetic defect in this membrane protein may be the cause of malignant hyperpyrexia (MH), which is triggered by depolarizing muscle relaxants and volatile anaesthetic agents. Caffeine enhances sarcoplasmic reticulum calcium release, and is employed in some laboratory muscle biopsy tests for MH susceptibility.

The response of striated muscle to repeated stimulation

As the muscle action potential and any refractory periods are over before contraction begins to fall, repetitive stimuli summate. If the stimuli are repeated at high frequency, the result is a smooth tetanus, with tension maintained at a high level (Fig. 1.38).

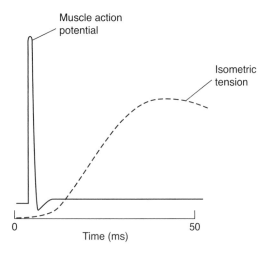

Figure 1.37 *Graphical representation of the muscle action potential.*

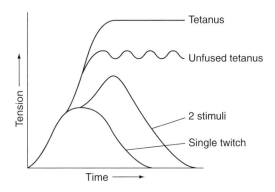

Figure 1.38 *Graph showing the response of striated muscle to repeated stimulation.*

CARDIAC MUSCLE

Anatomy

Cardiac muscle cells have one nucleus, are rich in mitochondria, and are made up of a network of branching fibres connected by intercalated discs which contain low-resistance gap junctions. Cardiac muscle is striated, but the pattern is not as ordered as in skeletal muscle. The pacemaker potential from the SA node is conducted throughout the heart, and the atria and ventricles contract in a coordinated fashion.

Excitation–contraction coupling

The cardiac action potential is prolonged with a plateau phase and involves inward calcium flow through the membrane. Contraction is achieved by a temporary release from the sarcoplasmic reticulum of calcium which binds to troponin, as in skeletal muscle. The opening of sarcoplasmic reticulum calcium-release channels is triggered by the inward flow of calcium during the action potential. Cardiac muscle does not contract if calcium is absent from the extracellular fluid. This form of excitation–contraction coupling may be described as 'calcium-triggered calcium release'.

The cardiac action potential and contraction

As the contraction lasts hardly longer than the action potential (Fig. 1.39), it is impossible to summate contractions or tetanize cardiac muscle.

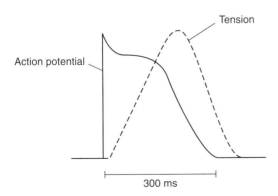

Figure 1.39   *The cardiac action potential.*

SMOOTH MUSCLE

Anatomy

There are two types of smooth muscles: single unit smooth muscle and multi-unit smooth muscle. Single-unit (also known syncytial or visceral) smooth muscles are joined by gap junctions that enable action potentials to travel from one fibre to another and cause the muscle to contract together as a single unit and are found in the walls of blood vessels, gastrointestinal tract and the genito-urinary system. Multi-unit smooth muscle fibres contain fibres that can contract independently of each other, such as those found in the ciliary muscle of the eye, the iris and the piloerector muscles. Smooth muscles contain actin and myosin, but are not striated. Actin filaments which form the contractile units are attached to dense bodies which are dispersed inside the cell and held together by a scaffold of structural proteins. Myosin filaments are interspersed among the actin filaments. The cells have one nucleus, are about $4\,\mu m$ in diameter and up to $400\,\mu m$ long, and are held in bundles by connective tissue.

Some smooth muscles resemble skeletal muscle as they are primarily controlled by motor nerves from the central nervous system; arteriolar vasoconstrictor smooth muscle and adrenergic nerves are an example. Other smooth muscles, including those of the viscera, are autorhythmic and contraction is only modified by nerve activity.

Unlike skeletal muscles, most smooth muscle contractions are prolonged and tonic in nature. The rate of myosin cross bridge attachment and release with actin ('cross-bridge' cycling) is much slower in smooth muscle. Consequently, smooth muscle require much less (1/10 to 1/300) energy for contraction which has a slower onset of contraction and relaxation. The maximum length and force of contraction of smooth muscle is often greater than that of a skeletal muscle because of the prolonged attachment of the myosin cross bridges to the actin filaments.

The resting membrane potential depends on the type of smooth muscle and is usually about -50 to $60\,mV$. Although action potentials do not occur in multi-unit smooth muscles, action or spike potentials occur in most single-unit smooth muscles either spontaneously or as a result of stretch, electrical stimulation, or the action of hormones or neurotransmitters (mediated by the flow of calcium ions into the cell). Slow-wave potentials (spontaneous slow-wave oscillations in membrane potentials) in single-unit smooth muscles occur as a result of rhythmical increases and decrease in the conductance of ion channels often when visceral muscle is stretched. When the slow wave rises above threshold ($-35\,mV$) action potentials are generated and spread over the muscle to cause muscle contraction. Spontaneous action potentials are frequently generated when visceral muscle is stretched and this response allows the gut wall to contract and resist stretch automatically.

Neural regulation of smooth muscle contraction depends on the type of innervation, neurotransmitters released, the distribution and types of receptors on the smooth muscle cell wall. In general, smooth muscle is innervated by the autonomic nervous system. Acetylcholine released by the parasympathetic nerves causes contraction (excitation) mediated by muscarinic receptors on the smooth muscle cell in some organs but may be an inhibitory mediator in others. When acetylcholine excites a smooth muscle cell, norepinephrine inhibits it. The smooth muscles of the uterus are not innervated.

Excitation–contraction coupling

Smooth muscles do not contain troponin but instead have calmodulin, a globular regulatory protein In smooth muscle, excitation–contraction coupling is achieved by calcium combining with calmodulin, which then activates myosin light chain kinase, a phosphorylating enzyme, which phosphorylates the regulatory light chain of myosin. The calcium ions may enter the cell via membrane channels during the action potential, but some smooth muscles also have a sarcoplasmic reticulum. Myosin phophatase splits the phosphate from the regulatory light chain of myosin, and the contraction ceases. The response to sustained or tonic stimulation is a rapid contraction followed by a sustained force maintained by a reduced cross-bridge cycling rate and energy consumption. This behaviour is called a 'latched state' and is advantageous in smooth muscles (e.g. blood vessels) that have to withstand continuous external forces.

MUSCLE SPINDLES, GOLGI TENDON ORGANS AND SPINAL REFLEXES

OVERVIEW

Muscle spindles detect muscle length and movement, whilst tendon organs sense tension. Muscle spindles are capsules of specialized fibres arranged in parallel with the muscle, and are innervated by primary and secondary afferents and γ motor nerves (Fig. 1.40). The γ motor fibres supply the contractile ends of the spindle and set the sensitivity of the afferent endings in the middle. Tendon organs lie in series with the muscle and are supplied by group Ib afferents.

Muscle spindle and tendon organ afferents respond differently to active skeletal muscle contraction (via α or γ nerves) or passive muscle stretching, permitting close monitoring of movement (Table 1.8).

MUSCLE SPINDLES

A muscle spindle is 3–10 mm in length and consists of three to twelve thin intrafusal striated muscle fibres which are attached at their distal ends to associated extrafusal skeletal muscle. The central

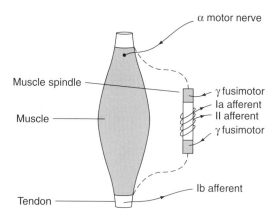

(NB. The spindle is embedded within the muscle, but drawn outside here)

Figure 1.40 *The relationship between muscle, the muscle spindle and the tendon organ.*

regions of each intrafusal fibre form a capsule containing several nuclei and is devoid of actin–myosin elements. When the nuclei are aggregated in the central region the fibre is called a nuclear bag fibre; whereas when the nuclei are are arranged linearly, the fibre is called a nuclear chain fibre. Typically a muscle spindle contains one to three nuclear bag fibres and three to nine nuclear chain fibres.

The sensory nerves are excited by any stretching of the non-contractile centre of the spindle muscle fibres. Two types of sensory fibres are associated with muscle spindle fibres. One is called the primary ending or the annulospiral ending. The

primary endings (Ia) are more sensitive to rate in change in length, and have been labelled 'dynamic'. The nuclear bag fibres are associated with the primary endings and therefore are responsible for the dynamic response. The second type of sensory fibre is the secondary or flower-spray ending which forms numerous small terminal branches that cluster around the the nuclear region of the nuclear chain fibres. The secondary endings (II) are more sensitive to absolute length of the fibre and are referred to as 'static' fibres.

The 'dynamic' and 'static' characteristics of the spindle are set by the different motor nerves and intrafusal fibres stretching the afferent endings. The γ-plate and some β nerves and nuclear Bag-1 fibres are 'dynamic'. The γ-trail nerves supplying nuclear Bag-2 and nuclear chain fibres are 'static' (Fig. 1.41).

The spinal effect of spindle excitation

Activity in muscle spindle afferents reflexly excites the motor neuron to the muscle group in which they lie, and inhibits any antagonists. When the central region of a spindle is slowly stretched, impulses in both the primary and secondary endings increase in proportion to the degree of stretch; this is called the static response. When the length of the spindle is suddenly increased, the primary sensory fibre

Table 1.8 *Response of muscle spindle and tendon organ afferents to active skeletal muscle contraction or passive stretching*

	Spindle afferents	Tendon organ	
α motor activity	↓	↑	– shortens spindle
γ motor activity	↑	–	– reflex ↑ α activates motor activity
Passive muscle stretching	↑	↑	– 'knee jerk' reflex

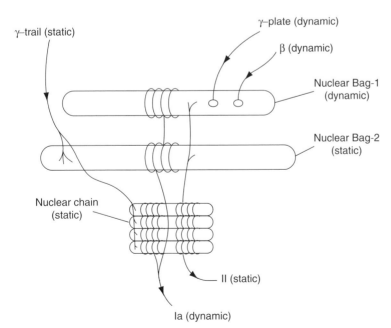

Figure 1.41 *The 'dynamic' and 'static' characteristics of the spindle.*

exhibits a vigorous response that signals the rate of change in length of the fibre and this is called a dynamic response.

The gamma loop

The extrapyramidal system controls γ motor neuron fibres, and activity in this can indirectly produce skeletal muscle contraction via spindle afferents and their reflex effect on α fibres (Fig. 1.42). There are two types of γ motor neuron: dynamic motor neuron fibres distributed to the nuclear bag that enhance the dynamic response when stimulated, and static motor neuron fibres distributed to nuclear chain fibres that enhance the static response.

About 30 per cent of the axons distributed to any muscle are from γ motor neurons. Both α and γ motor neurons are co-activated by signals from the motor cortex or other control centres. During a muscle contraction the stimulation of the γ motor neurons maintains the sensitivity of the muscle spindle. The γ motor neuron system is modulated by descending pathways from the reticular formation in the brainstem, which is under the influence of output from the cerebellum, basal ganglia and cerebral cortex as well as ascending spinoreticular pain fibres.

GOLGI TENDON ORGANS

The Golgi tendon organ is an encapsulated receptor through which some musle tendon fibres pass prior to their bony insertion. Some of the tendon

organ receptors respond vigorously when the tendon is stretched (dynamic response) but most tendon organ receptors monitor the degree of tension in the tendon at a steady state (static response).

SPINAL REFLEXES

Muscle spindle reflexes

Passive muscle stretching produces reflex contraction by exciting spindle primary afferents which synapse directly onto motor neurons in the anterior horn of the spinal cord (Fig. 1.43). This is the basis of the knee jerk, the only monosynaptic reflex in the body. Type Ia sensory fibres enter the spinal cord through the dorsal roots and their branches terminate in the spinal cord near the entry level or ascend to the brain. The fibres (from the primary sensory fibres) that terminate in the spinal cord synapse directly with α motor neurons in the ventral horn which innervate the extrafusal fibres of the same muscle. This neural pathway forms the circuitry for the stretch reflex. The stretch reflex has two components: a dynamic component

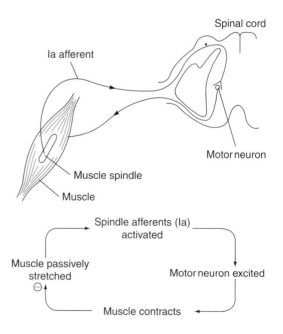

Figure 1.43 *Muscle spindle reflexes.*

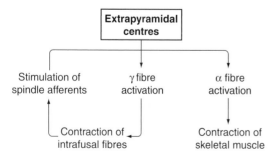

Figure 1.42 *The gamma loop.*

that is activated while the spindle is being stretched, and a static component that occurs when the muscle reaches a new static length. An important function of the dynamic component of the stretch reflex is its dampening effect on jerky or oscillatory movements. Clonus or unusual repetitive contractions of muscles occur when muscle spindle sensory function is abnormal.

Glutamate is the excitatory transmitter released and produces an excitatory postsynaptic potential (EPSP). EPSPs generate motor neuron action potentials by summating either spatially (many afferent cells synapse with different motor neurons), or temporally by repetitive activity in the same afferent fibre.

Opposing muscle groups are also inhibited by spindle spinal stretch reflexes. For example, during the knee jerk, flexor muscles are relaxed by muscle spindle afferents acting via spinal inhibitory interneurons (Fig. 1.44). The transmitter released by the inhibitory interneuron is glycine, which hyperpolarizes the motor neuron membrane by opening chloride channels, producing an inhibitory postsynaptic potential (IPSP).

Tendon organ spinal reflexes

If the muscle tension becomes excessive, high-threshold tendon organs are stimulated and reflexly inhibit the muscle and excite antagonists. Signals from the tendon organ are conducted through large myelinated type Ib fibres and inhibit the motor neurons innervating the muscle

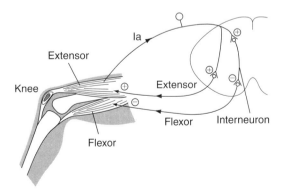

Figure 1.44 *The monosynaptic knee jerk reflex: nerve-muscle interaction.*

associated with the tendon organ. This negative feedback prevents injury to the muscle from excessive tension.

THE INITIATION OF SKELETAL MUSCLE CONTRACTION

Skeletal muscle contraction can be accomplished by the pyramidal system via α motor neurons, the extrapyramidal system acting via the gamma-loop, or by passive stretching producing a monosynaptic spindle reflex. Coordinated voluntary motor activity depends on coordination between the direct pyramidal and the indirect extrapyramidal motor pathways.

SENSORY RECEPTORS

CLASSIFICATION

Receptors can be classified by which stimulus they react to, and whether this originates from inside or outside the body (Table 1.9). Pain receptors must also be added to this classification. Such 'nociceptors' react to excessive, potentially harmful, mechanical, thermal, chemical and even light stimuli.

MECHANISMS OF RECEPTOR ACTIVATION

The simplest receptors are nerve endings which are depolarized by certain specific stimuli, often increased membrane sodium permeability. The resultant 'generator potential' is localized to the nerve terminal, but fires off propagated action potentials in the adjacent afferent axon (Fig. 1.45(a)).

The Pacinian corpuscle is a cutaneous receptor where the unmyelinated nerve ending is covered by a number of lamellae (Fig. 1.45(b)). Touch or pressure deforms the lamellae, and a generator potential is produced in the nerve ending. As is the case with most receptors, the Pacinian corpuscle is more sensitive to changes in applied pressure.

Table 1.9   *Classification of sensory receptors*

	Stimulus origin		
	Inside	Outside	
		Contact	Distance
Mechanoreceptors	Muscle length, tension Joint movement Arterial blood pressure	Touch	Hearing
Photoreceptors	–	–	Sight
Chemoreceptors	[H$^+$] body fluids	Taste	Smell
Thermoreceptors	Hypothalamic temperature receptors	Cutaneous temperature receptors	–

(a)

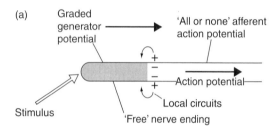

Generator potential varies as stimulus intensity
Action potential frequency varies as generator potential

(b)

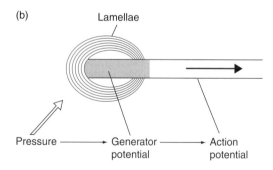

Figure 1.45   *Diagrammatic representation of the mechanisms of receptor activation. (a) Nerve endings; (b) the Pacinian corpuscle.*

In other cases the nerve ending is connected to a specialized receptor cell (e.g. vision, hearing, proprioception and muscle spindles). The stimulus produces a 'receptor potential' and depolarizes the free end of the afferent nerve, probably via chemical transmission (Fig. 1.46).

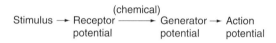

Figure 1.46   *Flow diagram from stimulus to action potential.*

SENSATION

Quality

The quality of sensation depends on the fact that different receptors are sensitive to a specific stimulus. Yet receptors can be stimulated by any incident stimulus of sufficiently high intensity (e.g. pressure on the eye produces the sensation of light).

Intensity

The receptor and generator potentials vary with stimulus intensity. In the afferent nerve, stimulus intensity is conveyed by the frequency of action potential firing.

Accommodation

Most receptors are more sensitive to changes in stimulus and reduce their response with continued application.

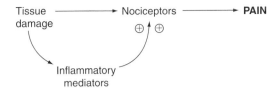

Figure 1.47 *Flow diagram from tissue damage to pain.*

PAIN RECEPTORS

Nociceptors can be divided into two main groups: C-polymodal and A-mechanoheat (AMH) nerve fibres. C-polymodal nociceptors (C-PMNs) respond to thermal, mechanical and chemical stimuli, and comprise the majority of fibres in many sensory nerves. A-mechano-heat (AMH) nociceptors are fine myelinated Aδ fibres, and respond mainly to thermal and mechanical stimuli.

After tissue damage, nociceptors do not accommodate, but they sensitize, so that even light touch can be painful. This is the basis for 'primary hyperalgesia' where the pain threshold is reduced in the localized area of tissue damage (Fig. 1.47). Sensitization is thought to occur because tissue damage releases many inflammatory factors that mediate the reduction of the stimulus threshold of nociceptors. Such tissue mediators include protons, neuropeptides (e.g. substance P), cytokines, prostaglandins, serotonin, histamine, kinins and ATP.

Reflections

1. An ion tends to flow across a membrane when there is a concentration difference of that ion or an electrical potential difference across the membrane. The electrochemical difference of an ion across the membrane represents the difference of chemical potential energy. The Nernst equation can be used to determine what the electrical potential difference across the membrane would have to be for a particular ion to be in equilibrium. An ion channel oscillates between an open (high conductance) and closed (low conductance) state.

2. All cells have a negative resting membrane potential with the cytoplasm electrically negative to the extracellular fluid. The resting membrane potential is established mainly by the high membrane permeability to potassium and the diffusion potentials due to potassium and sodium caused by the differences in the concentrations of both ions inside and outside the cell. The electrogenic nature of the Na^+–K^+ pump causes only a small negative charge of $-4\,mV$ inside the cell membrane and therefore has a minor contribution to the resting cell potential.

3. Nerve signals are transmitted by rapid changes in membrane potentials called action potentials. An action potential occurs with a sudden change from a negative resting potential to a positive membrane potential and almost equally returns to the resting membrane potential. During the course of an action potential, voltage-gated sodium and potassium channels are activated and inactivated. The voltage-gated sodium channels are instantaneously activated when the threshold potential ($-65\,mV$) is reached and cause a 5000-fold increase in sodium conductance. Then an inactivation process closes the sodium channel within a fraction of a millisecond. The onset of the action potential also activates the voltage-dependent potassium channels which begin to open more slowly. Shortly after the action potential is initiated, the sodium channels become inactivated and any amount of excitatory signal cannot open the inactivation gates – the absolute refractory period. The period that follows the absolute refractory period when stronger than normal stimuli can initiate an action potential and this is known as the relative refractory period.

4. Local currents produce electrotonic conduction and conduct subthreshold signals and action potentials. A subthreshold current decreases as it is conducted along a nerve fibre by electronic conduction (by 37% of its maximal strength over a distance of one length constant). However, an action potential is propagated rather than merely

conducted. It is regenerated as it moves along the cell so that an action potential remains the same size and shape as it is conducted. The velocity of conduction is determined by the electrical properties of the cell. A large-diameter nerve fibre has a faster conduction velocity. Myelination also dramatically increases the conduction velocity of a nerve axon. Action potentials are regenerated at the nodes of Ranvier, and the nerve impulse 'jumps' from node to node. This is known as saltatory conduction and is useful because it increases the conduction velocity of a nerve and conserves energy for propagation of the nerve impulse.

5. Many cellular regulatory substances exert their effects on cellular processes by signal transduction pathways. G proteins serve as intermediates between a receptor that is activated by an agonist and the cellular enzymes that are modulated in response to agonist binding. The GTP binding protein is activated and this stimulates or inhibits the activity of an enzyme or ion channel, and thereby alters the intracellular concentration of a second messenger such as cAMP, cGMP, inositol triphosphate, diacylglycerol or Ca^{++}. An increase in the levels of one or more second messenger increases the activity of a second messenger-dependent protein kinase. Many cellular processes are regulated by the phosphorylation of enzymes and ion channels. Certain membrane receptors for hormones and growth factors are tyrosine kinases or associated with tyrosine kinases that are activated by agonists.

6. Acetylcholine, various amines, glutamate, glycine and gamma-aminobutyric acid are important neurotransmitters in the central nervous system. Glycine and gamma-aminobutyric acid are major inhibitory neurotransmitters at synapses in the central nervous system. Glutamate is a major excitatory neurotransmitter in the central nervous system. There are five types of excitatory amino acid receptors. Neuropeptides function as neurotransmitters in the central nervous system.

7. Skeletal muscle fibres are striated because of the arrangement of thick myosin filaments and thin actin filaments in the myofibrils. The sarcomere, the contractile unit of the skeletal muscle, is bordered by two Z lines. Thin actin filaments extend from the Z line toward the centre of the sarcomere and the thick myosin filaments are located in the centre and overlap the actin filaments. Interaction between myosin and actin (dependent on Ca^{++}) causes muscle contraction, with myosin filaments pulling the thin actin filaments towards the centre of the sarcomere.

8. Motor centres in the brain control the activity of the α motor neurons in the ventral horns of the spinal cord. The motor neuron produces an action potential which initiates skeletal muscle contraction. The action potential passes down the T tubules of the sarcolemma and causes conformational changes in the dihydropyridine receptors that result in the opening of ryanodine receptors, which release Ca^{++} from the sarcoplasmic reticulum into the sarcoplasm. The sarcoplasmic Ca^{++} binds to troponin C and exposes the myosin-binding sites on the actin thin filaments. This causes tropomyosin to move toward the groove in the thin actin filament. Myosin cross-bridges pull the thin filaments toward the centre of the sarcomere, contracting the muscle fibre. Sarcoplasmic Ca^{++} is resequestered in the sarcoplasmic reticulum by Ca^{++}–ATPase and relaxation of the muscle occurs.

9. The roles of smooth muscle cells in hollow organs are to (a) develop force or shorten the muscle, and (b) contract tonically to maintain organ dimensions against imposed loads. Smooth muscles contain contractile units that consist of thick myosin filaments which interdigitate with large numbers of actin filaments attached to dense bodies and lack troponin. No striations are visible. Contraction of smooth muscle is dependent on both Ca^{++} release from the sarcoplasmic reticulum and Ca^{++} entry across the sarcolemma. The increase in sarcoplasmic Ca^{++} activates myosin light chain kinase which phosphorylates the cross-bridges and causes smooth muscle contraction. Dephosphorylation of the attached cross-bridges by myosin phosphatase causes smooth muscle relaxation. Shortening velocities and energy requirements of smooth muscles are very low compared with skeletal muscles.

Smooth muscles have variable velocity–stress relationships. The response to sustained stimulation is a rapid contraction followed by a force maintained with reduced cross-bridge cycling rates and energy (ATP) consumption. This is known as the latch behaviour. Smooth muscle activity is controlled by the autonomic nervous system, circulating hormones, locally generated signalling mediators, and stretch or shearing forces.

10. Sensory receptors include exteroreceptors, interoreceptors and proprioceptors. In general sensory transduction is accomplished by the production of a receptor potential, and encode modality, spatial localization, intensity, duration and frequency of stimuli. Sensory receptors may show accommodation. Muscle spindles are sensory receptors in skeletal muscles that lie parallel to the regular extrafusal muscle fibres. They consist of nuclear bag and nuclear chain intrafusal fibres. Ia afferent fibres form primary nerve endings on both nuclear bag and chain fibres. Group II afferent fibres form a secondary ending which is found chiefly on the nuclear chain fibres. Primary endings detect static (change in length) and dynamic (rate of change in muscle length) changes in the muscle, whereas secondary endings detect only static responses. The gamma efferent system controls the sensitivity of the muscle spindle. The muscle spindles also dampen jerky or oscillatory muscle contractions. The Golgi tendon organs, located in the tendons of the muscles, are arranged in series with the skeletal muscle. They are supplied by IIb afferent fibres and are stimulated by both stretch and contraction of the muscle. The stretch reflex includes a monosynaptic excitatory pathway from muscle spindle afferent (Ia and II) fibres to the α motor neurons to the same and synergistic muscle and a disynaptic inhibitory pathway to the motor neurons of the antagonist muscles.

2

Physiology of the nervous system

LEARNING OBJECTIVES

After studying this chapter the reader should be able to:

1. Compare and contrast the functional properties of voltage-gated, ligand-gated and non-gated ion channels
2. Describe the mechanisms involved in the process of synaptic transmission including inhibitory and excitatory postsynaptic potentials, and temporal and spatial summation
3. Discuss the major neurotransmitters, receptor subtypes, effects on the postsynaptic membrane and termination of action
4. Describe the formation and absorption of cerebrospinal fluid and explain its function and role in the regulation of intracranial pressure
5. Describe the physiological basis of somatic sensation: the senses of touch, pressure, vibration and temperature
6. Outline the visual and auditory pathways
7. Describe the organization of the vestibular system and its role in the maintenance of balance
8. Explain the role of the motor cortical regions in the execution of voluntary activity
9. Describe the role of the cerebellum in balance, refined coordinated movements and muscle tone
10. Outline the role of the basal ganglia in the planning and execution of defined motor activity
11. Describe the organization of the spinal cord and its role in reflex activity
12. Outline the descending pathways that modify the activity of the spinal cord
13. Compare and contrast non-REM sleep and REM (paradoxical) sleep
14. Describe the functions of the limbic system
15. Describe the anatomical organization of the autonomic nervous system and its separation into the sympathetic and parasympathetic divisions
16. Describe how the sympathetic nervous system regulates the cardiovascular system, visceral organs, and secretory glands
17. Describe how the parasympathetic nervous system regulates the gastrointestinal system, heart and secretory glands
18. Outline the functions of the hypothalamus

The nervous system coordinates the activities of many organs, and is responsible for modulating and regulating numerous physiological processes via a network of specialized cells. These cells form the brain and spinal cord (central nervous system, CNS), and sensory and motor fibres that enter and leave the CNS or are wholly outside the CNS (peripheral nervous system, PNS). The basic unit of the nervous system is a highly specialized cell for the reception and transmission of information, the neuron.

NEURONS

NEURON STRUCTURE AND PROPERTIES

Neurons are diverse in shape and size. A typical neuron has a cell body (or soma) with fibre-like processes called dendrites and axons emerging from it. The dendrites are branches that leave the cell body and receive information from adjoining neurons. The dendrites have knob-like extensions

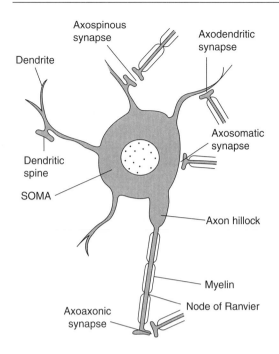

Figure 2.1   *Structure of a neuron.*

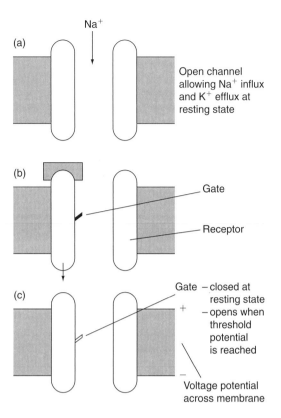

Figure 2.2   *Types of ion channels. (a) Non-gated ion channel (important for resting membrane potential); (b) ligand-gated ion channel; and (c) voltage-gated channel (important for initiation and propagation of action potential).*

called dendritic spines. The cell body has an extensive system of rough endoplasmic reticulum containing basophilic granules (Nissl substance) which synthesize proteins. The dendritic spines, dendrites and soma receive information from other cells.

The axon is a fibre-like structure that leaves the cell body and contains mitochondria, microtubules, neurofilaments and smooth endoplasmic reticulum, and its junction with the cell body is called the axon hillock (Fig. 2.1). The terminal end of the axon contains small vesicles packed with neurotransmitters.

When a neuron is activated, an electrical impulse called an action potential is generated at the axon hillock and then conducted along the axon. The action potential releases neurotransmitters from the nerve terminal, and these bind to receptors located on target cells, causing a flow of ions across the postsynaptic membrane.

Ion channels

Ions can flow across the nerve cell membrane through three types of channel: non-gated, ligand-gated and voltage-gated. The non-gated channels (Fig. 2.2(a)) are always open and are responsible for the efflux of potassium ions and the smaller influx of sodium ions when the neuron is in the resting state. In ligand-gated ion channels (Fig. 2.2(b)), the receptor may be part of the channel or may be coupled to the channel via a G protein and a second messenger. Voltage-gated channels (Fig. 2.2(c)) are sensitive to the voltage difference across the membrane. In the resting state, voltage-gated channels are closed, and they open when a critical membrane potential is reached during depolarization.

Non-gated ion channels are found throughout the neuron, and are important for the generation of the resting membrane potential. Ligand-gated channels are found predominantly on the dendritic spines, dendrites and cell bodies. Voltage-gated

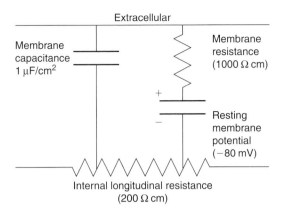

Figure 2.3 *Simplified electrical circuit of membranes.*

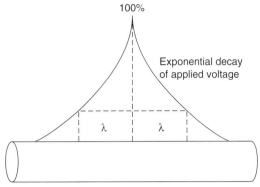

Length constant (λ) = distance from applied stimulus at which the voltage falls to 36.7% (1/e) of applied voltage

Figure 2.4 *Length constant (λ) of biological cable.*

channels are found mainly on the axons and axon terminals, and are important for the initiation and propagation of the action potential.

The electrical properties of the nerve membrane affect the flow of ions through the membrane, the generation and conduction of the action potential along the axon, and the integration of incoming information at the dendrites and the cell body. The movement of ions across the nerve membrane is driven by ionic concentrations and electrical gradients. The ease with which ions flow across the membrane is called the conductance, which is the inverse of resistance. Nerve membranes are able to store electrical charge, and this property is known as membrane capacitance. The amount of electrical charge a membrane can store is proportional to its surface area, and so the greater the surface area, the larger the capacitance (Fig. 2.3). A large-diameter dendrite can store more charge than a small-diameter dendrite of equal length.

Cable properties of a nerve fibre

The decay of electronic potential per unit length along the axon is given by the length or space constant (λ), defined as the distance from the stimulus to the point at which the voltage falls to 36.7 per cent (1/e) of the maximal potential (V_{max}) ($\lambda = \sqrt{r_m/r_a}$, where r_m is membrane resistance per unit length and r_a is the longitudinal resistance of the axoplasm per unit length). Thus, the length

or space constant increases as the membrane resistance increases and as the axoplasmic resistance decreases (Fig. 2.4).

Axons also behave like electrical cables. A long nerve fibre may be regarded as two parallel conductors with capacitors distributed along the cable (Fig. 2.5). One parallel (inner) conductor is the axoplasm of the nerve, and the other (outer) parallel conductor is the interstitial fluid bathing the nerve. The outer longitudinal resistance is low because the large volume of interstitial fluid carries the outside current. The inner longitudinal resistance is high because the cross-sectional area of the axoplasm is so small. The capacitance along the nerve fibre is due to the two-layered lipid matrix of the nerve membrane. As the membrane is not a perfect insulator, a potential applied to the nerve spreads longitudinally along the membrane and progressively leaks out across the membrane. The cable properties of the membrane determine the time course and voltage changes of an electrotonic potential. In turn, this will determine both the ability of a stimulus to a nerve to produce an action potential and the action potential propagation velocity. Synaptic and receptor potentials spread passively from the site of the membrane where they are produced to a part of the membrane that is able to produce an action potential. If the membrane resistance is reduced, the space constant falls, and the ability of the excitatory potential to spread passively to an action potential generating part of the membrane is reduced. An increase in the time constant will reduce the propagation velocity along the nerve.

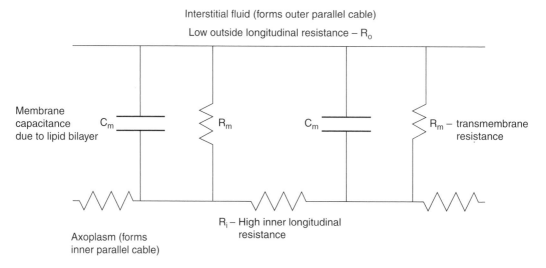

Figure 2.5   *The nerve fibre as a biological cable.*

SYNAPTIC TRANSMISSION

A synapse is the anatomical site where nerve cells communicate with other nerves, muscles and glands. Two types have been identified: electrical and chemical. Electrical synapses are formed by gap junctions that form low-resistance channels between the presynaptic and postsynaptic elements so that various ions can freely move between the two related neurons and mediate rapid transfer of signals that can spread through large pools of neurons. They may be found at dendrodendritic sites of contact between nerves that synchronize neuronal activity but they are uncommon in the mammalian nervous system.

Synaptic transmission in mammals usually occurs via chemical neurotransmitters (Fig. 2.6). The presynaptic terminal is depolarized by an action potential, which opens voltage-gated calcium channels; calcium ions flow into the presynaptic terminal, and cause neurotransmitter vesicles to fuse with the presynaptic membrane. The neurotransmitter is thus released into the synaptic cleft by exocytosis, it diffuses across the synaptic cleft and binds to receptors on the postsynaptic membrane and alters its permeability. The receptors in the postsynaptic membrane may be either ion channels or coupled with G proteins, which activate a second messenger system.

There is a synaptic time delay of less than 0.5 ms between the excitation of the presynaptic terminal

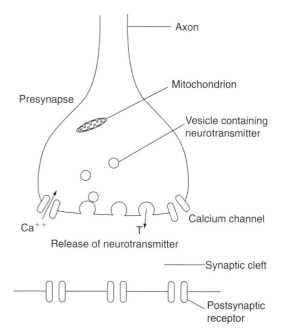

Figure 2.6   *Outline of a synapse. T, neurotransmitter.*

and the permeability change in the postsynaptic membrane. Most of the synaptic delay is accounted for by the presynaptic release of the transmitter. The transmitter activity is terminated by diffusion away from the cleft, by enzymatic breakdown, or by reuptake into the presynaptic terminal.

By altering the postsynaptic membrane permeability the transmitter produces a local potential change resulting in either depolarization or hyper-polarization. As depolarization leads to excitation of a neuron, it is called an excitatory postsynaptic potential (EPSP). An EPSP is a depolarization of a few millivolts resulting from an increased postsynaptic membrane conductance to Na^+ and K^+ ions. Na^+ ions move into the cell and K^+ ions move out. As the movement of Na^+ ions predominates, the net effect is a small depolarization of the postsynaptic membrane, bringing the membrane potential closer to the threshold required for opening of its voltage-gated channels so that an action potential is more likely to be triggered.

When the flux of ions ceases, the membrane will repolarize to its resting potential. The rate at which it repolarizes depends on the membrane resistance per unit area (R_m) and capacitance per unit area (C_m) . The time required for the membrane potential to decay to 37 per cent of its peak value is called the membrane time constant (T_M), which is the product of membrane resistance and capacitance, i.e. $T_M = R_m \times C_m$.

Some inhibitory transmitters increase K^+ and Cl^- conductance by causing a local increase in K^+ and Cl^- permeability. The postsynaptic membrane potential becomes hyperpolarized by an increased efflux of K^+ ions and influx of Cl^- ions. As hyperpolarization of the postsynaptic membrane prevents the cell from becoming activated, this is called an inhibitory postsynaptic potential (IPSP).

The neuronal membrane at rest is maintained at about $-65\,mV$ because it is more permeable to potassium ions than to sodium ions. As a result, positively charged potassium ions move out of the cell and the cell interior becomes negatively charged with respect to the extracellular environment. The interior of the soma and dendrites consist of a highly conductive cytoplasm with minimal resistance, and therefore, changes in one part of the neuron can easily spread to other parts of the neuron. In general, ligand-gated channels that allow sodium to enter the postsynaptic neuron are excitatory, whereas channels that allow chloride to enter (or potassium to exit) are inhibitory.

The passage of a current across a synapse requires a certain amount of time that varies from one neuronal pool to another. This is called synaptic delay, and it is determined by the time required to release the neurotransmitter, time needed for the transmitter to diffuse across the synaptic cleft,

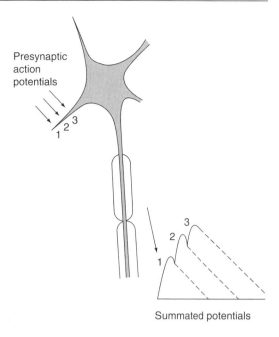

Figure 2.7 *Schematic illustration of temporal summation of potentials.*

transmitter–receptor binding time, and time required for ionic movement at the membrane and alter the membrane potential.

The excitability of the synapse is increased by acidic conditions in the extracellular synaptic environment, whereas more alkaline conditions decrease synaptic activity. A decrease on oxygen supply diminishes synaptic activity. When synapses are repetitively stimulated, the response of the postsynaptic neuron decreases over time as a result of an inability to replenish the supply of neurotransmitter rapidly at the presynaptic terminal. The synapse is said to be fatigued.

Properties of synapses

A synapse permits the transmission of a neural impulse in one direction from one nerve to another. A typical neuron in the CNS may receive inputs from several other neurons (convergence), or make synaptic contact with many other neurons (divergence). The postsynaptic potentials of neurons may be integrated by a process called summation.

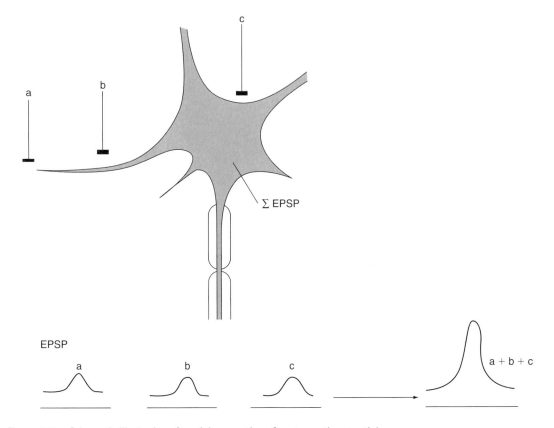

Figure 2.8   *Schematic illustration of spatial summation of postsynaptic potentials.*

Temporal summation therefore occurs when a second postsynaptic potential (excitatory or inhibitory) arrives before the membrane has returned to its resting level (Fig. 2.7). A typical postsynaptic potential may last about 15 ms and ion channels are open for less than 1 ms and there is usually sufficient time for several channels to open over the course of a single postsynaptic potential. The effects of these two potentials are additive over time. Spatial summation occurs when a number of axon terminals over the surface of a neuron are active simultaneously and their combined postsynaptic potential is greater than any one individual potential (Fig. 2.8). Commonly, the magnitude of a single EPSP may be 0.5–1 mV, far less than the 10–20 mV required to reach threshold. Spatial summation enables the combined EPSP to exceed threshold.

The amount of neurotransmitter released from an axon terminal may be reduced by presynaptic inhibition by the release of gamma-aminobutyric acid (GABA) from a neuron synapsing (axo-axonic synapse) with the presynaptic terminal (Fig. 2.9). GABA reduces the influx of calcium ions at the synaptic terminal during synaptic transmission.

Two types of GABA receptors are present on the presynaptic terminal. Activation of $GABA_A$ receptors opens chloride ion channels. The increased chloride conductance reduces the action potential, less calcium enters the nerve terminal and less neurotransmitter is released. When GABA binds to the $GABA_B$ receptor, a G protein that reduces the neurotransmitter release by two mechanisms is activated. First, a G protein may open K^+ channels, leading to hyperpolarization of the presynaptic terminal. Second, the G protein may directly prevent the opening of calcium channels when an action potential reaches the nerve terminal.

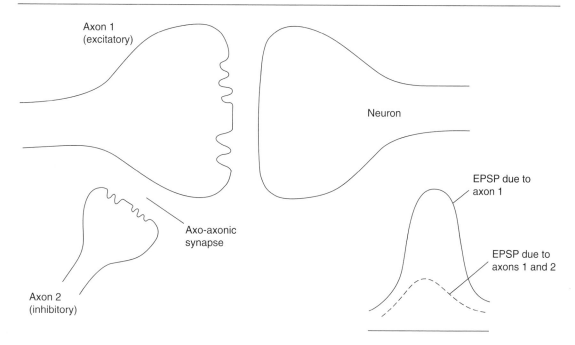

Figure 2.9 *Presynaptic inhibition.*

NEUROTRANSMITTERS

These include both excitatory and inhibitory amino acids, acetylcholine, catecholamines, indoleamines and neuropeptides.

Excitatory amino acids

Glutamate and aspartate are excitatory CNS neurotransmitters. Five types of glutamate receptors have been described. The N-methyl-D-aspartate (NMDA) receptor increases calcium conductance to produce an EPSP via an increase in intracellular diacylglyceraldehyde and inositol triphosphate (IP_3) activity. Activation of kainate and quisqualate receptors produces an EPSP by increasing Na^+ and K^+ conductance.

Inhibitory amino acids

GABA and glycine are inhibitory amino acids. The $GABA_A$ receptor is a ligand-gated Cl^- channel that produces an IPSP by increasing Cl^- ion conductance when GABA binds to the β subunits.

$GABA_B$ receptors activate G proteins and cause an increase in K^+ conductance, producing an IPSP. Glycine binds to specific receptors, causing hyperpolarization of the postsynaptic membrane.

Acetylcholine (ACh)

This is released at the presynaptic terminals of cholinergic neurons. The receptors for ACh may be either nicotinic or muscarinic. The nicotinic receptor consists of two α subunits, and a β, γ and δ subunit. ACh binds to the two α subunits leading to a conformational change, which increases Na^+ influx more than K^+ efflux, in turn producing a depolarization of the postsynaptic membrane (Fig. 2.10).

Muscarinic receptors are composed of seven membrane-spanning domains, and they work by G-protein activation. When ACh binds to the muscarinic M_1 receptor there is a decrease in K^+ conductance via phospholipase C activation and this produces membrane depolarization. When ACh activates an M_2 receptor, there is an increase in K^+ conductance by the inhibition of adenyl cyclase, causing hyperpolarization.

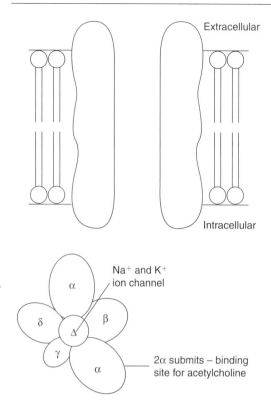

Figure 2.10 *Nicotinic acetylcholine receptor.*

Catecholamines

These are important neurotransmitters in the CNS and PNS, and include dopamine and epinephrine. Dopamine interacts with dopamine (D) receptors, which may be classified into two subfamilies, D_1 and D_2. D_1 receptors are coupled to G_1 proteins and stimulate adenyl cyclase. D_2 receptors reduce or do not change adenyl cyclase activity. Molecular cloning techniques have demonstrated the existence of five dopamine receptor subtypes, D_1, D_2, D_3, D_4 and D_5. The D_1 and D_5 receptors are classified as members of the D_1 subfamily because they have 80 per cent similarity of amino acid sequences in the transmembrane domains. The D_2, D_3 and D_4 receptor subtypes have substantial similarity in their amino acid sequences and are therefore classified as members of the D_2 subfamily.

Indoleamines (serotonin or 5-hydroxytryptamine [5-HT])

Serotonergic neurons are found in the CNS and PNS. 5-HT is stored in vesicles and released by exocytosis when the nerves are depolarized. 5-HT_3 receptor activation increases Na^+ and K^+ conductances, leading to an EPSP. Other subtypes of serotonin receptors are mediated by second messenger systems. Activation of 5-HT_{1A} receptors causes cyclic AMP activation and an increase in K^+ ion conductance, producing an IPSP.

Neuropeptides (met–enkephalin, substance P)

Several peptides act as neurotransmitters and interact with specific peptide receptors. Morphine receptors bind endogenous opioids (met-enkephalin, leu-enkephalin, dynorphin and β endorphin). Substance P is a neurotransmitter that depolarizes neurons in the spinal cord and the hypothalamus via IP_3 activation. Somatostatin is an important neuropeptide that inhibits electrical activity in the hypothalamus, hippocampus, the limbic system and the neocortex.

THE CENTRAL AND PERIPHERAL NERVOUS SYSTEMS

THE BLOOD–BRAIN BARRIER

Slow diffusion of many substances between the blood and the cerebrospinal fluid and the blood and the suggest the existence of a blood–brain barrier and a blood–cerebrospinal fluid (CSF) barrier (Fig. 2.11).

Morphologically, these barriers are formed by the capillary endothelium and the specialized ependymal cells at the brain–CSF interface. The blood–brain barrier is formed by the capillaries in the brain, and prevents the free diffusion of circulating substances into the brain interstitial space. The capillary endothelial cells have the following distinct features: (1) tight junctions (zona occludens) between adjacent cells; (2) the absence of fenestrations; and (3) a high content of mitochondria. Beyond the basement membrane of the capillary endothelium there is a perivascular area of closely applied foot processes of the astrocytes with intercellular clefts or channels. The blood–CSF

(a)

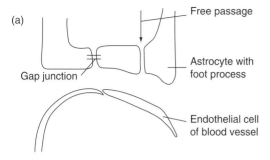

Free passage

Astrocyte with
foot process

Gap junction

Endothelial cell
of blood vessel

(b)   Cerebrospinal fluid in ventricle

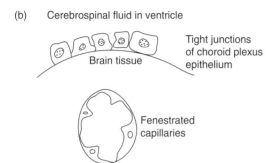

Tight junctions
of choroid plexus
epithelium

Brain tissue

Fenestrated
capillaries

Figure 2.11 *Barrier mechanisms in the brain.*
(a) Blood–brain barrier; (b) blood–cerebrospinal fluid
(CSF) barrier.

barrier is absent in the circumventricular organs, which abut on the third and fourth ventricles. These structures have capillaries that are porous with fenestrations, and include the median eminence, the area postrema, the subfornical organ and the pineal gland.

The blood–brain barrier provides a favourable environment for nervous tissue function by selective permeability to substances present in plasma. It is permeable to respiratory gases and substances of molecular weights less than 30 kDa, but is impermeable to large, polar or lipid-insoluble substances. In this manner, the brain is protected from potentially toxic substances, whilst metabolic substrates have free access.

Simple sugars (e.g. glucose) are transported across the blood–brain barrier by a specialized pump. A similar mechanism is thought to be responsible for the transport of hexoses, pentoses, serotonin, biogenic amines and some drugs.

The concentrations of calcium, magnesium and chloride ions in the CSF are controlled by active transport mechanisms at the blood–brain

barrier. A calcium influx pump is probably involved in the transport of calcium ion from the blood to the brain and then to the CSF. Magnesium and chloride ions are present in higher concentrations in CSF than in plasma, suggesting transport of these ions across the blood–brain barrier by both passive and active mechanisms. Potassium ions exchange very slowly between the blood and brain.

The bicarbonate ion (HCO_3^-) is involved in the regulation of CSF pH. Normally, the ratio of CSF to plasma bicarbonate concentration is 0.8. However, it has been demonstrated that HCO_3^- can be actively transported by either a primary HCO_3^- pump or a secondary ion pump (with H^+ or Cl^-) between brain interstitial fluid and blood.

There is no evidence that Na^+ is actively transported into the CSF. Na^+ diffuses passively across the blood–brain barrier, but the final Na^+ concentration in CSF is determined by the uptake of Na^+ by the brain and the exchange of water between the blood, the brain and the CSF.

CEREBROSPINAL FLUID

The CSF is a specialized extracellular fluid in the ventricles and the subarachnoid space. The functions of CSF are:

- *Mechanical protection by buoyancy*. As brain tissue has little rigidity and mechanical strength, flotation of the brain in the CSF prevents deformation and damage caused by acceleration or deceleration forces associated with head movements. Within the cranium and the spinal cord, counter-pressure of the CSF surrounding the blood vessels compensates for gravitational effects from postural changes and forces of external acceleration. The low specific gravity of CSF (1.007) produces buoyancy and reduces the effective weight of the brain from 1400 g to 47 g.
- *Maintenance of a constant ionic environment*. Neurons are highly sensitive to ionic changes, and the CSF provides a constant chemical environment for neuronal activity. The concentrations of Ca^{++}, K^+ and Mg^{++} ions in the CSF are important for the electrical activity of the brain. Ca^{++}, K^+, Mg^{++} and HCO_3^- ions are transferred into the CSF by active pumps, whereas H^+ and Cl^- are transferred by secondary

transport. The CSF also removes unwanted and toxic substances from the brain.

- *Acid–base regulation.* The acid–base status of the CSF is important for the control of respiration.
- *Nutritional and intracerebral transport.* Simple sugars, and basic and neutral amino acids are transported by active transport mechanisms between blood and the extracellular fluid compartment of the brain. Oxygen and carbon dioxide can diffuse freely across the blood–brain barrier. Neuropeptides secreted into the CSF are carried from one region to another.

CSF formation

CSF is formed in the brain and circulates through the subarachnoid space and the ventricular system. The ventricular system includes the two lateral ventricles, the third ventricle, the aqueduct of Sylvius, the fourth ventricle and the central canal of the spinal cord. In humans, the rate of CSF formation is 0.35–0.4 mL/min, or 500 mL per day. The total volume of CSF (120 mL) is exchanged approximately four times per day.

About 70 per cent CSF is derived from the choroid plexuses (the rich network of blood vessels covered by ependymal (epithelial) cells projecting into the ventricles) and the other 30 per cent from endothelial cells lining the brain capillaries. The composition of CSF differs from that expected if it were to be an ultrafiltrate of plasma, indicating that there may be some active secretion.

The capillary endothelium of the choroid plexus is fenestrated so that the plasma in the capillaries is filtered across the endothelium to form a protein-rich fluid in the choroid plexus stroma. As the choroid plexus epithelium is relatively impermeable because of the presence of apical tight junctions, ultrafiltration and secretion are important for the transport of selected constituents of the CSF. Hydrostatic pressure and bulk flow promote fluid entry into clefts between the epithelial cells of the choroid plexus. ATP membrane pumps at the basement membrane of the epithelial cells move Na^+ into the cells and K^+ and H^+ ions into the stroma. Cl^- ions coupled to Na^+ ions move into the epithelial cell of the choroid plexus. HCO_3^- and H^+ ions are formed within the epithelial cell

by the interaction of carbon dioxide and water, catalysed by carbonic anhydrase.

ATP-dependent membrane pumps at the secretory surface of the epithelial cells of the choroid plexus secrete Na^+ ions into the CSF and move K^+ ions into the cell. Water moves across the epithelial cells into the CSF because of the osmotic gradient. Cl^- and HCO_3^- ions also move passively along the electrochemical gradient into the CSF.

Glucose enters the CSF by a facilitated transport mechanism, which becomes saturated when the plasma glucose concentration is above 15–20 mmol/L. The CSF glucose concentration is approximately 60 per cent of that of plasma.

CSF protein concentrations are approximately 0.5 per cent of the respective plasma protein concentrations. In normal conditions protein enters the CSF at the choroid plexus by passage through the junctions of the epithelial cells and by vesicular transport across the epithelium. Some protein enters the CSF through extrachoroidal sites. Small amounts of protein in the extracellular fluid of the brain enter the CSF slowly by bulk flow. The CSF protein concentrations are lowest in the ventricles (26 mg/100 mL), intermediate in the cisterna magna (32 mg/100 mL), and highest in the lumbar sac (42 mg/100 mL). This is thought to be related to decreased clearance of protein by the dural sinuses in the lumbar sac.

In summary, the concentrations of magnesium ions (1.12 mmol/L; CSF/plasma ratio = 1.4) and chloride ions (124 mmol/L, CSF/plasma ratio = 1.14) are higher in cerebrospinal fluid than in plasma whereas the concentrations of potassium ions (2.9 mmol/L, CSF/plasma ratio = 0.67), calcium ions (1.15 mmol/L, CSF/plasma ratio = 0.7), glucose (3.7 mmol/L, CSF/plasma ratio = 0.82), and protein (0.18 g/L, CSF/plasma ratio = 0.002) are lower in the cerebrospinal fluid. The pH of cerebrospinal fluid is 7.32 and the concentration of sodium ions in the cerebrospinal fluid is 140 mmol/L which is similar to that of plasma.

When the intracranial pressure (ICP) increases to 20 mmHg (normal range 5–15 mmHg), there is minimal change in the rate of CSF production so long as the cerebral perfusion pressure (CPP) is greater than 70 mmHg. However, when the CPP falls below 70 mmHg, CSF formation falls as there is a reduction in cerebral and choroid plexus blood flow.

CSF circulation

Cerebrospinal fluid circulates through the ventricular system and the subarachnoid space from the formation sites to the absorption sites, and has a hydrostatic pressure of 6.5–20 cmH$_2$O (or 5–15 mmHg). Ciliary movements of the ependymal cells propel the CSF towards the fourth ventricle and the foramina of Luschka and Magendie into the cisterna magna. From the cisterna magna, CSF passes superiorly into the subarachnoid space around the cerebellar hemispheres, caudally into the spinal subarachnoid space, and cephalad to the basilar cisterns (around the premedullary pontine and interpeduncular cisterns). From the basilar cisterns, CSF flows through the prechiasmatic cistern and Sylvian fissure to the lateral and frontal cortical regions; and by a second route to the medial and posterior part of the cerebral cortex (Fig. 2.12). Respiratory oscillations and arterial pulsations of the cerebral arteries and choroid plexus provide additional momentum for the movement of CSF.

Reabsorption of CSF

The reabsorption of CSF into the venous blood is through the arachnoid villi located within the dural walls of the sagittal and the sigmoid sinuses, and in the spinal arachnoid villi located within the dural walls of the dural sinusoids on the dorsal nerve roots. The arachnoid villi function like one-way valves that allow CSF to flow into the cerebral venous sinuses when the CSF pressure is about 1.5 mmHg greater than the pressure in the blood in the venous sinuses. Mean CSF pressure in man is 15 cmH$_2$O (11.25 mmHg), while the pressure in the superior sagittal sinus is 9 cmH$_2$O (6.75 mmHg), providing a 6 cmH$_2$O (4.5 mmHg) pressure gradient for the passage of CSF across the arachnoid villi. Approximately 85–90 per cent of the CSF is absorbed by the intracranial arachnoid villi, and 10–15 per cent by the spinal arachnoid villi.

CSF is transported across the endothelium of the arachnoid villi primarily by pinocytosis and also by an extracellular route by the opening of

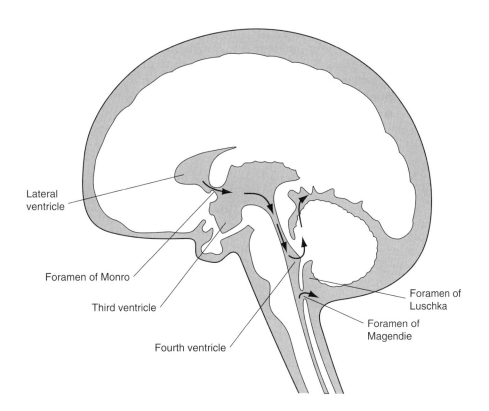

Figure 2.12   *Cerebrospinal fluid circulation.*

intercellular spaces. The rate of CSF reabsorption increases with CSF pressure. As intracranial pressure increases, the reabsorption of CSF rises by increased pinocytosis and opening of intercellular spaces. The resistance to reabsorption of CSF is normal until the CSF pressure is above 30 cmH$_2$O (22.5 mmHg), at which point the resistance decreases.

Under normal conditions there is an equilibrium between CSF formation and reabsorption. The rate of CSF formation remains normal and is relatively unaffected by an increase or decrease of ICP. However, if the ICP is sufficiently high to reduce cerebral perfusion pressure (CPP) to below 70 mmHg, CSF formation decreases. If the ICP is less than 7 mmHg, minimal reabsorption occurs. However, at an ICP greater than 7 mmHg, CSF reabsorption increases in a linear fashion with increased ICP up to 22.5 mmHg (30 cmH$_2$O).

CEREBRAL BLOOD FLOW AND OXYGENATION

Cerebral function is totally dependent on oxidative phosphorylation of glucose to provide ATP. Although the brain comprises only 2 per cent of the body weight, it uses 20 per cent of the total body's resting oxygen consumption. The lack of storage of substrates and the high metabolic rate of the brain accounts for the organ's sensitivity to hypoxia.

Regional cerebral blood flow (CBF) varies with the metabolic rates of local areas of the brain. CBF and cerebral metabolism are said to be coupled. Various regulatory mechanisms maintain the CBF at physiological levels. Local metabolic factors are involved in flow–metabolic coupling, and these include H$^+$, K$^+$, adenosine, phospholipid metabolites, glycolytic metabolites and nitric oxide.

Autoregulation refers to the phenomenon where CBF is kept constant over a mean arterial blood pressure range of 50–150 mmHg (6.7–20 kPa). Changes in perfusion pressure are thought to invoke a myogenic response (Bayliss effect) in the vascular smooth muscle. This myogenic response is sensitive to the mean blood pressure and to changes in pulse pressure. Nitric oxide appears to mediate basal vascular tone, but is unlikely to be directly involved in pressure autoregulation. Above a mean

arterial pressure of 150 mmHg, CBF passively increases with CPP and arterial pressure.

Carbon dioxide (CO$_2$) is an important factor in determining CBF. At a normal arterial blood pressure, CBF increases linearly by 2–4 per cent for every mmHg (0.13 kPa) increase in Pa_{CO_2} (between Pa_{CO_2} of 20 and 80 mmHg (2.7 and 10.7 kPa)). CO$_2$ diffuses rapidly across the blood–brain barrier, increases the concentration of extracellular fluid H$^+$ ions, and causes vasodilatation. However, the arteriolar tone modifies the effects of P_{CO_2} on CBF, and severe hypotension can abolish the ability of the cerebral circulation to respond to Pa_{CO_2} changes.

Within physiological limits, Pa_{O_2} does not affect CBF. However, CBF begins to rise when the Pa_{O_2} falls to 50 mmHg (6.7 kPa) and roughly increases two-fold at a Pa_{O_2} of 30 mmHg (4 kPa) because of tissue hypoxia and concomitant lactic acidosis.

The cerebral metabolic rate decreases by 7 per cent for each 1°C fall in body temperature. There is a parallel reduction in CBF with the decrease in the cerebral metabolic rate induced by hypothermia.

The cerebrovasculature is well innervated with serotonergic, adrenergic and cholinergic nerves, but their physiological role is unknown. The cerebral circulation has dense sympathetic innervation. When the sympathetic nervous system is activated, cerebral blood flow is maintained relatively constant by autoregulation. Under certain conditions, sympathetic stimulation can cause marked constriction of the cerebral arteries. During strenuous exercise or states of increased circulatory activity, sympathetic impulses can constrict the large and intermediate size arteries and prevent the high pressure from reaching the small blood vessels and prevent haemorrhage.

During hypercarbia, vasodilatation occurs in the normal cerebral vasculature. Cerebral steal refers to the phenomenon whereby there is a decreased blood flow in relatively ischaemic areas of the brain as a result of hypercarbia-induced vasodilatation in non-ischaemic areas. Conversely, vasoconstriction in the normal areas of brain induced by hypocarbia can redistribute blood to ischaemic areas. This is called the inverse steal, or Robin Hood, phenomenon.

INTRACRANIAL PRESSURE (ICP)

The normal intracranial pressure is between 5 and 15 mmHg. There are rhythmic variations in ICP, named A, B and C pressure waves (Fig. 2.13). Although B and C waves are associated with variations of respiratory movements and blood pressure, A (or plateau) waves are associated with raised ICP. Sneezing, coughing and straining can increase ICP by 45 mmHg (60 cmH$_2$O).

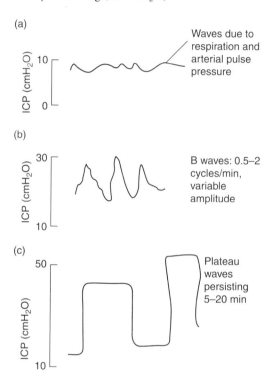

Figure 2.13 *Intracranial pressure (ICP) waveforms. (a) Normal; (b) B waves; and (c) plateau (A) waves.*

In adults the rigid cranium essentially forms a fixed brain volume. The contents of this fixed volume are the brain parenchyma, CSF and cerebral blood. Within the cranial vault, changes in the volume of any one component will alter the volume of one or more of the other components of the cranial contents. This is called the Monro–Kellie hypothesis (Fig. 2.14).

Therefore, the brain tissue mass, CSF volume and cerebral blood volume will determine ICP. The relationship between volume and pressure in the intracranial space is called compliance. Since

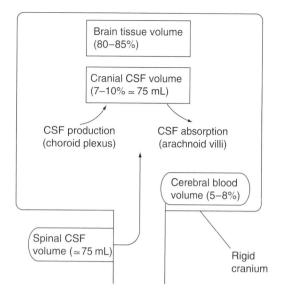

Figure 2.14 *Monro–Kellie doctrine.*

this is usually described as a change in ICP for a unit change in intracranial volume, it is strictly intracranial elastance that is measured or described (Fig. 2.15). (Compliance is strictly a change in volume per unit pressure.)

Various compensatory mechanisms that tend to minimize changes in ICP are invoked in the

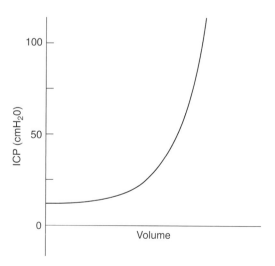

Figure 2.15 *Pressure–volume relationship (elastance) of brain.*

face of a rising ICP. Changes in CSF distribution and bulk flow are initial compensatory responses. Initially, CSF is displaced into the spinal subarachnoid space or the reabsorption rate may increase. A limited reduction in cerebral blood volume or cerebral tissue may follow.

BRAIN METABOLISM

Brain metabolism supports and regulates synaptic transmission, voltage-dependent and receptor-operated ion channels, and the synthesis, transport and packaging of transmitters. Normally, the brain uses D-glucose as its sole metabolic substrate. Glucose is transported into the brain by facilitated diffusion from the blood and this is independent of insulin. At rest the brain extracts about 10 per cent of the glucose delivered to it. In the neonate and during starvation in the adult, oxidation of ketone bodies becomes an important energy source for the brain. The brain uses 3–5 mL per minute of oxygen per 100 g brain tissue (about 15–20 per cent of resting total body oxygen requirements).

CLASSIFICATION OF SENSORIMOTOR NEURONS

Several factors influence the functional properties of peripheral nerves. The greater the axon diameter, the greater the conduction velocity. Roman numerals I, II and III are used to classify axons according to their size, while the letters A, B, and C are used to classify axons according to propagation velocity. A fibres include all the peripheral myelinated fibres and are subdivided into α, β, γ and δ fibres by their decreasing size and conduction velocities. B fibres are small myelinated preganglionic autonomic fibres. C fibres are small unmyelinated afferent and efferent fibres. However, the two classifications tend to overlap considerably (Table 2.1). The alphabetical A, B and C system is usually used for cutaneous afferents and the numerical I, II, III and IV system for muscle afferents. As the refractory period decreases with increasing diameter of the axons, the action potentials of the larger fibres are shorter in duration.

Table 2.1   *Classification of nerve fibres*

Type	Axon diameter (μm)	Conduction velocity (m/s)	Function
(a) Classification of motor nerves			
Aα	10–20	60–120	Motor
Aβ	5–10	40–70	Touch/pressure
Aγ	3–6	15–30	Proprioception
Aδ	2–5	10–30	Pain, temperature
B	1–3	3–15	Preganglionic (sympathetic)
C	0.5–1	0.5–2	Pain/temperature
(b) Classification of sensory nerves			
I	12–20	72–120	Touch, pressure, vibration
II	4–12	24–72	Position sense
III	1–4	6–24	Touch, pressure, cold
IV	0.5–1	0.5–2	Touch, pressure, cold, warm, dull ache

THE SENSORY SYSTEM

The sensory system detects, transmits and analyses stimulus information from inside and outside the body. This information is transmitted to the CNS where it is integrated to produce a conscious perception of the stimulus, alter behaviour, or elicit a reflex response. The sensory system consists of receptors, the afferent nerve fibres of the sensory neurons, and the central pathways activated by the stimuli.

Somatic senses can be divided into three main components: mechanoreception (tactile and position sensations), thermoreception (increases or decreases in temperature), and nociception (detection of tissue damage or release of specific pain mediators). The sensory modalities conveyed include discriminative (localized) touch, crude touch, pressure, vibration, and senses of static body position and movement which are collectively referred to as proprioception.

Receptors are structures that convert the various forms of energy produced by stimuli into nerve impulses, and they are classified according to their location in the body. Superficial somatic receptors are found on the body surface and mucosal membranes, and sense touch, pain and temperature. Deep somatic or proprioceptive receptors are found in muscles, tendons and joints, and these sense stretch and movement. Sensory signals that arise from internal organs (ectodermally derived structures) are called visceral sensations. Visceral receptors sensing stretch, chemical and pain stimuli are found in the walls of blood vessels, the gut, bladder and the peritoneum.

General properties of receptors

Receptors are usually specific or selective to one form of stimulus modality, and are activated in a limited area or receptive field. The nature and intensity of the stimulus may be important in evoking a transformation of the stimulus into a change in membrane potential. Temporal factors including discharge frequencies, any receptor adaptation and the duration of the stimulus may influence the change in membrane potential. Spatial factors such as the area stimulated and the number of receptors activated may also be important. Receptor excitability may be influenced by the CNS as well as by local factors.

A stimulus to a receptor causes a local change in membrane potential due to an increase in Na^+ conductance, and this produces a generator potential and then a propagated action potential along the sensory nerve to the CNS.

When a continuous stimulus is applied, adaptation or a decline in the receptor potential occurs. In slowly adapting receptors (muscle spindles and Golgi tendon organs) the receptor potentials decay slowly and are therefore prolonged. In rapidly adapting receptors (Pacinian corpuscles) the generator potentials fall rapidly below the threshold for action potential generation. Sensory adaptation is important for providing information about the rate of stimulus application and decreases the amount of afferent information to the brain. Sensory adaptation occurs by two mechanisms:

- A reduction in generator potential despite continuous stimulus application
- A diminished excitability of the receptor due to increased K^+ conductance, inactivation of Na^+

channels or increased activity of the Na^+/K^+ electrogenic membrane pump

The sensory system can code the intensity, location and quality of a stimulus. The modality of sensation depends on the specificity of the receptor and the location of a stimulus related to the receptive field. Regions of the body that are densely innervated (lips and fingers) can provide precise information about the shape, size and position of the stimulus. A large or suprathreshold stimulus can also increase the number of neurons responding by recruitment of more sensory units. Stimulus intensity may be coded by the frequency of impulse in each sensory unit, or by the number of active sensory units. There appears to be a linear relationship between the receptor generator potential amplitude and sensory nerve action potential frequency. The stimulus intensity may have a linear or a logarithmic (Weber–Fechner law) relationship with the frequency of action potentials in the sensory nerves.

Cutaneous sensors

Skin receptors sense touch, pain and temperature. Crude somatosensory mechanical sensations are mediated by unmyelinated nerve fibres. Fine touch, vibration and pressure sensation are transmitted by large myelinated fibres. Temperature and pain are transmitted by small myelinated Aδ fibres and unmyelinated C fibres.

The Pacinian corpuscle, present in the skin and fascia, is a rapidly adapting receptor with a large receptive field, that detects vibration. Meissner's corpuscle – another rapidly adaptive encapsulated receptor but with a small receptive field – is present in the skin of fingertips and lips which are particularly sensitive even to light touch. Merkel's disc is a slowly adapting receptor with a small receptive field and are thought to signal continuous touch of objects against the skin. Ruffini's corpuscle are slowly adapting encapsulated endings located in the skin and deeper tissues as well as joint capsules. They exhibit little adaptation and therefore signal continuous touch and pressure applied to the skin or movement around the joint where they are located (Fig. 2.16).

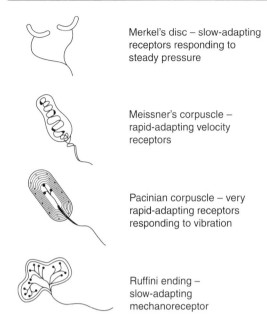

Figure 2.16 *Sensory receptors.*

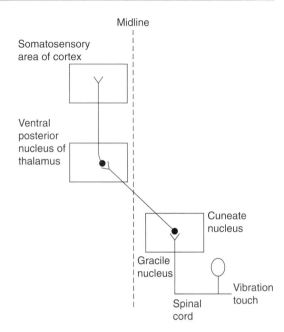

Figure 2.17 *Schematic diagram of the dorsal column.*

Thermoreceptors are specialized receptors located on the free endings of small myelinated (Aδ) and unmyelinated (C) fibres. There are two distinct types: (1) warm receptors active when the skin temperature is between 30 and 43°C; and (2) cold receptors active when the skin temperature is between 15 and 38°C. Specialized sensory receptors include the olfactory, taste and visual receptors.

Sensory pathways

The afferent nerve fibres enter the CNS via the spinal cord dorsal roots, the trigeminal nerve and the vagal nerve roots. After entering the spinal cord, afferent fibres ascend within the spinal cord via the dorsal column or the anterolateral spinothalamic pathways.

The primary or first-order neuron has its cell body in the dorsal root ganglion and its axon enters the dorsal column. The large myelinated Aβ fibres ascend in the dorsal column to the gracile and cuneate nuclei where they synapse with second-order neurons. These then cross over (decussate) to the opposite side of the medulla where they form the medial lemniscus, and end at the ventrobasal nucleus of the thalamus where they synapse with third-order neurons. The dorsal

column transmits discriminative touch, vibration and proprioception and forms the discriminative pathway (Fig. 2.17).

Fibres from the lateral division of the dorsal root are the smaller myelinated (Aδ) and unmyelinated C fibres. These fibres branch caudally and rostrally in the dorsolateral tract of Lissauer and then synapse with second-order neurons found in laminae I and V. These second-order neurons cross the midline to the contralateral side of the spinal cord and ascend in the anterolateral or spinothalamic tract, pass upwards in the lateral lemniscus, and terminate at the posterolateral nucleus of the thalamus by synapsing with third-order neurons, which project to the somatosensory cortex (Fig. 2.18). The fibres in the anterolateral pathway transmit touch, pressure, temperature and pain stimuli, and form the non-discriminatory (non-specific) pathway. Damage to the anterolateral pathway produces a dissociated sensory loss, as touch is preserved (via the pathway in the dorsal column) but with loss of pain and temperature sensation.

Somatosensory information from the face is carried in the trigeminal nerve and enter the brain stem at midpontine levels where the primary sensory fibres terminate in the principal trigeminal sensory nucleus, and then the axons cross the

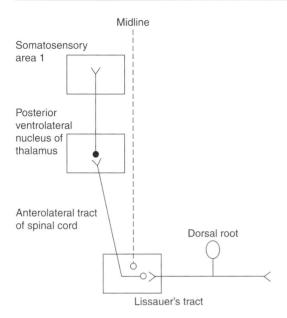

Figure 2.18 *Spinothalamic tract.*

midline and course rostrally and terminate in the ventral posteromedial nucleus. The fibres segregate into two pathways: one for processing pain, temperature and crude touch; and the other for discriminative touch, vibration, and proprioception.

Sensory information is also transmitted via two uncrossed neuron chains which terminate in the cerebellar hemisphere of the same side.

Visual pathway

The visual pathway is from the retina to the cerebral cortex. The rods and cones are the receptors in the retina. The rods are found throughout the retina (except for the fovea) and contain the photopigment rhodopsin. They are used in night vision and have high sensitivity, but do not transmit colour vision. The cones contain three pigments: erythrolabe (red-sensitive), chlorolabe (green-sensitive) and cyanolabe (blue-sensitive). They are concentrated in the fovea and are used in day and colour vision. The rods and cones are connected to ganglion cells, the axons of which are carried in the optic nerve and in the lateral geniculate body to the primary visual cortex. Fibres from the nasal halves of the retinae cross, while those of the temporal sides remain ipsilateral. Optic nerve fibres also pass to the midbrain

pretectal areas (pupillary reflexes) and to the superior colliculus (eye movements).

The direct light reflex is produced when the pupil constricts in response to light shone into that eye. The consensual light reflex occurs when the pupil in the opposite eye also constricts. The afferent pathway of the light reflex is the optic nerve to the pretectal region of the midbrain, and the efferent pathway is via the parasympathetic fibres of the third cranial nerve (Fig. 2.19).

Vestibular pathways

The vestibular system senses spatial orientation and movement and maintains body posture. The ampulla of the inner ear semicircular canals sense rotatory acceleration of the head, while the saccule and utricle sense the force of gravity and linear acceleration of the head.

The vestibular axons have their cell bodies in the vestibular ganglion and project to the vestibular nuclei in the pons and the dorsorostral part of the medulla. The lateral vestibular nucleus receives axons from the macula of the utricle, the cerebellum and the spinal cord. These axons facilitate α and γ motor neurons, which innervate skeletal muscles. The medial and superior vestibular nuclei control neck movements. The superior vestibular nuclei also have efferent outputs in the medial longitudinal bundle and the oculomotor nuclei, and mediate nystagmus (Fig. 2.20). The inferior vestibular nucleus integrates the vestibular apparatus and the higher brain centres.

Relay nuclei

Somatosensory pathways have several synaptic interruptions formed by relay nuclei in the dorsal column of the spinal cord: the gracile and cuneate nuclei of the dorsal column; the ventral posterior medial nuclei involving the trigeminal pathway and the taste pathway; and the ventral posterior lateral nucleus of the spinothalamic tract. The relay nuclei are important for the interaction and modification of the sensory input.

The mechanisms involved in the modification of input into the relay nuclei may be presynaptic, postsynaptic, or both. Interneurons are important for

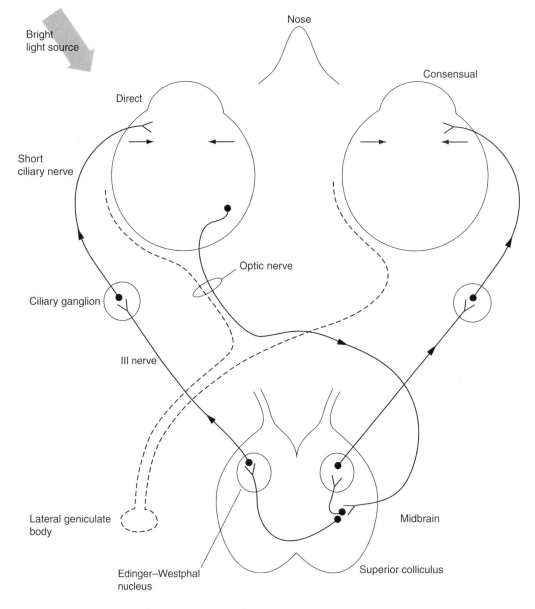

Figure 2.19 *Pupillary reflexes (direct and consensual).*

pre- and postsynaptic inhibition, producing lateral or afferent inhibition. This is particularly important in the visual and auditory systems. Functionally, lateral inhibition is important for spatial discrimination and the ability to localize stimuli.

The thalamic nuclei are the main somatosensory relay nuclei. The ventral posterolateral nucleus of the thalamus is important for nociceptive stimuli. The other thalamic nuclei may be involved in arousal, changes in affect, and learning and memory.

Somatosensory cortex

Sensory impulses project to the somatosensory cortex. The postcentral gyrus comprises the primary somatosensory cortex which corresponds to Brodmann's areas 3, 1 and 2. A smaller secondary somatosensory cortex lies along the lateral fissure. Within the primary somatosensory cortex, segregation of the body is maintained so that the face is located ventrally near the lateral fissure, the

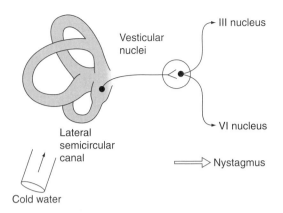

Figure 2.20 *Vestibulo-ocular reflex (caloric test).*

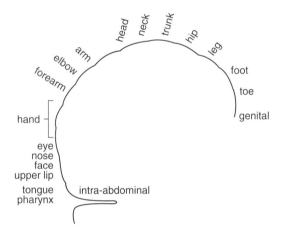

Figure 2.21 *Representation of the body in the somatosensory area.*

upper limb continues medially and dorsally from the face region and extends to the convexity of the hemisphere, and the lower extremity projects on to the medial surface of the hemisphere (Fig. 2.21). Those body surfaces with a higher density of sensory receptors are represented by larger areas in the cortex than those with a lower density of receptors.

The primary somatosensory cortex contains six horizontally arranged cellular layers numbered I to VI, with layer I at the cortical surface. Layer IV receives important projections from the ventrolateral posterior nucleus and the ventral posteromedial nucleus. From here information is conveyed to layer I to III and ventrally to layers V and VII.

Lesions that involve the primary somatosensory cortex result in the inability to localize precisely fine cutaneous stimuli on the body surface, the inability to assess the degree of pressure on the skin, and the inability to identify objects by touch or texture (astereognosis). Lesion that involve Brodmann's areas 5 and 7 damage the association cortex for somatic sensation resulting in the inability to recognize objects that have a relatively complex shape and the loss of awareness of the contralateral side of the body.

MOTOR FUNCTION AND ITS CONTROL

The cerebral cortex generates the neural impulses for voluntary movements of the skeletal muscles. These neural impulses are modulated by the basal ganglia, the cerebellum and the brainstem, and are then conveyed to the spinal cord, which contains the final common pathways to the skeletal muscles. Sensory feedback from muscle and joint receptors adjusts the motor commands during movement.

Spinal cord pathways

There are two types of motor neurons: α motor neurons, which are large cells with large axons ($10–20\,\mu$m in diameter) innervating the extrafusal fibres of skeletal muscle; and γ motor neurons, which are smaller cells with axons ($3–6\,\mu$m diameter) which innervate the muscle spindle intrafusal fibres. The cell bodies of the motor neurons are located in the ventral horn of the spinal cord, with those of the trunk and proximal limb lying in the medial part of the ventral horn, and those of the distal limb muscles lying in the lateral part of the ventral horn.

The descending motor pathways from the cerebral cortex excite the spinal cord motor neurons both directly and indirectly. There are six descending motor pathways, two from the cerebral cortex (the cortospinal and corticobulbar), and four from the brainstem (the rubrospinal, vestibulospinal, tectospinal and reticulospinal) as shown in Fig. 2.22.

The pyramidal (corticospinal) tract is the most important motor pathway for motor control, and consists of axons of pyramidal cells in layers III and V of the premotor, precentral and postcentral

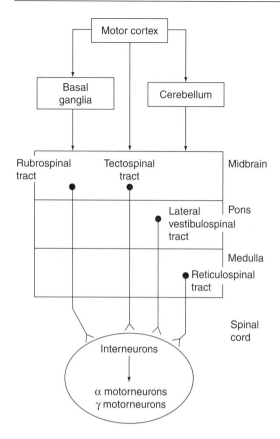

Figure 2.22 *Descending motor tracts from higher motor centres.*

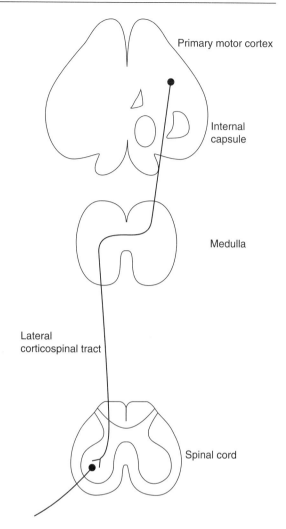

Figure 2.23 *Motor pathway – corticospinal tract.*

gyrus of the cortex (Fig. 2.23). The tract descends through the posterior part of the internal capsule and passes through the pons and medulla. In the ventral medulla the axons cross to the contralateral side and proceed to the spinal cord. The corticospinal tract controls muscles responsible for precise movements (fingers, hands) and the laryngeal muscles.

The rubrospinal, vestibulospinal and reticulospinal tracts form the extrapyramidal tract. These tracts maintain postural tone and direct voluntary movement. The rubrospinal tract arises from cells in the red nucleus, crosses contralaterally in the brainstem, receives input from the cerebral cortex, and runs down the spinal cord. The lateral vestibulospinal tract arises from the lateral vestibular nucleus and does not cross to the contralateral side. The reticulospinal tract originates from the reticular system in the pons and medulla.

The extrapyramidal system receives inputs from the cerebral cortex and cerebellum (Fig. 2.24).

Reticular formation

The reticular formation is a diffuse aggregation of cells (with a network of fibres that run in all directions) in the core of the brainstem. The reticular formation is concerned with somatic muscle control, and regulation of eye, neck, trunk and limb movements. They also receive somatic and proprioceptive sensory signals as well as descending inputs from the cerebral cortex and the limbic system. They are also interconnected with the fastigial and

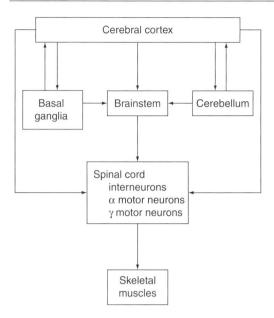

Figure 2.24 *Motor centres and connections.*

intermediate nuclei of the cerebellum. The reticular system can be divided into the pontine and medullary parts. The pontine part projects ipsilaterally down the spinal cord, whilst the medullary part sends axons down both sides of the spinal cord. The reticular fibres terminate on the ventromedial group of interneurons. The pontine part facilitates the antigravity reflexes and is important for the automatic maintenance of the erect posture whereas the medullary part suppresses spinal reflexes during sleep and may override spinal influences in voluntary movement via descending pathways.

Basal ganglia

The basal ganglia consist of five nuclei: the substantia nigra, the subthalamic nucleus, the globus pallidus, the putamen, and the caudate nucleus. They are extensively connected to the cortex and thalamus. The direct and indirect circuits that interconnect the structures comprising the basal ganglia are extremely complex. The direct pathways activate the muscles required to accurately perform movement whereas the indirect pathways inhibit muscles that would interfere with the intended movement. Lesions of the subthalamic nuclei produce disorders characterized by excessive, abnormal involuntary movements such as

athetosis (slow writhing movements), chorea (involuntary jerky movements of the face and extremities) and ballismus (violent flailing movements of one limb involving proximal muscles). Degeneration of dopaminergic cells of the substantia nigra causes Parkinson's disease.

Cerebellum

The cerebellum consists of the anterior, posterior and flocculonodular lobes. The cerebellar cortex consists of an outer layer containing basket and stellate cells, a middle layer containing Purkinje cells, and an inner layer containing granular cells with interneurons called Golgi cells. The cerebellum functions at several levels in motor control. At the spinal level it facilitates stretch reflexes so that the ability to manage an unexpected load change is enhanced. The cerebellum is connected to the vestibular system in the brainstem and this regulates posture, equilibrium and eye movements. The output of the cerebellum via the thalamus influences the cerebral cortex to provide accessory commands that control complex motor skills.

Input to the cerebellum is from three sources: directly from the cerebral cortex; from the brainstem via the extrapyramidal tracts and the vestibular system; and from the ascending spinal pathways via the dorsal and ventral spinocerebellar tracts. The spinocerebellar pathways form the major afferent pathway to the cerebellum and transmit proprioceptive information from the joints, muscles and skin. Neural impulses from the cerebellar nuclei are transmitted directly to the brainstem nuclei and then to the cerebral cortex via the thalamus, or project to the spinal cord (Fig. 2.25).

Functionally, the cerebellum can be divided into three parts: the archicerebellum consisting of the flocculonodular lobe; the paleocerebellum consisting of the anterior lobe; and the neocerebellum consisting of the posterior lobe (Fig. 2.26).

As the flocculonodular lobe is connected to the vestibular nucleus it is associated with the control of balance. The anterior lobe receives input from the spinocerebellar tracts and is responsible for controlling the tone of muscles maintaining posture and equilibrium. The posterior lobe is connected to the cerebral cortex and controls and coordinates motor function.

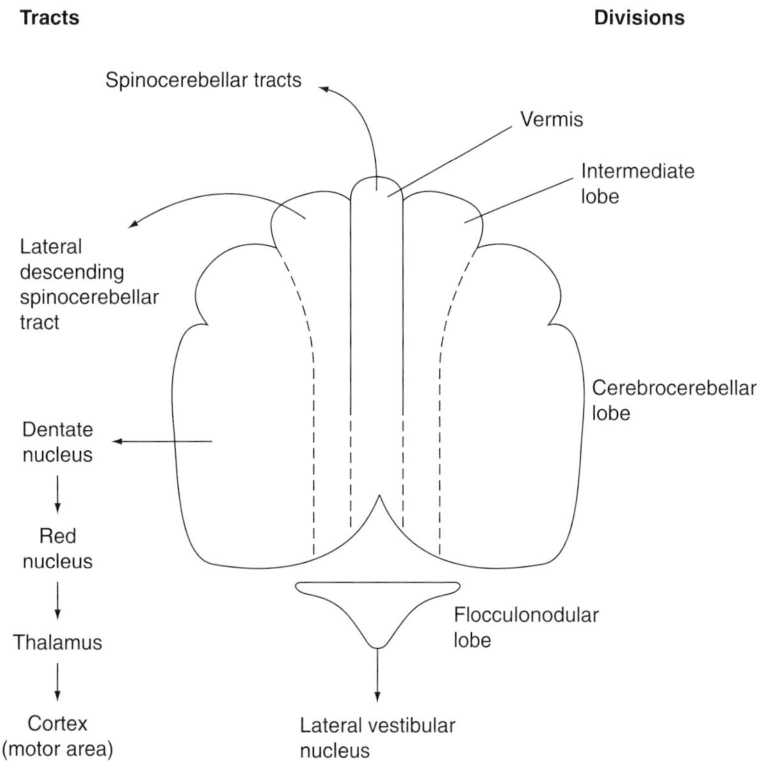

Tracts

Divisions

Spinocerebellar tracts

Vermis

Intermediate lobe

Lateral descending spinocerebellar tract

Cerebrocerebellar lobe

Dentate nucleus

↓

Red nucleus

↓

Thalamus

↓

Cortex (motor area)

Flocculonodular lobe

Lateral vestibular nucleus

Figure 2.25   *Tracts from the cerebellum.*

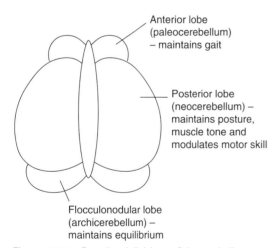

Anterior lobe (paleocerebellum) – maintains gait

Posterior lobe (neocerebellum) – maintains posture, muscle tone and modulates motor skill

Flocculonodular lobe (archicerebellum) – maintains equilibrium

Figure 2.26   *Functional divisions of the cerebellum.*

Motor cortex

The motor cortex generates and controls motor commands which are transmitted to the descending pyramidal and extrapyramidal tracts. The cerebral cortex contains three motor areas: the primary motor cortex, the supplementary motor

cortex, and the premotor cortex (Fig. 2.27). The primary motor cortex (Brodmann's area 4) is located in the frontal lobe anterior to the central sulcus, and contains Betz cells and is responsible for the excitation of spinal motor neurons, which generate muscular movement. Electrical stimulation of the primary motor complex produces only simple movements (flexion or extension) on the contralateral side. The primary motor cortex can

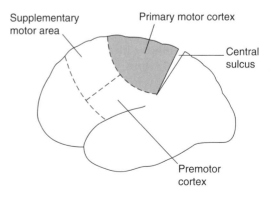

Supplementary motor area

Primary motor cortex

Central sulcus

Premotor cortex

Figure 2.27   *Motor areas of the cerebral cortex.*

modify motor commands because it receives sensory input from the spinal cord.

The supplementary motor area is located on the medial surface of each cerebral hemisphere above the cingulate gyrus, and controls and coordinates fine complex movements. The premotor cortex is located immediately anterior to the lateral part of the primary motor cortex. This cortex forms a portion of Brodmann's area 6 and stimulation of this area produces movements that involve groups of muscles. Broca's area (motor speech area) lies anterior to the primary motor cortex near the Sylvian fissure.

Limbic system

The limbic system is a large complex of brain structures composed of subcortical and cortical components. The subcortical components include the hypothalamus, septum, hippocampus, amygdala and parts of the basal ganglia. Surrounding the subcortical structures is the limbic cortex which is composed of the cingulated gyrus, orbitofrontal cortex, subcallosal gyrus and parahippocampal gyrus (Fig. 2.28).

The limbic system is the site of generation of emotion. It integrates many emotional and behavioural reflexes. Emotional expression arising from limbic system activity is mediated by the hypothalamus through the autonomic and somatic nervous systems. These include pressor and sympathoexcitatory responses triggered by an alerting stimulus, or a depressor response triggered by an emotional shock.

The hippocampus is concerned with learning, memory and behaviour. Stimulation of the hippocampus causes a range of somatovisceral responses associated with emotions and arousal (pupillary dilatation, defence movements of the body). Lesions of the hippocampus lead to anterograde amnesia and it is thought that the hippocampus is involved in the transformation from short-term to long-term memory. The amygdaloid nuclei are linked with fear and rage behaviour. The posterior orbital gyrus controls autonomic reactions, especially those affecting the cardiovascular, respiratory and gastrointestinal systems. The anterior cingulate gyrus is important for arousal reactions and control of movements. The periamygdaloid and prepyriform cortical

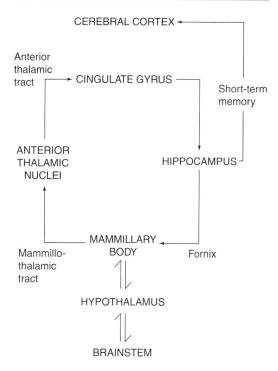

Figure 2.28 *Functional elements of the limbic system.*

areas are connected to the olfactory organs and are concerned with olfactory sensation.

Spinal control of movement

The spinal cord contains neural circuits that produce reflexes – stereotyped actions in response to peripheral stimuli. These reflexes may be simple or complex and produce purposeful movements rapidly.

A reflex can be defined as an automatic or involuntary stereotype response to a stimulus mediated by a receptor, an afferent pathway and an effector organ. The reflex pathway generally consists of a receptor, an afferent neuron, and is integrated by interneurons in the spinal cord. The final common efferent path is a motor neuron to the effector organ. Spinal cord ventral horn motor neurons may be either α motor neurons (14 μm in diameter and rapid conduction velocity) or γ motor neurons (5 μm in diameter and slower conduction velocity). There are also interneurons which are highly excitable and may have high spontaneous firing rates. The Renshaw cell is a special interneuron that receives collateral branches of collateral branches of motor neuron axons. The Renshaw cell provides inhibitory connections with the same or neighbouring motor neurons via its own axons.

Reflex actions have a number of characteristics which include stimulus specificity, a latent period of response due to synaptic delay, and facilitation by spatial or temporal factors.

The withdrawal reflex is an important cutaneous reflex which quickly removes the body from a painful or noxious stimulus. The receptors are the pain receptors on the free nerve endings of Aδ and C fibres, and the effector organs are skeletal muscles that withdraw the body from the stimulus. The withdrawal reflex is via a polysynaptic pathway with several interneurons linking the pain receptor with the α motor neuron of the limb, producing contraction of the flexor muscles. A continuation of the withdrawal reflex occurs even after the receptor stops firing because reverberating circuits produced by branches of the interneurons re-excite and prolong the motor neuron discharge, known as 'afterdischarge'. Interneurons from pathways that cross the spinal cord can stimulate the extensor motor neurons on the opposite side of the body to produce the cross-extensor reflex. Inhibition of antagonist muscles to the flexor muscles may occur when a reflex activates one group of motor neurons with a simultaneous inhibition of its antagonistic motor neurons – 'reciprocal innervation'.

The stretch or myotactic reflex is a monosynaptic reflex that causes contraction of a muscle when it is stretched (the knee-jerk reflex) (Fig. 2.29). The receptors activated by stretch are the primary muscle spindle afferent fibres carried by type Ia afferent fibres to the spinal cord. In the spinal cord, the Ia afferent axons synapse with motor neurons of the same muscle and also motor neurons of synergist muscles. Thus, the afferent Ia impulses produce contraction of both the stretched muscle and its synergist muscles. Collaterals of the type Ia fibre may synapse with inhibitory interneurons which inhibit antagonist muscles, causing reciprocal inhibition. The latency period of a stretch reflex is short because it is a monosynaptic pathway. Type IIa secondary afferent fibres can activate the synergist muscles to reinforce the reflex.

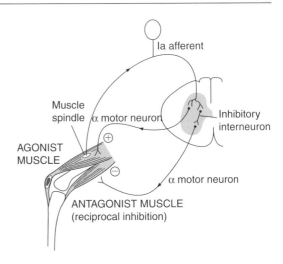

Figure 2.29 *Monosynaptic stretch (myotactic) reflex.*

The stretch reflex has two components: a phasic component, which is short and intense and produced by primary afferent (Ia) activity; and a tonic component, which is weaker and longer, and is produced by secondary (type II) afferent fibre activity.

The lengthening reaction is a protective reflex preventing muscle-fibre damage during a strong contraction by inhibiting activity in the α motor neuron. The lengthening reflex may have an important role in regulating tension during normal muscle activity. Thus, the force generated by a muscle during contraction is the stimulus for its own relaxation.

MUSCLE TONE

Muscle tone is the reflex resistance of a muscle to passive stretch, and is produced by tonic excitation by a steady level of motor neuron activity that keeps the muscle at a preset length. Muscle tone is necessary for the maintenance of posture, particularly against gravitational forces.

During a muscle contraction there would tend to be a reduced rate of Ia afferent discharge from the muscle spindles, preventing the CNS from detecting the rate and extent of muscle shortening because of the shortening of the intrafusal fibres. However, this is avoided by spontaneous γ motor neuron activity which causes the intrafusal muscle fibres to shorten along with the extrafusal fibres; as the central region of the intrafusal fibre does not contain sarcomeres, it remains stretched, thus maintaining afferent nerve activity. Thus coactivation of α and γ motor neurons prevents reduced Ia afferent discharge during muscle contraction. Direct stimulation of γ motor neurons can initiate movements, but this does not normally occur.

CONTROL OF POSTURE

Control of posture is the maintenance of the position (or attitude) of the head, limbs and trunk in space and is accomplished by reflexes acting on eye, neck, trunk and proximal limb muscles. The vestibulospinal and reticulospinal tracts control the motor neurons of the antigravity (postural) muscles. The postural reflexes involved are static reflexes against gravitational forces and vestibular reflexes in response to acceleration forces.

Static reflexes evoked by gravitational forces are essentially stretch reflexes combined into complex patterns. For example, the muscles extending the knee are continuously stretched during standing and therefore remain reflexly contracted. Segmental static reflexes may evoke responses on both sides of the body as a result of reciprocal innervation. General static reflexes can be produced by stimulation of the labyrinths. In addition, righting reflexes evoked by stimulation of skin receptors, labyrinthine receptors and muscle proprioceptors may be responsible for controlling muscle contraction required for adopting and maintaining posture. This is facilitated by visual reflexes.

Vestibular reflexes are evoked by linear and angular acceleration of the head. Receptors in the utricle and saccule of the inner ear respond to linear acceleration and evoke reflexes through the vestibulospinal tracts. Rotatory stimuli are detected by receptors in the semicircular canal and evoke reflex responses involving the eye, neck and proximal arm muscles. These mechanisms excite the α and γ motor neurons of the antigravity muscles responsible for muscle tone via the vestibulospinal and reticulospinal tracts.

There is close integration of the afferent visual and vestibular information about head position that is responsible for the control of posture. Both the visual and vestibular systems provide information about the position of the head in space, and signals from the vestibular system are continually checked against the visual signals.

ELECTROENCEPHALOGRAPHY

Electroencephalography is the recording of the spontaneous electrical activity of the brain. The electroencephalogram (EEG) is generated by the superficial layer of pyramidal cells by changes in postsynaptic potentials in the dendrites oriented perpendicular to the cortical surface. The current is the result of the summation of EPSPs and IPSPs. The recorded potentials range from 0 to $200\,\mu V$ and their frequency range from once every few seconds to 50 or more per second.

The electrical potentials of the EEG are the algebraic summation of the postsynaptic potentials of the pyramidal cells in response to rhythmic discharges from the thalamic nuclei. Input from the reticular formation interrupts these rhythmic potential changes and causes a desynchronization of the cortical waves.

The EEG is analysed by measuring the frequency and amplitude of the electrical activity and by recognition of wave patterns. There are three basic types of activity:

- Continuous and rhythmical
- Transient
- Background activity

The EEG is usually symmetrical in the two hemispheres. The normal EEG of the awake and alert state is of irregular (asynchronous) low-voltage $(30–80\,\mu V)$ waves of high frequency. The EEG of a deep sleep state is of regular (or synchronized), high-voltage (amplitude, i.e. microvolt range) waves of a low frequency.

The frequency spectrum of the EEG waves is: $<4\,Hz$, delta wave; $4–8\,Hz$, theta wave; $8–13\,Hz$, alpha wave; and $>13\,Hz$, beta wave. The alpha waves (approximately $50\,\mu V$) arise from the parieto-occipital area. The waves are symmetrical (left and right hemispheres) and are found in normal, awake but resting (eye closed) individuals. Activity or mental stimulation disrupts the alpha waves. The beta waves with a voltage of less than $50\,\mu V$, not affected by eye closure, are symmetrical and present in the frontal area. Intact thalamocortical cortical projections and functional ascending ascending reticular input to the thalamus are necessary for these waves to be recorded. Theta waves occur in the parietal and temporal areas in children but they can appear in adults during emotional stress. Delta waves occur during deep sleep, with serious organic brain disease and in infants. This sleep state is probably due to releasing the cortex from the influence of lower centres.

An abnormal EEG may reveal three types of abnormality:

- A generalized excess of slow-wave activity as seen in encephalopathy of metabolic or infective origin
- A focal excess of slow-wave activity, suggesting a focal abnormality
- An abnormal electrical discharge of high voltage, indicating an epileptiform disturbance

In cerebral ischaemia there is a loss of high-frequency (alpha and beta) waves and appearance of large-voltage (amplitude) slow (delta) waves. Prolonged ischaemia produces low-voltage slow waves, and the EEG may become isoelectric. The critical cerebral blood flow is defined as the flow below which ischaemic EEG changes will present within 3 min. Normal CBF is 54 mL/100 g per minute. The critical cerebral blood flow is 18–20 mL/100 g per minute for the awake individual. This critical cerebral blood flow varies from 10 mL/100 g per minute under isoflurane anaesthesia to 18 mL/100 g per minute under halothane anaesthesia.

EVOKED POTENTIALS

Stimulation of any sensory receptor evokes a minute electrical signal in the appropriate area of the cerebral cortex. When sensitive equipment is used to record cortical activity after multiple stimuli, the background 'noise' can be averaged out, leaving the signal evoked from the specific stimulus. An evoked potential has a smaller amplitude (1–5 μV) than the EEG (10–20 μV).

Sensory-evoked potentials are small waves produced in response to repetitive peripheral stimuli (Fig. 2.30). The waveforms are analysed for the latency (delay from stimulus to wave appearance) or amplitude (peak to trough voltage) of the characteristic waves. The waveforms are described as P (positive) or N (negative) and by their latency after stimulation.

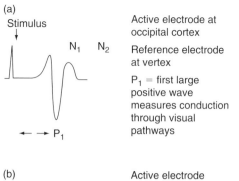

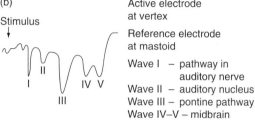

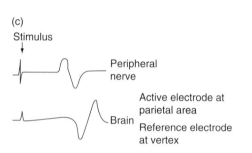

Figure 2.30   *Evoked potentials. (a) Visual-evoked potential (VEP); (b) brainstem auditory-evoked potential (BAEP); and (c) somatosensory evoked potential (SEP).*

Evoked potentials are classified according to the sensory pathway stimulated: visual-evoked potential (VEP); auditory-evoked potential (AEP); brainstem auditory-evoked potential (BAEP); and somatosensory evoked potential (SEP).

CONSCIOUSNESS

Consciousness may be described as a state of being aware of one's surroundings and of one's own thoughts and emotions, and is dependent on intact cerebral hemispheres interacting with the ascending reticular activating system in the brainstem, midbrain, hypothalamus and thalamus. In the conscious state the EEG shows irregular, high-frequency, low-voltage and asynchronous waves. Activity of the brainstem and the cerebral hemispheres determines the level of consciousness.

The reticular formation is located in the pons and medulla, and consists of small nuclei enmeshed in fibre tracts. The locus coeruleus, a group of neurons containing norepinephrine (noradrenaline) as a neurotransmitter, is found in the pontine reticular formation. Another group of neurons lying in the midline of the pons form the raphe nuclei. Some of the neurons in the pons and midbrain send axons to excite or inhibit the thalamus. Neural impulses from the reticular formation can modulate the activity of the thalamic pacemakers and thus influence cortical neuronal excitability. Non-specific nuclei of the thalamus (intralaminar and anterior thalamic nuclei) excite the cerebral cortex. Axons from the non-specific nuclei can also stimulate inhibitory interneurons linked to the thalamic nuclei to produce drowsiness. Lesions diffusely affecting the cerebral hemispheres or directly affecting the reticular activating system can impair consciousness.

SLEEP

Sleep may be described as a state associated with loss of reactivity to surroundings or unconsciousness from which one can be aroused by sensory stimulation. There are two different types of sleep: slow-wave sleep and rapid eye movement (REM) sleep. As one progresses from alert wakefulness to deep sleep, there is a gradual change in brain wave pattern from low-voltage/high-frequency alpha waves to high-voltage/low-frequency delta waves. There is also a progression from desynchronized activity (alert) to synchronous (deep sleep) patterns. REM sleep is also called paradoxical sleep because it is a sleep state, yet the brain exhibits asynchronous activity characteristic of the waking state. Sleep restores the normal balance of activity of many parts of the brain although the specific mechanism of this process is unknown. Sleep deprivation leads to irritability, sluggish thought processes and affects other body systems that regulate blood pressure, heart rate, peripheral vascular tone, muscle activity and basal metabolic rate. However, the mechanisms are not defined.

Slow-wave or non–rapid eye movement (non-REM) sleep

This is deep, restful sleep characterized by decreases in peripheral vascular tone, blood pressure, blood pressure, respiratory rate, and metabolic rate. Slow-wave sleep develops with maturation of the nervous system.

Slow-wave sleep is brought about by inhibition of the midline pontine and medullary nuclei (raphe nuclei). Three subcortical regions have been shown to be involved in the genesis of non-REM sleep. The diencephalic sleep zone, parts of the posterior hypothalamus and the surrounding intralaminar and anterior thalamic nuclei, produce synchronized low-frequency EEG waves associated with sleep. Stimulation of the forebrain sleep zone of the preoptic area also produces non-REM sleep. Various neurotransmitters are responsible for non-REM sleep, and release of prostaglandins, dopamine and muramyl peptide, and inhibition of serotoninergic neurons in the brain have all been implicated in slow-wave sleep.

Non-REM sleep is divided into four stages. In stage 1, as a person goes to sleep, the alpha waves are replaced by low-voltage theta waves of low frequency (4–6 Hz). This progresses to stage 2, which is characterized by occasional bursts of high-frequency waves (50 μV) called 'sleep spindles'. In stage 3 there are high-voltage delta waves with a frequency of 1–2 Hz, with bursts of rapid waves (called K complexes) superimposed on the delta waves. This then progresses to stage 4, when the large delta waves become synchronized.

Paradoxical or rapid eye movement (REM) sleep

It is called paradoxical sleep because the brain is quite active and skeletal muscle contractions occur and is associated with a rapid, low-voltage, irregular (desynchronized) EEG, which resembles the recording of cerebral activity seen in alert animals and humans. REM sleep lasts for 5–30 min and occurs at approximately 90 min. Babies at birth spend about 18 h per day asleep, of which 45–65 per cent is in the REM state. Adults aged over 50 years spend about 15 per cent of total sleep in REM sleep.

REM sleep is thought to be produced by sleep centres in the locus coeruleus and in the raphe nuclei of the pontine reticular formation. It is associated with large phasic waves called ponto-geniculo-occipital (PGO) spikes from the pons that pass rapidly to the geniculate body and then to the occipital cortex. REM sleep is mediated by norepinephrine (noradrenaline). There are several important features of REM sleep: dreaming occurs and can be recalled; muscle tone is substantially decreased; heart rate and respiration become irregular; muscle contractions such as rapid eye movements and bruxism occur; brain metabolism is increased by as much as 20 per cent; the electroencephalogram shows brain waves that are characteristic of the waking state; and glucocorticoid production is increased.

THE AUTONOMIC NERVOUS SYSTEM

FUNCTIONS AND STRUCTURE

The autonomic nervous system is the portion of the nervous system that controls the visceral functions of the body such as the involuntary control of cardiac output and vascular resistance, respiration, the iris, gastrointestinal function, micturition, body temperature and other neurochemical processes that maintain homeostasis. It is the motor system for the visceral organs, blood vessels and secretory glands.

The central portions of the autonomic nervous system are located in the hypothalamus, brain stem and spinal cord. The limbic system and parts of the cerebral cortex send signals to the hypothalamus and lower brain centres and this can influence the activity of the autonomic nervous system.

HYPOTHALAMUS

The hypothalamus is the centre of both neural and endocrine control of the internal organs of the body. The hypothalamus (which extends from the mammillary bodies posteriorly to the lamina terminalis anteriorly) can be divided into four functional regions: the anterior, medial, lateral and posterior hypothalamus.

- The *anterior hypothalamus* contains the supraoptic and paraventricular nuclei that give rise to the hypothalamo–hypophysial tract to the posterior lobe of the pituitary. The anterior hypothalamus controls the parasympathetic nervous system, heat-loss mechanisms, and the release of antidiuretic hormone (ADH) and oxytocin. The cells of the supraoptic nucleus have osmoreceptor functions and release ADH in the presence of increased plasma osmolality. The paraventricular nuclei release oxytocin. Releasing and inhibitory factors from the hypothalamus regulate the anterior pituitary gland. Stimulation of the anterior hypothalamus can also induce sleep.
- The *medial hypothalamus* contains the ventromedial and dorsomedial nuclei, and controls energy balance and sexual behaviour. The ventromedial nuclei inhibit appetite via the satiety centre (via sensing blood glucose levels).
- The *lateral hypothalamus* contains the tuberal nuclei and the medial forebrain bundle that carries the efferent pathways from the hypothalamus into the brainstem. It is associated with emotions and defence reactions. The thirst centre is located in the lateral hypothalamus. The lateral hypothalamus is responsible for the desire to seek food whereas the ventromedial nucleus is responsible for satiety.
- The *posterior hypothalamus* is the principal site of sympathetic nervous outflow. It controls the activity of the vasomotor centres of the brain, and is responsible for the sympathetic vasoconstrictor response to cold temperatures. When stimulated by the ascending reticular system, the posterior hypothalamus maintains wakefulness and triggers the alert mechanisms. The posterior and lateral hypothalamic areas increase blood pressure and heart rate whereas the preoptic area decreases blood pressure and heart rate. These effects are mediated by cardiovascular centres in the pontine and medullary reticular formation. Body temperature is controlled by neurons in the preoptic area that sense changes in the temperature of blood flowing through the area and activate body temperature-lowering or body temperature-elevating mechanisms.

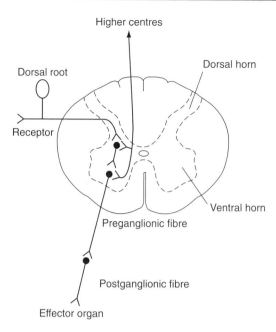

Higher centres

Dorsal root

Dorsal horn

Receptor

Ventral horn

Preganglionic fibre

Postganglionic fibre

Effector organ

Figure 2.31 *Autonomic reflexes.*

The autonomic nervous system consists of craniosacral and thoracolumbar outputs that innervate smooth muscle, the heart, the exocrine and endocrine glands as well as the enteric nervous system in the gastrointestinal tract, and the urogenital system. The main anatomical feature of the efferent limb of the autonomic nervous system is that it has two neurons in series, with one neuron cell body located along the efferent pathway to the visceral target organ. The cell bodies of these efferent pathways form the ganglia located in chains alongside the vertebral column, in plexuses in the abdomen, or within the target organ innervated. The output of the autonomic nervous system is continuous and tonic. The autonomic nervous system therefore plays an important role in homeostasis and is also concerned with non-homeostatic organs such as the reproductive system.

In summary, the autonomic nervous system consists of visceral receptors, afferent neurons, integrating centres in the CNS and efferent neurons to effector organs. The autonomic peripheral nerves consist of separate efferent neuron outflows from the CNS at the cranial, thoracolumbar and sacral levels which consist of preganglionic and postganglionic neurons in series (Fig. 2.31).

VISCERAL AFFERENT SYSTEM

The receptors of the autonomic nervous system are not well defined, but many are mechanoreceptors, chemoreceptors or osmoreceptors. The reflex pathways are multisynaptic, involving interneurons in the dorsal horn which synapse with preganglionic neurons located in the intermediolateral column of the thoracolumbar spinal cord and in the ascending spinal tract to the brainstem. Interneurons also synapse with the preganglionic vagal neurons. Autonomic afferents from blood vessels and visceral organs may enter the CNS by two different pathways. Afferent fibres from the aorta and its major branches (and from some of the viscera) travel along visceral nerves into the sympathetic trunk and reach the dorsal root ganglia via the white or grey rami communicantes. Other visceral afferent fibres join somatic spinal nerves and then travel to the dorsal root ganglia without entering the sympathetic chain.

AUTONOMIC GANGLIA

The autonomic ganglia consist of multipolar neuronal cell bodies with long dendrites. The preganglionic axons have several branches that synapse with cells in a number of ganglia. There are more postganglionic fibres than preganglionic nerves, and so stimulation of a single preganglionic neuron can activate many postganglionic nerves, resulting in divergence. In the human superior cervical ganglion numerous preganglionic fibres converge on a single postganglionic neuron, resulting in convergence.

The autonomic nervous system differs from the somatic nervous system in that all the final motor neurons lie completely outside the CNS. The preganglionic fibres are slow-conducting B or C fibres that release acetylcholine. The postganglionic fibres that originate from the ganglia and innervate target organs are largely slow-conducting, unmyelinated C fibres.

The autonomic nervous system is divided into the sympathetic and parasympathetic systems. Many organs are innervated by both systems, usually with opposing effects (Fig. 2.32).

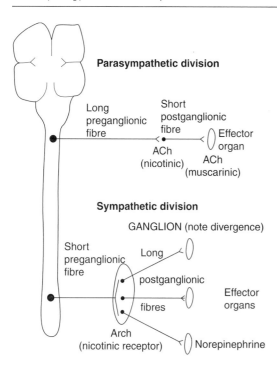

Figure 2.32 *Outline of the parasympathetic and sympathetic divisions of the autonomic nervous system.*

SYMPATHETIC NERVOUS SYSTEM

The sympathetic nerves originate from columns of preganglionic neurons in the grey matter of the lateral horn of the spinal cord from the first thoracic segment down to the second or third lumbar segment. The preganglionic fibres leave the spinal cord through the ventral roots with the spinal nerves and then leave the spinal nerves as white rami communicantes to synapse with the postganglionic neurons in the ganglia of the sympathetic chain. The white rami communicantes are myelinated B fibres.

The ganglia form the sympathetic chains. The postganglionic fibres leave the ganglia as grey rami communicantes and join the spinal nerves or visceral nerves to innervate the target organ. The grey rami are unmyelinated C fibres. The sympathetic chains extend down the length of the vertebral column and are divided into four parts:

1. A *cervical part* consisting of three ganglia (superior, middle and inferior) supplying the head, neck and thorax. The superior cervical ganglion sends postganglionic fibres to form the internal carotid plexus. The inferior cervical or stellate ganglion is fused with the first thoracic ganglion.
2. A *thoracic part* consisting of a series of ganglia from each thoracic segment. Branches from T_1–T_5 supply the aortic, cardiac and pulmonary plexuses. The greater and lesser splanchnic nerves are formed from the lower seven thoracic ganglia. The lowest splanchnic nerve arises from the last thoracic ganglion and supplies the renal plexus.
3. The *lumbar sympathetic ganglia* are situated in front of the vertebral column as the prevertebral ganglia. Some form the coeliac plexus.
4. The *pelvic part* of the sympathetic chain lies in front of the sacrum and consists of the sacral ganglia. The sacral ganglia contribute to the hypogastric and pelvic plexus distributed to the pelvic viscera and the arteries of the lower limbs.

The adrenal medullary tissues are considered to be modified sympathetic postganglionic neurons, and the nerves supplying these glands are thus equivalent to sympathetic preganglionic fibres. The postganglionic cells of the adrenal medulla release epinephrine. The postganglionic sympathetic nerves that innervate the blood vessels of muscles, sweat glands and the hair follicles in the skin release acetylcholine instead of norepinephrine (noradrenaline).

Effects of sympathetic nerve stimulation

Sympathetic nerve stimulation produces local and generalized effects. The local effects include:

*   Dilatation of the pupil and retraction of the eyelid (levator palpebrae).
*   Thoracic visceral effects of positive inotropic and chronotropic cardiac effects, pulmonary blood vessel vasoconstriction and bronchial smooth muscle relaxation.
*   Abdominal visceral effects of increased sphincteric tone and inhibition of peristalsis, leading to relaxation of the gut and reduced motility.

Vasomotor fibres in the splanchnic nerves constrict the splanchnic arterioles and redistribute the splanchnic blood volume.

- Pelvic visceral effects of relaxation of the bladder wall and the rectum with sphincter closure. Contraction of the smooth muscle of the seminal vesicles and prostate produces ejaculation.
- Cutaneous effects include piloerection, vasoconstriction and sweating.
- In the limbs, the arterioles to the skin constrict whilst the skeletal muscle arterioles vasodilate.

The general effects of sympathetic nerve activity are the maintenance of normal homeostatic mechanisms and mobilization of body reserves in stress conditions. The sympathetic system maintains homeostasis partly by a direct neural action on the target organs, and partly by releasing epinephrine into the bloodstream. Widespread activation of the sympathetic nervous system throughout the body called the 'fight or flight' reaction or stress response is brought about by fear, rage or severe pain resulting in increases in arterial pressure, muscle blood flow, metabolic rate, glycogenolysis, and mental alertness, and decreases in blood flow in the gastrointestinal tract and kidneys.

PARASYMPATHETIC NERVOUS SYSTEM

The parasympathetic nervous system arises from neurons in the brainstem and spinal cord sacral segments (S_2–S_4). As the parasympathetic ganglia are located near or within their effector organs, the parasympathetic postganglionic fibres are short, and they all release acetylcholine. The distribution of the parasympathetic outflow is restricted so that parasympathetic effects are more localized than sympathetic effects.

Cranial outflow

The parasympathetic fibres follow the distribution of the third, seventh, ninth and tenth cranial nerves. Preganglionic fibres of the third cranial nerve arise from the oculomotor nucleus and pass through the orbit to the ciliary ganglion. Postganglionic fibres from the ciliary ganglion supply the ciliary muscle and sphincter of the iris, and constrict the pupils.

Preganglionic fibres from the superior salivary nucleus of the seventh nerve form the chorda tympani and reach the submaxillary ganglion via the lingual nerve. The postganglionic fibres supply the submaxillary and sublingual salivary glands and cause salivary secretion. Preganglionic fibres arising from the inferior salivary nucleus of the ninth nerve form the lesser superficial petrosal nerve and reach the otic ganglion. The postganglionic fibres are distributed to the parotid gland via the auriculotemporal nerve and also cause salivary secretion.

The vagus nerve is the major part of the cranial parasympathetic outflow. The preganglionic fibres arise from the dorsal nucleus of the vagus in the medulla and terminate in the ganglia of plexuses or in the walls of visceral organs. Postganglionic fibres supply the sinoatrial node and atrial and junctional tissue of the heart and decrease cardiac excitability, contractility, conductivity and rate. Postganglionic fibres from the pulmonary plexus contract the circular muscles of the bronchi, producing bronchoconstriction. Vagal branches to the gastric plexus give rise to postganglionic fibres to the stomach, liver, pancreas and spleen. Stimulation of the vagus causes increased gastric motility and secretions, with relaxation of the pyloric sphincter. The intestinal branches of the vagus supply the small and large intestine down to the transverse colon. Parasympathetic stimulation increases peristalsis and relaxes the ileocolic sphincter.

Sacral outflow

The sacral outflow of the parasympathetic system arises from the second, third and fourth sacral segments of the spinal cord, and fibres enter the hypogastric plexus to innervate the descending colon, rectum, bladder and uterus. The sacral outflow of the parasympathetic system contracts the muscular wall of the rectum, relaxes the internal sphincter of the anus, and contracts the detrusor muscle of the bladder wall.

Enteric system

The autonomic nerves in the gastrointestinal tract form an extensive network in the wall of the intestinal tract in which many cells are not influenced by the central nervous system. The network is composed of ganglia and interconnecting bundles of axons that form the myenteric and submucosal plexuses in the intestinal wall. This system contains sensory neurons, intergrative interneurons, and excitatory and inhibitory motor neurons that generate coordinated patterns of activity. The excitatory interneurons and motor neurons release acetylcholine as a neurotransmitter, However, other interneurons may release serotonin, vasoactive intestinal peptide (VIP) or nitric oxide. Inhibitory neurons to smooth muscle that do not release acetylcholine or norepinephrone as a neurotransmitter are called non-adrenergic, non-cholinergic nerves.

NEUROTRANSMITTERS

Acetylcholine (ACh) is the neurotransmitter released at the synapse between the preganglionic and postganglionic neurons of both the sympathetic and parasympathetic systems. When the nicotinic receptors in the postganglionic neurons are stimulated by ACh, a fast EPSP is produced. In addition to the fast EPSP at the synapse, a slow IPSP due to inhibitory interneurons being connected to the preganglionic neurons can follow the fast EPSP. Acetylcholine is the transmitter at the postganglionic nerve (neuroeffector) endings of the parasympathetic system producing excitatory (e.g. salivary gland secretion) and inhibitory (e.g. sinoatrial node depression) actions. The neurotransmitter at the postganglionic nerve endings of the sympathetic system is norepinephrine (noradrenaline), and this produces both excitatory (e.g. vasoconstriction, positive chronotropic and inotropic cardiac actions) and inhibitory (e.g. relaxation of detrusor muscle) effects.

However, other neurotransmitters at the ganglionic synapses and neuroeffector junctions have been discovered. These chemical mediators include peptides (enkephalin, neuropeptide Y, vasoactive intestinal peptide, neurotensin), which modulate (excite or inhibit) synaptic excitability. They increase or decrease the efficacy of synaptic transmission without acting directly as neurotransmitters, a phenomenon known as 'neuromodulation'. In general, neuromodulation involves intracellular messengers, and therefore is slower (several seconds to days) than neurotransmission (milliseconds).

Cholinergic and norepinephrinergic (noradrenergic) nerve terminals also respond to other chemical mediators such as prostaglandins, adenosine, dopamine, 5-HT, peptides (enkephalin, neuropeptide Y) and GABA.

RECEPTORS

Receptors mediate actions of the neurotransmitters involved in the autonomic nervous system by activation of a second messenger system, or by a change in ion-channel permeability.

Acetylcholine receptors

Acetylcholine receptors are classified as nicotinic or muscarinic. The nicotinic receptor is a large protein consisting of five subunits (α_1, α_2, β, γ, δ) spanning the cell membrane. When the two alpha units are occupied by ACh, a conformational change occurs resulting in the ion channel opening with inward Na^+ ion and outward K^+ ion fluxes.

The muscarinic receptors are G protein-coupled receptors which activate phospholipase C, inhibit adenyl cyclase, or open K^+ ion channels. There are five subtypes of muscarinic receptors. The M_1 receptors are found in the CNS, autonomic ganglia and gastric parietal cells. The cellular effects produced by M_1 receptor activation include increased intracellular inositol triphosphate (IP_3) and diacylglycerol, causing a decrease in K^+ conductance and hence membrane depolarization. M_1 receptors are responsible for gastric acid secretion following vagal stimulation.

M_2 receptors are found in atrial and conducting tissues of the heart and decrease intracellular cyclic AMP, thus exerting an inhibitory effect on the heart by increasing K^+ conductance.

M_3 receptors produce mainly excitatory effects resulting in glandular secretion and visceral smooth contraction by an increase in inositol triphosphate production.

Adrenergic receptors

The adrenergic receptors are divided into α and β subtypes. The α receptors are subdivided into the

α_1 receptor, which activates phospholipase C, producing IP_3 and diacylglycerol as second messengers, and the α_2 receptor, which inhibits adenyl cyclase activity. The α_1 receptor mediates vasoconstriction, relaxation of gastrointestinal smooth muscle, salivary secretion and hepatic glycogenolysis. The α_2 receptor produces platelet aggregation and inhibits the release of norepinephrine (noradrenaline) from autonomic nerves.

The β receptor stimulates intracellular adenyl cyclase via a G protein. The β_1 receptor exerts positive inotropic and chronotropic cardiac effects, and also relaxes gastrointestinal smooth muscle. The β_2 receptor mediates bronchodilatation, vasodilatation, relaxation of gastrointestinal smooth muscle, hepatic glycogenolysis and muscle tremors. The β_3 receptor mediates lipolysis, especially in brown fat.

Reflections

1. The neuron is the functional unit of the nervous system. Neurons contain a nucleus, Nissl bodies (rough endoplasmic reticulum), Golgi apparatus, mitochondria, microtubules and microfilaments. The cell bodies give rise to dendrites (which receive information from other nerve cells) and axons (which transmit information to other neurons or non-neuronal cells).

2. Four types of non-neuronal cells exist in the central nervous system: astrocytes (which regulate the CNS microenvironment), oligodendroglia (form myelin in the CNS), Schwann cells (form myelin in the peripheral nervous system), microglia (CNS macrophages) and ependymal cells (line the ventricles).

3. Action potentials are generated along axons when a neuron is activated by a stimulus of a minimum strength to threshold potential. The action potential, an 'all or none' phenomenon, is caused by a large transient increase in membrane permeability to sodium ions via the opening of voltage-gated sodium channels. Following this, the sodium channels inactivate spontaneously and limit the action potential to 1 ms. The sodium channels cannot reopen until they are reprimed by spending a period at the resting membrane potential. After the passage of the action potential there is a period lasting 1–2 ms (absolute refractory period) when an action potential cannot be generated until sufficient sodium channels have returned to their resting state. Following the absolute refractory period, the axon is less excitable than normal and this period is known as the relative refractory period.

4. A synapse is a specialized junction between the axon of one neuron (presynaptic) and another (postsynaptic) neuron. The presynaptic neuron releases neurotransmitters into the synaptic cleft via calcium-dependent exocytosis of synaptic vesicles. The neurotransmitter binds to receptors in the postsynaptic membrane to cause a transient change in the membrane of the postsynaptic cell. Depolarization of the postsynaptic membrane results in an excitatory postsynaptic potential (EPSP) whereas hyperpolarization of the postsynaptic membrane leads to an inhibitory postsynaptic potential (IPSP). EPSPs and IPSPs are not 'all or none' potentials. As a result postsynaptic potentials can be superimposed on one another resulting in temporal and spatial summation.

5. The extracellular fluid in the CNS is regulated by the cerebrospinal fluid, the blood–brain barrier, and the astrocytes. Choroid plexuses form CSF. The CSF leaves the ventricles through the roof of the fourth ventricle, traverses the subarachnoid space, and is reabsorbed into the circulation via the arachnoid villi. CSF differs from plasma in having a lower concentration of glucose, protein and K^+ and a higher concentration of Cl^- and Mg^{++} ions.

6. The somatosensory system provides the CNS with information concerning touch, temperature, proprioception, and nociception (tissue damage). A receptive field is the region that, when stimulated, causes a response in sensory neurons. The size and the degree of overlap of receptive fields of adjacent nerves play an important role in the spatial discrimination of a

stimulus. Information from the somatosensory receptors reaches the cerebral cortex via the dorsal column–medial lemniscal pathway, and via the spinothalamic tract. The dorsal column is concerned with fine touch and position sense (proprioception), while the spinothalamic tract is concerned with crude touch, temperature and pain.

7. Pain is an unpleasant experience that is associated with actual or impending tissue damage. Nociception is triggered by thermal, mechanical or chemical stimuli and is conducted via Aδ- and C afferent fibres. These synapse within the spinal cord and project to the CNS via the spinothalamic and spinoreticular tracts. Following nerve injury or tissue damage, the nociceptive pathways can become hyperexcitable via the release of excitatory amino acids or prostaglandins causing the 'wind up' phenomenon and neuropathic pain.

8. The visual pathways are arranged so that each half of the visual field is represented in the visual cortex of the contralateral hemisphere. The fibres arising from the ganglion cells of the nasal retinal cross to the opposite side whereas those of the temporal retina do not. The right visual field is represented in the left hemisphere and vice versa. The axons of the ganglion cell reach the lateral geniculate bodies via the optic tracts. The axons of the geniculate body project to the visual cortex. The amount of light falling on the retina is controlled by the pupil. If light is shone into one eye, the pupil constricts (direct papillary response) and the pupil of the other eye also constricts (consensual papillary response). Sympathetic stimulation causes dilatation of the pupils as a result of contraction of the radial muscles of the iris. Parasympathetic stimulation causes papillary constriction as a result of contraction of the circular muscle of the iris.

9. The auditory system consists of the ear and the auditory pathways. The inner ear is the organ of hearing (via the cochlea) and balance (via the vestibular sytem). Sound waves evoke a travelling wave in the basilar membrane of the cochlea and this activates the hair cells via an increase in the frequency of action potentials in the nerve fibres of the cochlea nerve. These cochlea nerve fibres pass to the olivary nuclei. Fibres from the olivary nuclei project to the ipsilateral inferior colliculus (via the lateral lemniscus) and then to the medial geniculate body of the thalamus and end in the primary auditory cortex.

10. A reflex is an automatic or involuntary stereotypical response to a peripheral stimulus mediated by a receptor, an afferent pathway and an effector organ to produce purposeful movements rapidly. Interneurons may be present between the afferent and efferent neurons. The number of synapses in the reflex arc defines the reflex as monosynaptic or polysynaptic. The knee-jerk reflex is a monosynaptic reflex arc. Stretching of the quadriceps muscle stimulates the muscle spindles which excite the α motor neurons supplying the quadriceps muscle and cause it to contract. Stretch reflexes are important for the control of posture. Withdrawal reflexes are elicited by noxious stimuli and involve many muscles through polysynaptic pathways. Reciprocal inhibition ensures that extensor muscles acting on a joint will relax while flexor muscles contract. The Golgi tendon reflex is activated by discharge in the Golgi tendon organs and plays an important role in maintenance of posture.

11. Tone in the axial muscles is continuously adjusted to maintain and alter posture. This is achieved by control via basal ganglia and modified according to feedback from the vestibular system, the eyes, proprioceptors in the neck, and pressure receptors in the skin. A variety of reflexes such as righting reflexes, supporting reactions and stepping reactions help to adjust posture. The vestibular reflexes (labyrinthine righting, tonic and dynamic reactions) elicited by sudden changes in body position help to maintain posture.

12. The cerebral cortex contains primary, secondary and premotor motor areas in which stimulation of cells will elicit contralateral movements. Those areas of the body that control refined and complex movements have a disproportionately large area of representation. Outflow from the motor cortex is carried by the pyramidal (corticospinal) and

extrapyramidal tracts. The motor cortex also sends collaterals to the basal ganglia, cerebellum and brainstem. The motor areas receive inputs from the somatosensory system (via the thalamus), the visual system, cerebellum, and basal ganglia. The secondary motor area is important for programming voluntary movements.

13. The cerebellum is located dorsal to the pons and medulla. It is divided into anterior, posterior and a smaller flocculonodular lobe. The cerebellar cortex comprises of three layers of cells, the Purkinje layer, the granular layer and the molecular layer. The cerebellum is attached to the brainstem by nerves in the three cerebellar peduncles. Afferents from the vestibular nuclei, muscle proprioceptors and pons transmit impulses to the cerebellum. Efferent tracts from the cerebellum originate in the deep nuclei and pass to the thalamic nuclei, red nucleus, vestibular nuclei, pons and medulla. The two halves of the cerebellum control and receive inputs from ipsilateral muscles. The cerebellum plays an important role in coordination of postural mechanisms and the control of rapid muscular movements. It also supplements and correlates the activities of other motor areas.

14. The basal ganglia are deep cerebral nuclei, comprising the amygdala, caudate nucleus, globus pallidus and putamen (the last two form the corpus striatum). In association with the substantia nigra and red nucleus, they are involved in basic patterns of movements, representing motor programmes derived from the cortical association areas.

15. The EEG waves are waves of low amplitude (10–150 µV) and their frequency and amplitude depend on the state of arousal of the person. Sleep is initiated by changes in the activity of neurons in the diencephalon and brainstem. During sleep the EEG is dominated by slow-wave activity. This is associated with a slowing of heart rate and respiratory rate, decrease in blood pressure and somatic muscle relaxation. Bouts of REM sleep in which the EEG is asynchronous occur intermittently. During REM sleep the heart rate and respiration may become irregular and there are rapid eye movements, clonic movements of the limbs, and dreaming.

16. The autonomic nervous system is a division of the nervous system that controls the activity of the internal organs. Acetylcholine is the principal transmitter released by the preganglionic fibres of both the sympathetic and parasympathetic nervous systems. The parasympathetic postganglionic fibres secrete acetylcholine onto their target organs, while norepinephrine is principally secreted by the postganglionic sympathetic fibres. The sympathetic division prepares the body for 'fight or flight' reactions. Increased sympathetic activity is associated with increased heart rate, vasoconstriction of visceral organs, and vasodilatation in skeletal muscle. It is associated with glycogenolysis and gluconeogenesis in the liver. The parasympathetic system promotes 'rest and digest' (restorative) functions. Increased parasympathetic activity is associated with increased motility and secretion by the gastrointestinal tract and slowing of the heart rate.

3

Respiratory physiology

LEARNING OBJECTIVES

After studying this chapter the reader should be able to:

1. Understand cellular respiration with knowledge of the production of ATP, oxygen consumption and carbon dioxide production
   Describe:
   a. Aerobic and anaerobic metabolism
   b. Basic metabolic pathways for carbohydrates, fats and proteins
2. Understand the carriage of oxygen and carbon dioxide in the blood
   Describe:
   a. The functions of haemoglobin and carriage of oxygen and carbon dioxide in arterial and venous blood
3. Understand the mechanics of lung ventilation and describe the equilibrium between the mechanical forces of the lungs and thorax
   Describe:
   a. Pressures and flow during the breathing cycle
   b. The influence of elastic forces and expansion of the lung
   c. The influence of non-elastic forces and expansion of the lung
   d. Different types of gas flow in the lungs
   e. Lung volumes, the pressure–volume relationship of the lung, and an understanding of lung compliance and airways resistance
   f. The muscles and the work of ventilation

4. Understand gas exchange in the lungs
   Describe:
   a. The rate of pulmonary exchange of oxygen and carbon dioxide across the alveolar and pulmonary capillary wall
   b. The rate of transfer of oxygen and carbon dioxide in the oxygen and venous blood
   c. The functional anatomy of the airways, alveoli and pulmonary capillaries
   d. The diffusion of gases across the alveoli and pulmonary capillaries
   e. The importance of dead space in the lungs, and the calculation of each type
   f. The composition of alveolar gas and be able to use the alveolar air equation
   g. The pulmonary circulation, the effects of venous admixture (shunt), and the effect of the importance of ventilation perfusion inequality in the lungs
5. Understand the control of ventilation
   Describe:
   a. The respiratory centres, and the central and peripheral receptors influencing ventilation
   b. The reflex response to increased arterial carbon dioxide tension, reduced arterial oxygen tension and a rise in arterial blood hydrogen ion concentration
   c. The control of ventilation during exercise, and the effect of anaesthesia upon respiratory control

CELLULAR RESPIRATION

OVERVIEW

Cells need energy for work including muscle contraction, biosynthesis, active transport across membranes and generation of heat. Of these, the main consumer is the membrane sodium/potassium ATPase pump.

Energy is generated from metabolic fuels (carbohydrates, fats and proteins) and from reduced molecules which are oxidized to release energy. Oxidation involves removing electrons at high potential from the fuel molecules and transferring them to a lower potential, thus releasing energy. The removed electrons must be transferred to a suitable electron acceptor which has to be transportable, soluble in water, and generally available. In cells, oxygen is the electron acceptor used. Unfortunately, oxygen is too reactive to be the immediate oxidizing agent and so intermediates, nicotinamide adenine dinucleotide (NAD^+) and flavin adenine dinucleotide (FAD), are employed as carriers of electrons between the metabolic pathways and the site of oxygen consumption in the mitochondria. NAD^+ and FAD are reduced by the major metabolic pathways (e.g. glycolysis, citric acid (Krebs) cycle) to $NADH + H^+$ and $FADH_2$ and carry the electrons to the electron transport chain. In the electron transport chain, the electrons are transferred through a series of carriers of lower potential until they finally combine with oxygen to form water. In this process energy is released, and adenosine triphosphate (ATP) is formed from adenosine diphosphate (ADP) by the process of oxidative phosphorylation (Fig. 3.1).

Oxygen is not used until the end of the electron transport chain (carbon dioxide is produced in the citric acid cycle). If oxygen is not available, then NAD^+ and FAD are converted to their reduced forms, and oxidation stops.

Anaerobic metabolism can continue by $NADH + H^+$ (produced during glycolysis) transferring its electrons to pyruvate, thus producing lactate and regenerating NAD^+ so that glycolysis can continue. In most tissues, this is only a temporary solution and oxygen is still required to collect the electrons from lactate. Some cells that have no mitochondria – including erythrocytes – operate anaerobically by using lactate to carry electrons to a different, aerobic organ (e.g. the liver) for oxidation.

ATP: THE CURRENCY OF CELLULAR ENERGY

The immediate source of cellular energy is ATP which can lose one phosphate group producing ADP and usable energy. The quantity of ATP in the body is limited, and it is therefore an energy source which must be maintained by recycling ADP back to ATP using energy obtained from the catabolism of dietary carbohydrates, fats and proteins. In the presence of oxygen, ATP is replenished from ADP

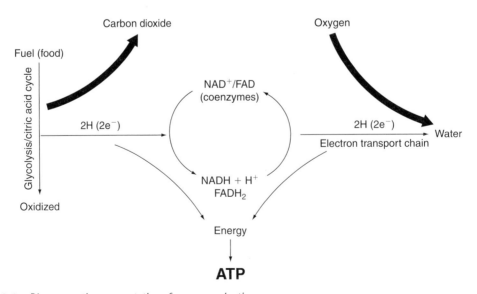

Figure 3.1 *Diagrammatic representation of energy production.*

by aerobic metabolism in the mitochondria. In the process of aerobic metabolism, carbon dioxide is produced in the citric acid cycle and oxygen is consumed by oxidative phosphorylation in the electron transport chain. In the absence of oxygen, or in cells with no mitochondria, some ATP is recycled by anaerobic metabolism in the cytoplasm.

The metabolism of organic molecules produces energy, but this cannot be used directly by cells. Instead, some of the energy produced by the breakdown of carbohydrates, proteins and fats is utilized to produce ATP from ADP (the rest of the energy is dissipated as heat). ATP is the energy 'currency' of the cell, and can be used directly to perform energy-requiring functions in the cell or to transfer energy to other intermediates.

The addition of a third phosphate group to ADP to produce ATP requires energy from the metabolism of organic molecules from food, and this is achieved by the two processes of substrate and oxidative phosphorylation during anaerobic and aerobic metabolism. The energy added to ADP is stored and can be released where necessary in the cell by the hydrolysis of ATP back to ADP. ATP is therefore a carrier, transferring energy in a storable form from the site of metabolism of organic molecules to cellular energy-requiring processes: membrane pumps, muscle contraction, and cellular synthetic processes. The amount of

ATP in the body is only sufficient to maintain resting functions for 1.5 min, and it must therefore be recycled continuously from ADP (Fig. 3.2).

ANAEROBIC OR AEROBIC METABOLISM

It is important to note that the processes of the citric acid cycle and the electron transport chain occur in the mitochondria of cells, and will only proceed in the presence of oxygen, but will continue to a mitochondrial P_{O_2} as low as 3 mmHg (0.4 kPa). Glycolysis takes place in the cytoplasm of the cell and can continue in the absence of oxygen by the transfer of electrons from NADH + H$^+$ to pyruvate with the production of lactate. To produce 1 mole of ATP from ADP, 7 kcal of energy is required. The aerobic metabolism of 1 mole of glucose produces 686 kcal of energy, of which 40 per cent is harnessed to produce up to 38 moles of ATP; the rest is lost as heat. The anaerobic metabolism of 1 mole of glucose produces only 2 moles of ATP, and is therefore much less efficient.

BASIC METABOLIC PATHWAYS

The processes of glycolysis, the citric acid cycle and the electron transport chain are described to show how they relate to cellular energy production.

Glycolysis

In glycolysis, sugars are metabolized by a series of 10 reactions that break down the 6-carbon molecules to produce two 3-carbon molecules of pyruvate. Two ATP molecules are used (reactions ① and ③), but four are produced (reactions ⑦ and ⑩), so there is a net gain of two ATP from each glucose molecule. In reactions ⑦ and ⑩, ATP is produced by substrate phosphorylation as the phosphate is transferred from a substrate molecule to ADP (in contrast to mitochondrial oxidative phosphorylation in the electron transport chain). All the reactions (Fig. 3.3) take place in the cytoplasm of the cell. Oxygen is not consumed, and carbon dioxide is not produced by glycolysis.

The end product of glycolysis is pyruvate, which under aerobic conditions enters the citric acid cycle. Under aerobic conditions the NADH + H$^+$ formed by reaction ⑥ is transferred indirectly to the mitochondrial electron transport chain to produce more ATP. This NADH + H$^+$ from glycolysis

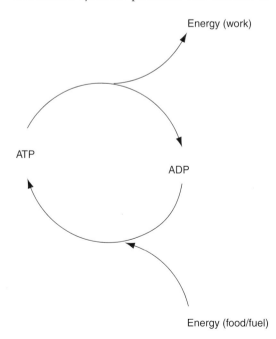

Energy (work)

ATP

ADP

Energy (food/fuel)

Figure 3.2   *ATP/ADP cycle.*

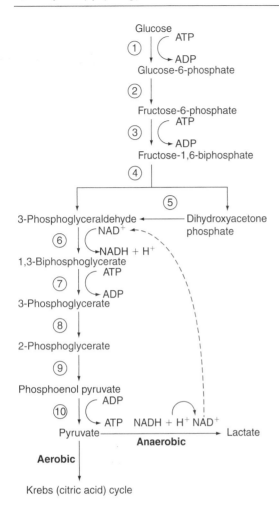

Production	ATP	NADH + H$^+$
Aerobic condition	2	2 (to electron transport chain)
Anaerobic condition	2	Nil (recycled to NAD$^+$ via lactate)

Figure 3.3 *Flow diagram showing the glycolysis pathway.*

in the cytoplasm is a charged molecule and cannot pass through the mitochondrial wall, but the electrons from the molecule can be transferred to any available NAD$^+$ or FAD inside the mitochondria. (Under aerobic conditions glycolysis can therefore continue, as NAD$^+$ used up in reaction ⑥ is regenerated by this process.) Under anaerobic conditions, glycolysis can still continue, as two electrons can be transferred from NADH + H$^+$ to pyruvate (to form lactate) to regenerate NAD$^+$ and maintain reaction ⑥, although the process is less efficient in

producing ATP than under aerobic conditions, and lactate is accumulated in the cell.

The citric acid cycle

The citric acid cycle (Fig. 3.4) is the next step in the production of cellular energy. It uses breakdown products of carbohydrate (pyruvate), fat and protein metabolism to produce carbon dioxide, some ATP and electrons in the form of hydrogen ions bound to intermediate carriers (NAD$^+$ and FAD) as NADH + H$^+$ and FADH$_2$, which then pass into the electron transport chain. The citric acid cycle takes place in the inner mitochondrial compartment, the matrix, and operates only in aerobic conditions as the electron transport chain is required to regenerate NAD$^+$ and FAD used in the cycle. Carbon dioxide is produced during the citric acid cycle, but oxygen is not consumed until the end of the electron transport chain.

The starting point of the citric acid (Krebs) cycle is acetyl coenzyme A. Coenzyme A is derived from vitamin B, and it transfers 2-carbon acetyl groups between molecules. When pyruvate (a 3-carbon chain, produced from glycolysis) enters the mitochondrion it reacts with coenzyme A to produce acetyl coenzyme A (2-carbon), carbon dioxide and NADH + H$^+$. This is the first point in the metabolic pathways where carbon dioxide is produced. In the first reaction of the citric acid cycle, the acetyl group of acetyl coenzyme A is transferred to the 4-carbon molecule oxaloacetate, producing citrate. During the citric acid cycle, two carbons are lost as carbon dioxide (reactions ③ and ④), electrons are donated to produce NADH + H$^+$ and FADH$_2$ (reactions ③, ④, ⑥ and ⑧), and one ATP is formed (reaction ⑤). At the end of the cycle, oxaloacetate is produced, ready to accept another acetyl group and recommence the process. The NADH + H$^+$ and FADH$_2$ formed in the cycle carry electrons to the transport chain where more ATP is produced by oxidative phosphorylation.

The electron transport chain

Oxidative phosphorylation is the primary mechanism by which ATP is produced in the body. Electrons are donated to the electron transport chain by the carriers NADH + H$^+$ and FADH$_2$, and are passed along a series of cytochromes until

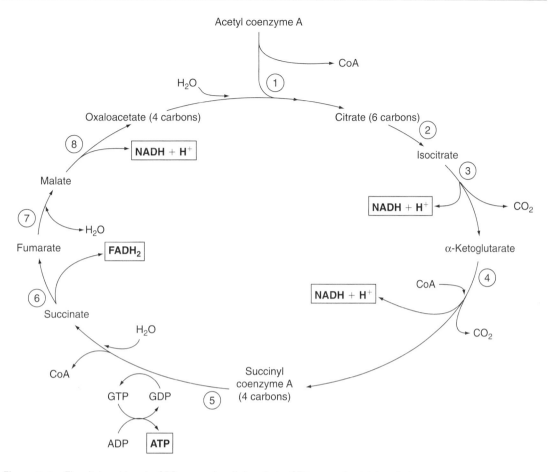

Acetyl coenzyme A

CoA

H_2O

①

Oxaloacetate (4 carbons)   Citrate (6 carbons)

②

⑧

$NADH + H^+$ Isocitrate

③

Malate

$NADH + H^+$ CO_2

⑦

H_2O

Fumarate α-Ketoglutarate

$FADH_2$

④

CoA

$NADH + H^+$

⑥

Succinate CO_2

H_2O

CoA

GTP   GDP   ⑤   Succinyl
                    coenzyme A
ADP   ATP   (4 carbons)

Figure 3.4   *The citric acid cycle. GDP, guanosine diphosphate; GTP, guanosine triphosphate.*

at the end of the chain they are accepted by oxygen to form water. The final cytochrome, which handles oxygen, is cytochrome a_3 – the site of action of the poison cyanide. As they pass down the transport chain, the electrons release energy which is used to form ATP – the process of oxidative phosphorylation. The mitochondria are therefore the primary site of cellular ATP production, carbon dioxide release and oxygen consumption.

The proteins forming the electron transport chain are found on the inside surface of the inner mitochondrial membrane (Fig. 3.5).

Electrons are transferred from NADH + H$^+$ and FADH$_2$ into the transport chain, which forms an energy gradient from high to low potential from beginning to end. As the electrons pass along the chain, down the gradient, they lose energy harnessed by the chain to form ATP at three specific points. The energy released from the electrons is used to pump hydrogen ions across the inner mitochondrial membrane to the cytoplasmic side,

producing a very high hydrogen ion concentration locally. At three points along the electron transport chain, there are molecular pores across the inner membrane so that the hydrogen ions can flow down their concentration gradient back into the matrix. As they do so, energy is released and is used to produce ATP from ADP and phosphate ions. There are therefore three points on the chain where ATP can be formed. As NADH + H$^+$ donates electrons into the beginning of the chain, three ATP molecules are formed. FADH$_2$ donates electrons further down the chain, and only produces two ATP molecules.

NADH + H$^+$ and FADH$_2$ release electrons into the transport chain, and NAD+ and FAD are regenerated to take part in earlier parts of the metabolic process. The H$^+$ released by NADH + H+ and FADH$_2$ flow along the inner mitochondrial membrane to react finally with oxygen and the transferred electrons to produce water at the end of the chain.

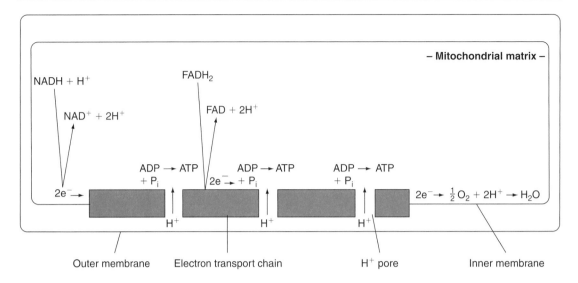

Figure 3.5 *Diagrammatic representation of a mitochondrion, showing the proteins forming the electron transport chain on the inside surface of the inner mitochondrial membrane.*

Carbohydrate, fat and protein metabolism

As described earlier, under aerobic conditions 38 moles of ATP can be produced from one molecule of glucose, but under anaerobic conditions only two ATP molecules are produced from glycolysis (only sugars can enter the glycolytic pathway). Fats represent 80 per cent of the stored energy in the body, and are broken down in the mitochondrial matrix by the removal of two carbon atoms at a time to form acetyl coenzyme A (enters the citric acid cycle) and hydrogen atoms combined with coenzymes (which enter the electron transport chain). By weight, the amount of ATP formed by fat is two and a half times that of carbohydrate, as fatty acids are more reduced molecules and can therefore donate more electrons. Proteins are broken down to amino acids from which the amino groups are removed to produce keto acids (these enter the citric acid cycle or are converted to glucose and fatty acids). The amino groups are removed by oxidative deamination (producing ammonia, which is converted to urea in the liver and excreted in the urine) or transamination (the amino group is transferred to a keto acid to form a new amino acid).

FUNCTIONS OF THE RESPIRATORY SYSTEM

The primary function of the lungs is gas exchange but there are also non-respiratory functions. These include (1) a blood filter, (2) an important reservoir of blood, (3) metabolic functions, (4) acid–base regulation, and (5) host defence functions. The pulmonary capillary bed acts as a blood filter preventing small clots and detached cells from reaching the systemic circulation. The pulmonary vessels are an important reservoir for blood. The pulmonary circulation, which has a high capacitance, accommodates about 16 per cent of the total blood volume in the supine position and about 9 per cent of the blood volume in the erect position. This blood volume can be redistributed to vital organs in hypovolaemic shock. The respiratory system has many metabolic functions such as the conversion of angiotensin I to angiotensin II; synthesis and breakdown of bradykinin; storage and release of serotonin and histamine; synthesis of peptides (like substance P), prostaglandins, surfactant and immunoglobulins; and inactivation of adrenaline and noradrenaline. Cytochrome P-450 isoenzymes are present in lung tissue. Mast cells in the lungs also secrete heparin. Acid–base regulation is accomplished by alterations of the arterial P_{CO_2} levels in the blood. Defence functions include the secretion of IgA, which provides innate immunity, and the removal of airborne particles by phagocytosis and mucociliary action. The respiratory system also has lymphoid tissue with T lymphocytes which appear to be a first line of defence as the lungs are exposed to the environment.

CARRIAGE OF OXYGEN AND CARBON DIOXIDE IN THE BLOOD

HAEMOGLOBIN

Within each of the biconcave disk-shaped erythrocytes there are two to three hundred million molecules of haemoglobin. Each haemoglobin molecule comprises four subunits of a polypeptide (globin) and a protoporphyrin ring containing a central iron atom in the ferrous state (Fe^{++}), haem. The haem group is attached to a histidine group on the globin protein to form one subunit of haemoglobin. One haem group can bind reversibly to one molecule of oxygen, and so each haemoglobin molecule can bind to four oxygen molecules (O_2). Adult haemoglobin (HbA) contains two α chains (141 amino acids) and two longer β chains (146 amino acids). The haem is attached to a histidine group at a constant distance from the end (N-terminal) on the globin chain.

In the erythrocyte, haemoglobin is folded into a complex, convoluted, shape with the haem enclosed in a cleft formed by the protein, forming loose bonds with amino acids within the globin chain. The conformation is affected by bonds formed between amino acids in the globin chains. The access to, and affinity of, oxygen to the haem group is affected by the conformation of the haemoglobin molecule, which is altered by physical and chemical factors (including temperature, hydrogen ion concentration, carbon dioxide and 2,3-diphosphoglycerate concentration). Such physicochemical factors can therefore change the affinity of haemoglobin for oxygen.

The Bohr effect

For the reason given above, increased hydrogen ion concentration, temperature, 2,3-diphosphoglycerate

concentration or carbon dioxide tension alters the access of oxygen to the haem groups, reducing the affinity of haemoglobin for oxygen, and therefore increases the unloading of oxygen to the tissues. Hydrogen ions bind to the α-amino and imidazole groups of haemoglobin and alter the allosteric conformation of the haemoglobin which reduces the affinity of oxygen to haem. Carbon dioxide binds to the N-terminal amino acid residues and this reduces the accessibility of the haem groups to oxygen and causes the oxyhaemoglobin curve to shift to the right. 2,3-Diphosphoglycerate binds to the β chains of one tetramer of deoxyhaemoglobin, causing a conformational change in the haemoglobin molecule which reduces oxygen affinity.

The Haldane effect

The increased capacity of deoxyhaemoglobin to carry carbon dioxide is referred to as the Haldane effect. In contrast to the dissociation of oxygen from haemoglobin, the dissociation of CO_2 from blood is directly related to the P_{CO_2} and therefore the dissociation curve for CO_2 is linear. The attachment of oxygen to haem reduces the capacity of haemoglobin to carry carbon dioxide. Oxygen increases the ionization of nitrogen groups, reducing both the capacity of the globin chain to carry carbon dioxide as carbamino compounds. Therefore deoxyhaemoglobin can carry more carbon dioxide in the form of carbamino compounds which account for about one third of the arterial venous difference of carbon dioxide carried in blood. In addition, deoxyhaemoglobin is more basic and therefore has an increased buffering capacity for carbon dioxide. Therefore deoxygenated venous blood freely takes up and transports more carbon dioxide than oxygenated arterial blood.

2,3-Diphosphoglycerate

This is produced by a side-shunt from glycolysis and is present in large quantities in erythrocytes. 2,3-Diphosphoglycerate binds with the β chains of haemoglobin, changing the protein conformation and reducing the oxygen affinity. When the 2,3-diphosphoglycerate concentration rises in the erythrocyte, there is increased unloading of oxygen from haemoglobin, increasing the tissue supply. 2,3-Diphosphoglycerate production is increased by anaemia and exposure to high altitudes.

Types of haemoglobin

Haemoglobin exists as several different forms:

- *Fetal haemoglobin* (*HbF*) comprises two gamma chains and two alpha chains, and has greater affinity for oxygen than HbA, thus enhancing oxygen transfer to the child from the maternal circulation in the placenta.
- *Haemoglobin S* (*HbS*). This is much less soluble in the reduced form than HbA, and can lead to abnormal 'sickle-shaped' erythrocytes which are fragile and can block small blood vessels, resulting in anaemia and sickle crises. The only difference between HbS and HbA is that, in the former, valine replaces glutamic acid as the sixth amino acid along the beta chain.
- *Methaemoglobin*. A small portion of the iron ions in haemoglobin exist as the trivalent, ferric form (Fe^{+++}) and are unable to carry oxygen. Deficiency of the enzyme converting ferric ions to the ferrous state, or some drugs (including prilocaine), can increase the amount of methaemoglobin and impair oxygen carriage in the blood.

Functions of haemoglobin

Haemoglobin binds reversibly to oxygen, and carries it in the blood from the lungs to the tissues for aerobic metabolism in the mitochondria. Other functions include the carriage of carbon dioxide from the tissues to the lungs as carbaminohaemoglobin, and acting as a buffer due to the large number of imidazole groups present in the histidine moieties.

ARTERIAL AND VENOUS BLOOD OXYGEN AND CARBON DIOXIDE PARTIAL PRESSURES AND CONTENTS

The partial pressures and contents of oxygen and carbon dioxide in arterial and venous blood are indicated in Table 3.1.

The normal oxygen consumption per minute is 250 mL; the total amount of oxygen in the body is only approximately 1.5 L, and less than half is immediately available for use. Carbon dioxide production is 200 mL/min, and the total body content amounts to 120 L.

OXYGEN CARRIAGE IN THE BLOOD

Oxygen in the blood

The presence of haemoglobin in the erythrocyte allows the blood to carry much more oxygen than would dissolve in it at atmospheric pressure. Oxygen, when bound to haemoglobin, does not contribute directly to the P_{O_2} of the blood: only dissolved oxygen has such an effect.

OXYGEN DISSOLVED IN THE BLOOD

The amount of oxygen carried in physical solution in the plasma and erythrocytes is directly proportional to the P_{O_2} of the blood and is 0.003 mL/100 mL blood per mmHg (of P_{O_2}) at 37°C. This amounts to 0.3 mL of oxygen per 100 mL blood at the normal arterial P_{O_2} of 100 mmHg (13.3 kPa).

OXYGEN CARRIAGE BY HAEMOGLOBIN

At normal atmospheric pressure, 98 per cent of the oxygen in blood is carried by haemoglobin. The oxygen-combining capacity of haemoglobin in adult blood is 1.306 mL/g (1.306 mL of oxygen carried by each gram haemoglobin when fully saturated). For fetal blood, the oxygen combining

Table 3.1 *Partial pressures and oxygen and carbon dioxide content in arterial and venous blood*

	P_{O_2} (mmHg/kPa)	O_2 content (mL/100 mL blood)	P_{CO_2} (mmHg/kPa)	CO_2 content
Arterial blood	100/13.3	20	40/5.3	48
Mixed venous blood	40/5.3	15	46/6.1	52

capacity is 1.312 mL/g. Increasing the P_{O_2} of blood increases the amount of oxygen combined with haemoglobin. As the P_{O_2} increases, haemoglobin becomes more saturated with oxygen: consequently, the ratio of oxygenated to deoxygenated haemoglobin increases. The relationship between P_{O_2} and percentage saturation of haemoglobin is 'S'-shaped because the reaction of each of the four subunits of haemoglobin with oxygen occurs sequentially, with each facilitating the next. The kinetics of the reaction of the fourth unit is faster than the previous reactions, so that oxygen continues to combine with haemoglobin, despite the haemoglobin molecule having fewer available binding sites. The oxygen capacity of blood is defined as the maximum amount of oxygen that can be carried by haemoglobin.

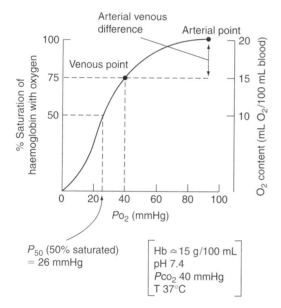

P_{50} (50% saturated) = 26 mmHg

| Hb $\simeq$ 15 g/100 mL |
| pH 7.4 |
| P_{CO_2} 40 mmHg |
| T 37°C |

Figure 3.6 *The oxygen–haemoglobin dissociation curve.*

The oxygen–haemoglobin dissociation curve

A diagram can be constructed relating P_{O_2} with per cent saturation of haemoglobin with oxygen (amount of oxygen combined with haemoglobin $\times$ 100 [i.e. a percentage] divided by the oxygen capacity) (Fig. 3.6). A third axis can be included for the oxygen content (i.e. the amount of oxygen carried by haemoglobin, in 100 mL of blood).

The saturation of haemoglobin with oxygen increases markedly on the steep part of the curve, from a P_{O_2} of 10–60 mmHg (1.3–8 kPa). Any further increase in P_{O_2} produces a smaller increase in saturation (at a P_{O_2} of 60 mmHg (8 kPa), the saturation is 90 per cent). In the lungs venous blood (P_{O_2} 40 mmHg (5.3 kPa), saturation 75 per cent) takes up oxygen quickly and the saturation rises, but with a minimal rise in blood P_{O_2} so that the pressure gradient for oxygen uptake into the blood from the alveoli is maintained. (Conversely, in the tissues the P_{O_2} gradient from blood to the tissues is maintained whilst oxygen is given up.) The plateau of the upper part of the curve means that a slight fall in alveolar and arterial P_{O_2} will only result in a modest reduction in haemoglobin oxygen saturation.

Under normal conditions, arterial blood has a P_{O_2} of 100 mmHg (13.3 kPa), is almost 100 per cent saturated with oxygen, and carries approximately 20 mL of oxygen per 100 mL blood in combination with haemoglobin. As the blood passes through the tissue capillaries, 5 mL of oxygen per 100 mL blood is removed to maintain aerobic metabolism. Mixed venous blood has a P_{O_2} of 40 mmHg (5.3 kPa), but is still 75 per cent saturated with oxygen, having 15 mL of oxygen per 100 mL blood in combination with haemoglobin.

Various factors can change the position of the curve, moving it to the left or right. The P_{50} of blood can be used to define this: P_{50} is the P_{O_2} at which the blood is 50 per cent saturated with oxygen, and is normally 26 mmHg (3.5 kPa).

THE BOHR EFFECT

Changes in blood P_{CO_2}, hydrogen ion concentration and temperature influence the affinity of haemoglobin for oxygen. When these factors increase (as in a working muscle), haemoglobin releases oxygen more easily, and this is revealed by a shift to the right of the oxygen dissociation curve. In the pulmonary capillaries, when P_{CO_2} and hydrogen ion concentration fall, the affinity of haemoglobin for oxygen increases, favouring oxygen uptake into the blood. This Bohr shift of the oxygen–haemoglobin dissociation curve therefore favours haemoglobin oxygen uptake in the lungs and oxygen delivery in the tissues. As described, 2,3-diphosphoglycerate also reduces the affinity of haemoglobin for oxygen, shifting the dissociation curve to the right.

The consequences of left and right shifts of the dissociation curve can be observed by examining the effects on the partial pressure at the venous point, i.e. the P_{O_2} when 5 mL of oxygen has been removed from arterial blood; normally 40 mmHg (5.3 kPa), which can be thought of as the P_{O_2} gradient maintaining oxygen flow into the tissues when oxygen delivery has normally finished. A shift to the right of the dissociation curve increases the P_{O_2} of the venous point, enhancing oxygen delivery. A shift to the left of the dissociation curve reduces the P_{O_2} at the venous point, curbing oxygen delivery to the tissues (Fig. 3.7).

OXYGEN IN ARTERIAL BLOOD

Arterial blood has a partial pressure of oxygen (P_{O_2}) of 100 mmHg (13.3 kPa), although this falls with age:

$$P_{O_2} = 102 - \text{age[in years]}/3, \text{ in mmHg or}$$
$$= 13.6 - 0.044 \times \text{age in years, in kPa),}$$

Arterial blood is 100 per cent saturated with oxygen and contains 20 mL of oxygen per 100 mL blood when the haemoglobin concentration is

14 g per 100 mL (at normal temperature, P_{CO_2} and 2,3-diphosphoglycerate concentration).

OXYGEN IN VENOUS BLOOD

After the tissues have removed sufficient oxygen for aerobic metabolism (only 5 mL oxygen per 100 mL blood), venous blood still contains 75 per cent of the oxygen contained in arterial blood (75 per cent saturated with oxygen). That is, the oxygen content of venous blood is 15 mL per 100 mL at a partial pressure of oxygen in venous blood ($P_{V_{O_2}}$) of 40 mmHg (5.3 kPa). There is therefore a significant reserve of oxygen in venous blood available for increases in tissue oxygen requirements.

Oxygen-carrying capacity of the blood and oxygen delivery

The oxygen-carrying capacity of blood is dependent upon the P_{O_2} and the haemoglobin concentration of blood. The oxygen delivery to the tissues per minute by the cardiovascular system is the product of the oxygen content in arterial blood (20 mL per 100 mL of blood; $\times$ 10 for a litre of blood) and the cardiac output (5 L/min), and is normally 1000 mL oxygen per minute (the tissue oxygen usage is 250 mL/min). Calculation of the oxygen delivery can be expanded to include the factors affecting the oxygen content of arterial blood (the majority being carried by haemoglobin, and a small amount dissolved):

Oxygen delivery to tissues =

$$\left[\begin{array}{l} (\text{Hb (g/100 mL)} \times 1.306 \times \%Sa_{O_2}) + \\ (0.003 \times Pa_{O_2} \text{ mmHg}) \times 10 \end{array} \right] \times$$
$$\text{cardiac output (L/min)}$$

$$= 1000 \text{ mL/min (under normal conditions)}$$

Calculating oxygen delivery

Oxygen transfer into blood in the pulmonary capillaries

Although mixed venous blood is 75 per cent saturated with oxygen, it has a low P_{O_2} of 40 mmHg

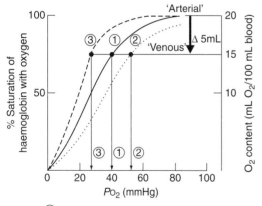

① Normal P_{O_2} venous point (40 mmHg)

② ↑P_{CO_2}, ↑T°C, ↑H⁺, ↑2,3-DPG
   'RIGHT' SHIFT
      ⇒P_{O_2} venous point increased

③ ↓P_{CO_2}, ↓T°C, ↓H⁺, ↓2,3-DPG, HbF
   'LEFT' SHIFT
      ⇒P_{O_2} venous point reduced

Figure 3.7   *Graphical representation of the Bohr effect. 2,3-DPG, 2,3-diphosphoglycerate.*

(5.3 kPa); oxygen combined to haemoglobin does not directly contribute to the P_{O_2}. In the pulmonary capillary, venous blood is exposed to the higher P_{O_2} across the blood gas barrier in the alveolus (100 mmHg (13.3 kPa)). Oxygen diffuses down the partial pressure gradient from the alveolus, and dissolves in the plasma and in the erythrocyte. Most of this oxygen does not remain in physical solution, but combines with haemoglobin. The blood P_{O_2} remains below alveolar P_{O_2} until haemoglobin is almost fully saturated. The P_{O_2} gradient from the alveoli to the blood is therefore maintained by the combination of oxygen with haemoglobin, encouraging transfer. Removal of carbon dioxide in the lung also favours the uptake of oxygen by haemoglobin.

Oxygen transfer from blood in the peripheral capillaries

As the tissue P_{O_2} is low, oxygen diffuses down the partial pressure gradient and out of the blood. The blood P_{O_2} falls, and oxygen is released from haemoglobin into physical solution. The process continues until adequate oxygen has been released from the blood for aerobic metabolism (each 100 mL of blood loses 5 mL of oxygen, and the P_{O_2} falls from 100 to 40 mmHg (13.3 to 5.3 kPa)). The P_{O_2} gradient from the blood to the tissues is maintained by the steep part of the oxygen haemoglobin dissociation curve as the blood P_{O_2} tends to be maintained while oxygen is lost. The higher tissue P_{CO_2}, temperature and hydrogen ion concentration encourage oxygen release by haemoglobin via the Bohr effect.

CARBON DIOXIDE CARRIAGE IN THE BLOOD

Carbon dioxide in the blood

Carbon dioxide exists in three forms in blood: dissolved in physical solution, as bicarbonate, and as the carbamino form in combination with proteins, mainly haemoglobin. In arterial blood, bicarbonate accounts for 90 per cent of the carbon dioxide, 5 per cent is dissolved and 5 per cent is present as carbamino compounds. In the capillaries of the peripheral circulation, carbon dioxide is added to the blood from the tissues, so that the carbon dioxide content of venous blood is higher than arterial. Venous blood, with a lower oxygen content, has a higher capacity to carry carbon dioxide. Of the carbon dioxide transferred from the tissues and evolved in the lung, 30 per cent is transferred in the venous blood as carbamino compounds, 10 per cent is dissolved, and 60 per cent exists in the form of bicarbonate.

CARBON DIOXIDE IN SOLUTION

Carbon dioxide is more soluble than oxygen, and dissolved carbon dioxide accounts for 10 per cent of the carbon dioxide evolved in the lungs.

CARBON DIOXIDE, CARBONIC ACID AND BICARBONATE

In solution, carbon dioxide combines with water to produce carbonic acid, which then dissociates to form bicarbonate and hydrogen ions. The first reaction is slow, but is catalysed by the enzyme carbonic anhydrase present in erythrocytes and the endothelium, but not in the plasma.

CARBAMINO COMPOUNDS

In blood, amino groups in proteins can combine with carbon dioxide to form carbamates, mostly with haemoglobin.

THE HALDANE EFFECT

Deoxygenated haemoglobin carries more carbon dioxide than the oxygenated form, and increased formation of carbamino compounds accounts for most of the Haldane effect. The secondary reason is that deoxygenated, or reduced, haemoglobin is more basic and is a better buffer for the hydrogen ions released by carbonic acid when bicarbonate is produced.

The carbon dioxide–blood dissociation curve

The relationship between blood P_{CO_2} and carbon dioxide content is more linear than that for oxygen. The CO_2 content is also higher than the oxygen content. As venous blood can carry more CO_2 than arterial blood (the Haldane effect), different carbon dioxide content curves can be drawn at different blood oxygen saturations. In the physiological range, blood CO_2 content increases from the arterial point (P_{CO_2} 40 mmHg (5.3 kPa) and 100 per cent oxygen saturation) to the mixed venous point (P_{CO_2} 46 mmHg (6.1 kPa) and 75 per cent oxygen saturation) (Fig. 3.8).

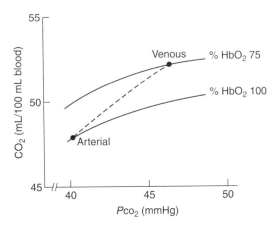

Figure 3.8 *The carbon dioxide–blood dissociation curve.*

Carbon dioxide transfer into the blood in the systemic capillaries

Carbon dioxide is produced in the mitochondria and diffuses down a partial pressure gradient into the interstitial fluid, across the capillary wall, and into the plasma. Some dissolves in the plasma, and some in the erythrocyte. Most of the carbon dioxide forms either carbamino compounds or bicarbonate (carbonic anhydrase) inside the erythrocyte. As oxygen is also moving out of the erythrocyte into the tissues, the storage of carbon dioxide is enhanced by the increased buffering of hydrogen ions (to HHb) released from carbonic acid, thus allowing increased bicarbonate production, and the formation of carbamino haemoglobin (HbNHCOOH). Bicarbonate ions diffuse out of the cell (hydrogen ions cannot do this and must be buffered inside the erythrocyte) and in order to maintain electrical neutrality, chloride ions diffuse in – as does water (Fig. 3.9). This is called the 'chloride shift' or Hamberger effect.

Carbon dioxide transfer from blood in the pulmonary capillaries

The reverse of the process described above takes place when the erythrocytes are in the pulmonary capillaries. The P_{CO_2} of venous blood (46 mmHg (6.1 kPa)) is higher than alveolar P_{CO_2} (40 mmHg (5.3 kPa)), and carbon dioxide diffuses out of the blood into the alveolus (there is therefore a partial

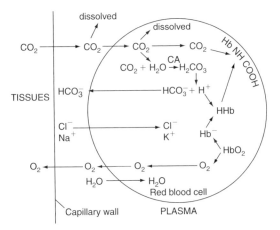

Figure 3.9 *Diagrammatic representation of the uptake of CO_2 and liberation of O_2 in systemic capillaries. CA, carbonic anhydrase. (Reproduced, with permission, from Respiratory Physiology – the essentials, 5th Edition. John B West, 1995, Williams and Wilkins.)*

pressure gradient from the mitochondria in tissue cells to the alveoli), while oxygen diffuses into the erythrocyte. As described, the loss of carbon dioxide from the erythrocyte in the lung enhances oxygen uptake by haemoglobin. The alveolar P_{CO_2} is determined by the balance between carbon dioxide output from the blood and alveolar ventilation. The fractional concentration of carbon dioxide in the alveoli equals the rate of carbon dioxide output (normally 200 mL/min) divided by the minute alveolar ventilation. At the end of the pulmonary capillary, the blood P_{CO_2} is very close to that of alveolar gas.

THE MECHANICS OF LUNG VENTILATION

EQUILIBRIUM BETWEEN THE LUNG AND THORAX

The lungs lie within the thorax, covered by the visceral pleura and separated from the parietal pleura on the inside of the chest wall by the (potential) intrapleural space. The diaphragm separates the lungs from the abdominal contents. The elastic forces of the lung and the chest wall are in equilibrium; the tendency of the lung is to contract down, and the tendency of the chest wall is to expand (muscle tone in the diaphragm also contributes to this), resulting in a negative intrapleural pressure (Fig. 3.10). At the end of a normal expiration, the two opposing forces balance, and the lung volume is at the functional residual capacity (FRC). The transpulmonary pressure is the difference between the alveolar and intrapleural pressures.

Inspiration is an active process. Increased inspiratory muscle activity shifts the equilibrium between the lung and chest wall to favour expansion, and a volume of air (a tidal volume) enters the airways through the nose and mouth. Expansion of the chest wall reduces intrapleural pressure and increases transpulmonary pressure, expanding the lung. The work done by inspiratory respiratory muscles against the elastic forces opposing expansion of the lung is stored as potential energy, and expiration is normally passive during quiet breathing. When the inspiratory muscle activity stops, the balance of the elastic forces returns the lung and chest wall down to the FRC and gas leaves the airways via the mouth.

PRESSURES AND FLOW DURING THE BREATHING CYCLE (FIG. 3.11)

At rest

At rest, the pressure in the alveoli of the lung is the same as at the mouth, i.e. zero with respect to atmospheric pressure. The intrapleural pressure is $-5\,cmH_2O$. The volume of gas within the lungs is the FRC, and there is no gas flow into or out of the airways.

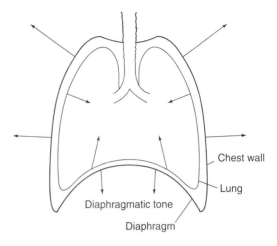

Figure 3.10 *Diagrammatic representation of the forces exerted within the thorax.*

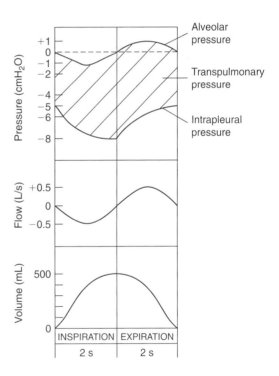

Figure 3.11 *The relationship between pressure, flow and volume during the breathing cycle.*

Inspiration

Intrapleural pressure falls because of the activity of the inspiratory muscles expanding the chest wall. This is transmitted across the lung, and alveolar pressure falls towards $-1\,cmH_2O$ and, as this is less than atmospheric pressure at the mouth, air flows into the lung. At the end of inspiration, intrapleural pressure is $-8\,cmH_2O$, alveolar pressure is again zero (atmospheric), gas flow into the lungs has ceased, and the lung volume has increased by about 500 mL (tidal volume).

Expiration

When the activity of the inspiratory muscles stops, the lung and chest wall reduce in volume as the equilibrium between the expanding chest wall and the elasticity of the lung moves back down to the resting volume. Intrapleural pressure increases (becomes less negative) and alveolar pressure becomes positive (equal to $+1\,cmH_2O$) and, as this is more than mouth pressure, gas flows out of the airways and is exhaled. At the end of expiration, intrapleural pressure is again $-5\,cmH_2O$, alveolar pressure is zero, expiratory flow has ceased, and the lung has returned to the resting volume (FRC). Expiration is normally a passive process, facilitated by the potential energy stored in the elastic expanded lung, but during more forceful breathing the expiratory muscles are also activated.

ELASTIC FORCES AND EXPANSION OF THE LUNG

The lung behaves as an elastic body. Factors involved in this include the surface tension developed by the very large air–water surface area of the fluid lining the alveoli and the elasticity of the lung tissues.

Surface tension

Surface tension develops at air–water interfaces where the forces of attraction between the water molecules are much greater than those between water and gas molecules. (The surface tension (T) is the force acting along a 1-cm line in the surface of the liquid.) The result is that the liquid surface area becomes as small as possible, as in a bubble of water where a spherical shape is assumed, and

pressure (P) is produced inside the bubble because of the surface tension. The pressure inside the bubble is determined by the surface tension and the radius (R) of the bubble, according to the law of Laplace (Fig. 3.12).

As the surface tension of water is constant, the pressure within small bubbles will be greater than that in larger ones, and the former will empty into the latter (Fig. 3.13).

The alveoli are lined with fluid over a large area, and the surface tension of the lung is therefore a considerable force resisting inflation. This was confirmed by the finding that the pressure required to inflate the lung is less when water is used instead of air, removing the effect of surface tension. In addition, as the pressure in small

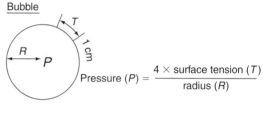

Bubble

$$Pressure\ (P) = \frac{4 \times surface\ tension\ (T)}{radius\ (R)}$$

Fluid-lined alveolus

$$P = \frac{2T}{R}$$

Figure 3.12 *The law of Laplace.*

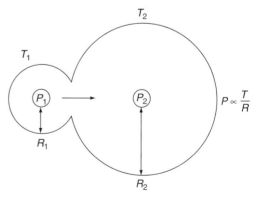

$$P \propto \frac{T}{R}$$

$T_1 = T_2$

As $R_1 < R_2$

$\Rightarrow P_1 > P_2$

$\Rightarrow$ small bubble empties

Figure 3.13 *Pictorial representation of the law of Laplace.*

alveoli should be higher, instability is produced as small alveoli would tend to empty into larger ones, producing collapse.

Surfactant

In the lung, the effect of the fluid lining the alveoli is minimized by the production of surfactant, which acts as a detergent, reducing the force of attraction between water molecules and thus reducing surface tension. Surfactant preferentially reduces surface tension in small alveoli and as the alveolar diameter is reduced (as during expiration), so that the pressure in small alveoli is less (law of Laplace), thus preventing collapse. This may be because the surfactant molecules become more closely packed together, and have a greater effect, with the reduction in alveolar surface area during expiration (Fig. 3.14).

Because of this effect of surfactant, the lung mimics the behaviour of an elastic body (although it is not made up primarily of elastic tissue). That is, the surface tension and therefore recoil pressure of the lung each decreases with lung volume.

The effect of surfactant on smaller alveoli is also important with respect to the change in alveolar diameter during inspiration and expiration. As alveolar diameter falls during expiration, surface tension is reduced. Surfactant also prevents pulmonary oedema, as the surface tension of the fluid lining the alveoli produces transudation of water from the pulmonary capillaries into the alveoli.

Type II alveolar cells produce surfactant and store it in the cytoplasmic lamellated bodies. Phospholipids are present in abundance in surfactant; the charged ends lie within the fluid lining the alveolus, while the hydrophobic fatty acid chains project into the alveolar gas. Some 90 per cent of surfactant is lipid, with most being phospholipid, but some cholesterol. The main lipid content is dipalmitoyl phosphatidyl choline, with significant amounts of phosphatidyl glycerol. Proteins comprise 10 per cent of surfactant.

Tissue elastic forces

Elastin fibres are present in the airways and alveoli, and they also contribute to the behaviour of the lung as an elastic body.

NON-ELASTIC FORCES AND EXPANSION OF THE LUNG

The major 'non-elastic' forces opposing ventilation include resistance to air flow and deformation of the lung tissues, both of which are related to the rate of flow of gas. Work done by the inspiratory respiratory muscles against these forces is not stored, but is lost, as heat.

LAMINAR, TRANSITIONAL AND TURBULENT GAS FLOW (FIG. 3.15)

Laminar flow

In a smooth-walled tube at low gas flow rates, the flow can be described as laminar where the flow rate has a velocity profile with gas at the centre of the tube flowing more quickly, while the lamina near the wall is almost stationary. That is, laminar

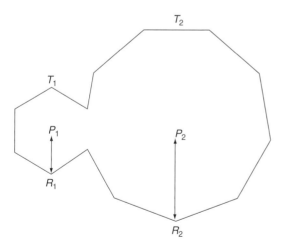

But, surfactant preferentially reduces T_1

$$T_1 << T_2$$
$$\Rightarrow \quad P_1 \approx P_2$$

Figure 3.14 *The effect of surfactant on pressure and surface tension.*

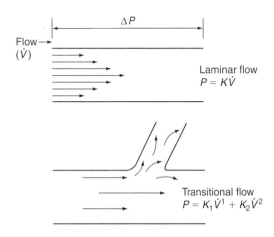

Figure 3.15   *Diagrammatic representation of laminar, transitional and turbulent flows.*

gas flow occurs as a series of concentric tubes parallel to the wall of the tube, and where the velocity is greatest at the centre and the velocity profile across the tube is parabolic. The pressure required to produce laminar flow is related linearly to the flow rate, and can be calculated from the Poiseuille equation:

$$\text{Flow rate} = \frac{P\pi r^4}{8\,\eta L}$$

$P = \Delta P$ (driving pressure)
r = radius of tube
η = viscosity
L = length of tube

The Poiseuille equation

   With laminar flow, the resistance to flow is affected markedly by the radius of the tube as it is inversely related to the fourth power of the radius;

any reduction in bronchial calibre increases airway resistance greatly:

$$R = \frac{8\eta L}{\pi r^4}$$

Calculation of the resistance to laminar flow

Transitional flow

Branches in airways may cause localized, disorganized eddies in the gas flow.

Turbulent flow

At higher flow rates, or in branching tubes, the gas flow is turbulent, and pronounced mixing of gas occurs. For a given flow velocity, the necessary driving pressure (and work done by the respiratory muscles) is greater for turbulent than for laminar flow as the former is related to the square of the flow rate. In contrast to laminar flow, the viscosity of the gas is not important, but the required pressure is instead proportional to the density of the gas for turbulent flow. The required driving pressure is also inversely proportional to the fifth power of the radius of the tube (Fanning equation).

Reynolds number

The factors determining whether flow will be laminar or turbulent include the velocity of flow, the radius of the tube, and the density and viscosity of the gas. Turbulence is more likely at high velocity, in tubes of large diameter, with a gas of high density and low viscosity. An equation can be described for these factors, to calculate the Reynolds number:

$$Re = \frac{\rho D \nu}{\eta}$$

D = diameter of tube
ν = velocity of flow
ρ = density of gas
Re = Reynolds number
η = viscosity of gas

Calculation of the Reynolds number

In smooth tubes, flow is likely to be turbulent when the Reynolds number exceeds 2000.

The pattern of gas flow in the lung

In the lung, which consists of a series of branching tubes, laminar flow probably only occurs in very small conducting airways where the Reynolds number is low. Flow tends to be turbulent in the trachea and in the larynx, especially when the velocity is high. Flow is transitional in most of the bronchial tree, and the pressure required is related both directly to the flow rate (as with laminar flow) and to the square of the flow rate (as with turbulent flow).

LUNG VOLUMES

Many of the lung volumes described can be measured by observing ventilation through a simple water volumetric spirometer. The residual volume (and therefore FRC and total lung capacity) cannot be measured in this way, requiring the use of gas dilution techniques. The volumes given in the following text are average values for adults (Fig. 3.16).

- *Functional residual capacity (FRC)*. This is the 2500 mL of air in the lungs at the end of a normal expiration (when the subject is standing). The FRC is the volume of the lungs at which the elastic outward force of chest wall expansion is balanced by the inward recoil of the lungs; muscular tone in the diaphragm is also involved (when this is lost, FRC falls by 400 mL). This volume maintains oxygenation of blood passing through the lungs during expiration or during breath-holding, and acts as an oxygen reserve so that alveolar oxygen partial pressure falls only by 3 mmHg during expiration. FRC is directly related to height and is 10 per cent less in females than males, but is not correlated with age (lung elasticity is governed by surface tension which does not change with age). The FRC is reduced by up to 1000 mL when the supine position is adopted, probably due the weight of the abdominal contents shifting towards the chest.

- *Tidal volume*. A normal resting breath, usually about 500 mL. Tidal volume is measured by spirometry as the volume difference between the resting inspiratory volume to the FRC.

- *Inspiratory reserve volume*. This is the volume of air that can be inspired over and above the resting tidal volume, and is normally about 3000 mL.

- *Inspiratory capacity*. This is the total volume (3500 mL) that can be inspired from the resting expiratory state (from the FRC).

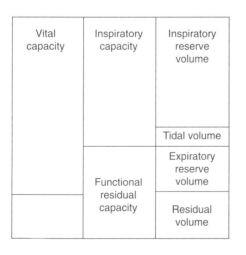

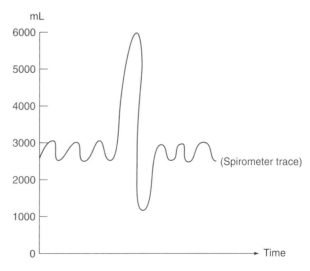

Figure 3.16 *Lung flow volumes (adult average).*

- *Vital capacity.* This is the maximal volume (4500–5000 mL) expired after a maximal inspiration.
- *Total lung capacity.* The total volume (6000 mL) of air in the lungs after a maximal inspiration (this cannot be measured by simple spirometry).
- *Expiratory reserve volume.* The additional volume (1500 mL) that can be expired at the end of a normal expiration (from the FRC).
- *Residual volume (RV).* The volume of air (1000–1200 mL) remaining in the lungs after a maximal expiration. The RV and FRC cannot be measured by simple spirometry because the lungs cannot be emptied completely after a forced expiration. They can be measured indirectly using a dilution technique involving 10 per cent helium. The helium dilution technique is an excellent technique for the measurement of FRC and RV in normal individuals. In patients with diseased lungs, the helium dilution technique gives a falsely low FRC value because of trapped gas in the lungs. This problem can be overcome by using the body pletysmograph technique where the lung volume is determined by applying Boyle's law.

THE PRESSURE–VOLUME RELATIONSHIP OF THE LUNG

The lung does not expand in a linear fashion with decreasing intrapleural pressure. Instead, there is a sigmoid relationship between lung volume and the intrapleural pressure. The curves for inspiration and expiration are different; higher expanding pressures are required for a given lung volume during inspiration than expiration (hysteresis) (Fig. 3.17). Even when intrapleural pressure is zero, or positive, a volume of air remains in the lungs because some small airways close (as there is then no transmural pressure to keep them open) and trap air in the alveoli. The lung volume, the closing capacity, at which this trapping occurs, rises with age.

The shape of the curve in Fig. 3.17 means that, at lower (less negative) intrapleural pressures, the lung will be relatively compressed, but the increase in volume with the inspiratory fall in intrapleural pressure will be marked as it occurs on the steep

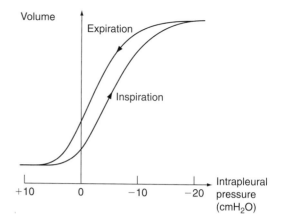

Figure 3.17   *The pressure–volume relationship of the lung.*

part of the pressure–volume curve. Conversely, at more negative intrapleural pressures, the lung is relatively expanded, but the increase in volume during inspiration with the fall in intrapleural pressure will be relatively less.

In the standing position, because of the weight of the lung in the thorax, the intrapleural pressure is not constant over the surface of the lung from the apex to the base. At the resting expiratory point (FRC), the apex is exposed to a more negative intrapleural pressure than the base of the lung. This means that, in the standing position, the apex of the lung is more expanded than the relatively compressed base, but the base expands more on inspiration. When standing, the base of the lung is therefore better ventilated (change of volume) than the apex (Fig. 3.18).

If breathing takes place at low lung volumes (from lower than the FRC), the intrapleural pressures are less negative. The intrapleural pressure at the lung base can exceed the atmospheric pressure in the airways, which close. Breathing at low lung volumes is therefore associated with airway closure at the base of the lungs and relatively more favourable ventilation of the upper areas of the lungs.

The closing capacity of the lungs

As described earlier, alveoli and airways at the dependent parts of the lung are smaller than those at the top. During expiration to volumes below

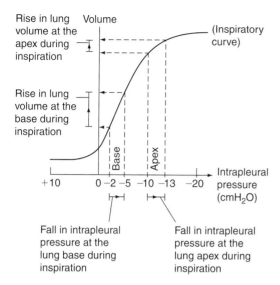

Figure 3.18 *Regional variation in lung ventilation.*

the FRC, such dependent small airways can collapse and close, at the 'closing capacity' of the lung, before expiration reaches the residual volume. When airways closure occurs, the oxygenation of blood passing through the lungs is impaired as the closed airways and unventilated alveoli are still perfused with pulmonary capillary blood. In young adults, closure of airways does not occur during normal tidal breathing, as the closing capacity is below the FRC. With age, the volume at which closure occurs rises, so that by the age of 65 years the closing capacity exceeds the FRC in the standing position. This means that, in the elderly, closure of basal airways happens during normal tidal breathing, impairing oxygenation of the blood. The rise in closing capacity is one of the main reasons why arterial P_{O_2} falls with age. This effect is important in the supine position as the FRC is less than when standing; by the age of 45 years, the closing capacity encroaches on the supine FRC so that airways closure occurs in normal tidal breathing.

LUNG COMPLIANCE

Lung compliance is the change in volume produced by a unit change in the expanding, transpulmonary,

pressure. As described earlier, the most important determinant of the elastic behaviour of the lung is the surface tension of the fluid lining the alveoli. The compliance of the lung is therefore much increased by surfactant.

If a lung is expanded by a given volume, and that volume is then held steady, the initial pressure required for the inflation falls to a lower value necessary to maintain the new volume. This is known as time-dependent elastic behaviour of the lung, and is produced by stress relaxation and redistribution of gas within the lung. Stress relaxation is demonstrated by many elastic tissues; for instance, a spring, where the initial tension required during stretching falls when the new length is maintained.

Redistribution of gas may occur in the lung after gas flow has ceased at the mouth, because differing airways and alveoli have differing resistances and compliances which determine how quickly, and by how much, they fill. The importance of this time-dependent elastic behaviour of the lung is that it will alter the calculated compliance of the lung depending on whether the calculation is made under static or dynamic conditions. As a result, static lung compliance exceeds dynamic compliance.

Static compliance

This is the change in volume of the lung produced by a unit pressure change, the flow of gas having ceased. For example, a subject can inspire a known volume of gas, and then relax against a closed airway for as long as possible, while intrapleural pressure is measured (usually by observing oesophageal pressure).

The compliance of the normal lung is $200\,mL/cmH_2O$, and the compliance of the thoracic cage has a similar value. As the lung and thoracic cage compliances are in series (and are analogous to electrical capacitances), the reciprocal of the total compliance of the respiratory system is equal to the sum of the reciprocals of the lung and the thoracic cage (1/total compliance = 1/200 + 1/200). Thus, the total compliance of the respiratory system is of the order of $100\,mL/cmH_2O$.

Dynamic compliance

Dynamic compliance is a measure of the volume and pressure changes which occur during normal breathing, without any breath-holding. The calculation of dynamic compliance is made from the volume and pressure values at times of no gas flow at the mouth at the end-expiratory and end-inspiratory points.

AIRWAYS RESISTANCE

The airways resistance to flow is low, of the order of $2 \, cmH_2O/L$ per second. The air flow of a normal tidal volume only requires a pressure difference between the mouth and the alveoli of $1 \, cmH_2O$. Airways resistance is influenced by lung volume, and decreases with inflation. Physical factors involved include the lateral traction exerted on the bronchi by the surrounding elastic tissues in the lung, and the transpulmonary pressure, which maintains the patency of small airways and increases with lung volume. Conversely, airways resistance rises when lung volume falls. In the lung, the main resistance is found in the bronchi of medium diameter, not the very small airways in the periphery. The very large number of the small airways and their large combined cross-sectional area are the reasons why these very small bronchi account for very little of the airways resistance.

Control of bronchial muscle tone

The parasympathetic system predominates in the neuronal control of bronchial tone and produces bronchoconstriction by the release of acetylcholine from vagal efferents which acts on smooth muscle muscarinic receptors. Sympathetic innervation of the lung is less important than the parasympathetic outflow, but circulating epinephrine can produce bronchodilation via β_2 adrenergic smooth muscle receptors. Non-cholinergic, non-adrenergic nerves also supply the bronchial muscles and produce bronchodilation. Histamine, released from mast cells, causes bronchial muscle constriction and mucosal swelling. Leukotrienes and some prostaglandins – products of arachidonic acid metabolism – also produce bronchospasm.

Additional factors involved in uneven regional ventilation of the lung: compliance, resistance and time constants

The preferential ventilation of dependent areas of the lung has been described earlier. In addition, differences between the resistances and compliances of airways and alveoli can also produce regional differences in ventilation. This is because of the effect of airways resistance and alveolar compliance on the rate of movement of air into and out of regions of the lung.

A lung unit can be considered to be an airway with the alveoli it supplies. If an airway has higher resistance than normal, then the movement of air into or out of that lung unit will be slower. If a lung unit has a lower compliance than normal, then the flow of air into that unit will cease sooner than in other units. The resistance and compliance of a lung unit therefore affect the time-dependent filling or emptying of that unit. This can be expressed by the time constant of the lung unit, which is the product of resistance and compliance.

The time constant is used to describe the rate of change of an exponential process, and is the time at which the process would have been complete had the initial rate of change continued. An exponential process, as the flow of air into a lung unit, is 95 per cent complete after three time constants.

The resistance for normal lung tissue is $2 \, cmH_2O/L$ per second, and the compliance (lung and chest wall) is $100 \, mL/cmH_2O$, giving a time constant of $0.2 \, s$. For normal lung tissue, filling (or emptying) is therefore 95 per cent complete within $0.6 \, s$. Lung units with high resistance will have longer time constants and will take longer to fill, but units with low compliances will fill and empty quickly. The converse is true for units with low resistances and high compliances. The presence of lung units with varying time constants means that there can still be flow of gas within the lung between units at the end of expiration and inspiration (when gas flow has ceased at the mouth), and that regional ventilation will depend on the frequency of ventilation.

Frequency-dependent compliance

The presence of lung units with differing time constants means that the measured dynamic compliance will decrease as respiratory rate increases, being 'frequency-dependent'.

Flow-related airways compression

Airways can collapse when the surrounding (intrapleural) pressure exceeds their own intraluminal pressure. This does not occur in normal tidal volume breathing as the airway intraluminal pressure exceeds the surrounding pressure throughout the breathing cycle. During a forced expiration, the intrapleural pressure becomes positive, and may exceed the airway intraluminal pressure which declines along the length of the airway because of the pressure fall due to the rapid expiratory flow and the airways resistance. The airway remains open to the equal pressure point (a point along the conducting airways between the alveoli and the trachea) where intrapleural and intraluminal pressures are equal; distal to this point, the airway collapses. Such flow-related airways compression takes place in the larger bronchi in a forced expiration, and limits the rate of expiratory flow. At normal lung volumes the expiratory flow rate is effort-independent, being limited by airways compression rather than by expiratory effort. Flow-related airways collapse is more likely when breathing at low lung volumes, when airways resistance is increased (bronchospasm), or if the elastic recoil of the lung is diminished (emphysema).

THE MUSCLES OF VENTILATION

The thorax, containing the lungs, is separated from the abdomen by the diaphragm. The cross-sectional area of the thorax (and lung volume) is increased by the lateral and anterior movement of the ribs. Contraction of the diaphragm increases the vertical dimension of the chest by pushing the abdominal contents down. Inspiration is an active process, but expiration is passive during normal quiet breathing. When breathing is increased, or during expiration to below the FRC, the expiratory muscles are recruited.

Inspiratory muscles

The diaphragm, supplied by the phrenic nerve (C3,4,5), is the principal muscle of ventilation. Contraction results in the descent of the domes of the diaphragm by 1–2 cm during quiet breathing, and by up to 10 cm in forced inspiration. The external intercostal muscles slope down and anteriorly, and move the ribs upwards and forwards. The scalene muscles are active during even quiet breathing, and elevate the rib cage. The sternocleidomastoid muscles are recruited when breathing is increased, when they also elevate the rib cage.

Expiratory muscles

Expiration is normally passive, the expiratory muscles only being recruited when ventilation increases. The important expiratory muscles are those of the abdominal wall (rectus abdominis, internal and external obliques and transversalis), which reduce the vertical dimension of the thorax by pushing the diaphragm up. The internal intercostal muscles are also expiratory, and they move the ribs down and in.

THE WORK OF VENTILATION

In normal tidal breathing, the inspiratory muscles do all the work of ventilation as expiration is passive. The inspiratory muscles work against both elastic and non-elastic (frictional) forces as described earlier. During inspiration, half of their work is expended against frictional forces and is dissipated as heat. The other half of the work of inspiration is against elastic forces, and this is stored as potential energy in the expanded elastic lung tissues. During expiration, as the inspiratory muscles relax and the elastic tissues return to their resting length, the stored potential energy is released and is employed to overcome frictional airway and tissue forces.

The work of ventilation can be analysed using pressure–volume curves of the lungs during inspiration and expiration (Fig. 3.19). The intrapleural pressure change shown is that required to produce a normal tidal volume breath, from the FRC. During inspiration, work is performed against both elastic (area ①) and non-elastic forces (area ②). The work against elastic forces is stored in the lung tissue as potential energy and increases with rises in tidal volume, or if the compliance of the lung is low. The work done against non-elastic forces is lost as heat, and is increased by faster flow rates during rapid breathing or by rises in airways resistance. The total work of inspiration, and therefore the total work of normal breathing, is the sum of the work done against elastic and non-elastic forces (area ② plus area ②).

The total work of breathing is low, and the oxygen requirement of the respiratory muscles is only 3 mL/min. The oxygen requirement of the respiratory muscles rises with increased minute volume, and can be greatly increased by lung diseases, including emphysema.

During expiration (Fig. 3.20), work must be done to overcome non-elastic forces (tissue and airways resistance, area ③), but in normal breathing

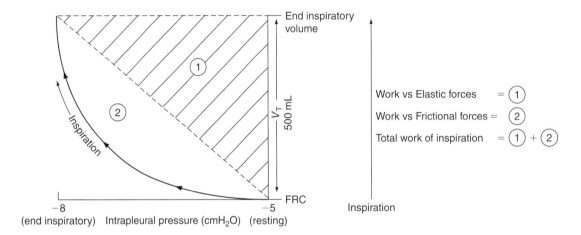

Figure 3.19 Analysis of work during inspiration.

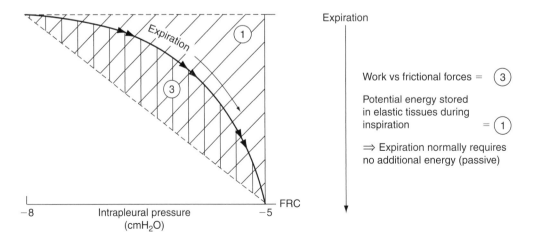

Figure 3.20 Analysis of work during expiration.

this is accounted for by the potential energy stored in the lung elastic tissues stretched during inspiration. If airway resistance increases, or if the expiratory flow rate rises with faster breathing, the required work for expiration may exceed the stored potential energy, the expiratory muscles must be recruited and expiration becomes an active process.

GAS EXCHANGE IN THE LUNGS

RATE OF PULMONARY EXCHANGE OF OXYGEN AND CARBON DIOXIDE

The minute pulmonary ventilation of 5 L/min is the product of the tidal volume (500 mL) and the respiratory rate, normally 10 breaths per minute. Because of the effects of dead space, the alveoli are ventilated by only 4 L each minute of fresh gas during inspiration, containing 840 mL of oxygen and little or no carbon dioxide. The pulmonary circulation (5 L/min) removes 250 mL of oxygen (for tissue consumption) and delivers 200 mL of carbon dioxide (product of tissue metabolism) per minute into the alveolar gas volume. As a result, 590 mL of oxygen and 200 mL of carbon dioxide are exhaled per minute from the alveolar gas during expiration.

RATE OF TRANSFER OF OXYGEN AND CARBON DIOXIDE IN ARTERIAL AND VENOUS BLOOD

Each minute, 5 L of venous blood containing 15 mL of oxygen per 100 mL (750 mL oxygen) passes through the pulmonary capillaries, where 250 mL of oxygen is gained by oxygenation. Thus, arterial blood carries 1000 mL of oxygen to the tissues each minute, where 250 mL is consumed (this can increase with tissue metabolism), leaving 750 mL oxygen in the blood of the venous return. From the carbon dioxide dissociation curve, the carbon dioxide content of arterial blood is 48 mL per 100 mL blood. Thus, in the cardiac output of 5 L/min, 2400 mL of carbon dioxide is carried to the tissues in arterial blood, where 200 mL

are added; the venous blood carbon dioxide content is 52 mL per 100 mL of blood. Venous blood therefore carries 2600 mL of carbon dioxide to the lungs per minute, where 200 mL are expired, leaving 2400 mL of carbon dioxide in the arterial blood per minute.

Oxygen and carbon dioxide partial pressures in the circulation

In inspired gas, the $P_{I_{O_2}}$ is 160 mmHg (21.3 kPa), and $P_{I_{CO_2}}$ is almost zero. In alveolar gas, the $P_{A_{O_2}}$ is 105 mmHg (14 kPa) and $P_{A_{CO_2}}$ 40 mmHg (5.3 kPa). In arterial blood, the $P_{A_{O_2}}$ is 100 mmHg (13.3 kPa) and $P_{A_{CO_2}}$ 40 mmHg (5.3 kPa). In the tissues, the P_{O_2} can be very low and the P_{CO_2} in excess of 46 mmHg (6.1 kPa). In mixed venous blood, the $P_{\bar{v}_{O_2}}$ is 40 mmHg (5.3 kPa) and $P_{\bar{v}_{CO_2}}$ is 46 mmHg (6.1 kPa).

FUNCTIONAL ANATOMY OF THE AIRWAYS, ALVEOLI AND PULMONARY CAPILLARIES

The airways branch and divide from the trachea into smaller generations, and these can be separated into two functional zones according to whether they have alveoli and therefore can take part in gas exchange with the pulmonary capillaries. The conducting zone comprises the airways without alveoli, from the trachea (generation 1) down to the terminal bronchioles (generation 16). Cartilage disappears from the airway walls from the 11th generation; the diameter of airways beyond this point is mainly determined by lung volume. The respiratory zone begins with the respiratory bronchioles (generations 17–19), the first airways to have alveoli in their walls, through the alveolar ducts (generations 20–22) to the alveolar sacs (generation 23). The airways and alveoli distal to a single terminal bronchiole make up a functional unit: the pulmonary lobule. The function of the conduction zone (volume 150 mL) is to permit bulk flow of air to and from the respiratory zone during inspiration and expiration, while also warming and adding moisture to the inspired air. The respiratory zone comprises the majority of the lung volume, being 3000 mL in volume, and exchange of oxygen and carbon dioxide with the pulmonary capillary blood occurs here, not by bulk flow, but by diffusion of gas that takes place rapidly within the pulmonary lobule because of the very short distances involved.

There are 200 to 600 million (average 300 million) alveoli in the lungs, and their diameter varies with lung volume; at FRC, the mean diameter is 0.2 mm. Exchange of oxygen and carbon dioxide takes place across the alveolar wall between the gas in the respiratory zone and the pulmonary capillary blood. The alveolar walls form septae between adjacent alveoli, which are polyhedral in shape rather than spherical as the septae are flat. The air-facing surfaces of the walls are lined by a one-cell-thick layer of flat, type I cells. Type II cells are also present and are rounded with large nuclei and microvilli and cytoplasmic granules containing stored surfactant, which they produce.

The interstitial space between alveolar walls contains the pulmonary capillaries, diameter 10 μm and endothelial thickness 0.1 μm, which form dense networks passing over a number of alveoli. An erythrocyte passes through this network (and two or three alveoli) in 0.75 s. On one side, the capillary is closely applied to the alveolar wall, so that the thickness from alveolar gas to pulmonary capillary blood is only 0.3 μm, representing a very short diffusion barrier for oxygen and carbon dioxide between blood and alveolar gas (in comparison, erythrocyte diameter is 7 μm). There is a very large total surface area of alveoli in contact with pulmonary capillaries, estimated to be 50–100 m^2.

DIFFUSION OF OXYGEN AND CARBON DIOXIDE ACROSS THE ALVEOLI AND PULMONARY CAPILLARIES

The flow of oxygen and carbon dioxide across the blood gas barrier of the alveolar wall, interstitial fluid and pulmonary capillary endothelium is governed by Fick's law of diffusion. This relates the flow of gas across a membrane to the area and thickness of the membrane, the partial pressure difference of the gas across the membrane,

and the diffusion constant (D) of the individual gases:

$$\text{Flow of gas} \propto \frac{A}{T} \cdot D \, (P_1 - P_2)$$

where:
A = lung area (50–100 m^2)
D = diffusion constant of the gas
T = lung thickness (0.3 μm)
$(P_1 - P_2)$ = partial pressure gradient across the membrane

and diffusivity:

$$D_{gas} \propto \frac{\text{solubility of gas}}{\sqrt{\text{mol. wt. of gas}}}$$
$$(D_{CO_2} \simeq 20 \times D_{O_2}).$$

Fick's law of diffusion

The pressure gradient for oxygen is from the alveolar $P_{A_{O_2}}$ of 105 mmHg (14 kPa) down to the mixed venous $P\bar{v}_{O_2}$ of 40 mmHg (5.3 kPa) – a partial pressure difference of 65 mmHg (8.7 kPa). The pressure gradient for carbon dioxide is from the venous $P\bar{v}_{CO_2}$ of 46 mmHg (6.1 kPa) down to the alveolar $P_{A_{CO_2}}$ of 40 mmHg (5.3 kPa) – a partial pressure difference of only 6 mmHg (0.8 kPa). Each minute, the P_{O_2} gradient of 65 mmHg (8.7 kPa) is associated with the diffusion of 250 mL of oxygen, while the much smaller P_{CO_2} gradient still transfers 200 mL of carbon dioxide. This is because of the influence of the diffusion constant in Fick's law: carbon dioxide is much more soluble and has a diffusion constant 20 times greater than that of oxygen.

The diffusion process is rapid. When an erythrocyte passes through the lungs, equilibration with alveolar oxygen and carbon dioxide partial pressures is accomplished within 0.25 s – only a third of the total time that the cell spends in the pulmonary capillaries at the normal resting cardiac output. During exercise, the cardiac output rises, and the time spent by an erythrocyte in the pulmonary capillaries may be reduced to only 0.25 s, but this is still sufficient for the transfer by diffusion of oxygen and carbon dioxide across the blood–gas barrier.

ALVEOLAR VENTILATION AND DEAD SPACE

Part of each tidal volume, the dead space, does not reach regions of the lung involved in gas exchange, and is therefore exhaled unchanged. The minute alveolar ventilation, the delivery of air by bulk flow to the respiratory zone where exchange of O_2 and CO_2 takes place, is therefore the product of respiratory rate and the tidal volume minus the dead space volume. Dead space can be thought of as the wasted portion of each tidal volume, as it does not contribute to O_2 and CO_2 exchange in the lungs.

The minute alveolar ventilation is the product of the tidal volume minus the dead space volume times the respiratory frequency (of the order of 4 L/min). Alveolar CO_2 is inversely related to the alveolar minute ventilation.

ANATOMICAL, ALVEOLAR AND PHYSIOLOGICAL DEAD SPACES

- The *anatomical dead space* is the volume of air contained in the conducting airways, that is, the volume of gas in airways without alveoli.
- The *alveolar dead space* is that part of the inspired gas, which passes through the anatomical dead space to mix with alveolar gas, but is not perfused by blood and does not contribute to gas exchange (i.e. ventilated, but unperfused, alveoli).
- The *physiological dead space* is the sum of the anatomical and the alveolar dead spaces.

Within one expiratory tidal volume, there will be anatomical dead space gas, and alveolar dead space gas mixed with gas from alveoli that are both ventilated and perfused and actually take part in gas exchange – 'ideal alveolar' gas (Fig. 3.21). Only this ideal alveolar gas contains carbon dioxide.

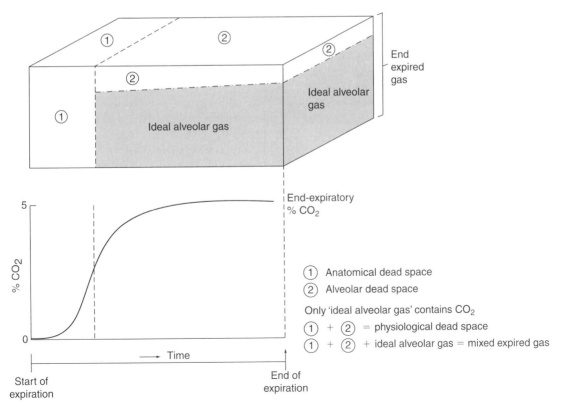

Figure 3.21   *The components of an expired tidal volume.*

The mixed expired CO_2 concentration is the CO_2 content of the expired breath divided by the total volume (measured by collecting an expired volume and measuring the CO_2 concentration). End-expired CO_2 concentration is measured at the end of expiration; this may not be the same as the ideal alveolar CO_2 concentration, as end-expired gas may contain alveolar dead space gas.

The anatomical and physiological dead spaces can be measured by Fowler's method and the Bohr equation, respectively. The alveolar dead space is obtained by subtracting the anatomical from the physiological dead space.

MEASUREMENT OF ANATOMICAL DEAD SPACE BY FOWLER'S METHOD

In 1948, Fowler described a method of estimating anatomical dead space based on the washout of nitrogen from the lungs after a normal size breath of 100 per cent oxygen, starting from a lung volume of FRC. The anatomical dead space can be defined, and measured, as the gas volume exhaled before the measured nitrogen concentration reaches the alveolar plateau using a rapid nitrogen analyser. By simultaneously measuring a single expired volume and nitrogen concentration, anatomical dead space can be calculated by a geometric method. This method has been modified using carbon dioxide as the tracer gas and measuring expired carbon dioxide concentration (Fig. 3.22).

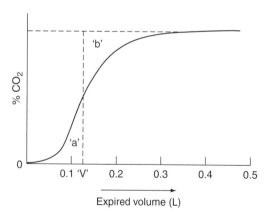

If area 'a' = area 'b'
Then 'V' = anatomical dead space

Figure 3.22 *Fowler's method for measuring anatomical dead space.*

In general, the volume of the anatomical dead space, in millilitres, is approximately equivalent to the weight of the subject in pounds. It is reduced by the subject lying down, and also by endotracheal intubation. Extending the neck and protruding the jaw increases anatomical dead space.

MEASUREMENT OF PHYSIOLOGICAL DEAD SPACE BY THE BOHR EQUATION

The basis of the Bohr equation is that only ideal alveolar gas contains carbon dioxide (i.e. the alveolar and anatomical dead spaces do not contain carbon dioxide), and that the carbon dioxide content of mixed expired gas must be equal to the product of alveolar ventilation and the concentration of carbon dioxide in ideal alveolar gas:

Volume of CO_2 eliminated from 'ideal alveolar' gas = volume of CO_2 in mixed expired gas

$\Rightarrow$ %CO_2 alveolar gas $\times$ alveolar ventilation = %CO_2 mixed expired gas $\times$ minute volume of ventilation

For one breath:
%CO_2 ideal alveolar gas $\times (V_T - V_D)$ = %CO_2 mixed expired gas $\times V_T$

Now:
V_D = physiological dead space
V_T = tidal volume
%CO_2 mixed expired gas $(P\bar{E}_{CO_2})$
%CO_2 ideal alveolar gas $\equiv Pa_{CO_2}$ (arterial P_{CO_2})

Therefore:
$$\frac{V_D}{V_T} = \frac{(Pa_{CO_2} - P\bar{E}_{CO_2})}{Pa_{CO_2}}$$

(the ratio of physiological dead space to the tidal volume can be estimated by measurement of the P_{CO_2} in arterial blood and mixed expired gas)

The Bohr equation for physiological dead space

The ratio of V_D/V_T is normally from 0.2 to 0.3. In healthy individuals, the alveolar dead space is small and the estimated values for the anatomical and physiological dead spaces (from Fowler's method and the Bohr equation, respectively) are similar. In lung diseases, where the distributions of alveolar ventilation and pulmonary blood flow are abnormal, the alveolar dead space can be significantly increased, and the physiological dead space will then exceed the anatomical dead space.

THE OXYGEN AND CARBON DIOXIDE COMPOSITION OF ALVEOLAR GAS

As described, blood passing through the pulmonary capillaries equilibrates with the alveolar oxygen and carbon dioxide partial pressures: P_{AO_2} 105 mmHg (14 kPa) and P_{ACO_2} 40 mmHg (5.3 kPa). Any factors that change alveolar oxygen and carbon dioxide partial pressures will therefore affect arterial partial pressures. Given a normal inspired P_{O_2} of 160 mmHg (21.3 kPa) and P_{CO_2} of close to zero, the primary factors affecting alveolar gas composition are alveolar ventilation and tissue metabolism (Fig. 3.23).

Alveolar ventilation delivers oxygen to, and removes carbon dioxide from, the alveolar gas. Tissue metabolism (via the pulmonary blood flow) delivers carbon dioxide to, and removes oxygen from, alveolar gas. The balance between the two processes of alveolar ventilation and tissue metabolism will determine the alveolar, and arterial, oxygen and carbon dioxide partial pressures. Increased alveolar ventilation, with a constant metabolic rate, will increase P_{AO_2} and reduce P_{ACO_2}. Conversely, reduced alveolar ventilation reduces P_{AO_2} and increases P_{ACO_2}. If the tissue metabolic rate rises without any change in ventilation, then P_{AO_2} will be reduced and P_{ACO_2} increased. A fall in metabolic rate will increase P_{AO_2} and reduce P_{ACO_2}. Such changes in alveolar partial pressures

will produce parallel changes in arterial blood partial pressures of oxygen and carbon dioxide. In contrast, the result of simultaneous increases in tissue metabolism and alveolar ventilation, as in normal aerobic exercise, is that alveolar (and blood) partial pressures of oxygen and carbon dioxide remain unchanged.

The carbon dioxide content of alveolar gas

Hyperventilation is present when the P_{ACO_2} (and P_{aCO_2}) is reduced below the normal value of 40 mmHg (5.3 kPa). As described earlier, this depends on the balance between alveolar ventilation and metabolism, and could be produced either by increased alveolar minute ventilation, or a reduced metabolic rate. Hypoventilation is present when the P_{ACO_2} is increased by reduced alveolar ventilation or increased metabolism. The alveolar carbon dioxide concentration is therefore determined by the ratio of the rate of carbon dioxide output (dependent upon tissue metabolic rate and the carriage of carbon dioxide by the cardiac output to the lungs) divided by the rate of alveolar ventilation. P_{ACO_2} is therefore inversely related to alveolar ventilation:

$$\% \text{ Alveolar CO}_2 = \frac{\text{CO}_2 \text{ output}}{\text{alveolar ventilation}}$$
$$= \frac{200 \text{ mL/min}}{4000 \text{ mL/min}} = 5\%$$

The P_{ACO_2} can be calculated if the atmospheric pressure, carbon dioxide output, alveolar ventilation and inspired carbon dioxide concentration are known:

$$\text{Alveolar } P_{CO_2} (P_{ACO_2}) =$$
$$\text{barometric pressure (dry) (mmHg)}$$
$$\times \left[F_{ICO_2} + \frac{\text{CO}_2 \text{ output}}{\text{alveolar ventilation}} \right]$$
$$(\text{normally} \simeq 0)$$
$$= P_{ACO_2} \simeq 35\text{--}40 \text{ mmHg} (4.7\text{--}5.3 \text{ kPa})$$

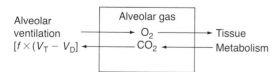

Figure 3.23 *The oxygen and carbon dioxide composition of alveolar gas (f is breathing frequency).*

As there is little difference between alveolar and arterial P_{CO_2}, it is clear that, given stable tissue metabolism and carbon dioxide output, PA_{CO_2} is a useful clinical measurement of changes in the rate of alveolar ventilation.

The oxygen content of alveolar gas and the alveolar air equation

By the combined effects of venous blood which reaches the arterial circulation without passing through pulmonary capillaries and areas in the lung of mismatch between pulmonary ventilation and perfusion, arterial P_{O_2} may differ significantly from the P_{O_2} of the ideal alveolar gas described above (i.e. alveolar gas which is both ventilated and perfused normally). As ideal alveolar gas cannot be sampled directly (end-expired gas may also contain gas from alveolar dead space), the P_{O_2} of ideal alveolar gas must be calculated by the alveolar air equation. The basis for this equation is that the effects of shunt and ventilation perfusion mismatch do not produce significant differences between alveolar and arterial P_{CO_2}. In addition, the respiratory quotient (the ratio of carbon dioxide production to oxygen use, normally 0.8), which is determined by the dietary substrates used for metabolism, must be taken into account. By convention, the P_{CO_2} of ideal alveolar gas is accepted as being equal to Pa_{CO_2}, and the respiratory quotient of ideal alveolar gas taken to be the same as that of expired air. A useful simple form of the alveolar air equation only requires knowledge of the inspired P_{O_2}, Pa_{CO_2}, and the respiratory quotient (assumed to be 0.8):

$$\text{Alveolar } P_{O_2}(\text{mmHg}) \approx$$
$$\text{inspired } P_{O_2} - \frac{\text{arterial } P_{CO_2}}{\text{respiratory quotient}}$$

Alveolar air equation for PA_{O_2} (simple form)

This form of the alveolar air equation produces a small error as it does not account for an additional effect of the respiratory quotient: the expired gas volume differs from the inspired volume. The inspired and expired gas volumes may also differ if inert gases (nitrogen) are not in equilibrium – as may be the case during anaesthesia. A more complex form of the alveolar air equation accounts

for these factors, and only requires knowledge of inspired P_{O_2}, Pa_{CO_2}, and collection of mixed expired gas for measurement of mixed expired P_{O_2} and P_{CO_2} (the respiratory quotient is not required):

$$\text{Alveolar } P_{O_2} =$$
$$\text{inspired } P_{O_2} -$$
$$\text{arterial } P_{CO_2} \left(\frac{\text{inspired } P_{O_2} - P\bar{E}_{O_2}}{P\bar{E}_{CO_2}} \right)$$

$$\left[\begin{array}{l} P\bar{E}_{O_2}: \text{ mixed expired } P_{O_2} \\ P\bar{E}_{CO_2}: \text{ mixed expired } P_{CO_2} \end{array} \right]$$

Alveolar air equation (complex form)

The alveolar air equation permits calculation of the ideal alveolar–arterial P_{O_2} difference, changes of which usually reflect altered pulmonary ventilation perfusion matching. Increased mismatch will lower arterial P_{O_2}, increasing the calculated ideal alveolar–arterial P_{O_2} difference.

THE PULMONARY CIRCULATION

The pulmonary and systemic circulations are in series, and have almost identical blood flows. Unlike the high-pressure and high-resistance systemic circulation which controls and changes perfusion to individual organs, the pulmonary circulation operates at much lower pressures, with very distensible and low resistance vessels receiving and distributing the venous return over the large area of thin alveolar walls available for gas exchange. By operating at low pressures, the pulmonary circulation minimizes the transudation of fluid into the interstitial spaces between alveolar and capillary walls (this would increase the diffusion barrier for gas exchange, as occurs in pulmonary oedema). The low-pressure pulmonary circulation is the reason that right ventricular work (and muscle thickness) is less than that of the left ventricle. A consequence of the low-pressure nature of the pulmonary circulation is that the effect of hydrostatic pressure is significant with the lung in the upright position, and perfusion pressure then decreases from the base to the apex so that mismatch of perfusion with alveolar ventilation can occur, impairing gas exchange.

Vessels of the pulmonary circulation

The volume of blood in the lungs is 450 mL, of which 70 mL is contained in the capillaries. The pulmonary artery is much shorter, thinner-walled and more distensible than the aorta, and the pulmonary arterioles contain little smooth muscle, although the muscle present can produce effective vasoconstriction. The pulmonary venules and veins have thin walls that are almost devoid of smooth muscle and are very distensible. The distensible vessels of the pulmonary circulation can accept the output of the right ventricle with little rise in pressure. The pulmonary capillaries are arranged around the small airways so that blood flows in thin sheets over the alveolar walls, maximizing the surface area for gas exchange (50–70 m^2). The pulmonary capillaries are applied directly to the alveolar walls so that flow through them is governed not only by pulmonary arterial and venous pressure, but also by the alveolar pressure. The pulmonary arteries and veins are expanded by the elastic pull of the surrounding lung tissues, and increase in diameter when the lung expands. When pulmonary blood flow and pulmonary vascular pressures increase, the increased blood volume is accommodated by both distension of patent vessels and recruitment of previously closed vessels so that pulmonary vascular resistance falls.

The bronchial circulation

The blood supply of the airways down to the terminal bronchioles is from the bronchial arteries – branches of the thoracic aorta. The bronchial circulation is normally 1 per cent of the cardiac output. A proportion of the bronchial venous drainage is not into the systemic venous return, but is a shunt into the pulmonary venous blood, which is fully saturated with oxygen having passed through the pulmonary capillaries. This, together with some coronary venous blood draining directly into the left atrium or ventricle, produces a small reduction of aortic arterial P_{O_2}. This small amount of venous blood returning to the left side of the heart means that the output of the left ventricle must be slightly more than that of the right ventricle.

Pressures in the pulmonary circulation

Pressures in the pulmonary circulation are only 20 per cent of systemic pressures, because the resistance of the former is much lower. Pulmonary artery systolic pressure is 25 mmHg and diastolic pressure 10 mmHg, with a mean pulmonary artery pressure of 15 mmHg. Left atrial pressure is 5 mmHg, and the average pressure drop across the pulmonary circulation is therefore 10 mmHg. Pulmonary capillary flow is markedly pulsatile because the low pulmonary arteriolar resistance does not damp the pressure waveform produced in the pulmonary artery by the systolic output of the right ventricle. A value of 8 mmHg has been suggested for the mean pulmonary capillary pressure.

Distribution of blood flow in the lung

In the upright position, hydrostatic pressure has significant effects on the pattern of blood flow in the lung as the pressures at the apex of the lung may be 10 mmHg less, and those at the base may be 10 mmHg more, than the mean pulmonary artery (15 mmHg) and venous (5 mmHg) pressures quoted above, which are measured at heart level. At the lung apex, the mean arterial pressure may only be 5 mmHg (15 minus 10 mmHg) and the venous pressure −5 mmHg (5 minus 10 mmHg). At the base of the lung, mean pulmonary arterial pressure may be 25 mmHg (15 plus 10 mmHg) and venous pressure 15 mmHg. As described, pulmonary capillaries are thin-walled structures directly applied to the alveolar walls. If alveolar pressure exceeds the capillary pressure, the vessel closes and flow stops. During normal breathing, alveolar pressure varies from atmospheric pressure by plus or minus 1 cmH$_2$O, but may be much higher during positive-pressure ventilation. The patency and blood flow through pulmonary capillaries therefore depend on the balance of vascular and alveolar pressures. The upright lung can be divided into three zones on the basis of the relationship of pulmonary arterial, venous and alveolar pressures:

- *Zone 1.* At the top of the lung, if alveolar pressure was greater than pulmonary artery pressure, there would be no blood flow, and ventilation to this area would be wasted, or dead space. In the normal lung, this zone is not present, but it may appear if pulmonary artery pressure falls, or alveolar pressure is increased.

- *Zone 2.* In the normal lung, from the apex down to 10 cm above the level of the heart, there is an area of intermittent capillary blood flow. In zone 2, the pulmonary artery pressure is greater than alveolar pressure, and venous pressure is less than alveolar. The capillary will flutter between being opened and closed. When it is closed there is no flow, the capillary pressure rises towards the arterial value until it exceeds the alveolar pressure, the vessel opens and blood flow starts. As flow takes place, the hydrostatic pressure falls along the capillary length (due to the effect of vessel resistance), until alveolar pressure exceeds the capillary pressure at a point close to the venous end, the vessel closes, flow stops, and the cycle then repeats. The important pressure gradient in zone 2 is the difference between arterial and alveolar pressures; venous pressure is unimportant. This can be described as a 'waterfall' effect, since just as flow over a waterfall is not affected by the height of the fall, flow in this lung zone is unaffected by venous pressure. Descending from the upper to the lower border of zone 2, blood flow increases as the arterial pressure rises whilst the alveolar pressure is constant.
- *Zone 3.* From 10 cm above the level of the heart down to the base of the upright lung, both arterial and venous pressures are more than the alveolar pressure. The capillaries remain open, and flow is determined by the difference between the arterial and venous pressures. The vessel transmural pressure between the inside and the outside of the distensible pulmonary vessels increases progressively towards the base of the lung, so that vessel diameter increases, resistance to blood flow decreases, and blood flow increases from the top to the bottom of zone 3.

Autonomic control of the pulmonary circulation

The pulmonary blood vessels have sympathetic nerve endings, although the density of receptors is less than in peripheral vessels. Both α and β receptors are present, producing vasoconstriction and vasodilatation, respectively. The pulmonary circulation also has a vagal nerve supply, and acetylcholine acts at muscarinic receptors to produce vasodilatation. The role of the autonomic nervous system in modulating pulmonary blood flow is unclear, as the vessels are normally maximally dilated, and local factors, specifically alveolar P_{O_2}, may be more important.

Hypoxic pulmonary vasoconstriction

Blood passing through the pulmonary circulation is diverted away from poorly ventilated areas of the lung by the constriction of small pulmonary arteries in the presence of low alveolar P_{O_2} (not low blood P_{O_2}). The mechanism involves a potent vasodilator, nitric oxide, which is continuously synthesized by the pulmonary artery endothelium in the presence of a normal alveolar P_{O_2}. If the alveolar P_{O_2} falls below 70 mmHg (9.3 kPa), endothelial nitric oxide synthesis is reduced, producing vasoconstriction. This is important for matching local lung perfusion to ventilation. Poorly ventilated lung areas have a low alveolar P_{O_2}, which induces local vasoconstriction that diverts blood flow to better ventilated areas, promoting gas exchange. This mechanism is important at birth as hypoxic pulmonary vasoconstriction is diminished by the first breath, and pulmonary vasodilatation and increased lung blood flow occur.

VENOUS ADMIXTURE (SHUNT)

The normal circulation of venous deoxygenated blood drained from individual organs is via systemic veins to the right side of the heart (mixed venous blood), through the pulmonary circulation where it is oxygenated before it leaves the pulmonary capillaries (pulmonary end-capillary blood), to the left side of the heart via the pulmonary veins, and then to systemic arteries (arterial blood). However, a small proportion of venous blood, normally from the bronchi and the left ventricle, returns directly to the left side of the heart, mixing some deoxygenated blood with the much larger amount of pulmonary circulation oxygenated blood.

This small proportion of venous blood can be described as venous admixture, or shunt. Blood passing into the aorta from the left ventricle is the sum of the normal venous return to the right side of the heart producing oxygenated blood from the pulmonary capillaries, and the much smaller amount of shunt that bypasses oxygenation in the lungs and drains to the left side of the heart. Because of this, arterial blood has a lower P_{O_2} than pulmonary end-capillary blood. As the P_{O_2} of pulmonary end-capillary blood is taken to be identical to the ideal alveolar P_{O_2}, the result of this mixing is revealed by the alveolar–arterial P_{O_2} difference. The definition of venous admixture is the amount of mixed venous blood required to mix with pulmonary capillary blood to produce the observed alveolar–arterial P_{O_2} difference. The venous admixture may not be equal to the actual amount of shunted blood, which may not have the same oxygen content as mixed venous blood.

Sources of venous admixture

There are two normal anatomical sources of shunt. Part of the bronchial circulation drains into the pulmonary veins; this represents less than 1 per cent of the cardiac output, but can increase in lung diseases, including emphysema. Thebesian veins, from the walls of the left ventricle, drain into the left side of the heart, representing some 0.3 per cent of the cardiac output. Disorders associated with increased venous admixture include congenital heart disease with right-to-left shunting, and the increased venous drainage from pulmonary neoplasms and areas of infection. Areas of airway collapse, which are perfused but not ventilated, also produce shunt, being extremes cases of mismatch of ventilation and perfusion, where the ratio of ventilation to perfusion is zero.

The shunt equation

The ratio of the shunt flow to the total blood flow from the left ventricle (the cardiac output) can be calculated by considering the different oxygen contents of shunted, pulmonary end-capillary,

and arterial blood. The basis for the shunt equation is that the oxygen carried by the cardiac output (arterial blood) must be the sum of the oxygen carried in the pulmonary end-capillary blood flow and the oxygen in the shunt blood flow. The assumption made is that shunted blood has the same oxygen content as mixed venous blood, although, for example, the bronchial and thebesian venous blood oxygen contents may be different:

$$\begin{pmatrix} \text{Cardiac output} \\ O_2 \text{ content} \end{pmatrix} =$$

$$\begin{pmatrix} \text{shunt} \\ O_2 \text{ content} \end{pmatrix} + \begin{pmatrix} \text{pulmonary blood flow} \\ O_2 \text{ content} \end{pmatrix}$$

Now:

Cardiac output = total blood flow = $\dot{Q}t$
Shunt flow   = $\dot{Q}s$
Pulmonary flow = $\dot{Q}t - \dot{Q}s$

and:

Ca_{O_2} = arterial O_2 content ⎫ measured from
$C\bar{v}_{O_2}$ = mixed venous O_2 ⎬ blood samples
          content ⎭

Cc'_{O_2} = pulmonary end-capillary O_2 content (obtained from the ideal alveolar P_{O_2}, from the alveolar gas equation and the oxygen–haemoglobin dissociation curve):

then:

$$\dot{Q}t \times Ca_{O_2} = (\dot{Q}s \times C\bar{v}_{O_2}) + (\dot{Q}t - \dot{Q}s)\, Cc'_{O_2}$$

$$\Rightarrow \frac{\dot{Q}s}{\dot{Q}t} = \frac{Cc'_{O_2} - Ca_{O_2}}{Cc'_{O_2} - C\bar{v}_{O_2}}$$

The effect of venous admixture on arterial oxygen and carbon dioxide contents and partial pressures of gases

Venous admixture reduces arterial blood oxygen content below that of pulmonary end-capillary blood. As the amount of shunted blood is relatively small in comparison to the pulmonary blood flow, the effect on arterial oxygen content is small, but the resultant depression of arterial P_{O_2} is significant. The P_{O_2} of pulmonary end-capillary blood lies at the upper flat part of the oxyhaemoglobin dissociation curve, and the reduction in oxygen content produced by the shunt results in an appreciable fall in arterial blood P_{O_2}. Arterial blood carbon dioxide content and P_{CO_2} are relatively unaffected by venous admixture, because of the steeper carbon dioxide–haemoglobin dissociation curve and the reflex ventilatory response to any increase in P_{CO_2}.

VENTILATION–PERFUSION RATIO

The prime function of the lung is to promote gas exchange between air and venous blood, oxygenating and removing carbon dioxide from the blood. Ventilation delivers oxygen to, and removes carbon dioxide from, the alveolar gas, and pulmonary perfusion provides venous blood with a high P_{CO_2} and a low P_{O_2}, ready for gas exchange. Mismatching of ventilation and perfusion impairs the transfer of oxygen and carbon dioxide in the lungs. The ratio of ventilation to perfusion determines the concentrations and partial pressures of oxygen and carbon dioxide of alveolar gas, and of arterial blood. If the ventilation–perfusion ratio is high, then the alveolar P_{O_2} is high and the P_{CO_2} low. If the ratio is low, then the alveolar P_{O_2} is low and the P_{CO_2} high. The overall ventilation–perfusion ratio in the normal lungs is 0.8 (alveolar ventilation 4 L/min and pulmonary blood flow 5 L/min), but this is not constant over the whole lung as the fall in perfusion towards the top is much more than the reduction of ventilation. Within the normal lung there are therefore regional differences in the ventilation–perfusion ratio, which reduce the efficiency of gas exchange. Lung disease affecting airways and the pulmonary blood vessels can produce gross mismatching of ventilation and perfusion.

The ventilation–perfusion (V/Q) ratio and composition of alveolar gas

The consequences of changes in the ventilation–perfusion ratio can be considered by the effects on the gas composition of a single alveolus. With a normal ventilation–perfusion ratio of 0.8, breathing air, the alveolar P_{O_2} is 105 mmHg (14 kPa) and the P_{CO_2} 40 mmHg (5.3 kPa) (as described earlier for 'ideal alveolar' gas). If the ventilation to this alveolus is progressively reduced, then the P_{O_2} would fall and the P_{CO_2} rise until, when there is no ventilation, the alveolus would contain gas at the same partial pressures as venous blood (P_{O_2} 40 mmHg (5.3 kPa), P_{CO_2} 46 mmHg (6.1 kPa)). Alternatively, if the blood flow to the alveolus was progressively reduced, then the alveolar P_{O_2} would rise and the P_{CO_2} fall until, when there is no perfusion, the alveolus would contain gas at the same partial pressures as inspired gas (P_{O_2} 160 mmHg (21.3 kPa), P_{CO_2} 0 mmHg).

The oxygen and carbon dioxide partial pressures of all the alveoli in the lung lie between the values of venous blood and inspired air, depending on their ventilation–perfusion ratio. Most alveoli have a ventilation–perfusion ratio close to the lung average of 0.8, with a P_{O_2} of 105 mmHg (14 kPa) and P_{CO_2} of 40 mmHg (5.3 kPa). However, alveoli with very high ventilation–perfusion ratios exchange little gas as their blood flow is low, and their ventilation is wasted, dead space. Alveoli with very low ventilation–perfusion ratios also exchange little gas because their ventilation is so low, and the venous blood perfusing them passes almost unchanged into the arterial side of the circulation, as a form of venous admixture.

Effects of ventilation–perfusion mismatching in the upright lung

At the top of the lung, alveoli have a high ventilation–perfusion ratio of 3, with a high P_{O_2} of 130 mmHg (17.3 kPa) and a low P_{CO_2} of 28 mmHg (3.7 kPa). At the bottom of the lung, alveoli have a low ventilation–perfusion ratio of 0.6, with a low P_{O_2} of 88 mmHg (11.7 kPa) and a high P_{CO_2} of 42 mmHg (5.6 kPa).

The main effect of this mismatch of ventilation–perfusion ratio over the upright lung is to reduce the arterial partial pressure of oxygen below that of the mixed alveolar P_{O_2}. For oxygen transfer, the effect of the high ventilation–perfusion ratio area at the top of the lung cannot compensate for the low P_{O_2} alveoli at the lung bases. This is because most of the pulmonary blood flow passes through the bases at the lower alveolar P_{O_2}, and the apical alveoli with high ventilation–perfusion ratios cannot add significantly more oxygen to the haemoglobin of the blood perfusing them as the oxygen–haemoglobin dissociation curve is flat at the upper end.

The reduction of arterial P_{O_2} by ventilation–perfusion inequality can be corrected by increasing the inspired concentration of oxygen. That is, when 100 per cent oxygen is breathed, ventilation–perfusion mismatch has little effect on arterial blood P_{O_2}.

The P_{CO_2} of arterial blood is less affected by mismatch as areas of high ventilation–perfusion ratio can compensate more effectively, because the carbon dioxide blood-dissociation curve is linear in the physiological range. In addition, any rise in the P_{CO_2} of arterial blood reflexly stimulates increased ventilation. For these reasons, arterial P_{CO_2} is normally not increased by ventilation–perfusion inequality. If the degree of ventilation–perfusion inequality is increased, the arterial P_{CO_2} may instead be reduced by increased ventilation driven by hypoxia.

Assessment of ventilation–perfusion inequality by the ideal alveolar–arterial P_{O_2} difference, the physiological dead space and the shunt equation

Ventilation–perfusion inequality produces a difference between the P_{O_2} of arterial blood and alveolar gas. Unlike arterial blood, alveolar gas cannot be sampled directly, but the alveolar gas equation can be used to calculate the ideal alveolar P_{O_2}, and the ideal alveolar–arterial P_{O_2} difference then determined. Areas of the lung with high ventilation–perfusion ratios produce wasted ventilation, or dead space, and this can be assessed by using the Bohr equation to calculate the physiological dead space. Areas of the lung with low ventilation–perfusion ratios have the same effect as the admixture of venous blood, and this can be estimated using the shunt equation.

CONTROL OF VENTILATION

In the early 1920s, Lumsden postulated that respiratory control was mediated by the pneumotaxic centre in the upper pons, the apneustic centre in the lower pons, and the inspiratory and the expiratory centres in the medulla oblongata. The pneumotaxic centre was considered to promote expiration and the apneustic centre promoted expiration, and the two together finely tuned normal breathing. However, no specific group of neurons has ever been identified as the so-called apneustic centre and theories involving it have been abandoned. The pneumotaxic centre is now referred as the pontine respiratory group and comprises of expiratory neurons in the nucleus parabrachialis medialis and the inspiratory neurons in the nucleus parabrachialis lateralis and the Kolliker–Fuse nucleus.

THE RESPIRATORY CENTRE IN THE MEDULLA

The respiratory centre in the medulla oblongata drives the rate and volume of ventilation via the motor nerves to the respiratory muscles, and is modulated by feedback loops from central and peripheral chemoreceptors sensing blood P_{CO_2}, P_{O_2} and hydrogen ion concentration (Fig. 3.24).

The medullary respiratory centre also has neuronal input from the cerebral cortex, pons, hypothalamus, lung stretch receptors, pharyngeal mechanoreceptors, baroreceptors and joint and muscle receptors. It may be that the spontaneous pacemaker activity of neurons in the respiratory centre is responsible for the automatic pattern of inspiration and expiration. It is more likely that the pattern of inspiration and expiration is governed by a much more complex system, involving the respiratory centre in the medulla and the many sources of afferent activity impinging upon it, as well as higher centres in the CNS. For example,

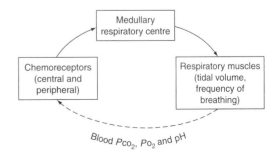

Figure 3.24 *Ventilatory control by feedback loops.*

the pneumotaxic centre in the pons modifies ventilation, and may determine the lung volume at which inspiration stops.

The respiratory centre in the medulla comprises three groups of neurons, dorsal and ventral, respiratory groups and the Botzinger complex in the reticular substance of the medulla oblongata. The dorsal group which is a part of the nucleus tractus solitarius lies near the termination of afferents of the IXth and Xth cranial nerves, is inspiratory, and affects the timing of the respiratory cycle. The ventral group which is located in the nucleus ambiguous and nucleus retroambigualis is mainly expiratory. The Botzinger complex is located rostral to the nucleus ambuguus and consists of entirely expiratory neurons. Neuronal activity in the respiratory centre during inspiration and expiration can be described in three phases. Within these anatomical groups, at least six types of cell have been described according to when they are active in inspiration and expiration as early inspiratory, inspiratory ramp, late-onset inspiratory, early expiratory, early peak whole expiratory, and expiratory ramp neurons:

- *Inspiratory phase.* A distinct beginning then a gradual increase ('ramp') of inspiratory nerve and muscle activity, with activation of dilator muscles of the pharynx.
- *Expiratory phase I.* A reduction in inspiratory motor nerve discharge, with a declining tone of the inspiratory muscles.

- *Expiratory phase II.* The inspiratory nerves (and muscles) are inactive. During normal quiet breathing, the expiratory neurons and muscles are quiescent, but are recruited with increased ventilation.

CENTRAL CHEMORECEPTORS

These neurons lie in the medulla, 200–400 μm below the ventral surface, and are separate from the respiratory centre. The central chemoreceptors are stimulated by an increase in the hydrogen ion concentration of the cerebrospinal fluid (CSF), which is determined, in the short term, by arterial blood P_{CO_2}. Carbon dioxide passes across the blood–brain barrier easily and, with water, produces carbonic acid which dissociates to hydrogen ions in the CSF. Changes in blood pH have less effect on the central chemoreceptors as the blood–brain barrier slows the passage of hydrogen ions from blood into the CSF (Fig. 3.25).

The central chemoreceptors are not stimulated by blood P_{O_2}. Hypoxia (in the absence of input from the peripheral chemoreceptors) depresses the respiratory centre.

Stimulation of the central chemoreceptors increases their afferent output, increasing the activity of the medullary respiratory centre and the respiratory muscles. A rise in arterial P_{CO_2} will therefore produce a reflex increase in ventilation (Fig. 3.26). Conversely, a reduction in hydrogen

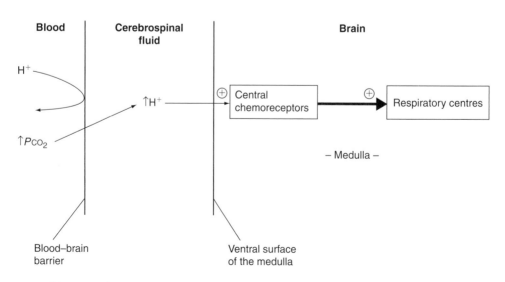

Figure 3.25 *Diagrammatic representation of the effect of the blood–brain barrier.*

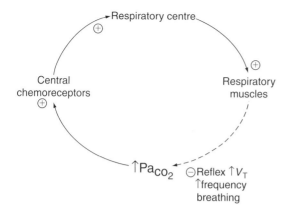

Figure 3.26   *The effect of raised arterial P_{CO_2} on the respiratory centre.*

ion concentration inhibits the central chemoreceptors. Most of the ventilatory response to carbon dioxide is via the central, not the peripheral, chemoreceptors.

The central chemoreceptors are particularly sensitive to changes in blood P_{CO_2} tension, as the buffering capacity of the CSF is less than that of blood, with a normal pH of 7.32. If the P_{CO_2} tension is chronically elevated by lung disease, the CSF pH can be reset by movement of bicarbonate ions across the blood–brain barrier into the CSF, so that the central chemoreceptor sensitivity to blood carbon dioxide tension is lost.

PERIPHERAL CHEMORECEPTORS

The carotid bodies, supplied by the glossopharyngeal nerve, lie at the bifurcation of the common carotid artery, and the aortic bodies, supplied by the vagus nerve, lie in the aortic arch (the carotid bodies are the more important in humans). The peripheral chemoreceptors have a high blood flow (much in excess of their weight), so that their oxygen requirement is met mainly by dissolved oxygen, and they are sensitive to reductions in arterial rather than venous P_{O_2}. The peripheral chemoreceptors are stimulated by low oxygen tension, not low oxygen content, of the blood. Conditions with low blood oxygen content, but relatively normal oxygen tension (anaemia, carboxyhaemoglobin), do not stimulate ventilation via the peripheral chemoreceptors.

The afferent nerve activity from the peripheral chemoreceptors increases as blood P_{O_2} falls, and

Table 3.2   *Stimuli to peripheral chemoreceptors*

Arterial blood	Carotid	Aortic
↓ Pa_{O_2}	+	+
↑ Pa_{CO_2}	+	+
↓ pH	+	0

reflexly stimulates ventilation via the respiratory centre in the medulla. In experiments using high inspired concentrations of oxygen it can be demonstrated that afferent nerve activity from the peripheral chemoreceptors begins when the P_{O_2} is reduced below a threshold of 500 mmHg (67 kPa). Both the carotid and aortic bodies are sensitive to changes in P_{CO_2} (Table 3.2). The carotid bodies are, in addition, stimulated by increases in blood hydrogen ion concentration.

The mechanism by which the peripheral chemoreceptors are activated is unclear, but it may involve neurotransmitter production by glomus cells. Intracellular ATP production in glomus cells falls with any reduction in arterial P_{O_2}, and the production of norepinephrine, dopamine and acetylcholine increases. Hypoxia may then excite activity in afferent nerves from the peripheral chemoreceptors by the release of neurotransmitters from glomus cells. Raised P_{CO_2} also stimulates glomus cell neurotransmitter release, perhaps by raising intracellular hydrogen ion concentration.

The response of the peripheral chemoreceptors to a fall in arterial blood P_{O_2}, rise in P_{CO_2}, or rise in hydrogen ion concentration is to stimulate a reflex compensatory increase in ventilation via the medullary respiratory centre (Fig. 3.27).

The response rate of the peripheral chemoreceptors is very rapid, and they can detect the oscillations of arterial blood P_{CO_2} tension which occur during the inspiration and expiration cycle. A fall in blood pressure, with low perfusion of the metabolically active aortic and carotid bodies, stimulates ventilation.

OTHER FACTORS INVOLVED IN CONTROL OF VENTILATION

Cortical and hypothalamic influences

The hypothalamic input to the respiratory centre provides a continual background excitatory drive

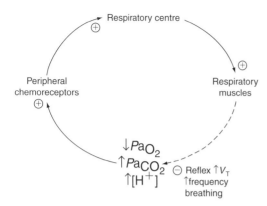

Figure 3.27 *The reflex ventilatory response via the peripheral chemoreceptors.*

to the dorsal respiratory group. Ventilation is partly under voluntary control, and this may be important in the initiation of increased ventilation at the beginning of exercise, as well as the coordination of breathing with speaking and coughing. The cerebral cortex and the limbic system integrate emotional and painful stimuli and have both excitatory and inhibitory effects on the respiratory centres via the hypothalamus.

Lung receptors

The Hering–Breuer inflation reflex is produced by stretch receptors in the smooth muscles of the trachea and lower airways and stimulated by lung inflation. Lung inflation inhibits directly and via the pontine respiratory group the inspiratory neurons of the dorsal respiratory group via the vagus nerve, terminating inspiration. A deflation reflex is also present, which stimulates inspiration. The importance of the Hering–Breuer reflex in humans is not clear, and it may only be important when the tidal volume exceeds 1 L.

Head's paradoxical reflex is transient stimulation of inspiration by a passive inflation of the lung, and can be demonstrated in the neonate, and at times during anaesthesia. Juxtacapillary receptors (J receptors) or pulmonary C fibres, in the alveolar walls close to the capillaries, are activated by an increase in interstitial fluid or mechanical distortion, and may be involved in the ventilatory response to pulmonary oedema. They do not play

a role in normal breathing. Irritant receptors in the airways detect cold air, dusts and noxious gases, producing bronchoconstriction via a vagal reflex.

Joint and muscle receptors

Afferent activity to the medulla from joint and muscle receptors, monitoring limb movement, reflexly stimulates ventilation. Muscle spindles in the intercostal muscles and the diaphragm sense the force necessary for ventilation, and may be involved in the reflex ventilatory response to airway obstruction.

Baroreceptors

Acting via the aortic arch and the carotid sinus baroreceptors, hypotension stimulates and hypertension depresses ventilation.

Temperature

Increased body temperature stimulates the respiratory centre.

Hormones

Both epinephrine and norepinephrine stimulate ventilation.

REFLEX VENTILATORY RESPONSES

Response to a raised arterial P_{CO_2}

Under normal circumstances, the arterial blood P_{CO_2} is the most important factor in the control of ventilation. By the feedback mechanisms described earlier, arterial P_{CO_2} is kept within ±3 mmHg (0.4 kPa) of the normal value of 40 mmHg (5.3 kPa). An elevated P_{CO_2} increases ventilation via both the central and the peripheral chemoreceptors, with the former being the more important. When P_{CO_2} rises, the peripheral chemoreceptors are also stimulated by the rise in arterial blood hydrogen ion concentration from the dissociation of carbonic acid (Fig. 3.28).

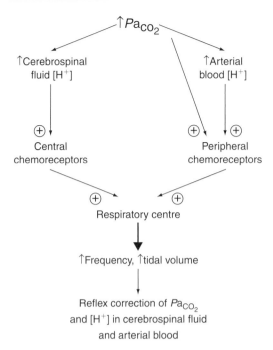

Figure 3.28   *The ventilatory response to carbon dioxide.*

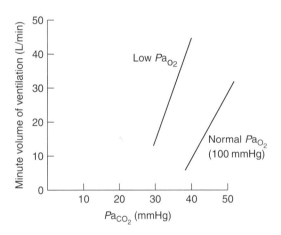

Figure 3.29   *The ventilatory response to carbon dioxide is enhanced by hypoxia.*

With a normal arterial P_{O_2}, the minute volume of ventilation is reflexly increased by 2–3 L/min for a 1 mmHg (0.13 kPa) rise in arterial blood P_{CO_2}. A low arterial P_{O_2} amplifies the ventilatory response to carbon dioxide (Fig. 3.29).

Response to a fall in arterial P_{O_2}

The response to falling arterial P_{O_2} is driven by the peripheral chemoreceptors. Hypoxia depresses the

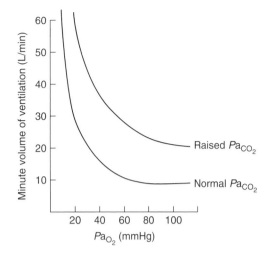

Figure 3.30   *The ventilatory response to hypoxia.*

ventilatory centre in the medulla. In the absence of peripheral chemoreceptors, hypoxia depresses ventilation, although prolonged hypoxia can produce mild cerebral acidosis, which stimulates ventilation.

The peripheral chemoreceptors are exquisitely sensitive to the partial pressure of oxygen, and increase their afferent neuronal activity in response to any fall in arterial P_{O_2}, although the reflex increase in ventilation produced by this via the respiratory centre only becomes significant when arterial P_{O_2} falls below 60 mmHg (8 kPa). The shape of the P_{O_2}–ventilation response curve is a rectangular hyperbola, asymptotic to ventilation at high P_{O_2} (i.e. zero hypoxic drive) and to the P_{O_2} at which ventilation becomes (theoretically) infinite at a P_{O_2} of 32 mmHg (4.3 kPa). The ventilatory response to hypoxia is amplified by elevations of arterial blood P_{CO_2} (Fig. 3.30).

In chronic lung conditions with elevated arterial P_{CO_2} the ventilatory response to carbon dioxide is lost by resetting of the central chemoreceptors by correction of the CSF hydrogen ion concentration by uptake of bicarbonate ions. In addition, renal compensation reduces the effect of elevated carbon dioxide on the peripheral chemoreceptors. Ventilation is then controlled by arterial P_{O_2}, not P_{CO_2}.

Response to a rise in arterial blood hydrogen ion concentration

The carotid bodies are stimulated by blood hydrogen ion concentration, and reflexly stimulate

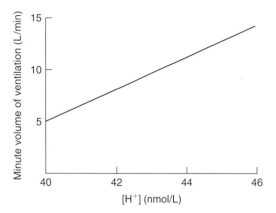

Figure 3.31 *The ventilatory response to a raised blood [H⁺].*

ventilation. The hydrogen ions may be formed from carbon dioxide, or from metabolic production of acids, including lactic acid. Hydrogen ions do not pass easily through the blood–brain barrier, but some may cross if the blood concentration is markedly elevated and stimulate the central chemoreceptors (Fig. 3.31).

Integrated ventilatory response to carbon dioxide, oxygen and hydrogen ions

The minute-to-minute control of ventilation is primarily by the effect of arterial P_{CO_2} on the central chemoreceptors. Significant reduction of ventilation, or a low inspired oxygen concentration, will stimulate the peripheral chemoreceptors and amplify the response to carbon dioxide by reducing arterial P_{O_2} and, if aerobic metabolism cannot be maintained, increasing lactic acid production.

EXERCISE AND THE CONTROL OF VENTILATION

During exercise, the increased metabolic requirement of the active skeletal muscles means that oxygen consumption can rise from the basal rate of 250 mL/min up to 3–4 L/min in athletes. With increased oxygen consumption, the production of carbon dioxide by skeletal muscle increases. The minute volume of ventilation increases to 120 L/min during exercise, delivering more oxygen, and removing more carbon dioxide.

It is not clear what controls ventilation during exercise. In moderate exercise, ventilation rises in step with the increased tissue metabolic needs, and this response occurs without significant changes in arterial blood P_{CO_2}, P_{O_2} or hydrogen ion concentration. During normal aerobic exercise, arterial blood P_{CO_2} falls slightly, and P_{O_2} and hydrogen ion concentration do not change. The systemic venous blood P_{CO_2} does rise and the venous P_{O_2} does fall during exercise. In severe exercise, the arterial P_{O_2} may fall, and the skeletal muscles obtain some energy by anaerobic metabolism, releasing lactic acid and hydrogen ions into the blood. Arterial blood P_{CO_2} falls during strenuous exercise because of the additional stimulus to ventilation of the peripheral chemoreceptor response to the low P_{O_2} and increased hydrogen ion concentration.

Descending neural influences from motor areas of the cerebral cortex excite the medullary respiratory centre, and this may explain the abrupt increase in ventilation at the onset of exercise. Joint and muscle receptors, sensing limb movement, send afferent information to the medulla and increase ventilation. The peripheral chemoreceptors may detect an increase in the arterial P_{CO_2} fluctuation within the respiratory cycle, due to the raised tidal volume during exercise. Other factors which may stimulate ventilation during exercise are the release of epinephrine and norepinephrine, and a rise in blood temperature.

ANAESTHETIC AGENTS AND THE CONTROL OF VENTILATION

Volatile anaesthetic agents, barbiturates and opioid analgesics depress the ventilatory response to hypercarbia, hypoxia and acidosis. The rate of rise of minute ventilation with elevated P_{CO_2} is reduced, and the 'apnoeic threshold' (the arterial P_{CO_2} at which spontaneous breathing resumes after hyperventilation) is elevated. The slope of the CO_2 response curve is also reduced. Volatile anaesthetic agents depress the hypoxic (P_{O_2}/ ventilation) response curve even at very low concentrations. These effects persist into the early postoperative period and increase the risk of hypoxia developing.

BREATH HOLDING

When the breath is held after breathing air, alveolar gas reaches equilibrium with mixed venous blood within a few minutes and the $P\mathrm{CO_2}$ will increase at a rate of 3–6 mmHg/min (0.4–0.8 kPa/min). What actually happens to the arterial blood gases depends on the patency of the airway and the composition of the ambient gas if the airway is patent. If the airway is occluded the alveolar $P\mathrm{O_2}$ decreases to approximately the mixed venous $P\mathrm{O_2}$ within a minute. If the airway is patent and air is the ambient gas, the initial changes are as described above but ambient air is drawn in to replace the decrease in lung volume (difference between oxygen uptake and carbon dioxide output) during the apnoea. Nitrogen in the air accumulates in the alveoli and hypoxia occurs after about 2 min.

If the airway is patent and oxygen is the ambient gas, then oxygen that continuously removed from the alveolar gas is replaced by oxygen drawn in by mass movement. As no nitrogen is added to the alveolar gas, the alveolar $P\mathrm{O_2}$ falls (about 3–6 mmHg (0.4–0.8 kPa) as fast as the $P\mathrm{CO_2}$ rises). The patient will not become seriously hypoxic for several minutes.

The 'break point' for breath holding is the point at which the stimulus to breathe overwhelms any voluntary effort to hold the breath. When the breath is held after breathing air the break point occurs at a $P\mathrm{CO_2}$ of 50 mmHg (6.7 kPa). This does not mean that $P\mathrm{CO_2}$ is the dominant factor and concomitant hypoxia is probably more important. Preliminary oxygen breathing prolongs breath holding as it delays the onset of hypoxia. Breath-holding time is also proportional to the lung volume at the onset of breath holding. This is because of the increase in oxygen stores with larger initial lung volumes and other effects mediated by nervous afferents arising from the lungs and chest wall. Prolonged durations of breath holding can be achieved after hyperventilation and preoxygenation.

Reflections

1. The functions of the lungs primarily involve gas exchange. The non-respiratory functions of the lungs include filtration, synthesis and metabolism of endogenous (surfactant, bradykinin and other vasoactive mediators) and exogenous substances, a blood reservoir and host defence. The basic physiological unit of the lung is the terminal respiratory unit which consists of the respiratory bronchiole, alveolar ducts, and the alveoli. The lungs have two separate blood supplies: the pulmonary circulation which brings deoxygenated blood from the right ventricle to the respiratory units for gas exchange, and the bronchial circulation which provides nourishment to the lung parenchyma. The pulmonary circulation has a high capacitance and can accommodate large volumes of blood. The lungs have host defence functions via IgA secreted by the epithelium, macrophages and T lymphocytes in the lymphoid tissue of the lung. The lungs are supplied by the autonomic system and do not have voluntary motor innervation fibres. Pain fibres are only found in the pleura. Stimulation of the parasympathetic system leads to airway constriction, vasodilatation and increased glandular secretions. Sympathetic stimulation causes airway relaxation, vasoconstriction, and inhibition of glandular secretion.

2. Lung volumes are determined by the balance between the inward elastic recoil of the lungs and the outward elastic recoil of the chest wall. The functional residual capacity (FRC) is the resting volume of the lung which is determined by the balance of lung elastic recoil and chest wall recoil forces. Lung compliance is a measure of the distensibility of the lung and is determined both by the elasticity of the lung parenchyma and the surface tension of the air–liquid interface of the alveoli. The surface tension is reduced below that of water by pulmonary surfactant that is secreted by type II alveolar cells. Pulmonary fibrosis is associated with a decrease in lung compliance whereas lung compliance increases in emphysema because of a loss in elastic recoil. Factors that

influence lung volumes include age, gender, height, body weight, posture and obesity. The work of breathing is equal to the change in pressure times the change in volume.

3. A positive transpulmonary pressure is required to increase lung volume. At end inspiration and end expiration (points of zero air flow) the pressure across the respiratory system is zero. The first eight airway generations are the major sites of airway resistance. Airway resistance decreases with increases in lung volumes and with decreases in gas density. It is also determined by changes in airway calibre (varies inversely to the fourth power of the radius). The equal pressure point is the point at which the pressure inside and surrounding the airway are equal. As the lung volume and elastic recoil decrease, the equal pressure point moves closer towards the alveolus. In patients with chronic obstructive disease the equal pressure point is closer to the alveolus.

4. The partial pressure of oxygen in the alveolus is given by the alveolar air equation and this can be used to calculate the alveolar–arterial oxygen difference. There is an inverse relationship between the partial pressure of carbon dioxide in the alveolus and alveolar ventilation. The relationship between carbon dioxide production and alveolar ventilation is given by the alveolar carbon dioxide equation. The volume of air in the conducting airways is called the anatomical dead space. The physiological dead space is that part of the tidal volume that does not participate with gas exchange and includes the anatomical dead space and the alveolar dead space (alveoli that are ventilated but not perfused).

5. The pulmonary circulation is a low-resistance, low-pressure system with a high capacitance. The arteries of the pulmonary circulation are thin walled and distensible. The pulmonary vessels are seven times more compliant than systemic blood vessels. The recruitment of new capillaries is an unique feature of the pulmonary circulation. The extra-alveolar vessels are arteries, arterioles, veins, and venules that are not subjected to alveolar pressures, but are affected by intrapleural and interstitial pressure changes.

Hypoxic vasoconstriction occurs in small arterial vessels in response to low alveolar P_{O_2}.

6. Regional differences in ventilation and perfusion in the lungs are due to the effects of gravity. In the upright subject, blood flow increases linearly from the apex to the base of the lung where flow is greatest. The lung is divided into three zones: zone 1 represents the apex where blood flow does not occur (dead space); zone 2, the cascade or starling resistor zone where $P_a > P_A > P_v$; and zone 3 where $P_a > P_v > P_A$, and blood flow follows the vascular pressure gradients. The ventilation/perfusion (V/Q) ratio at the apex of the lung is high (3.3) whereas the V/Q ratio at the bottom of the lung is very low (0.6). In the normal lung the overall V/Q ratio is about 0.8. There are four mechanisms for hypoxaemia: anatomical shunt, physiological shunt, V/Q inequalities, and hypoventilation. Hypercarbia occurs as a result of an increase in dead space or hypoventilation.

7. Gases such as nitrous oxide, helium and ether have a rapid rate of air to blood equilibration and are described as perfusion limited. Gases such as carbon monoxide (CO) have a slow air-to-blood equilibration rate and are diffusion limited. Normally, oxygen transport is perfusion limited, but under certain conditions it can be diffusion limited. The diffusion capability across the alveolar capillary membrane is measured using CO gas, and is expressed as D_{LCO}. Although the diffusing capacity for carbon dioxide is 20 times that for oxygen, the overall rate of equilibration is similar for both gases.

8. Oxygen binds rapidly and reversibly to the haem groups of the haemoglobin molecule. The oxygen dissociation curve is sigmoid-shaped. The steep part of the curve (20–60 mmHg) indicates that oxygen is readily released from haemoglobin during low P_{O_2} states. In the plateau portion of the curve, increasing the P_{O_2} has only a minimal effect on haemoglobin saturation. Increases in Pa_{CO_2}, hydrogen ions, 2,3-DPG and temperature decrease the affinity of haemoglobin for oxygen. This is called the 'Bohr effect' which facilitates the delivery of oxygen to the tissues and uptake of oxygen in the lungs.

9. The major source of carbon dioxide is the mitochondria during aerobic respiration. CO_2 interacts with H_2O to form H_2CO_3 in the red blood cell catalysed by carbonic anhydrase. H_2CO_3 dissociates to H^+ and HCO_3^-. HCO_3^- diffuses out of the red blood cell across a concentration gradient and is replaced by chloride ions to maintain electrical neutrality. This is known as the chloride shift. The CO_2 dissociation curve from blood is linear and directly related to P_{CO_2}. Deoxygenated haemoglobin has an increased capacity to carry more CO_2 and this is called the 'Haldane effect'.

10. The control of respiration is primarily automatic via the respiratory centres in the brainstem, although voluntary control can also occur via the cerebral cortex and limbic system. Ventilatory control is mediated by the respiratory control centre, central chemoreceptors, peripheral chemoreceptors, and pulmonary mechanoreceptors. The respiratory control centre consists of the dorsal respiratory group and the ventral respiratory group. Rhythmic breathing is mediated by tonic or continuous inspiratory drive from the dorsal inspiratory group, and by intermittent or phasic expiratory inputs from the ventral respiratory group including the Botzinger complex, cerebrum, hypothalamus and ascending spinal sensory tracts. The central chemoreceptors are located near the ventral surface of the medulla and are responsible for most of the chemical stimulus to breathing, They respond to changes in the pH of CSF brought about by alterations in arterial P_{CO_2}. The peripheral chemoreceptors respond mainly to a low P_{O_2} and to a lesser extent pH and P_{CO_2}. Lung mechanoreceptors can also influence breathing. Juxtacapillary C fibre receptors are stimulated by increases in interstitial pressures near the pulmonary capillaries.

4

Cardiovascular physiology

LEARNING OBJECTIVES

After studying this chapter the reader should be able to:

1. Understand the functions and layout of the cardiovascular system
   Describe:
   a. General functions of the pulmonary and systemic circulations, the pumps and circuitry, with an understanding of the pressure, area and velocity of flow in the vessels
   b. The distribution of blood volume in the cardiovascular system and the organ distribution of the cardiac output

2. Understand the functions of the heart
   Describe:
   a. The functional anatomy of the heart, structure of the muscular walls, the coronary arteries, cardiac energy production, and the nerve supply
   b. The ionic basic of the cardiac action potentials, the relationship between the cardiac muscle contraction and refractory periods, and excitation–contraction coupling in cardiac muscle cells
   c. The generation and conduction of the cardiac action potential, the progress of the potential through the chambers of the heart, and electrocardiography
   d. The mechanical events of the cardiac cycle, with knowledge of the biophysical determinants of cardiac muscle contraction from isolated preparations, studies of performance of the whole heart and ventricular pressure–volume relationships
   e. The effect of the autonomic nervous system on cardiac performance

3. Understand the physical factors governing blood flow through vessels
   Describe:
   a. Poiseuille's equation, and observed physiological deviations from Poiseuille's equation
   b. The relationship between pressure and flow in vessels of the systemic circulation with a description of the total fluid energy comprising potential and kinetic energy

4. Understand the role of the vessels of the systemic circulation
   Describe:
   a. The role of arteries, arterioles, capillaries, lymphatics and veins
   b. The determination of arterial blood pressure
   c. The relationship between the return of blood to the heart and cardiac output

5. Understand the control of the cardiovascular system
   Describe:
   a. The organization of the autonomic nervous system and the central and peripheral control of the cardiovascular system
   b. The integration of the control of the cardiovascular system
   c. The efferent pathways and effectors of the cardiovascular system, sensors and measured variables, and effects of the arterial baroreceptor reflex upon arterial blood pressure
   d. The control of the special circulations including the heart, skeletal muscle, splanchnic, kidneys, the brain, the skin and the lungs

6. Understand the integrated cardiovascular responses to haemorrhage, the Valsalva manoeuvre and exercise

FUNCTIONS AND LAYOUT OF THE CARDIOVASCULAR SYSTEM

GENERAL FUNCTIONS

The cardiovascular system (Fig. 4.1) circulates blood through the vessels of the pulmonary and systemic capillaries for the purpose of exchange of oxygen, carbon dioxide, metabolic nutrients, waste products and water in the tissues and in the lungs.

The right and left ventricles are two pumps in series that contract simultaneously to perfuse blood through the pulmonary and systemic circulations. As described in Chapter 3 (p. 102), the right ventricle pumps blood at low pressure through the pulmonary capillaries, where haemoglobin takes up oxygen and loses carbon dioxide. The oxygenated blood then passes to the left atrium via the pulmonary veins. The left ventricle pumps blood to the organs and tissues of the body, and exchange of oxygen, carbon dioxide, water, nutrients and waste products takes place in the systemic capillaries.

The cardiovascular system can be considered to have as many roles as the diverse tissues and organs it supplies. The transfer of oxygen and carbon dioxide between the lungs and the tissues is the system's prime function. The gastrointestinal vessels absorb nutrients from the gut, and perfuse the liver. The cardiovascular system then delivers these nutrients (glucose, amino acids, fatty acids) to the tissues for cellular metabolism, and removes waste products. The renal circulation is essential for water and electrolyte homeostasis and the excretion of waste products. The cardiovascular system also has an important role in the distribution of body water between the intravascular, extracellular and intracellular spaces: a reduction in the hydrostatic pressure in the systemic capillaries facilitates the movement of interstitial and intracellular water into the intravascular space. The hormonal roles of the cardiovascular system include the delivery of endocrine hormones from their release sites to the target tissues, and the production of atrial natriuretic factor (ANF) within the heart. The cardiovascular system has an immune

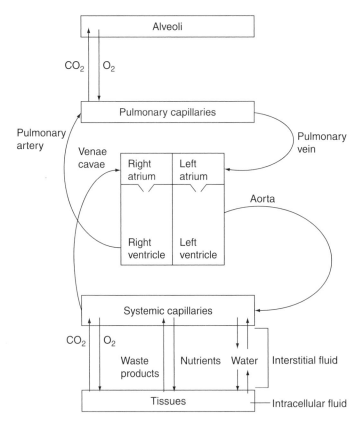

Figure 4.1 *General functions of the cardiovascular system.*

role, transporting antibodies and immune cells around the body. The cardiovascular system is involved in temperature regulation, as skin blood flow and thus heat loss can be varied according to body temperature.

PUMPS AND CIRCUITRY

The low-pressure pulmonary and high-pressure systemic circulations operate in series. The different vascular beds of the systemic circulation are in parallel, and the flow to them is determined by the resistance vessels, the peripheral arterioles (Fig. 4.2).

The function of the left ventricle is to pump blood around the systemic circulation. The pulsatile output of the left ventricle is converted to continuous flow by the elastic properties of the aortic wall, the presence of resistance in the peripheral vessels, and the prevention of retrograde flow by the aortic valve (the Windkessel effect). The high-pressure systemic circulation allows the cardiac output to be rapidly distributed and redistributed between different organs, and provides sufficient hydrostatic pressure for filtration in the renal glomeruli. The systemic arterial blood pressure is the fundamental monitored and regulated variable of the cardiovascular system. The arterioles

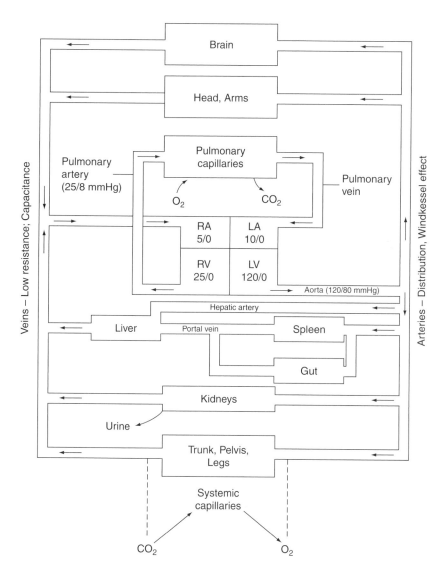

Figure 4.2 *The pumps and circuitry of the cardiovascular system.*

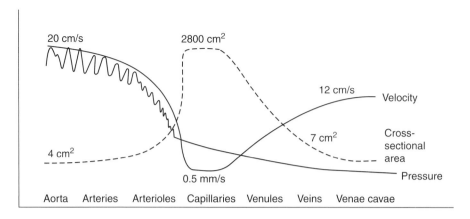

Figure 4.3   *Pressure, cross-sectional area and velocity of flow.*

determine the distribution of the cardiac output around the various organs by their variable resistance, determine arterial blood pressure by influencing total peripheral resistance, and reduce intravascular pressure upstream of the thin-walled capillaries. The capillaries consist of only a single layer of endothelium (which lines the entire cardiovascular system), and provide a very large surface area for the exchange of oxygen and nutrients between the tissues and the blood that flows slowly through them. Thin-walled lymphatic vessels are permeable to protein, collect any excess fluid and protein from the interstitial spaces and drain into veins in the thorax. The small venules and larger veins are low-resistance conduits for the return of blood to the right atrium. The large veins normally contain some 60 per cent of the blood volume, and their capacitance can be altered by sympathetic nerve activity. The function of the right ventricle is to pump blood through the pulmonary circulation, which receives all of the output of the that ventricle.

Pressure, cross–sectional area and velocity of flow in the systemic vessels

There is significant resistance to blood flow in the small arteries, and this increases in the arterioles. As a result the pressure developed in the left ventricle drops significantly as blood flows through the small arteries and arterioles, and the pulsatile flow present in the large arteries is damped and becomes continuous before it reaches the systemic

capillaries. There is an inverse relationship between vessel cross-sectional area and velocity of blood flow in the systemic circulation. At the resting cardiac output (the product of the heart rate and the stroke volume ejected by each ventricular contraction), blood flows through the aorta (cross-sectional area $4\,cm^2$) at 20 cm/s (Fig. 4.3). Although only one-quarter of the capillaries are patent at rest, this still represents a very large cross-sectional area of $2800\,cm^2$, and blood flow slows to 0.5 mm/s. In the veins the vessel cross-sectional area decreases and the velocity of blood flow increases again to 12 cm/s in the venae cavae. Blood flow in the capillaries is therefore continuous and slow, conditions favouring diffusional exchange of nutrients and waste products between the tissues and blood.

DISTRIBUTION OF BLOOD VOLUME IN THE CARDIOVASCULAR SYSTEM

The normal total blood volume is 5–6 L in males, and 4–5 L in females. When a subject is supine, 75 per cent of the blood volume is in the systemic circulation, 16 per cent in the pulmonary circulation and 8 per cent in the heart. Some 6 per cent of the total blood volume is in the systemic capillary exchange vessels, 60 per cent in the veins and 15 per cent in the arteries and arterioles. In the pulmonary circulation, 3 per cent of the total blood volume is in the pulmonary capillaries, with 8 per cent in the arteries and 5 per cent in the veins. When standing, the volumes in the heart and pulmonary circulations fall to 6 and 9 per cent of total blood volume respectively, and the amount in the veins increases.

The organ distribution of the cardiac output

The normal resting cardiac output (product of the heart rate and the left ventricular stroke volume) is 5–6 L/min, and this can increase to 20–30 L/min during exercise, depending on the level of fitness of the individual. As all the systemic vascular beds are exposed to the same mean arterial pressure, the distribution of the cardiac output to individual organs is determined by the local organ vascular resistance, which varies from tissue to tissue. The arteriolar diameter determines the resistance to flow through the organs, and this varies with tissue metabolic demands and sympathetic vasoconstrictor nerve activity. Approximate organ blood flows at rest are: brain 750 mL/min; coronary 250 mL/min; kidneys 1100 mL/min; skeletal muscle 1200 mL/min; abdominal organs 1400 mL/min; skin 500 mL/min; other tissues 600 mL/min. During exercise, the blood flow to the skeletal muscles can increase to 20 L/min.

THE HEART

FUNCTIONAL ANATOMY

The cardiac chambers and valves

The heart is a muscular pump with four chambers: the right atrium and ventricle, and the left atrium and ventricle. Blood flowing through the chambers of the right side of the heart has no direct connection with the chambers of the left side; the right and left sides of the heart are two pumps in series, separated by the pulmonary and systemic vessels. Blood from the systemic veins is carried by the superior and inferior venae cavae to the right atrium, is pumped by the right ventricle at low pressure into the pulmonary artery, passes through the pulmonary capillaries, is carried by the four pulmonary veins to the left atrium, and is pumped by the left ventricle at high pressure into the aorta to perfuse the systemic tissues once more. The

walls of the cardiac chambers are made of the cardiac muscle, the myocardium, which is lined inside by endothelium and outside by mesothelial epicardium. The thin but fibrous pericardium encloses and limits sudden overdistension of the heart chambers. The pericardial space contains a small amount of lubricating pericardial fluid.

Cardiac valves ensure unidirectional flow of blood within the cardiac chambers, movement of the valve flaps being passive. The tricuspid (right) and mitral (left) atrioventricular (AV) valves lie between the atria and ventricles, and prevent reflux of blood into the atria during ventricular contraction. Chordae tendineae connect the edges of the AV valves to papillary muscles within the ventricles. When atrial pressure exceeds ventricular pressure, the AV valves open and ventricular filling takes place. When the ventricles contract, ventricular pressure exceeds atrial pressure and the AV valves close. The papillary muscles and the chordae tendineae limit any eversion or bulging of the AV valves into the atria during ventricular contraction. Semilunar valves lie between the right ventricle and pulmonary artery (pulmonary valve) and the left ventricle and aorta (aortic valve). They open during ventricular contraction, when ventricular pressure exceeds pulmonary arterial and aortic pressure, and close passively when ventricular pressure falls during diastole, preventing reflux of blood into the ventricles.

During ventricular relaxation, blood flows continuously into the right atrium from the systemic veins and into the left atrium from the pulmonary veins, the atrioventricular valves are open, and the ventricles fill from the atria until they are more than three-quarters full. The atria then contract and complete ventricular filling (to 130 mL blood volume). Shortly afterwards, the ventricles contract, the AV valves close, the aortic and pulmonary valves open and ventricular ejection takes place (70 mL ejected). When ventricular contraction stops, the aortic and pulmonary valves close, the AV valves open, and ventricular filling starts again.

The structure of the muscular walls of the heart chambers

The pressures developed within the cardiac chambers determine the thickness of the muscle wall. The atria operate at low pressures and have thinner walls than the ventricles. The left ventricle has a much thicker muscle wall than the right ventricle, and develops much higher pressures.

Although the chambers of the right and left sides of the heart function as two separate pumps in series, the muscle fibres of the atria are continuous, as are the fibres of the ventricles. The fibrous atrioventricular ring separates and prevents direct electrical coupling of the atria and ventricles, and conduction of the cardiac action potential from the atria to the ventricles is through the specialized conducting tissues of the AV node and bundle of His. This means that when a cardiac action potential is generated in the sinoatrial (SA) node, the atria contract simultaneously (to complete ventricular filling), and then there is a short delay (for transmission through the AV node) before the ventricles contract together.

Cardiac muscle cells have striations similar to those of skeletal muscle cells, being made up of sarcomeres (between Z lines) containing myosin thick filaments (in the A band) and thin actin filaments (in the I band) which are attached to the Z lines. When compared with skeletal muscle, cardiac muscle cells are shorter and thicker, and form branching networks with intercalated disks between the ends of adjacent fibres that contain low electrical resistance gap junctions. Cardiac muscle operates as a functional syncytium, although it is not a true syncytium (a mass of protoplasm with many nuclei forming one cell), because each myocardial cell has its own nucleus within its own membrane. Cardiac muscle functions as a syncytium due to the presence of low-resistance connections between adjacent cells, and when an action potential is generated, the atria or the ventricles contract together.

The cardiac muscle cells are arranged in spiral layers anchored to the fibrous ring at the base of the heart and which encircle the chambers. During ventricular contraction, the base of the heart is pulled downwards and the heart rotates to the right, and this can be palpated as the 'apex beat'. The left ventricle contracts in a concentric, squeezing, fashion, whilst right ventricular contraction is more of a bellows action.

The coronary arteries

The left and right coronary arteries arise from the aortic root behind the cusps of the aortic valve. The right coronary artery perfuses the right ventricle and atrium. The left coronary artery divides into anterior descending and circumflex branches and perfuses the left ventricle and atrium. In humans there is overlap between the right and left coronary arteries: in 50 per cent of individuals the right coronary artery is dominant; the left coronary is dominant in 20 per cent. Myocardial capillary density is very high, facilitating the delivery of oxygen and nutrients. Most of the venous blood from the coronary circulation drains into the right atrium via the coronary sinus, but some capillary beds drain directly into the cardiac chambers via the anterior coronary and the thebesian veins. The normal resting coronary blood flow is 250 mL/min, with a myocardial oxygen consumption of 8–10 mL/min per 100 g of heart tissue. The oxygen extraction of cardiac muscle is very high at rest (coronary venous blood oxygen content is only 5 mL/100 mL), and blood flow must increase when oxygen consumption increases. Coronary blood flow increases several-fold with exercise and decreases moderately during hypothermia and hypotension.

Cardiac substrate utilization and efficiency

The heart is versatile in consuming various substances for energy, and in general the use of substrates is in proportion to their blood concentrations. Up to 40 per cent of the total cardiac oxygen consumption is by carbohydrate metabolism: glucose and lactate are used in equal proportion, and insulin enhances cardiac glucose uptake. The main substrate used by the heart is esterified and non-esterified fatty acid, which accounts for 60 per cent of cardiac oxygen consumption. In normal circumstances the heart produces energy by oxidative pathways. During hypoxia, anaerobic glycolysis does occur, but in ischaemia lactic acid is not washed out, the hydrogen ion concentration rises, glycolysis stops and the myocardial cell dies. The efficiency of the heart (work done/energy used) is low at 14–18 per cent, but comparable with that of man-made pumps. Cardiac efficiency improves during exercise as the cardiac output increases considerably without a proportional increase in oxygen use (as there is little change in mean arterial blood pressure).

The nerve supply to the heart

The heart has dual and opposing nerve supplies: parasympathetic (acetylcholine transmitter, slows heart rate) and sympathetic (catecholamine transmitter, increases heart rate and force of contraction). Both the parasympathetic and sympathetic nerve supplies originate in cardiovascular control centres in the medulla. The parasympathetic supply is via the right and left vagus nerves. The right vagus mainly slows depolarization of the SA node, and the left vagus slows conduction through the AV node. The released acetylcholine has an immediate effect on the heart, but the cardiac response to vagal nerve discharge is transient as both nodes are rich in cholinesterase. Therefore, the vagus nerves have a beat-by-beat effect on the heart. The cardiac sympathetic fibres originate from the intermediolateral columns of the upper thoracic spinal cord, synapse in the middle or stellate ganglia, and then form a complex nerve plexus (mixed with parasympathetic fibres) to the heart. The transmitter released is norepinephrine which has a slower onset but longer duration of action than acetylcholine.

CARDIAC ACTION POTENTIALS

Nerve action potentials last only 1 ms, and are produced by changes in membrane permeability to sodium and potassium ions. However, cardiac action potentials last much longer (250 ms) because changes in membrane calcium permeability produce a prolonged plateau phase. The duration of cardiac action potentials varies inversely with the heart rate. The autorhythmicity of the SA and AV nodes is explained by cyclical changes in membrane ionic permeability.

Fast- and slow-response cardiac action potentials

There are two main types of cardiac action potentials. The 'fast response' type is seen in normal atrial and ventricular muscle cells and in the Purkinje fibres. The 'slow response' cardiac action potential is seen in the SA and AV nodes.

The ionic basis of the fast-response cardiac action potential

Atrial and ventricular muscle and Purkinje fibre action potentials differ from those in nerves as they are much longer in duration, with a distinct plateau phase when depolarization is maintained (Fig. 4.4).

The membrane potential comprises five phases:

- *Phase 0.* The cell is depolarized from the resting membrane potential by a rise in membrane sodium permeability. Fast sodium channels open; these are similar to those in nerves and are sensitive to tetrodotoxin. Potassium conductance decreases.
- *Phase 1.* Partial repolarization is produced by a rapid decrease in sodium permeability.
- *Phase 2 – plateau.* The cell membrane permeability to calcium rises, maintaining depolarization. Sodium conductance continues to decline slowly.
- *Phase 3 – repolarization.* Potassium, sodium and calcium permeability return towards normal.
- *Phase 4 – the resting potential* (-90 mV). The membrane potential is mainly governed by potassium permeability.

These phases are not separate events. Depolarization to the threshold potential affects consecutively sodium, calcium and potassium channels in

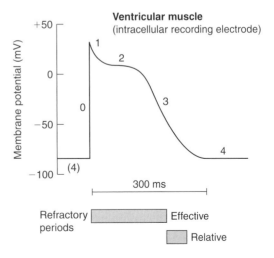

Figure 4.4 *Intracellular electrode recording of membrane potential in ventricular muscle.*

the membrane and produces depolarization, the plateau and eventual repolarization of the ventricular cells.

CARDIAC CALCIUM CHANNELS

Two types of calcium channels are involved in the cardiac action potential: long-lasting (L-type) and transient (T-type). The L-type channels produce a long-lasting calcium current, are the predominant calcium channels in the heart, and begin to open during the action potential upstroke (phase 0) when the membrane is depolarized to $-10\,mV$. The calcium channel antagonists verapamil and nifedipine block L-type calcium channels. Catecholamines increase the activation of L-type calcium channels via β-receptor stimulation of membrane adenyl cyclase and increased intracellular cyclic AMP production. The T-type channels open briefly at more negative membrane potentials during phase 0 (-70 to $-40\,mV$) than the L-type channels and are not affected by catecholamines.

REFRACTORY PERIODS

The ventricular muscle action potential lasts for 250 ms. Of this, the absolute refractory period accounts for the first 200 ms and the relative refractory period for the other 50 ms. The absolute refractory period extends into phase 3, and during this period the cell will not respond to further excitation.

The ionic basis of the slow–response cardiac action potential

THE SA NODE

The SA node has no resting state; rather there is a pacemaker potential which generates cardiac autorhythmicity. Phases 1 and 2 (of the fast response cardiac action potential) are absent in the SA node as there is no depolarization plateau (Fig. 4.5).

- *Phase 4.* The pacemaker potential is produced by a fall in membrane potassium permeability and an increase in a slow inward current. The slow inward current consists of a voltage-gated increase in calcium permeability (via T-type channels) and activity of the electrogenic sodium–calcium exchange system, driven by inward movement of calcium ions. This pacemaker activity brings the cell to the threshold potential.
- *Phase 0.* Depolarization is produced by opening of voltage-gated calcium channels (L-type) and inward movement of positive ions. There is no sodium current involved in the SA node potential.
- *Phase 3.* Repolarization is accomplished by a late increase in potassium permeability and outward flow of ions.

Again, a process of changing membrane permeabilities causes the SA node cardiac action potential. A cycle of reduced potassium, increased calcium and then increased potassium membrane permeability produces the SA node autorhythmicity.

Parasympathetic nerve stimulation increases the membrane potassium permeability of the SA node, hyperpolarizes the cell, and inhibits spontaneous cardiac activity. Sympathetic nerve activity has the opposite effect by opening calcium channels.

THE AV NODE

The ionic basis of the AV node is the same as that for the SA node, but the rate of depolarization of the pacemaker potential during phase 4 is slower.

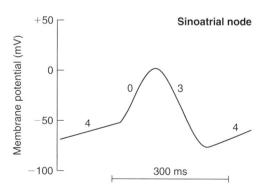

Figure 4.5 *Membrane potential in the sinoatrial node.*

RELATIONSHIP BETWEEN CARDIAC ACTION POTENTIAL, MUSCLE CONTRACTION AND REFRACTORY PERIODS

As the contraction lasts hardly longer than the action potential, and the cell is in the absolute refractory period for most of the duration of the action potential, it is impossible to summate contractions or tetanize cardiac muscle (Fig. 4.6).

As described earlier, the absolute refractory period accounts for 200 ms of the cardiac action potential, and the relative refractory period 50 ms. During the absolute refractory period the cardiac cell is inexcitable, but a supramaximal stimulus can generate an action potential during the relative refractory period. Such an action potential has a slower rate of depolarization, smaller amplitude, and is shorter than a normal action potential; moreover, the contraction produced is weaker. Because of the long duration of the absolute refractory period, any contraction generated during the relative refractory period can only occur when the muscle is already relaxing after the previous normal contraction. It is impossible to produce tetanic contraction of cardiac muscle with high-frequency stimuli, as can be done with skeletal muscle. The long cardiac refractory periods prevent the ventricles from being activated again before they have had time to relax and fill with blood after a normal contraction.

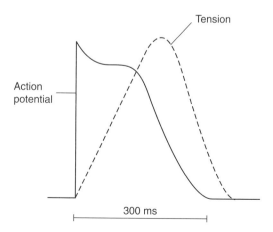

Figure 4.6 *The cardiac action potential and developed tension.*

EXCITATION–CONTRACTION COUPLING IN CARDIAC MUSCLE CELLS

Contraction in cardiac fibres is by interaction of actin and myosin filaments in the presence of calcium. Tropomyosin consists of two α-helical chains, and lies in the groove between the two actin polymers, preventing the interaction of myosin with actin – an effect modulated by troponin. The globular protein troponin is present with tropomyosin at regular intervals on the thin actin filaments, with one molecule of each for every seven actin monomers. There are three subunits: troponin-T binds tropomyosin; troponin-I inhibits actomyosin ATPase; and troponin-C binds calcium. In the presence of calcium, the configuration of the troponin–tropomyosin complex alters and myosin can interact with actin.

As in skeletal muscle, contraction in cardiac muscle is achieved by a temporary release from the sarcoplasmic reticulum of calcium, which binds to troponin-C, permitting interaction of actin and myosin filaments. The cardiac muscle relaxes when the sarcoplasmic reticulum takes up and sequesters calcium, again by active transport. Unlike skeletal muscle, the opening of sarcoplasmic reticulum calcium-release channels is triggered by the inward flow of calcium across the cell membrane and the T-tubules during the action potential. (In skeletal muscle, the action potential itself releases calcium from the sarcoplasmic reticulum.) The membrane calcium channels are opened by phosphorylation by a cyclic AMP-dependent protein kinase. Cardiac muscle does not contract if calcium is absent from the extracellular fluid. This form of excitation–contraction coupling may be described as 'calcium-triggered calcium release', and is an amplification process whereby the movement of a small amount of calcium into the cell causes the temporary release of a much larger amount of calcium from the sarcoplasmic reticulum. The amount of calcium released inside the cardiac cell is an important determinant of the force of contraction. In the cardiac cell, the amount of calcium normally released is not sufficient to combine with all the troponin available (which is the case in skeletal muscle), so factors which increase the intracellular calcium concentration can increase the force of contraction.

The calcium which entered the cell to open the sarcoplasmic reticulum channels and induce contraction is removed during diastole by a membrane exchanger which removes one calcium in exchange for the entry into the cell of three sodium ions (countertransport), and by active transport of calcium out of the cell.

Catecholamines (norepinephrine), acting via β_1 receptors and a G protein, which activates membrane adenyl cyclase and increases the activity of a cyclic AMP-dependent protein kinase, augment the force of cardiac contraction by phosphorylation and opening of membrane calcium channels. They also phosphorylate myosin, increasing the rate at which cross-bridge cycling occurs with actin. Catecholamines also phosphorylate phospholamban, which stimulates the active uptake of calcium into the sarcoplasmic reticulum, and troponin-I, which inhibits calcium interaction with troponin-C. Catecholamines therefore augment contraction (by opening calcium channels and phosphorylation of myosin) and relaxation (via phospholamban and troponin-I) of the cardiac muscle. Sympathetic stimulation therefore speeds both emptying and filling of the heart.

Digoxin increases the intracellular calcium ion concentration by inhibiting the active membrane Na, K-ATPase, increasing intracellular sodium concentration and thus impairing the activity of the membrane countertransport mechanism that excretes calcium in exchange for sodium entry. Digoxin therefore increases intracellular sodium and calcium concentrations and increases the force of contraction of the cardiac muscle.

GENERATION AND CONDUCTION OF THE CARDIAC ACTION POTENTIAL

Cardiac pacemaker cells generate the inherent automaticity and rhythmicity of the heart; the ionic basis of pacemaker potentials was described earlier. The SA node, other atrial centres, the AV node and the bundle of His all have inherent pacemaker activity, but the SA node is the most active and suppresses the others. The cardiac action potential generated by the SA node spreads directly through atrial muscle by gap junctions between adjacent fibres, producing simultaneous contraction of both atria. There is no direct electrical connection between atrial and ventricular muscle, and the cardiac action potential reaches the ventricles through the AV node and then the bundle of His. AV node transmission is associated with a delay of 0.1 s, which allows the atria to finish filling the ventricles before they contract.

The SA node

The normal pacemaker for the heart, the SA node, lies in the right atrium close to the entry of the superior vena cava, is 2 mm thick and 8 mm long, and is perfused with blood via the sinus node artery. Two cell types are present: small round cells which are probably the pacemakers; and longer, elongated cells. The natural SA node discharge rate (in the absence of any autonomic influence) is 100 per minute. If the activity of the SA node is depressed, one of the other pacemaker centres can take over and determine the heart rate.

Atrial conduction

From the SA node the cardiac impulse spreads through atrial muscle at a rate of 1 m/s. The cardiac action potential is conducted from the SA node to the left atrium by a special pathway, Bachmann's bundle, and to the AV node by the anterior, middle and posterior internodal pathways.

The AV node

The AV node lies at the base of the right atrium on the right side of the interatrial septum, near the opening of the coronary sinus. It contains the same two types of cells as the SA node, but with fewer round cells. The AV node is the only conducting pathway from the atria to the ventricles through the fibrous atrioventricular ring, and conduction is slow (0.05 m/s). The AV node has three zones: the AN zone which is transitional between the atria and the node; the N zone; and the NH zone which is the origin of the bundle of His. The AN and N zones are responsible for the 0.1 s delay in transmission of the cardiac action potential from the atria to the ventricles.

The AV node cannot transmit more than 220 impulses per minute, much less than the theoretical maximal rate of atrial discharge of 400 per minute. At high rates of atrial discharge, many cardiac impulses therefore fail to pass through the AV node. The cells of the N zone have a prolonged relative refractory period, and AV node conduction slows as the rate of atrial firing increases. The heart cannot beat in a coordinated fashion more than 220 times a minute, because the AV node protects the ventricles from high rates of atrial depolarization.

Vagal activity delays or stops AV node transmission because acetylcholine increases membrane potassium permeability and hyperpolarizes the N zone cells. Norepinephrine released by sympathetic nerves speeds AV conduction by increasing the membrane calcium permeability of the N zone cells.

Ventricular conduction

After passing down the right side of the interventricular septum for 1 cm, the bundle of His splits into right and left bundle branches to the right and left ventricles (the left bundle branch divides into anterior and posterior divisions). The bundle branches supply the dense network of Purkinje fibres, which innervate the ventricles. Purkinje cells are the largest diameter cells in the heart, and their conduction velocity is fast (1–4 m/s), so that ventricular activation is rapid. The endocardial surfaces of the ventricles are activated first; early interventricular septum and papillary muscle contraction provides a firm base for ventricular contraction and prevents AV valve eversion. The cardiac action potential then spreads out to the epicardial surfaces and, as the right ventricle is thinner than the left, the outside of the right ventricle is activated first. The apical regions of the ventricles are activated before the bases, propelling blood up and out of the ventricular chambers.

The Purkinje fibres have a long refractory period and block many premature atrial impulses that may have passed through the AV node, and this is especially effective at low heart rates. If the function of the AV node is impaired and impulses from the SA node are not being transmitted, bundle of His cells can take over as the pacemaker to the ventricles, maintaining a ventricular rate of 30–40 per minute.

ELECTROCARDIOGRAPHY

All the cardiac action potentials from each heartbeat sum to produce a voltage that can be measured as a potential difference between two electrodes placed on the surface of the body (approximately 1 mV): the electrocardiogram (ECG). The electrical changes in the heart are complex three-dimensional vectors of potential, but recording the potential difference between two electrodes on the skin reflects only the magnitude, not the direction, of the potential. Therefore, standard ECG recording is known as scalar electrocardiography.

The ECG has P, QRS and T waves (Fig. 4.7). The P wave is produced by atrial depolarization, the QRS complex by ventricular depolarization, and the T wave by ventricular repolarization. The ST segment lies on the isoelectric line as the whole of the ventricular myocardium is depolarized during that time. Atrial repolarization takes place during the QRS complex, and is not seen on the ECG.

The PR interval is the time taken for excitation to spread through the atria, AV node, and the bundle of His (0.12 to 0.21 s). AV node transmission (0.1 s) accounts for most of this delay between atrial and ventricular depolarization. The QRS interval is the time taken for excitation to spread through the ventricles (0.06 to 0.12 s). The QT interval is the duration of ventricular depolarization, and varies inversely with the heart rate (0.3 to 0.4 s).

The Einthoven triangle and the standard limb leads

Einthoven described the electrical activity of the heart as being at the centre of an equilateral triangle formed by the shoulders and the pubic symphysis. Electrodes on the right and left arms and on the left foot approximate the corners of Einthoven's triangle, and record cardiac electrical activity in the vertical plane. Each of the three standard limb leads records the ECG from two corners of the triangle. Lead I records between the left and right arms. Lead II is between the left foot and the right arm. Lead III is between the left foot and the left arm (a fourth electrode acts as an earth electrode). By convention, if a wave of cardiac depolarization is travelling towards an

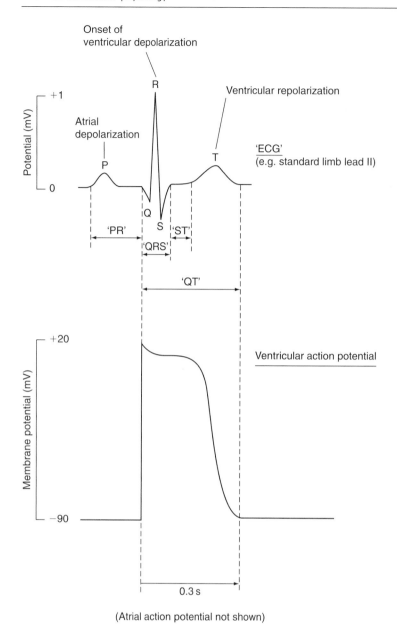

Figure 4.7 *Standard nomenclature of the ECG recording and what each section represents.*

electrode, a positive deflection is recorded on the ECG. If a wave of depolarization is travelling away from an electrode, a negative deflection is recorded. For example, in standard limb lead II, if the depolarization wave is travelling down towards the left foot, a positive wave is recorded on the ECG. If a wave of repolarization is travelling towards an electrode, a negative deflection is recorded. If the repolarization wave is travelling away from the electrode, a positive deflection is recorded on the

ECG. This arrangement was chosen so that the QRS complex would normally be upwards in recordings from standard leads I, II and III.

For example, the P, QRS and T waves in standard limb lead II are produced by the changing course over time of waves of depolarization and repolarization of the cardiac chambers during each heartbeat. Atrial depolarization commences in the SA node and spreads down and to the left to the AV node; lead II records a positive (upwards)

P wave. Ventricular depolarization starts in the interventricular septum and spreads down and to the right; lead II records a small negative Q wave. Ventricular depolarization then spreads epicardially, and the larger bulk of the left ventricle means that the net effect is for the depolarization to spread down and to the left; lead II records a large positive R wave. During activation of the remaining areas of the ventricle the depolarization spreads upwards; lead II records a small negative S wave. The wave of ventricular repolarization moves from the epicardial to the endocardial surface from the ventricle; lead II records a positive T wave. The predominant direction of the vector through the heart during depolarization of the ventricles is from base to apex.

MECHANICAL EVENTS OF THE CARDIAC CYCLE

The cardiac cycle comprises two phases defined by ventricular muscle mechanical activity: systole (contraction) and diastole (relaxation). During systole the ventricles contract, the AV valves close (first heart sound), and intraventricular pressure rises until the aortic and pulmonary valves open (isovolumetric ventricular contraction); ventricular ejection of blood then takes place (stroke volume). During diastole the ventricles relax, the aortic and pulmonary valves close (second heart sound), intraventricular pressure falls until the AV valves open (isovolumetric ventricular relaxation), and the ventricles then fill with blood again. Atrial contraction completes ventricular filling before the onset of the next ventricular systole.

At an average resting heart rate of 72 beats per minute, the cardiac cycle lasts 0.8 s; systole is 0.3 s and diastole 0.5 s. Therefore, at rest, ventricular filling accounts for two-thirds of the cardiac cycle. At the maximum heart rate of 200 beats per minute – as may be seen with heavy exercise – each cycle only lasts 0.3 s; the ventricular action potential is shortened so that systole is 0.15 s, but only 0.15 s remains for ventricular filling during diastole.

Figure 4.8 demonstrates the relationship between the atrial and ventricular components of the ECG, the first and second heart sounds, the pressures in the left atrium and ventricle, aortic pressure, left ventricular volume, the position of the AV and aortic and pulmonary valves, and the phases of the cardiac cycle.

The sequence of events in the cardiac cycle can be described by beginning from before the onset of atrial contraction, in mid-diastole, through late diastole, isovolumetric contraction, ventricular ejection, to isovolumetric ventricular relaxation and rapid ventricular filling in early diastole.

Mid-diastole (slow ventricular filling)

Atrial and ventricular pressures are low and, as the former is slightly higher, the ventricles fill slowly from the atria through the open AV valves. The ventricles are already 80 per cent full because of rapid filling during early diastole. The aortic and pulmonary valves are closed. During diastole, the aortic and pulmonary pressures fall as blood moves out into the vascular system, whilst the ventricular pressures rise slightly as blood enters the ventricles.

Late diastole

The SA node discharges (ECG P wave) and atrial contraction (atrial pressure 'a' wave) forces most of the blood in the atria into the ventricles, producing a small rise in ventricular pressure. Atrial contraction accounts for one-fifth of the end-diastolic ventricular volume, which is 130 mL when the subject is standing, and 160 mL when lying down.

Isovolumetric ventricular contraction (early systole)

The ventricles contract and ventricular pressure increases rapidly, closing the AV valves (first heart sound). The onset of ventricular contraction is at the same time as the R wave of the ECG. The contracting ventricles cannot eject any blood yet because the aortic and pulmonary valves remain closed, and this phase is described as isovolumetric ventricular contraction (the semilunar valves only open when the ventricular pressures rise above the aortic and pulmonary artery pressures). The AV valves bulge into the atria with ventricular contraction (atrial pressure 'c' wave) and the pressure rises to 10 mmHg in the left atrium and 5 mmHg in the right atrium.

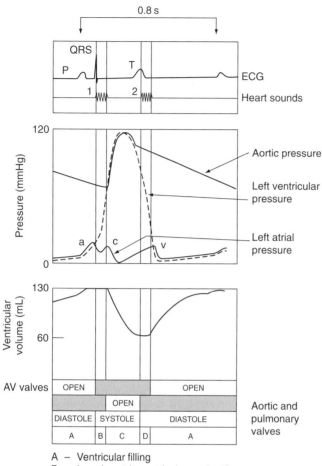

A – Ventricular filling
B – Isovolumetric ventricular contraction
C – Ventricular ejection
D – Isovolumetric ventricular relaxation

Figure 4.8 *The events of the cardiac cycle.*

Ventricular ejection

The semilunar valves open and ventricular ejection takes place when ventricular pressure exceeds aortic and pulmonary artery pressure. Aortic pressure rises from a diastolic low of 80 mmHg up to 120 mmHg during left ventricular ejection. Pulmonary artery pressure rises from 8 mmHg diastolic to 25 mmHg systolic during right ventricular ejection. Ventricular ejection consists of an early short rapid ejection phase and a prolonged reduced ejection phase.

In late systole the ventricles repolarize (T wave of the ECG) and begin to relax, and ventricular pressure drops slightly below aortic. Ventricular ejection continues slowly because of the momentum added

to the blood during the rapid ejection phase. The elasticity of the stretched walls and the presence of peripheral resistance to flow in the arterioles maintain aortic pressure.

At the end of systole, 60 mL of blood remains in each ventricle, 70 mL having been ejected as the stroke volume. The ejection fraction is the ratio of the stroke volume over the end-diastolic ventricular volume (70/130 when standing).

Atrial pressure falls sharply, to zero or negative values, during the rapid ejection phase as ventricular contraction pulls the atrioventricular fibrous ring downwards, lengthening and increasing atrial volume. Thereafter, atrial pressure rises during systole as blood continuously returns from the systemic and pulmonary veins.

Isovolumetric ventricular relaxation (beginning of diastole)

Closure of the aortic and pulmonary valves marks the end of ventricular systole. The closure of the semilunar valves produces the second heart sound, which may be split as the aortic valve closes before the pulmonary valve. The incisura is produced in the aortic pressure waveform by closure of the aortic valve. The ventricular muscle relaxes, and, as the AV and semilunar valves are closed, this phase is known as isovolumetric ventricular relaxation.

During isovolumetric ventricular relaxation atrial pressure rises to 5 mmHg in the left atrium and 2 mmHg in the right atrium.

Early diastole (rapid ventricular filling)

The AV valves open when ventricular pressure falls below atrial pressure. The ventricles fill rapidly, and ventricular and atrial pressures both decline (atrial pressure 'v' wave). The rapid filling of the ventricles during the early part of diastole is important, as the time available for ventricular filling is shortened when the heart rate is high; at rates above 200 per minute the filling time becomes inadequate.

BIOPHYSICAL DETERMINANTS OF CARDIAC MUSCLE CONTRACTION

Three main factors – preload, afterload and autonomic activity to the heart – determine the volume of blood ejected by the ventricles during systole and the ejection pressure. The preload is the stretching of the ventricular muscle fibres at the end of ventricular filling, as represented by the end-diastolic volume. This is the basis of Starling's law of the heart – that the diastolic length of the ventricular fibres determines their force of contraction. The afterload is the pressure that the ventricle has to eject against during systole, after isovolumetric ventricular contraction and aortic and pulmonary valve opening. The term afterload is used as the ventricle does no work against this load until the aortic valve opens and ejection begins. Up to a point, increases in afterload can produce an increase in the tension developed by the ventricle.

Sympathetic activity increases the rate and force of ventricular contraction at any given fibre length. Preload and afterload are intrinsic factors influencing the muscle of the heart, whereas sympathetic nervous activity is an extrinsic factor.

STUDIES OF ISOLATED CARDIAC MUSCLE PREPARATIONS

Isolated papillary muscles have been used to examine the effects of preload, afterload and sympathetic activity on cardiac muscle performance. Electrodes are used to stimulate the papillary muscle and the contraction is observed either when the muscle length is fixed (isometric contraction) or when it is allowed to shorten against a load (isotonic contraction).

A useful mechanical model for cardiac muscle is of a contractile element (CE) in series with an elastic element (SE) and in parallel with a second elastic element (PE). The PE represents the elasticity of the resting muscle, and the SE the elastic behaviour of contracting myocardium. When the muscle fibres are activated, CE shortens, SE is stretched, and force is developed. Figure 4.9 illustrates the changes in the contractile and elastic elements of the model during an isometric contraction.

The length–tension relationship of isometric contraction

The tension developed by the isolated papillary muscle depends on the initial resting length of the fibre, which can be altered. As the resting length is increased, the papillary muscle twitch tension increases in magnitude. When the muscle is over-stretched, the developed tension falls (Fig. 4.10).

This effect of resting length on isometric contraction tension depends firstly on the degree of overlap of the contractile actin and myosin fibres. At the normal resting length of the papillary muscle the overlap of actin and myosin is not optimal. An increase in the resting length increases the effective overlap of actin and myosin filaments, and the isometric twitch tension is increased. If the muscle is over-stretched, the twitch tension falls because the overlap of actin and myosin is reduced.

Rest   **Isometric contraction**

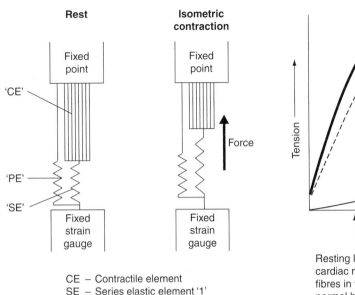

CE – Contractile element
SE – Series elastic element '1'
PE – Parallel elastic element '2'

Figure 4.9   *Mechanical model for cardiac muscle.*

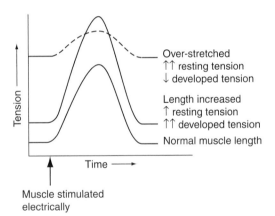

Figure 4.10   *The length–tension relationship of isometric contraction.*

There are other reasons for the length–tension relationship of papillary muscles. Lengthening the muscle increases the sensitivity of troponin to calcium and induces an increase in the concentration of free calcium within the cell, increasing the force of contraction.

For isolated papillary muscle, there is therefore a relationship between initial resting length and developed isometric contraction tension. The elastic elements in the muscle produce a resting tension, and this increases as the muscle is lengthened.

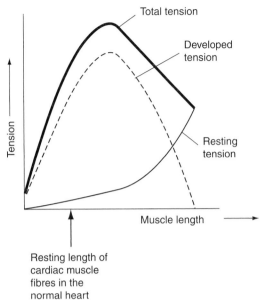

Figure 4.11   *Length–tension curves for cardiac muscle: resting, isometric developed and total tensions.*

The total tension during the isometric twitch is the sum of the resting and the developed tensions. The over-stretched cardiac muscle has a high resting tension, but isometric contraction is weak. The relationship between the resting muscle length, resting tension and isometric developed and total tensions is shown in Figure 4.11. Muscle fibres in the normal heart lie on the ascending part of this curve, so that increases in length produce more forceful contractions.

The effect of catecholamines on papillary muscle isometric contractions

The addition of norepinephrine to the fluid perfusing a papillary muscle preparation increases both the strength and rapidity of the isometric twitch (Fig. 4.12). Catecholamines increase the contractility (force of contraction at a constant resting fibre length) of cardiac muscle. This is known as the positive inotropic effect of the sympathetic nervous system on the heart. In contrast, catecholamines have no effect on skeletal muscle contraction. Figure 4.12 also illustrates that norepinephrine speeds both the onset and relaxation of the muscle twitch. As discussed earlier under excitation–contraction coupling, catecholamines augment both contraction (by opening membrane

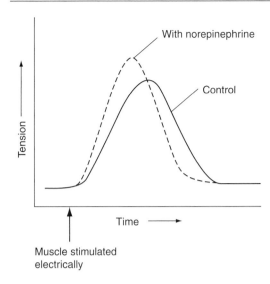

Figure 4.12   *The effect of catecholamines on cardiac muscle contraction.*

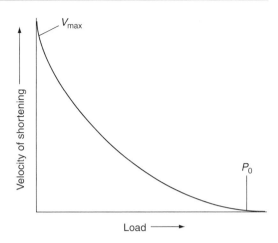

Figure 4.13   *Isotonic papillary muscle contraction: velocity of shortening and load.*

calcium channels and phosphorylation of myosin) and relaxation (via phospholamban and troponin-I) of cardiac muscle. Sympathetic stimulation therefore speeds both the emptying and filling of the heart.

Isotonic contraction of isolated papillary muscle

In this case, the papillary muscle is attached to a pivot and allowed to shorten against a load when stimulated. In an isotonic contraction, CE shortens and develops a force that stretches SE. The developed force increases until it matches the load against which the muscle is working, then exceeds it and the muscle shortens. A constant force that equals the load stretches SE, and the velocity of shortening is therefore an indicator of the performance of CE. The variables of particular interest in studies of isotonic contractions are the velocity of contraction and the maximum load that the muscle can move. Attaching a small preload weight to the free end of the pivot sets the initial length of the muscle. By increasing the preload, the effect of resting muscle length on the velocity of shortening of the muscle can be studied. Once the preload has set the resting length, the pivot is fixed so that the muscle cannot stretch further, and an additional afterload is added to the free end of the pivot. By increasing the afterload, the

effect of the load the muscle is working against on the velocity of shortening can be studied. The preload represents the resting end-diastolic length of the cardiac muscle fibres (end-diastolic volume). The aortic pressure when the aortic valve is open is the equivalent of the afterload for the left ventricle (in the intact ventricle, afterload is defined as the tension operating on the muscle fibre to resist change, after the onset of contraction).

The relationship between the velocity of contraction and the force the muscle is working against is hyperbolic (Fig. 4.13). The velocity of contraction is maximal when the load is zero (V_{max}). As the load is increased, the velocity of contraction falls until it reaches zero, at which point the developed force is equal to the maximal isometric force the muscle can produce (P_0).

Increasing the initial resting muscle length by increasing the preload increases the velocity of contraction and increases the maximal isometric force (P_0), but does not change the maximum velocity at zero load, V_{max}. In contrast, the addition of norepinephrine to the bath perfusing the muscle does increase V_{max}, as well as increasing the velocity of contraction at other loads and P_0 (Fig. 4.14).

Summary of findings from studies of isolated papillary muscle

Studies of isometric and isotonic contractions demonstrate that both the force and velocity of contraction of cardiac muscle cells are influenced by the

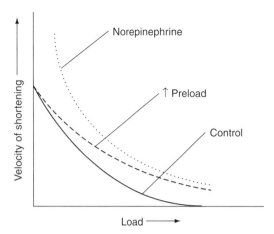

Figure 4.14 *Isotonic contraction: preload and catecholamines.*

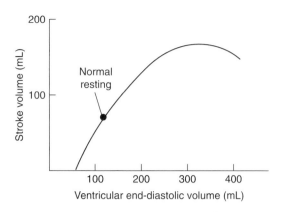

Figure 4.15 *A Starling curve showing the relationship between ventricular end-diastolic volume and stroke volume.*

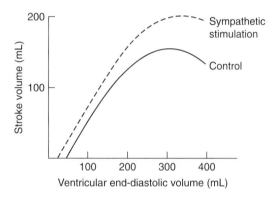

Figure 4.16 *Starling curve showing the positive inotropic effect of the sympathetic nervous system.*

intrinsic factors of preload and afterload and the extrinsic factor of sympathetic autonomic activity.

STUDIES OF MECHANICAL PERFORMANCE OF THE WHOLE HEART

The heart–lung preparation

Starling developed an animal heart–lung preparation in order to study the performance of the ventricles and the effects of changes in preload, autonomic activity and afterload. The lungs of the animal are artificially ventilated, the right atrium is supplied with blood from a reservoir, the output of the left ventricle is passed via a variable resistance and then back to the venous reservoir, and the sympathetic and parasympathetic nerves can be stimulated electrically. Ventricular volume is measured and stroke volume can be determined. Raising or lowering the venous reservoir changes the preload, and altering the arterial resistance changes the afterload to the left ventricle.

PRELOAD

In the heart–lung preparation, raising the venous reservoir increases the right atrial pressure, end-diastolic ventricular volume and stroke volume. Starling's law of the heart is that the force of contraction of ventricular muscle is proportional to the initial resting length. A Starling curve can be drawn, relating ventricular end-diastolic volume

to the stroke volume (Fig. 4.15). Normally, the heart lies on the ascending part of the curve and an increase in end-diastolic volume will produce an increase in stroke volume.

SYMPATHETIC AND PARASYMPATHETIC NERVE ACTIVITY

Stimulating the sympathetic nerves to the heart–lung preparation increases the stroke volume at any given end-diastolic volume. The Starling's curve is therefore shifted to the left (Fig. 4.16), demonstrating the positive inotropic effect of the sympathetic nervous system.

In the heart–lung preparation, sympathetic stimulation alone reduces the end-diastolic volume

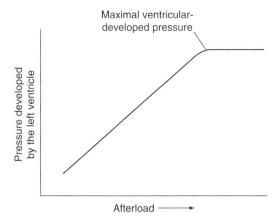

Figure 4.17 *Graphical representation showing the effect of increased afterload (aortic pressure) until maximal ventricular-developed pressure is reached.*

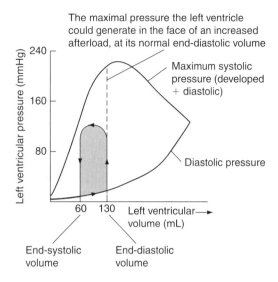

Figure 4.18 *Normal left ventricular–pressure loop.*

and right atrial pressure because of enhanced ventricular emptying. If the venous reservoir is raised simultaneously, the combined effect with sympathetic stimulation is then an increase in stroke volume, with little change in the end-diastolic volume, as may be found during exercise.

AFTERLOAD

Increases in afterload (aortic pressure) can produce higher peak systolic pressures, until the point is reached at which the ventricle cannot generate sufficient pressure to open the aortic valve. Ventricular contraction is then isometric and there is no ejection of blood (Fig. 4.17). The pressure the ventricle produces at this point is the maximal isometric pressure it can produce at a given preload. An increase in preload can produce a further rise in the maximal isometric force, if the heart lies on the ascending part of the Starling curve.

THE NORMAL PRESSURE–VOLUME LOOP OF THE LEFT VENTRICLE

The volumes and pressures at which the left ventricle operates can be related to a diagram showing ventricular volume, resting diastolic pressure, and total systolic pressure (Fig. 4.18). This is the equivalent for the whole heart of Figure 4.11 for the isolated papillary muscle of length, and resting, maximal developed and total tensions.

In diastole, the left ventricular volume rises from 60 mL at the end of systole to the end-diastolic volume of 130 mL, and the ventricular pressure rises from 5 mmHg to 10 mmHg. During isovolumetric ventricular contraction, the pressure increases to 80 mmHg before the aortic valve opens. During the ejection phase the ventricle contracts and ejects the stroke volume of 70 mL blood against the afterload of increasing aortic pressure. Ventricular pressure rises during the ejection phase to 120 mmHg, and then falls to 100 mmHg at the end of systole when the aortic valve closes again. During isovolumetric relaxation the ventricular pressure falls once more to the end-systolic point. The area enclosed by the pressure–volume curve (the shaded area in Fig. 4.18) reflects the external work done by the left ventricle in each cardiac cycle.

The left ventricle normally operates on the ascending part of the curve relating volume to pressure. Also, the point at which the aortic valve opens is normally at a much lower pressure than the maximum the ventricle could generate. As described earlier, a rise in afterload, as with an increase in peripheral resistance, results in an increase in pressure generated by the ventricle up to the maximal isometric point. However, ventricular work (area enclosed by the pressure–volume curve) and oxygen requirement rise significantly when the ventricle develops high pressures in response to an increased afterload.

VENTRICULAR PRESSURE–VOLUME RELATIONSHIPS

The analysis of the instantaneous relationships between pressure and volume of both the left and right ventricles allows the workload of the heart to be assessed. The area of the pressure–volume loop of the left ventricle represents the stroke work. In addition, the pressure–volume relationships at the end of systole and at the end of diastole allow cardiac contractility and ventricular stiffness to be determined. The pressure–volume relationships at the end of systole reflect the active contractile properties of cardiac muscle, whereas the pressure–volume relationships at diastole depend on the passive properties or stiffness of the ventricular muscle.

The ventricular pressure–volume loops consist of four segments corresponding to isovolumetric contraction, ejection, isovolumetric relaxation, and ventricular filling. In the normal heart the pressure–volume loops are approximately rectangular for the left ventricle, and more triangular for the right ventricle. During the isovolumetric contraction the left ventricle becomes spherical and the endocardial surface area tends to decrease, while during isovolumetric relaxation the ventricle becomes ellipsoidal and the area tends to increase (Fig. 4.19). The contraction of the right ventricle is not synchronous, starting in the inflow tract, and in a peristaltic manner reaches the outflow tract after a delay of 50 ms (Fig. 4.20).

During ischaemia, the ventricular pressure–volume loops appear to lean to the right (Fig. 4.21). This is caused by lengthening of ventricular muscle during isovolumetric contraction (systolic lengthening) and its shortening during isovolumetric relaxation (postsystolic shortening). Systolic lengthening represents displacement (bulging) of the ischaemic muscle segment during the contraction of the normal ventricular muscle. Postsystolic shortening during isovolumetric relaxation may be caused by active shortening of the ventricular muscle or elastic recoil of the ventricle with profound ischaemia. Abnormal pressure–volume loops in the absence of ischaemia can be caused by altered excitation–contraction coupling of ventricular muscle produced by the interaction between calcium antagonists and halogenated anaesthetic agents.

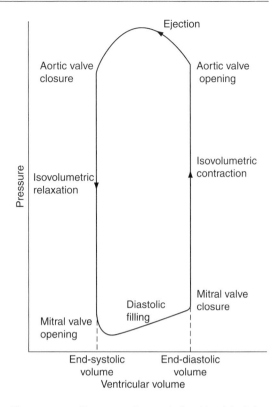

Figure 4.19   *Pressure–volume relationship of the left ventricle.*

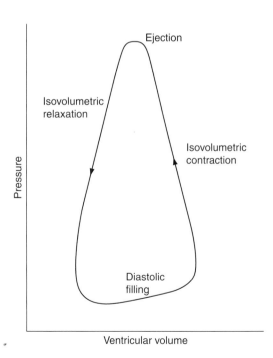

Figure 4.20   *Pressure–volume relationship of the right ventricle.*

Changes in preload

In initial studies using isolated papillary muscles, physiologists defined preload as the initial stretch (diastolic fibre length) of cardiac muscle before contraction. In an intact heart, ventricular preload is defined as the end-diastolic volume – that volume producing the initial passive stretch of the myocardium prior to active contraction. An increase in preload causes an increase in end-diastolic volume, resulting in an increase in end-diastolic fibre length of the ventricular muscle. This increases the velocity of ventricular muscle shortening for a given level of afterload via the Frank–Starling mechanism. Thus, with a constant afterload and myocardial contractility, increasing the preload increases stroke volume by the Frank–Starling mechanism. The end-systolic ventricular volume is unchanged (Fig. 4.22).

Changes in afterload

Afterload, as defined in an isolated muscle, is the external load or force that has to be generated before the muscle is shortened. In the intact heart, it represents the force opposing the cardiac muscle shortening and ejection of blood during ventricular contraction. It is affected by the tension or instantaneous force within the ventricular wall and the arterial impedance (forces outside the heart that oppose ventricular ejection) during ventricular contraction. Systolic wall tension (pressure radius/wall thickness) or stress has been used to define afterload. Thus, afterload can be defined as stress (force per unit area) encountered by the myocardial fibres of the ventricle after the onset of shortening. Increasing afterload results in an increase in the end-systolic ventricular volume. With a constant end-diastolic ventricular volume (if the preload is not changed), this results in reduction of the stroke volume (Fig. 4.23).

End–systolic ventricular pressure–volume relationships and changes in contractility

When resistance to ventricular ejection is altered, pressure–volume loops of the left ventricle at

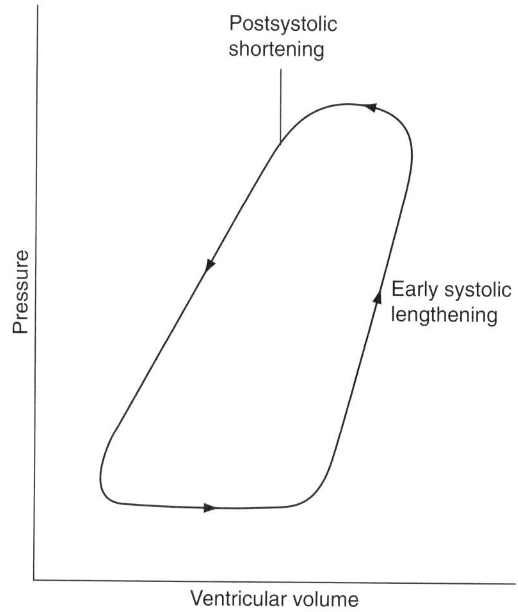

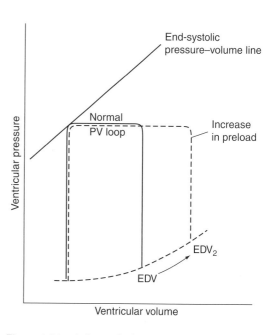

Figure 4.21 *Pressure–volume relationship of the left ventricle during moderate ischaemia.*

Figure 4.22 *Left ventricular pressure–volume (PV) relationships with increased preload. EDV, end-diastolic volume; EDV2, end-diastolic volume with increased preload.*

end-systole extend along a straight line which is called the end-systolic pressure–volume (ESPV) line, the gradient of which provides an index of contractility. Contractility is defined as the intrinsic ability of the myocardial fibre to shorten (that determines

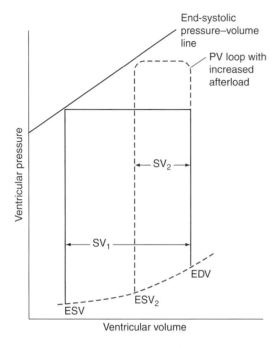

Figure 4.23 *Pressure–volume relationships with increased afterload. SV_1, stroke volume (normal); SV_2, stroke volume with increased afterload; ESV, end-systolic volume; ESV_2, end-systolic volume with increased afterload; EDV, end-diastolic volume.*

the velocity and extent of shortening) independent of preload and afterload. An increase in myocardial contractility (inotropy) increases the gradient of the slope of the ESPV, associated with a widening of the pressure–volume loop because end-systolic shortening is enhanced (Fig. 4.24(a)). Extrapolation of the ESPV to zero pressure at the x-axis defines the volume of the ventricle if the ventricular pressure is zero, V_0. Negative inotropy may cause either a depression of the slope of the ESPV line without a change in V_0 or a parallel shift of the ESPV line without a change of its slope, but with a change in V_0 to V_0^1 (Fig. 4.24(b)).

Diastolic function

Normal diastolic function of the ventricle is dependent on ventricular diastolic compliance, distensibility and relaxation. Extrinsic and intrinsic factors affect ventricular diastolic function. Examination of the ventricular diastolic pressure–volume diagrams can differentiate the diastolic dysfunction caused by altered ventricular compliance, distensibility and relaxation (Fig. 4.25).

Compliance is defined as the ratio of a volume change to the corresponding pressure change ($\Delta V/\Delta P$) or the slope of the pressure–volume relationship. Stiffness is the inverse of compliance. Therefore, the slope of the diastolic pressure–volume relationship of the left ventricle would indicate the compliance of the left ventricle. An

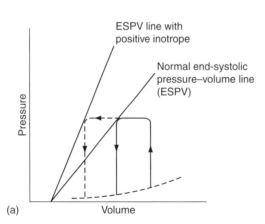

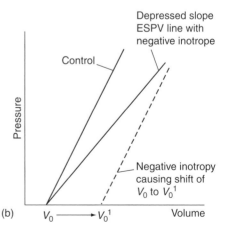

Figure 4.24 *Pressure–volume relationships with changes in contractility. (a) Increased contractility; (b) decreased contractility.*

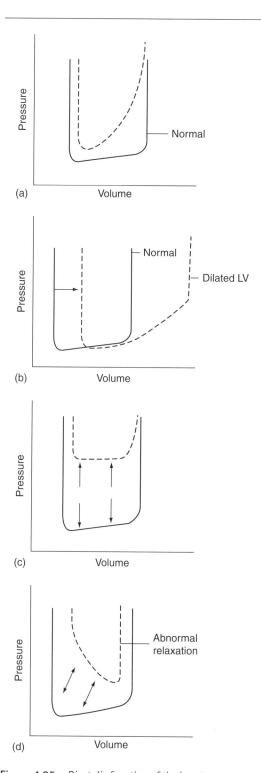

increase in the steepness of the diastolic pressure–volume relationship of the left ventricle indicates reduced compliance or increased stiffness of the left ventricle. Normally, the left ventricle does not exhibit constant compliance. Left ventricular compliance is high at low end-diastolic volumes, but this progressively decreases at higher end-diastolic volumes. Decreased left ventricle compliance may be produced by increased chamber stiffness resulting from left ventricular hypertrophy – as in systemic hypertension or aortic stenosis. It can also be caused by increased cardiac muscle stiffness caused by restrictive cardiomyopathies such as amyloidosis.

An increase in diastolic pressure at a given ventricular volume indicates decreased ventricular distensibility. This produces a parallel upward shift of the diastolic pressure–volume relationship. Reduced distensibility may occur from either intrinsic causes such as myocardial ischaemia, or extrinsic restrictions to ventricular filling such as constrictive pericarditis or pericardial effusion.

For ventricular relaxation, energy in the form of ATP is required for reuptake of calcium ions into the sarcoplasmic reticulum and for the detachment of actin–myosin cross-bridges in cardiac muscle. When isovolumetric relaxation is delayed, early diastolic filling is reduced, whereas filling is reduced throughout diastole when ventricular relaxation is incomplete. Relaxation is impaired in myocardial ischaemia, and in hypertrophic and congestive cardiomyopathies.

Systolic function

Systolic function refers to the ability of the left ventricle to generate an adequate stroke volume or to perform external work under varying conditions of preload, afterload and contractility. The ejection fraction is commonly used as an index of global systolic function. It is defined as:

$$\frac{\text{LVEDV} - \text{LVESV}}{\text{LVEDV}}$$

Figure 4.25 *Diastolic function of the heart.*
(a) Increased chamber stiffness (diminished compliance);
(b) increased chamber volume; (c) pericardial restriction,
e.g. effusion; and (d) abnormal relaxation, e.g. ischaemia.

where LVEDV is left ventricular end-diastolic volume and LVESV is left ventricular end-systolic volume.

An increase in afterload with a constant pre-load and contractility produces an increased LVESV with LVEDV unchanged, causing a fall in stroke volume and ejection fraction. With preload and afterload constant, an increase in contractility augments the ejection fraction as there is a marked decrease in the LVESV.

The gradient of the ESPV line (as described earlier) provides an index of myocardial contractility independent of preload and afterload. Qualitative assessment of regional wall motion by the left ventricle is another index of systolic function. This may be assessed by left ventriculography or by echocardiography. Normal areas of the myocardium exhibit concentric inward movement during systole, whilst hypokinetic areas caused by ischaemia show reduced concentric inward motion during systole. Akinetic areas do not exhibit any motion during systole, and are usually composed of ischaemic myocardium with a fibrous scar. Dyskinetic areas exhibit a paradoxical bulging during systole and usually represent regions with little or no viable myocardium. Regional wall motion abnormalities are more sensitive indicators of coronary artery disease than indices of global systolic function. This is because global systolic function can be maintained in the presence of regional wall dysfunction by compensatory increases in wall shortening in areas of normal wall motion, as long as large areas of the ventricular wall do not have abnormal wall motion.

The pressure–volume area: index of mechanical energy and work

The area within the pressure–volume loop represents the stroke work (mechanical energy) done by the heart during a single contraction. This represents the external work of the ventricle. During isovolumic contraction, pressure within the ventricle develops. Although there is no ejection and therefore no external work, energy is expanded to generate the potential energy, and this is converted to heat during diastole. The triangle formed by the ESPV line, the end-diastolic pressure–volume (EDPV) line and the line representing isovolumic relaxation represents the amount of

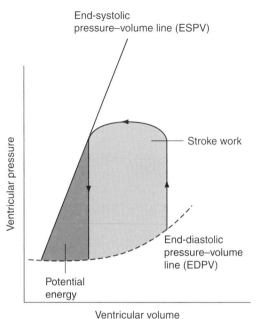

Figure 4.26 *Pressure–volume area.*

potential energy available during a contraction (Fig. 4.26). This correlates with the heat generated by the heart during contraction. The total mechanical work plus the heat generated by the heart during contraction is represented by the pressure–volume area (PVA) (Fig. 4.26). The PVA correlates well with the amount of oxygen consumed by the myocardium during a single contraction. When the PVA is extrapolated to zero pressure, myocardial oxygen consumption is still present. This basal oxygen consumption in the absence of pressure development is required to keep the cells alive, and activation energy is required for the biochemical processes associated with excitation–contraction coupling.

PHYSICAL FACTORS GOVERNING BLOOD FLOW THROUGH VESSELS

POISEUILLE'S EQUATION

The pressure developed by contraction of the left ventricle produces the flow of blood around the systemic vessels of the cardiovascular system. This flow of the blood is determined by the pressure gradient applied, the vessel length and diameter, and by the viscosity of blood. Using rigid, uniform and unbranching glass capillary tubes, Poiseuille studied the steady, laminar, flow of a Newtonian (homogeneous) fluid, and described the mathematical relationship between flow, pressure, vessel length and radius, and the fluid viscosity. In laminar flow all the elements of the fluid move in streamlined layers parallel to the vessel wall that slip over one another; the layer in contact with the wall does not move, and the central layer has the highest velocity. Laminar flow in a tube consists of concentric fluid cylinders sliding over each; the central cylinder has the highest velocity, whilst the outermost cylinder adjacent to the vessel wall is stationary, and the profile of flow velocity across the tube is parabolic. From Poiseuille's equation, laminar flow is proportional to the hydrostatic pressure gradient acting along the vessel and to the fourth power of the vessel radius, and is inversely proportional to the fluid viscosity and the vessel length:

$$\text{Flow} = \frac{\pi (P_i - P_o)r^4}{8\eta L}$$

where $(P_i - P_o)$ is the hydrostatic pressure gradient acting along the length of the vessel, η is the blood viscosity, L is the vessel length, and r is the vessel radius. The term $\pi/8$ is a constant obtained from the mathematical relationship of volume passing per unit time.

Viscosity

In laminar flow, fluid layers slip over and interact with each other, and the ease with which they move is determined by the fluid 'viscosity' or 'lack of slipperiness'. The higher the viscosity, the greater the interaction there is between adjacent layers so they slip over one another less easily, and the flow is reduced for a given pressure gradient. A Newtonian fluid, as studied by Poiseuille, is a homogeneous fluid with a viscosity that is not changed by the flow rate.

Newton defined fluid viscosity as the ratio of 'shear stress' to 'shear rate', where 'shear stress' is the force applied to a plate moving on the surface of a liquid in a container divided by the plate area (with units dyn/cm^2), and the 'shear rate' is the velocity of movement of the liquid beneath the plate divided by the depth of the liquid in the container [with units (cm/s)/cm]. (In laminar flow the fluid close to the plate moves quickest, deeper layers are slower, and the layer adjacent to the vessel wall is still.) For a 'slippery' liquid, the velocity of movement is high for a given applied force, and the ratio shear rate/shear stress – the viscosity – is low. Given the same applied force and plate area (the same shear stress), with a liquid of 'low slipperiness' the velocity of liquid movement will be lower, and the viscosity is high. Fluids of high viscosity flow less easily. Another definition of a Newtonian fluid is that the shear rate and shear stress are proportional; this is true for water, but not blood.

Units of viscosity

The unit of viscosity is the poise (equal to 1 dyn · s/cm^2). At 20°C the viscosity of water is 0.01 poise, or 1 centipoise. Temperature changes the viscosity of fluids inversely. At 37°C the viscosity of water is reduced to 0.695 centipoise, but at 0°C it is increased to 1.8 centipoise. The term 'relative viscosity' of a fluid to that of water may be used. For example, at 37°C plasma has a viscosity of 1.2 centipoise, or a relative viscosity to water of 1.7. Blood is a non-Newtonian fluid, and the measured viscosity varies with the rate of flow. Blood has a relative viscosity of 4–5 when measured at moderate or high shear rates, but the viscosity is much higher at low shear rates.

Hydraulic resistance to blood flow

The hydraulic resistance to blood flow of the cardiovascular system vessels is the ratio of the pressure drop/flow (separate resistances are calculated for the systemic and pulmonary circulations). By substituting this into Poiseuille's equation, we can obtain an hydraulic resistance equation, which reveals that the resistance to blood flow depends on the vessel size (radius and length) and the viscosity of blood:

$$\text{Resistance} = \frac{8\eta L}{\pi r^4}$$

where η is the blood viscosity, L is the vessel length, and r is the vessel radius.

From the hydraulic resistance equation it can be seen that resistance varies directly with liquid viscosity and vessel length, and varies inversely with vessel radius. In laminar flow, resistance arises from the viscous forces between the moving layers of fluid, and this is determined by the liquid viscosity. An increase in viscosity increases hydraulic resistance. In laminar flow there is no friction between the vessel wall and the liquid because the lamina adjacent to the wall is stationary. However, the vessel length and radius are important determinants of hydraulic resistance as the dimensions of the tube alter the frictional forces between the laminae that are moving. The radius is particularly important as resistance is inversely proportional to the fourth power of the radius. For example, halving the vessel radius increases the hydraulic resistance 16-fold.

UNITS OF HYDRAULIC RESISTANCE

Resistance is calculated by the division of pressure (mmHg) by flow (L/min), and so the unit of resistance is mmHg/L per minute (Wood units). Often, organ blood flow is expressed as mL per minute, and the unit of resistance is then mmHg/mL per minute (the unit used in this text for 'determinants of mean arterial pressure'; see pages 148–149). An adult with a mean aortic pressure of 100 mmHg, right atrial pressure 0 mmHg, and a cardiac output of 5000 mL/min has a total peripheral resistance equal to the pressure drop across the system (100 − 0 mmHg) divided by the flow of 5000 mL/min giving 100/5000 = 0.02 mmHg/mL per minute (equivalent to 20 Wood units). When comparing different organ circulations, resistance may be expressed per 100 g tissue. [NB. In clinical practice, the total

peripheral resistance (or systemic vascular resistance, SVR) and pulmonary vascular resistance are expressed in absolute units as dyn · s/cm^5. These absolute resistance units may be converted to Wood units by dividing by 80. For example, the normal SVR is 700 to 1600 dyn · s/cm^5, or 9–20 Wood units.]

The resistance of vessels in series and in parallel

In the cardiovascular system, arteries, capillaries and veins lie in series, but many individual organ circulations (and vessels within them) lie in parallel (see Figs 4.1 and 4.2). The effect of adding hydraulic resistances is quite different if vessels are in series or in parallel (as with electrical resistance). When different vessels lie in series, the total resistance of all the vessels taken together is the sum of the individual resistances. However, if the vessels lie in parallel, the reciprocal of the total resistance equals the sum of the reciprocals of the individual resistances. That is, when vessels lie in parallel, the combined total hydraulic resistance is less than any one of the individual vessel resistances.

Individual organ circulations lie in parallel, and so the hydraulic resistance of an individual organ is greater than the value calculated earlier for total peripheral resistance (0.02 mmHg/mL per minute). For example, the blood flow of one kidney is of the order of 550 mL/min, and the renal hydraulic resistance can be calculated as 100/550 = 0.18 mmHg/mL per minute.

Capillaries are of smaller diameter than arterioles, yet the main fall in arterial blood pressure (and therefore the main site of resistance) is seen in the larger arterioles, not the narrower capillaries (see Fig. 4.3). This is because there is such a great number of capillaries in parallel that the combined hydraulic resistances of the capillary beds is much lower than that of the arterioles. Arterioles also lie in parallel in organs, but the number of these vessels is insufficient to compensate for their small diameter, so that arteriolar resistance is the main source of hydraulic resistance in the peripheral vascular system.

Changes in hydraulic resistance in the systemic circulation

The hydraulic resistance equation demonstrates that the important factors are blood viscosity and vessel length and diameter. Of these, vessel length is constant and blood viscosity is relatively so. However, the radius of the arterioles can change considerably by contraction or relaxation of their smooth muscle walls. Therefore, changes in arteriolar diameter are the main way by which peripheral resistance is altered in the systemic circulation.

OBSERVED PHYSIOLOGICAL DEVIATIONS FROM POISEUILLE'S EQUATION

Poiseuille's equation was derived from studies of steady laminar flow of a Newtonian fluid, under a constant pressure head, in rigid glass tubes. Conditions in the cardiovascular system differ considerably from this as the output of the heart is pulsatile, the vessels are sometimes distensible or contractile, blood viscosity changes with vessel size and the flow rate, and the pattern of flow may be turbulent.

Blood vessels are not rigid tubes

In rigid tubes flow is proportional to the applied pressure head. Blood vessel walls, however, have both elastic and muscular components that confer them, respectively, with the properties of distensibility and contractility that can alter the relationship between pressure and flow. The pressure–flow relationships of different blood vessels vary considerably, but may be considered as being of three types. In the first type, the vessels are distensible and are stretched by increased pressure so that flow increases more than would occur in a rigid tube (the adventitia limits the distensibility so at high pressures flow increases linearly with pressure, as in a rigid tube). In the second type, the vessels are distensible but stretching of their wall stimulates a myogenic muscular contraction by which the effect of increasing pressure on flow is reduced. In the third type, the vessels have a

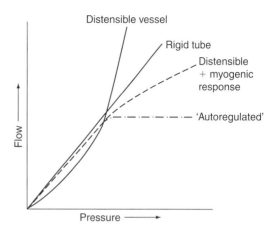

Figure 4.27   *The pressure–flow relationships of different blood vessels.*

myogenic mechanism so well developed that flow is kept constant, or 'autoregulated', over a range of increasing pressure (Fig. 4.27).

Pulsatile blood flow

Whereas Poiseuille's equation was obtained with steady flow, the output of the left ventricle is intermittent. However, one function of the large elastic arteries is to convert the intermittent ventricular output to more continuous pulsatile flow. Because of this, flow in the capillaries is relatively steady.

The anomalous viscosity of blood

Unlike water, blood viscosity is not constant at a given temperature, but changes with the haematocrit, vessel diameter and flow rate. Blood viscosity also tends to be lower when measured *in vivo* than *in vitro*. In addition, exposure to changes in ambient temperature affect blood viscosity in the skin. The terms 'anomalous' or 'apparent' viscosity of blood can be used to describe these effects.

The relative viscosity of blood to that of water rises with the haematocrit. For example, at moderate flow rates, with an increased haematocrit of 0.60 (normal 0.45), the relative viscosity of blood is doubled to 8–10.

The apparent viscosity of blood falls progressively when the vessel diameter is reduced below 300 μm. Arterioles are smaller than this, so this effect reduces the frictional forces resisting flow through these small, high-resistance vessels. In tubes of 30–40 μm diameter (the size of many arterioles), the relative viscosity of blood is only 2.5 (normally 4–5 in larger vessels). In capillary-sized tubes (6 μm diameter) the viscosity of blood is as low as that of plasma. This effect explains why blood viscosity measured *in vivo* is lower than *in vitro*. It can be seen that this effect reduces the pressure required to perfuse the vessels of the microcirculation. One reason for the effect of vessel diameter on apparent blood viscosity is by a change in the composition of blood when it is flowing through small vessels. The red cells tend to occupy the axial, central, fast-moving stream, whilst plasma flows slowly through the small vessel in the marginal streams. The result is that the ratio of red cell volume to blood volume (haematocrit) is effectively reduced when blood is in small vessels of less than 300 μm diameter. It is not clear why red cells tend to move away from the vessel wall and occupy the axial stream, but erythrocyte flexibility is important and this is enhanced by blood fibrinogen (erythrocytes must also be flexible in order to pass through capillaries, as they have larger diameters than the vessels). In capillaries, another reason for the effect of vessel diameter is that the erythrocytes flow in a single-file pattern that also reduces the viscosity.

Axial streaming is also seen with changes in blood flow rate. The apparent viscosity of blood decreases as the shear rate is increased – an effect known as 'shear thinning'. At high shear rates, the red cells lie in the axial stream with their long axis in the direction of the flow, where the flow rate is fastest. Thus, the viscosity of blood is less at high flow rates and greater at low flow rates. In addition, at low flow rates the erythrocytes tend to aggregate, increasing the viscosity. Fibrinogen increases the tendency of red cells to aggregate at low flow rates.

The vessels of the skin and subcutaneous tissues are exposed to significant ambient temperature changes, and cooling increases blood viscosity and encourages stasis.

Turbulent flow

If the driving hydrostatic pressure across a tube is gradually increased, a point is reached at which the flow is no longer proportional to driving pressure, but increases more slowly only with the square-root of pressure, and this is caused by a change from laminar to turbulent flow. In turbulent flow, fluid does not move in discrete laminae, but there are disorganized eddies with rapid radial and circumferential mixing, and the frictional resistance to flow is increased. The heart has to do more work to produce a given flow when turbulence is present. To double turbulent flow, the hydrostatic pressure must be increased four-fold. Turbulence tends to develop in large-diameter vessels with high flow velocity, low fluid viscosity and high fluid density. The Reynolds number (Re) can be used to predict the presence of turbulent flow in a long straight tube:

$$\text{Reynolds number } (Re) = \frac{\rho D v}{\eta}$$

where ρ is the fluid density, D is the vessel diameter, v is the mean velocity, and η is the fluid viscosity.

Flow is usually turbulent if Re is more than 2000, and laminar if the calculated Re is less than 1000. In the cardiovascular system, turbulence occurs normally in the ventricles and the aorta (aortic root $Re > 3000$: peak velocity 70 cm/s; diameter 2.2 cm; blood density 1.06 g/cm^3; blood viscosity at 37°C is 0.03 poise), but not in the small vessels (arteriolar $Re = 0.5$). Turbulent flow may be audible as a murmur. For example, functional murmurs may be heard in severe anaemia because of the reduced blood viscosity and the high flow velocities produced by the high cardiac output.

PRESSURE AND FLOW

Whilst the hydrostatic pressure gradient is often described as producing flow, it is more correctly the total fluid energy that determines liquid movement between two points in a tube. Total fluid energy comprises three factors: pressure energy (potential energy), gravitational potential energy, and kinetic energy. In a fluid moving horizontally in a tube the gravitational energy component can be ignored, and thus:

Total fluid energy =
    potential energy + kinetic energy ($\frac{1}{2}mv^2$)

where *m* is the mass and *v* is velocity. *P* is the lateral hydrostatic pressure, measured by a side-tube in a vessel. The total fluid energy is taken to be constant, and if the velocity of flow changes, the kinetic energy changes, and the lateral hydrostatic pressure, *P*, is altered. If a tube containing flowing fluid narrows, the velocity of flow, and therefore the kinetic energy component, increases. As the total fluid energy is taken to be constant, the increase in the kinetic energy component results in a decrease in the potential energy component, so that the measured lateral pressure (*P*) decreases in the narrowed section. If the tube widens again, the importance of the kinetic component diminishes once more. Conversely, if a tube suddenly widens, the velocity of flow and the kinetic energy component falls, so *P* is increased. Because of the constancy of total fluid energy, the pressure measured by a side-tube falls in narrowed vessels and increases in widened vessels.

This effect must be remembered when the pressure in vessels or cardiac chambers is measured by catheterization, as side-facing catheters only measure the lateral pressure *P* (catheters with a lumen facing the flow can be used to measure the kinetic and potential energy components). In the arterial and venous systems the effect of kinetic energy is only significant in the aorta and vena cava during heavy exercise. However, the kinetic energy factor is important in the atria and ventricles, even at the normal resting cardiac output. In aortic stenosis the lateral pressure at the aortic root (coronary perfusion pressure) is reduced by the kinetic effect of the increased blood velocity passing through the narrowed valve. The lateral pressure within an aneurysm is increased because of the reduced flow velocity in the widened vessel.

VESSELS OF THE SYSTEMIC CIRCULATION

GENERAL DESCRIPTION OF THE VESSELS

Vessel walls, apart from capillaries, have three layers: the intima, the media and the adventitia. The intima is a sheet of endothelial cells, the media contains smooth muscle cells embedded in a matrix of elastin and collagen, and the adventitia is an inelastic connective tissue sheath. The diameter, wall thickness and relative proportions of elastic, smooth muscle and connective tissues are shown in Table 4.1.

Large arteries tend to be elastic, arterioles have much smooth muscle, capillaries are thin-walled endothelial structures, and the large veins have more inelastic connective tissue in their walls. The wall structure reflects the function of the vessels: the aorta is a low-resistance conducting vessel that is stretched during systole and maintains blood flow during diastole by elastic recoil; arterioles control organ flow, total peripheral resistance and arterial blood pressure; capillaries are exchange vessels; and the large veins are low-resistance capacitance

Table 4.1   *The structure and properties of the different types of blood vessel (average dimensions)*

	Aorta	Artery	Arteriole	Capillary	Venule	Vein	Vena cava
Internal diameter	25 mm	4 mm	30 μm	8 μm	20 μm	5 mm	30 mm
Wall thickness	2 mm	1 mm	6 μm	0.5 μm	1 μm	0.5 mm	1.5 mm
Velocity of blood flow	20 cm/s			0.5 mm/s			12 cm/s
Endothelium	+	+	+	+	+	+	+
Elastic tissue	+ + + +	+ + +	+ +	0	0	+ +	+ +
Smooth muscle	+ +	+ + +	+ + + +	0	0	+ +	+ +
Connective tissue	+ + +	+ +	+ +	0	+	+ +	+ + + +

vessels. The velocity of blood flow is high in the arteries, is low in the much larger volume of all the capillaries, and increases again in the veins.

Endothelium

The whole of the cardiovascular system (heart and blood vessels) is lined with a single layer of endothelial cells forming the surface in contact with the blood. In addition to their capillary functions of nutrient and waste product diffusion and fluid filtration, endothelial cells produce vasoactive substances throughout the cardiovascular system. Prostacyclin is produced by the endothelium from arachidonic acid, and inhibits platelet adhesion and aggregation and vessel constriction. Endothelial cells generate the vasodilator nitric oxide (NO, endothelium-derived relaxing factor) from L-arginine. NO increases vascular smooth

muscle cyclic GMP, which decreases the intracellular calcium concentration, producing muscle relaxation and vasodilatation. Endothelial NO production is stimulated by acetylcholine, ATP, bradykinin, serotonin, substance P and histamine. NO production may also be enhanced by the effect on the endothelial cell surface of the shear stress of blood flow. Endothelial cells also make the potent peptide vasoconstrictor endothelin, which increases peripheral vascular resistance and arterial blood pressure. Vascular endothelial cells, when stimulated by angiogenic factors, are able to form new capillary networks by cell division and movement.

The aorta

The elastic nature of the aorta and large arteries converts the intermittent ventricular output to a

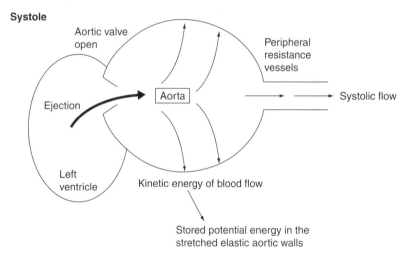

Systole

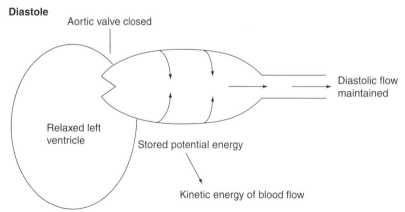

Diastole

Figure 4.28 *The Windkessel effect.*

continuous, pulsatile, blood flow to the periphery. This effect is known as the hydraulic filter, or the Windkessel effect (after a similar principle used to produce continuous water flow in early fire-engine pumps) (Fig. 4.28).

During ventricular systole, blood is ejected into the aorta. One-third of the ejected blood flow passes through the arteries to the tissues, but because of the impedance to flow of the arteriolar peripheral resistance, most remains within the aorta and large arteries and distends their elastic walls. That is, some of the kinetic energy of the blood ejected during systole is converted to potential energy stored in the stretched aortic wall elastic tissues (Fig. 4.28). In diastole, the aortic valve closes (preventing flow of blood back into the ventricle) and the stretched elastic aortic walls contract and maintain blood flow. In diastole, the stored potential energy is converted back again to the kinetic energy of flowing blood (Fig. 4.28). If this mechanism was not present, arterial blood pressure would be very low during diastole, and blood flow would stop.

For this system to work effectively, three factors are important. The first is the elasticity of the aorta; if this is reduced (as with age), then the system becomes less efficient. The second factor is the presence of a peripheral resistance; if this is reduced, as by the administration of peripheral vasodilators, flow tends to be less constant. The third factor is the presence of a functioning aortic valve that closes to prevent retrograde flow; in aortic incompetence peripheral blood flow tends to be less constant.

In essence, the elastic nature of the aorta allows it to work as a second pump in series with the ventricle during diastole.

The Windkessel effect reduces the cardiac workload, as excess work must be done to pump a given flow intermittently rather than continuously. One consequence of the loss of arterial elasticity with age is that this aortic hydraulic filter becomes less effective.

ARTERIES AND ARTERIAL BLOOD PRESSURE

Arteries are low-resistance conduits of blood with elastic walls that contribute to the Windkessel effect of the aorta. The arterial system is one of low volume but high pressure, permitting rapid distribution and redistribution of the cardiac output to all the different organs of the body. The arteries (with the arterioles) contain approximately 15 per cent of the total blood volume (i.e. 750 mL) at a mean pressure of around 100 mmHg. Aortic and arterial pressures rise during systole to 120 mmHg and fall to 80 mmHg in diastole. The arterial pulse pressure is the difference between the systolic and diastolic pressures:

$$\text{Pulse pressure} = \text{systolic} - \text{diastolic pressure}$$

The mean arterial pressure is the average arterial pressure acting during the cardiac cycle, and can be measured by averaging the pressure recorded under an arterial pressure curve over time. The mean arterial pressure can be calculated approximately as being equal to the diastolic pressure plus one third of the pulse pressure:

$$\text{Mean arterial pressure} = \text{diastolic} + \tfrac{1}{3}(\text{systolic} - \text{diastolic})$$

The systolic stretching of the aortic wall travels peripherally as a pressure wave along the arterial walls much faster (metres per second) than the blood flows (only centimetres per second); it is this pressure wave that is palpated at peripheral arteries. As the arterial walls stiffen with age, the velocity of this pressure wave increases.

The aortic pressure waveform rises in systole to 120 mmHg, has an incisura at the point of aortic valve closure, and falls in diastole to 80 mmHg. The waveform changes as blood flows through the arteries. As blood passes down the arterial tree, the diastolic and mean pressures fall gradually. In healthy young adults there are other important changes in the shape of the arterial pressure wave as blood flows peripherally: the systolic wave becomes elevated and narrowed; the incisura and other high-frequency components are damped out; and another wave (the dicrotic wave) appears on the diastolic portion. In elderly individuals with stiff arteries these latter changes are diminished, and the shape of the aortic and peripheral artery pressure waveforms are then similar.

Determinants of mean arterial blood pressure

The mean arterial blood pressure is determined by the amount of blood in the arterial system at any point in time. The volume of blood in the arteries is determined by the amount of blood entering the aorta as the cardiac output, and the blood volume leaving the arteries as peripheral run-off into the capillaries. The mean arterial blood pressure is stable when the cardiac output and the peripheral run-off are equal. The cardiac output is the product of the heart rate and the stroke volume. The peripheral run-off is determined by the mean arterial blood pressure acting across the total peripheral resistance of the arterioles (Fig. 4.29).

The relationship between mean arterial blood pressure, total peripheral resistance and peripheral run-off from the arteries can be described using the equation:

$$Resistance = \frac{pressure}{flow}$$

and:

$$Total\ peripheral\ resistance = \frac{mean\ arterial\ pressure}{peripheral\ run\text{-}off}$$

or:

$$Peripheral\ run\text{-}off = \frac{mean\ arterial\ pressure}{total\ peripheral\ resistance}$$

Normally,

$$Peripheral\ run\text{-}off = \frac{100\ mmHg}{0.02\ mmHg/mL\ per\ minute}$$
$$= 5000\ mL/min$$

If the cardiac output increases (and the total peripheral resistance does not change), there will be an increase in the arterial blood volume that increases the mean arterial blood pressure, until the peripheral run-off rises to equal the new cardiac output. For example, normally the cardiac output is equal to the peripheral run-off at 5000 mL/min, the mean arterial blood pressure is 100 mmHg, and the total peripheral resistance is 0.02 mmHg/mL per minute. If the cardiac output increases to 6000 mL/min (and the total peripheral resistance is unchanged), the mean arterial blood pressure will rise until the peripheral run-off also increases to 6000 mL/min. The new mean arterial blood pressure can be calculated:

$$Peripheral\ run\text{-}off = \frac{mean\ arterial\ pressure}{total\ peripheral\ resistance}$$

$$6000\ mL/min = \frac{mean\ arterial\ pressure}{0.02\ mmHg/mL\ per\ minute}$$

and:

$$Mean\ arterial\ pressure = 6000 \times 0.02$$
$$= 120\ mmHg$$

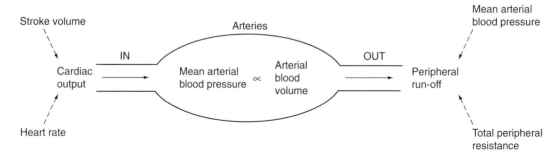

Figure 4.29 *Diagrammatic representation showing the determinants of mean arterial blood pressure.*

That is, with a rise in cardiac output the equilibrium between flow into and out of the arteries is restored by the increased mean arterial blood pressure increasing the peripheral run-off.

Changes in total peripheral resistance also alter the mean blood pressure. A rise in total peripheral resistance will tend to reduce peripheral run-off, and arterial blood volume and mean arterial pressure increase until the rise in pressure restores the peripheral run-off to its previous value (equal to the cardiac output). For example, if the total peripheral resistance increases to 0.03 mmHg/mL per minute and the cardiac output is unchanged, the mean arterial pressure will increase until the peripheral run-off is restored to 5000 mL/min, to equal the cardiac output once more. The new mean arterial pressure value can be calculated:

$$\text{Peripheral run-off} = \frac{\text{mean arterial pressure}}{\text{total peripheral resistance}}$$

$$5000 \text{ mL/min} = \frac{\text{mean arterial pressure}}{0.03 \text{ mmHg/mL per minute}}$$

and:

$$\begin{aligned}\text{Mean arterial pressure} &= 5000 \times 0.03 \\ &= 150 \text{ mmHg}\end{aligned}$$

That is, with a rise in peripheral resistance the equilibrium between flow into and out of the arteries is restored because the increased mean arterial pressure returns the peripheral run-off to its normal value.

The main determinants of mean arterial blood pressure are therefore the cardiac output and the total peripheral resistance, acting via changes in the arterial blood volume. Arterial wall compliance is not a primary determinant of mean arterial blood pressure. However, arterial compliance does determine the rate at which a new mean arterial pressure is attained upon cardiac output or total peripheral resistance changes. An individual with stiff, low-compliance arteries (as found in the elderly) will show more rapid changes in mean arterial pressure with alterations in cardiac output or total peripheral resistance.

Determinants of arterial pulse pressure

Stroke volume and arterial compliance are the main determinants of arterial pulse pressure. During systole, the arterial pressure rises from the diastolic value of 80 mmHg up to 120 mmHg because of the increase in arterial blood volume produced by the rapid ventricular ejection phase of the cardiac cycle. If the stroke volume is increased, the systolic pressure increases, but the diastolic pressure rise is less, and so the arterial pulse pressure increases. A fall in arterial compliance increases the pulse pressure for two reasons: (1) the ventricular ejection of blood into the stiff arteries produces a greater increase in systolic pressure, and (2) the hydraulic filter (Windkessel), which normally maintains diastolic pressure, is impaired by the reduced arterial elasticity.

ARTERIOLES

Arterioles are small-diameter vessels which have abundant smooth muscle in their walls and provide the main site of resistance to blood flow in the cardiovascular system. After passing through the high-resistance arterioles, the pulsatile arterial blood flow at a mean pressure of 100 mmHg is converted to steady flow in the capillaries at a pressure of 35 mmHg. The importance of arteriolar resistance to blood flow is that it can be varied by alterations in arteriolar size. According to the resistance equation:

$$\text{Resistance} = \frac{8\eta L}{\pi r^4}$$

where η is the blood viscosity. Therefore,

$$\text{Arteriolar resistance} = \frac{8\eta \times \text{arteriolar length}}{\pi (\text{arteriolar radius})^4}$$

Arteriolar length and blood viscosity cannot be altered, but the radius can change and this affects arteriolar resistance markedly as the resistance is inversely proportional to the fourth power of the radius. Because of spontaneous contractile myogenic activity in the smooth muscle of their walls and tonic sympathetic output to these vessels, arterioles are normally under a state of resting partial vasoconstriction. Changes in the factors controlling arteriolar tone can therefore lead to vasodilatation (an increase in radius) or vasoconstriction (a decrease in radius), with resultant effects on arteriolar resistance. Arteriolar dilation tends to reduce the hydrostatic pressure drop across these vessels: arterial pressure decreases and capillary pressure increases. Arteriolar constriction tends to increase the hydrostatic pressure drop: arterial pressure increases and capillary pressure decreases.

Arteriolar functions

The functions of arterioles are reflected in the haemodynamic consequences of changes in arteriolar resistance on mean arterial blood pressure, the blood flow to individual organs, and capillary hydrostatic pressure. Arterioles have three main functions:

- Alterations of total peripheral resistance
- Alterations of the vascular resistances of individual organs
- Alterations of capillary hydrostatic pressure

ARTERIOLES AND TOTAL PERIPHERAL RESISTANCE

As explained earlier, the total peripheral resistance is one of the main factors affecting mean arterial blood pressure. As the arterioles are the main component of peripheral resistance, alterations in their radius affects mean arterial blood pressure. Generalized arteriolar constriction, as by an increase in sympathetic nervous system outflow, increases total peripheral resistance, and mean arterial blood pressure rises as a consequence. Generalized arteriolar dilation, as by a reduction in sympathetic outflow, decreases total peripheral resistance, and mean arterial blood pressure falls as a consequence.

ARTERIOLES AND THE VASCULAR RESISTANCES OF INDIVIDUAL ORGANS

The blood flow to an organ depends upon the pressure perfusing it (the mean arterial blood pressure) and the organ vascular resistance. That is,

$$Flow = \frac{pressure}{resistance}$$

and:

$$Organ\ blood\ flow = \frac{mean\ arterial\ pressure}{organ\ vascular\ resistance}$$

The arterioles of an individual organ bed can dilate or constrict independently of other tissues. For example, the arterioles in an exercising muscle dilate (under the control of systemic and local factors described later), so the vascular resistance falls and blood flow to the muscle rises, delivering more oxygen and nutrients as required. The arteriolar resistance of a given organ in comparison with other organs will determine the proportion of the cardiac output it receives.

ARTERIOLES AND CAPILLARY HYDROSTATIC PRESSURE

Arteriolar resistance influences the capillary hydrostatic pressure. The importance of this is that the capillaries are the interface between the intravascular (3 L plasma) and the larger interstitial (11 L) and intracellular (28 L) fluid volumes, and arteriolar dilation or constriction affects the distribution of the total body water (42 L) between these compartments. The capillary hydrostatic pressure is an important determinant of the bulk flow of water between the intravascular and interstitial fluid volumes. If capillary hydrostatic pressure falls, fluid moves by bulk flow from the interstitial volume (and then out of the intracellular volume) into the capillaries. If capillary hydrostatic pressure rises, fluid moves by bulk flow from the capillaries into the interstitial volume. During haemorrhage and intravascular volume depletion, reflex sympathetic discharge produces arteriolar constriction, the capillary hydrostatic pressure falls, and fluid moves by bulk flow from the interstitial volume into the capillaries to replenish the depleted intravascular volume. Arteriolar dilation increases capillary hydrostatic pressure, and fluid moves by bulk flow out of the capillaries into the interstitial volume. Arterioles therefore affect the distribution of total body water between the intravascular, interstitial, and intracellular compartments. It is important to note, however, that venous resistance also significantly affects capillary hydrostatic pressure.

The control of arteriolar smooth muscle tone

A number of local and systemic factors are involved in the control of arteriolar smooth muscle tone and radius; the relative importance of these factors varies between organs.

LOCAL FACTORS

Local myogenic control of arteriolar smooth muscle

Arterioles contain smooth muscle fibres that contract spontaneously in response to a rise in pressure within the vessel. As the pressure rises, the arterioles constrict, arteriolar resistance increases, and the organ blood flow is kept constant in the face of the rising perfusion pressure. With a fall in pressure, the arterioles dilate and organ blood flow is again kept constant. This myogenic activity is dominant in the brain and the kidneys, and is the mechanism of the autoregulation of blood flow seen in those organs over a wide range of mean arterial blood pressure. Myogenic control is much less important in skeletal muscle, and the skin.

Local metabolic control of arteriolar smooth muscle

Arteriolar smooth muscle has a resting tone that can be reduced by vasodilatory tissue metabolites. The production of such metabolites increases (and oxygen tension falls) with organ metabolic activity, so they increase tissue blood flow according to metabolic requirements. Many factors have been proposed as being responsible for this metabolic control of arteriolar tone, including a fall in oxygen tension, a rise in carbon dioxide tension, a rise in temperature, hydrogen ions, potassium, lactic acid, pyruvate, inorganic phosphate, interstitial fluid osmolarity, adenosine and ATP, ADP and AMP. It is not clear which of the proposed factors is involved, and it may be that their relative importance varies from tissue to tissue. Metabolic regulation

of arteriolar tone may be responsible for the hyperaemia seen after the blood flow to an organ is stopped temporarily (reactive hyperaemia), and that observed with increased tissue activity (active hyperaemia). The metabolic control of arteriolar tone is especially important in the heart, skeletal muscle and brain.

Local tissue vasoactive chemical control of arteriolar smooth muscle

The salivary, sweat and intestinal glands produce the enzyme kallikrein, which converts inactive plasma kininogens to active kinins, such as bradykinin, that relax arteriolar smooth muscle. When these glands are active, blood flow is increased by the local release of bradykinin. Endothelial cells also affect arteriolar smooth muscle by the local release of vasoactive substances: prostacyclin and NO are potent vasodilators; endothelin is a vasoconstrictor that acts locally and systemically. Other tissues can produce local factors affecting arteriolar tone. For example, if a vessel wall is damaged or cut, platelets aggregate and release the powerful vasoconstrictor thromboxane A_2.

SYSTEMIC FACTORS

Extrinsic sympathetic nerve control of arteriolar smooth muscle

Arterioles have a profuse sympathetic nerve supply, and norepinephrine released from the nerve endings acts at α receptors to produce vasoconstriction (and also at β_2 receptors to produce vasodilatation, but this effect is weaker). Because of the tonic sympathetic output from the medullary cardiovascular centres, there is a basal sympathetic nervous discharge to the arterioles. Therefore, arteriolar resistance can be either raised or lowered by an increase or decrease in sympathetic nerve activity, respectively. The extrinsic sympathetic nervous control of arteriolar tone is important in the skin, kidneys and gut. Extrinsic sympathetic nervous control is less important in the brain and heart.

Skeletal muscle has a second 'sympathetic cholinergic' nervous supply, under cortical control. These sympathetic nerve endings release acetylcholine that produces arteriolar relaxation and vasodilatation. This sympathetic cholinergic nerve supply may increase skeletal muscle blood flow at the onset of exercise, or with anger or fear.

Extrinsic parasympathetic nerve control of arteriolar smooth muscle

The parasympathetic control of arterioles is much less important. The vessels of the external genitalia have a dual autonomic nerve supply, with parasympathetic dilator nerves as well as a sympathetic constrictor supply. Other vessels, in the heart, brain and lungs, do have a parasympathetic nerve supply, but the role of these fibres is unknown.

Extrinsic hormonal control of arteriolar smooth muscle

Epinephrine (adrenaline) is released from the adrenal medulla by sympathetic activity controlled by the hypothalamus (norepinephrine is also released, but epinephrine accounts for 80 per cent of the adrenal medullary catecholamine production in humans). In the arterioles, epinephrine acts at both α (vasoconstrictor) and β_2 (vasodilator) receptors. The effect of circulating epinephrine on the blood flow of an organ depends on the relative proportion of α and β_2 receptors present in the arterioles. In the heart and skeletal muscle there are relatively more β_2 receptors, and circulating epinephrine produces vasodilatation. In contrast, circulating epinephrine produces arteriolar vasoconstriction in the gut and the skin because of the greater proportion of α receptors present. Other hormones affect arteriolar smooth muscle tone: angiotensin II and vasopressin are vasoconstrictors, and atrial natriuretic factor (ANF) is a vasodilator.

CAPILLARIES

There are an estimated 25 000 million capillaries in the body, and they perform the essential cardiovascular system function of nutrient and metabolite exchange between blood and tissues. Normally, 6 per cent of the total blood volume is in the systemic capillaries and 3 per cent is in the pulmonary capillaries. Capillaries are thin-walled vessels made up of tubes of endothelial cells lying on a basement membrane. The capillary wall has channels connecting the inside and outside of the vessel. (NB. This is not present in the brain where the endothelial cells of the blood–brain barrier are joined by tight junctions.) There are narrow intercellular clefts between adjacent endothelial cells, and also there are fused-vesicle channels formed by amalgamation of some of the many endocytotic and exocytotic vesicles present. Fluids and solutes move across the capillary wall by diffusion, filtration and by pinocytotic transport of vesicles (carrier-mediated transport is also important in the capillaries of the brain). Capillary wall permeability varies between tissues (high in the liver, low in the brain), and is higher at the venous than at the arterial end of the vessel.

Capillaries range in diameter from 5 to 10 μm, and are 1 mm long. Erythrocytes (diameter 7 μm) have to be flexible to deform and squeeze through the smaller capillaries. The capillary network provides a vast surface area for diffusion of nutrients to tissues, and the diffusion distance between the blood and cells is short (maximum 50 μm). More capillaries open with increased tissue activity, and the diffusion distance is then decreased to facilitate the nutrient and metabolite movement. Capillary density varies between organs, and tissues with high metabolic demands such as cardiac and skeletal muscle have many capillaries, whereas less active tissues (such as cartilage) have fewer. Only one-quarter of all the body's capillaries are open at rest, but this represents a very large capillary bed cross-sectional area, and so blood flow slows to 0.5 mm/s. At rest, a red cell may take 2 s to pass through a capillary. During increased tissue activity, this transit time can fall to 1 s, but this is still sufficient for diffusional exchange to take place.

Capillary blood flow depends on the dilation or constriction of the arterioles feeding them. Arteriolar vasoconstriction reduces capillary blood flow, and vasodilatation increases capillary flow. In some tissues capillaries arise from metarterioles connecting arterioles and venules. Metarterioles contain smooth muscle and the point of origin of a capillary is encircled by a smooth muscle precapillary sphincter under the control of local tissue metabolites. Contraction of the precapillary sphincter closes the capillary completely. Precapillary sphincters may also be present at the point of origin of capillaries from arterioles. In certain tissues there are also arteriovenous shunts from arterioles to venules that are under autonomic control and which subserve functions unrelated to tissue nutrition (e.g. arteriovenous shunts in the skin and temperature regulation).

The capillary wall is only one cell thick (0.5 μm), yet it is able to withstand hydrostatic pressures of around 35 mmHg. The wall tension is the force per unit length tangential to the vessel wall that opposes a distending force that tends to pull apart a theoretical split in the wall (i.e. prevents vessel rupture), and is related to the distending force (the product of transmural pressure and radius), by the law of Laplace:

Wall tension = pressure × radius

Because of their small radius, the wall tension required to prevent capillary rupture is low and so these thin-walled vessels can withstand relatively high intravascular pressures.

Capillary functions

Capillaries have two main functions: (1) the delivery and removal of nutrients and metabolites to tissues by diffusion across the capillary wall, and (2) the distribution of fluid between the intravascular and extravascular compartments by bulk flow of fluid and solutes across the semipermeable capillary wall (impermeable to large protein molecules).

NUTRIENT AND METABOLITE EXCHANGE BY DIFFUSION

Diffusion is the main way by which gases, nutrients, metabolites and water move between blood, the interstitial fluid and the cells.

Lipid-soluble substances – including oxygen and carbon dioxide – pass across the capillary wall with ease through the endothelial cell lipid membranes. Indeed, oxygen and carbon dioxide are so lipid-soluble that some exchange may take place in the arterioles, before the blood reaches the capillaries. The process of diffusion is affected by the tissue metabolic state. If a tissue (e.g. a skeletal muscle) increases activity, then intracellular oxygen tensions falls and this favours oxygen diffusion from the capillary blood. As a consequence, intracellular carbon dioxide tension rises, favouring diffusion into the capillary blood.

Water-soluble substances diffuse across the capillary wall through the water-filled channels of the intercellular clefts between adjacent endothelial cells, but this depends on the size of the molecule. Small molecules such as water, sodium and chloride ions, glucose and urea diffuse easily across the capillary wall. Large molecules with a molecular weight of 60 000 Da or more (including albumin) cannot pass through the intercellular clefts and so diffusion across the capillary is minimal (some proteins can diffuse through the larger fused-vesicle channels). The size of the water-filled channels explains the variation in capillary permeability observed between organs. Brain capillaries have no intercellular clefts, and even small water-soluble molecules must cross these vessels by carrier-mediated transport. In contrast, liver capillaries have large intercellular channels that even large proteins can pass through.

Diffusion is also the main way by which water moves to and fro between capillaries and cells. In total, 300 mL water per 100 g of tissue moves across the capillary wall by diffusion each minute – a value which exceeds capillary blood flow. In comparison, less than 1 mL of water per 100 g of tissue is filtered each minute by the capillaries.

BULK FLOW FLUID EXCHANGE BY FILTRATION

The capillary wall is a semipermeable membrane as it is permeable to water and solutes, but impermeable to large proteins (including albumin). A plasma ultrafiltrate, free of protein, is filtered by bulk flow through the capillary wall by the action of opposing hydrostatic and oncotic pressures. The function of this bulk flow of fluid across the capillary wall is fluid distribution between the intravascular and extravascular compartments (not nutritional). Four 'Starling forces' (Fig. 4.30) are involved in this filtration process: capillary hydrostatic pressure; interstitial hydrostatic pressure; plasma oncotic pressure, and interstitial fluid oncotic pressure. The net filtration pressure (NFP) moving fluid across the capillary wall is the balance of the opposing capillary and hydrostatic pressures minus the balance of the opposing plasma and interstitial oncotic pressures.

The capillary hydrostatic pressure is the prime force moving fluid into the interstitial space. It is opposed by the hydrostatic pressure of the interstitial

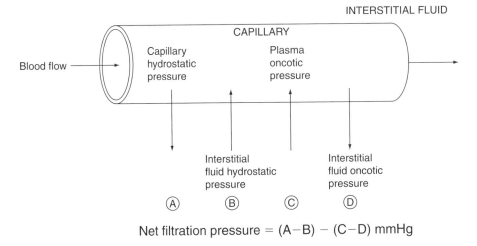

Net filtration pressure = (A−B) − (C−D) mmHg

Figure 4.30   *The Starling forces.*

fluid, but this is normally very low. The capillary hydrostatic pressure falls as blood flows along from the arterial to the venous end of the vessel. Albumin has a molecular weight of 60 000 Da, and cannot pass across the capillary wall. The plasma oncotic pressure is produced by the higher albumin concentration inside the capillary. This produces an osmotic gradient across the capillary wall – the plasma oncotic pressure of 28 mmHg, which opposes the filtration of fluid out of the capillary. There is some protein present in the interstitial fluid, and this produces a smaller interstitial fluid oncotic pressure, which tends to move fluid out of the capillary.

The NFP – the resultant pressure of the four Starling forces – can be calculated at the arterial and venous ends of a capillary as:

$$NFP = (P_c - P_{if}) - (\pi_p - \pi_{if})$$

where P_c is the capillary hydrostatic pressure, and:

P_c at the arterial end of the capillary = 35 mmHg
P_c at the venous end of the capillary = 15 mmHg
P_{if} interstitial fluid hydrostatic pressure = 0 mmHg
π_p plasma oncotic pressure = 28 mmHg
π_{if} interstitial fluid oncotic pressure = 3 mmHg
NFP at arterial end of the capillary = 10 mmHg
NFP at venous end of the capillary = −10 mmHg
Bulk flow across the capillary wall = $k \times$ NFP, where k is the capillary membrane filtration constant.

Therefore:

- At the arterial end of the capillary, bulk flow is out of the vessel into the interstitial fluid – filtration.
- At the venous end of the capillary, bulk flow is into the vessel from the interstitial fluid – absorption.

The result is that the balance of the Starling forces tends to filter fluid out of the capillary at the arterial end, and absorb fluid into the vessel at the venous end. The net fluid loss from filtration in the capillaries is only 4 L per day, and this is reabsorbed by the lymphatics.

Depending on the hydrostatic pressure present, capillaries may show net filtration or reabsorption along their entire length. The glomerular capillaries show net filtration because of the high glomerular hydrostatic pressure. In the lungs, the mean capillary hydrostatic pressure is only 8 mmHg and absorption of fluid occurs in these vessels. Net absorption also takes place in intestinal capillaries because of the balance of hydrostatic and oncotic forces.

Factors determining capillary hydrostatic pressure

Of the four Starling forces, only capillary hydrostatic pressure is under immediate physiological control. The factors controlling capillary hydro-static pressure therefore determine the distribution of body fluid between the intravascular and extravascular compartments.

Capillary hydrostatic pressure is determined by the ratio of the resistances of the arterioles (precapillary resistance) and the venules and veins (postcapillary resistance). A reduction in arteriolar resistance increases capillary hydrostatic pressure, and an increase in precapillary resistance decreases capillary hydrostatic pressure. A rise in venous resistance increases capillary hydrostatic pressure, and a fall in postcapillary resistance reduces the hydrostatic pressure. It can be seen that capillary hydrostatic pressure tends to vary as the ratio of the postcapillary resistance over the precapillary resistance. Capillary hydrostatic pressure is proportional to:

$$\frac{\text{postcapillary resistance}}{\text{precapillary resistance}}$$

The smooth muscle of the postcapillary vessels (venules) is controlled by the same systemic and local factors described earlier for the arterioles. The precapillary sphincters are subject to the same controls, and there is evidence for the presence of distinct postcapillary sphincters that are subject to the influence of systemic and local factors.

The arterial and venous blood pressures can also influence capillary hydrostatic pressure, but a given change in venous pressure has a greater effect than the same change in arterial pressure. Capillary hydrostatic pressure is increased by an elevated venous pressure, as seen in the legs upon standing, or in the presence of cardiac failure.

THE LYMPHATICS

The lymphatic capillaries arise in the tissues and drain lymph – a fluid derived from interstitial fluid – through the lymph nodes and progressively larger vessels that open into the right and left subclavian veins. Only central nervous system tissues, cartilage, bone and epithelium do not have lymph vessels. The lymphatic vessels contain valves that ensure that the flow of lymph is in one direction, from the interstitial fluid back into the cardiovascular system. The lymphatic capillaries are the first vessels of the lymphatic system; no other vessels drain into them. The walls of lymphatic capillaries consist of one layer of endothelium and are permeable to fluid and protein. The lymphatics are the only way that protein lost from the vessels can be returned to the cardiovascular system. Flow of lymph is promoted by the contraction of smooth muscle in the walls of the lymphatics, by skeletal muscle contraction, and by the one-way valves. Each day the lymphatics return to the cardiovascular system the net 4 L of interstitial fluid filtered from the capillaries and any albumin lost from the systemic vessels (one-quarter to one-half of the circulating plasma proteins).

VEINS AND VENOUS RETURN

Blood flows from the large cross-sectional area of the capillary bed into the venules, the larger veins, the venae cavae, and then the right atrium. In the venules some metabolic exchange continues to take place with the interstitial fluid. Smooth muscle, elastic and connective tissue fibres appear in the walls as the veins increase in size (see Table 4.1). The smooth muscle is innervated by sympathetic nerve fibres that release norepinephrine. When the transmural pressure falls below $6\,\text{cmH}_2\text{O}$ the veins collapse and become elliptical, reducing their cross-sectional volume. Veins of the limbs have many one-way valves that encourage drainage into the larger veins (there are no valves in the venae cavae, cerebral and portal veins).

As the large veins join, the venous cross-sectional area is reduced, and the velocity of blood flow increases. In the venae cavae the velocity of blood flow is 12 cm/s – only slightly less than the aortic value of 20 cm/s. The mean pressure is 10–15 mmHg in the venules, 4–8 mmHg in the larger veins, and 0–2 mmHg in the venae cavae. Right atrial contraction produces pressure pulsations in the venae cavae. The venous system is 25 to 30 times more compliant than the arterial system, the systemic veins contain 60 per cent of the total blood volume, and the veins can accommodate large volumes of blood with little rise in pressure. Any volume of blood lost from or added to the cardiovascular system is in the ratio of 25:1 to 30:1 (venous:arterial) because of the high venous compliance.

Sympathetic nerve stimulation, via an α-adrenoreceptor-induced contraction of the vein wall smooth muscle, decreases venous compliance, increases venous pressure, and moves blood out of the capacitance vessels into the right side of the heart. Circulating vasoactive hormones including epinephrine, angiotensin and vasopressin have similar venous effects. Normally, there is a basal sympathetic nerve outflow to the veins producing a tonic smooth muscle contraction, the venomotor tone. Cardiovascular reflexes, such as the arterial baroreceptor system, increase or decrease the venomotor tone to maintain a constant arterial blood pressure.

Functions of the veins

The venous system is therefore a low-pressure, high-volume capacitance system, in contrast to the high-pressure, low-volume arteries. The functions of the veins are:

- To serve as low-resistance pathways for the return of blood to the heart
- To act as capacitance vessels that maintain the filling pressure of the heart

DISTENSIBILITY OF VEINS

The shape of a vein is altered by the hydrostatic pressure within the vessel (Fig. 4.31). At low pressures, veins collapse and become elliptical, but with a small rise in hydrostatic pressure, they become circular, with a much larger internal volume. At higher venous pressures, the circular veins are stretched

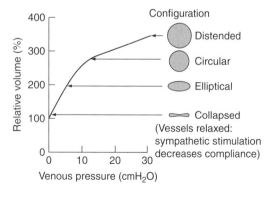

Figure 4.31 *Configurational changes in veins.*

and their compliance is reduced. Such changes occur when venomotor smooth muscle tone is low: increased sympathetic nerve activity reduces the compliance and the volume of blood in the veins.

There are two main consequences of the configurational change in veins. The first is that veins can accommodate large volumes of blood with only a small change in internal hydrostatic pressure (by changing from elliptical to circular). The second consequence is that if the hydrostatic pressure falls and the transmural pressure approaches zero, veins collapse to the elliptical form and their flow resistance increases. For example, in normal individuals (sitting or standing) the veins of the neck are collapsed because they are 5–10 cm above the heart and the transmural pressure is zero.

THE EFFECT OF POSTURE ON VEINS

In the supine position the mean blood pressure in the vessels of the foot will be (approximately) arterial 100 mmHg and venous 15 mmHg. In the absence of muscle activity in the leg, all the leg vein valves will be open and there will be a continuous flow of blood back to the heart.

When standing, the foot is 120 cm below the level of the heart, and the gravitational hydrostatic consequence is an increase of the vascular pressures, both arterial and venous, by 85–90 mmHg (120 cmH$_2$O). The right atrial pressure is unchanged. If the veins were rigid tubes, this would have no effect as neither the arterial to venous nor the venous to right atrial pressure differences alter. However, the resultant elevated venous transmural pressure expands the leg veins and increases the volume of blood in them. This would tend to reduce the return of blood to the heart, the cardiac output, and the arterial blood pressure. The cardiovascular response to this is mediated by the arterial baroreceptors that reflexly increase sympathetic outflow to the veins to reduce their compliance and maintain cardiac filling pressures (the sympathetic outflow to the heart and resistance vessels is also increased). In addition, the combined effect of skeletal muscle activity in the legs, together with the action of the venous valves, tends to minimize these postural changes.

Standing reduces the arterial and the venous pressures in the brain by 50 mmHg, but again the arterial to venous perfusion gradient is unchanged.

When standing, the cerebral venous sinus pressure may fall to $-40\,mmHg$, but the vein walls do not collapse because a simultaneous fall in cerebrospinal fluid pressure maintains a relatively normal cerebral venous transmural pressure. Furthermore, the cerebral venous sinuses are held open by extravascular tissues. For these reasons, during neurosurgery with the patient in the sitting position, large air emboli can enter into the cerebral venous sinuses if they are inadvertently opened.

Determinants of venous return to the heart

The factors that determine the rate of venous blood flow back to the heart are:

- The pressure gradient for venous return
- Venous valves
- The skeletal muscle pump
- The respiratory pump
- The effect of ventricular contraction and relaxation
- Venomotor tone

MEAN SYSTEMIC FILLING PRESSURE AND THE PRESSURE GRADIENT FOR VENOUS RETURN

Guyton proposed the concept of a 'mean systemic filling pressure' – the average of all the pressures in the different vessels weighted according to their relative compliances. This is a single hydrostatic assessment of the filling of the cardiovascular system, and indicates the pressure moving blood back toward the right atrium to maintain the cardiac output. If the heart of an animal is arrested experimentally, the mean systemic filling pressure (MSFP) is then the static pressure in the cardiovascular system after rapid equilibration of the arterial and venous pressures. (In this experimental condition the relative venous and arterial compliances lead to a transfer of blood from the arteries to the veins, and the fall in arterial pressure is much greater than the rise in venous pressure.) In the functioning cardiovascular system, the MSFP approximates to the mean venous pressure. The normal value of the MSFP is $7\,mmHg$, with a range of 0 to $20\,mmHg$ (the mean pulmonary filling pressure is $2\,mmHg$).

The pressure gradient through the veins to the right atrium is the main factor determining the rate of venous return, and is equal to the difference between the MSFP ($7\,mmHg$) and the mean right atrial pressure ($1\,mmHg$).

The pressure gradient for venous return:
= MSFP − mean right atrial pressure
= $6\,mmHg$

The MFSP increases with venomotor tone or blood volume, and falls with venodilation or blood loss. Changes in total peripheral resistance do not affect the MFSP.

VENOUS VALVES

The limb veins have one-way valves that prevent retrograde flow.

THE SKELETAL MUSCLE PUMP

Alternating contraction and relaxation of limb skeletal muscle squeezes blood out of the veins towards the heart. During contraction, the veins are compressed and blood is expelled from them towards the heart, and when the muscles relax the veins fill again (one-way flow is ensured by the venous valves). During standing, rhythmical skeletal muscle contractions in the veins of the leg reduce venous pressure and volume. During exercise, the skeletal muscle pump increases the venous return.

THE RESPIRATORY PUMP

The respiratory cycle changes in intrathoracic pressure facilitate venous return during inspiration. During inspiration the intrapleural pressure falls from $-5\,cmH_2O$ down to $-8\,cmH_2O$, and descent of the diaphragm increases intra-abdominal pressure: both increase the movement of blood from extrathoracic veins to the right atrium. During expiration, the effects are reversed. During inspiration the thoracic blood volume increases by $250\,mL$ and the right ventricular stroke volume rises by $20\,mL$. The effect on the left ventricle is different, as during inspiration the increased capacity of the pulmonary vessels decreases left ventricular stroke volume. The respiratory variation in left ventricular stroke volume between inspiration and expiration is only of the order of 5 per cent. The effect of the respiratory pump increases during exercise, but the effect is limited by the development of negative pressure and the collapse of veins as they enter the chest.

EFFECT OF VENTRICULAR CONTRACTION AND RELAXATION

During the rapid ejection phase of ventricular systole, the atrial pressure falls sharply, to zero or negative values, as ventricular contraction pulls the atrioventricular fibrous ring downwards, lengthening and increasing atrial volume. This increases the flow of blood into the atria from the venae cavae and pulmonary veins. During early diastole the ventricles fill rapidly, and ventricular and atrial pressures both decline (atrial pressure 'v' wave), and this effect may facilitate the flow of blood into the atria.

VENOMOTOR TONE

An increase in venomotor tone reduces the compliance and capacity of the veins, and increases the MSFP. Venomotor tone is reflexly increased via the sympathetic nerve system during exercise, or by blood loss. Venomotor tone has more effect on venous return when the venous pressure is normal and the veins are in their circular configuration, containing large volumes of blood (*see* Fig. 4.31).

THE RELATIONSHIP BETWEEN VENOUS RETURN AND CARDIAC OUTPUT

Normal circulatory function is brought about by an interaction between the peripheral circulation and the heart. The heart performs the work of pumping the blood, and the cardiac output is regulated by the autonomic nervous system which regulates changes in heart rate and myocardial contractility. However, the cardiac output is often limited by the amount of blood returning to the heart from the systemic circulation – the venous return.

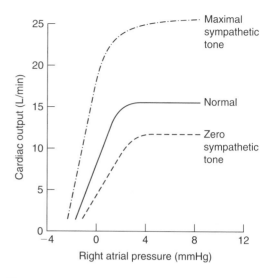

Figure 4.32   *Effect of sympathetic tone on cardiac output curves.*

The cardiac output curve

Cardiac output curves that relate cardiac output to right atrial pressure (Fig. 4.32) graphically illustrate the Frank–Starling mechanism, whereby raising the cardiac preload results in an increase in cardiac output. The cardiac output curves described by Guyton were obtained (in an experimental animal preparation) by cannulating the right atrium and varying the height of an attached reservoir of blood.

The normal curve has a very steep upstroke so that small changes to the right atrial pressure produce a large change in cardiac output. The curve reaches a plateau, when the cardiac output is independent of the right atrial pressure. The height of the plateau is dependent on the maximum cardiac pumping capacity. At a normal resting sympathetic tone the normal human heart can produce a cardiac output of 5 L/min at a right atrial pressure of zero (Fig. 4.32). The output rises to a maximum of 10–15 L/min at a right atrial pressure of 4 mmHg. With increased sympathetic stimulation (as in exercise) the maximum pumping capacity rises to 20–30 L/min (Fig. 4.32). This shows that the normal heart has a significant reserve pumping capacity. Since the venous return to the heart is substantially less than the maximum cardiac pumping capacity, the heart normally adjusts its output to match venous return, using the

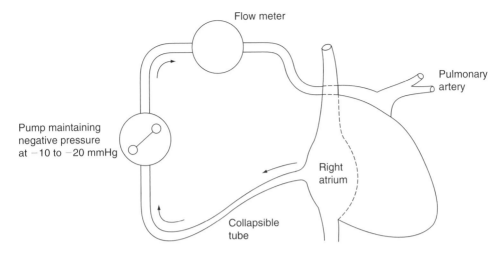

Figure 4.33 *Right heart bypass preparation.*

Frank–Starling mechanism. Therefore cardiac output is determined by the peripheral circulatory factors that regulate venous return.

The normal venous return curve

The tendency for blood to return to the heart from the peripheral circulation can be characterized by the venous return curves described by Guyton. In an experimental animal preparation (Fig. 4.33), a bypass pump replaces the right ventricle and blood enters the pump via a collapsible tube and is returned to the pulmonary artery. The pump maintains a pressure of −10 to −20 mmHg. The right atrial pressure is adjusting by raising or lowering the segment of the collapsible tubing. When the right atrial pressure in this preparation is raised to 7 mmHg, venous return falls to zero, the circulation stops, and pressure equalizes throughout the systemic blood vessels. This occurs because blood collects in the compliant venous capacitance vessels. The pressure at which venous return ceases is the MSFP. This corresponds to the degree of filling of the systemic circulation and is determined by the blood volume and venous capacitance. The MSFP is normally about 7 mmHg in humans. As the right atrial pressure is reduced to below MSFP, venous return increases up to about 5 L/min in humans when the right atrial pressure is zero. Therefore venous return decreases linearly with

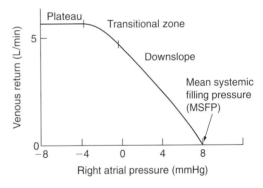

Figure 4.34 *Normal venous return curve.*

rises in right atrial pressure between zero and the MSFP (Fig. 4.34). Venous return is proportional to the difference between MSFP and right atrial pressure, that is the hydrostatic pressure gradient promoting venous return. Between this range of right atrial pressure (i.e. 0 to 7 mmHg), venous return can be defined by the equation:

$$\text{Venous return} = \frac{\text{mean systemic filling pressure} - \text{right atrial pressure}}{\text{resistance to venous return}}$$

The 'resistance to venous return' is a term that reflects the resistance and capacitance of the venous circulation. However, this relationship is not valid when the right atrial pressure is below zero. As the right atrial pressure becomes more negative, venous return reaches a plateau at a pressure 20 per cent above that at a right atrial pressure of zero. The plateau occurs because the extrathoracic veins collapse and act as Starling's resistors, thus preventing any further increase in venous return. The pressure in the extrathoracic veins cannot be reduced below -4 mmHg. This explains why increased pump function that merely reduces right atrial pressure cannot produce a large increase in cardiac output unless there is a concomitant increase in venous return.

Effects of changes in MSFP

Changes in the MSFP cause a parallel displacement of the venous return curve. An increase in the MSFP shifts the venous return curve to the right and enhances venous return at any given right atrial pressure (Fig. 4.35). Conversely, a decreased MSFP shifts the curve to the left and reduces venous return. A 15 per cent increase in blood volume will double the MSFP, whereas a 15 per cent decrease will reduce it to zero. Therefore, venous return is dependent on the blood volume. Sympathetic tone will influence the MSFP by effects on the venous capacitance vessels. Maximal sympathetic discharge will increase the MSFP to about 17 mmHg. This is important in producing the enhancement of venous return during compensation for hypovolaemia.

Effects of changes in resistance to venous return

Changes in the resistance to venous return alter the slope of the venous return curve. A decrease in resistance to venous return rotates the venous return curve clockwise, increasing venous return (Fig. 4.36). A rise in resistance to venous return rotates the venous return curve counter-clockwise and decreases venous return (Fig. 4.36). Alterations in resistance to venous return are from changes in peripheral vascular resistance caused by autoregulatory vasoconstriction or vasodilatation. For example, during exercise, metabolites in the muscle cause marked vasodilatation in muscle that reduces the venous resistance and enhances venous return.

The maintenance of a stable circulation requires an equilibrium between venous return and cardiac output. These two variables must be equal and cannot differ for more than a few heartbeats before a new equilibrium state is reached. The point of intersection of the cardiac output and venous return curves is known as the 'equilibrium point' (Fig. 4.37). The normal curves intersect at a right atrial pressure of zero and a cardiac output or venous return equal to 5 L/min. Changes in either cardiac output or venous return will shift the equilibrium point and change the right atrial pressure.

The circulatory effects of alterations in sympathetic tone illustrate how changes in cardiac function and venous return affect the equilibrium point

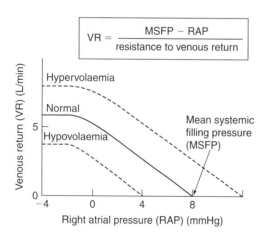

Figure 4.35 *Effect of mean systemic filling pressure (MSFP) on venous return curve.*

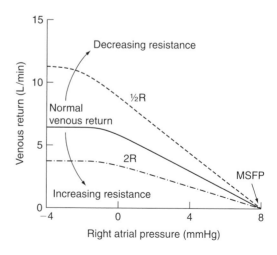

Figure 4.36 *Effect of changes in resistance (R) to venous return.*

(Fig. 4.38). In humans, the normal curves equilibrate at a right atrial pressure of zero, and a cardiac output and venous return equal to 5 L/min. A total sympathectomy, such as that caused by a high spinal anaesthesia, impairs cardiac contractility and causes vasodilatation of capacitance vessels. The impaired cardiac function shifts the cardiac function curve to the right, whilst the vasodilatation decreases the MSFP and shifts the venous return

curve to the left. This shifts the equilibrium point, resulting in a 40 per cent decrease in cardiac output. Sympathetic stimulation enhances cardiac function, and vasoconstriction of the capacitance vessels increases the MSFP and enhances venous return (Fig. 4.38). The new equilibrium point indicates enhanced cardiac output with a reduced right atrial pressure. Maximal sympathetic stimulation can double the cardiac output whilst reducing right atrial pressure. Most of the increased cardiac output results from the shift in the venous return curve caused by the vasoconstriction of the capacitance vessels, and not from the enhanced cardiac output.

Cardiac output is regulated by changes in heart rate and stroke volume mediated by the autonomic system as well as changes in the peripheral circulation that alter venous return. These principles may be applied to analyse clinical derangements of circulatory function.

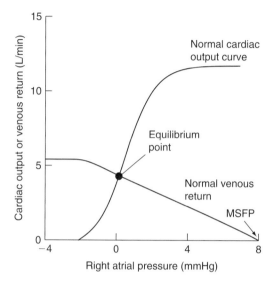

Figure 4.37 *Interaction of cardiac output and venous return curves.*

CONTROL OF THE CARDIOVASCULAR SYSTEM

OVERVIEW

The cardiovascular system is controlled by central nervous control and integration; efferent autonomic nerves and hormones; effectors in the heart,

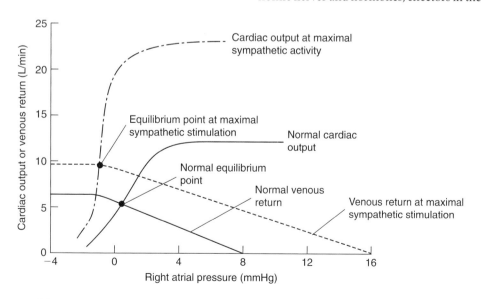

Figure 4.38 *Effect of sympathetic activity on the 'equilibrium point'.*

vessels and kidneys; the measured and controlled variables (e.g. arterial blood pressure); receptors (e.g. stretch receptors in the carotid and aortic bodies); and afferent pathways carrying information back to the control centres (Fig. 4.39).

The arterial blood pressure is the primary controlled cardiovascular variable. Short-term changes in arterial blood pressure are minimized by feedback loops that reflexly produce compensatory changes in autonomic nerve discharge to the heart and vessels. For example, a rise in arterial blood pressure reflexly increases vagal discharge and reduces sympathetic output from the brain, producing compensatory falls in cardiac output and peripheral resistance. Although the heart and peripheral resistance vessels have their own intrinsic activity, the central nervous system influence descending upon them is essential for the maintenance of a normal arterial blood pressure. Disruption of this efferent autonomic influence, as by transection of the cervical spinal cord, produces a marked fall in arterial blood pressure.

FUNCTIONAL ORGANIZATION OF CARDIOVASCULAR SYMPATHETIC AND PARASYMPATHETIC NERVES IN THE MEDULLA

Many centres in the central nervous system influence the cardiovascular system, including the medulla, the cerebellum, the hypothalamus and the cerebral cortex. The central control of the cardiovascular system involves the processing of afferent information in the medulla and higher brain centres, modulation of the activity of the medullary sympathetic and parasympathetic neurons, and alteration of the outflow of autonomic nervous activity. In the medulla and spinal cord, the neuronal cells involved in the control of the cardiovascular system are:

- Premotor sympathetic and preganglionic parasympathetic nerves in the medulla
- Preganglionic sympathetic nerves in the spinal cord intermediolateral column

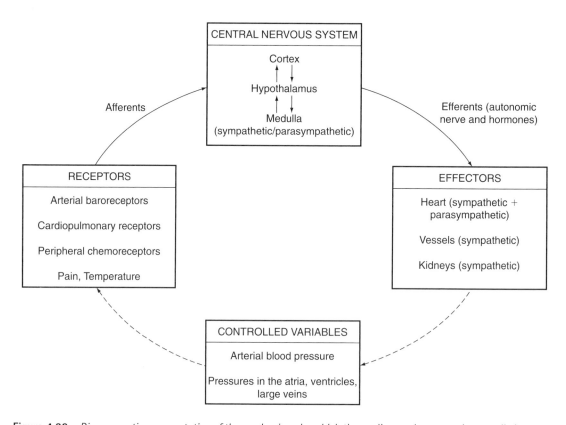

Figure 4.39 *Diagrammatic representation of the mechanisms by which the cardiovascular system is controlled.*

- Afferent fibres in the glossopharyngeal and vagus nerves
- Interneurons in the medulla

Efferent sympathetic nerve activity is via the preganglionic nerve cells in the intermediolateral columns of the thoracic and lumbar spinal cord (T1 to L2), and fibres run out to the sympathetic chain from where postganglionic fibres supply the heart, vessels and the kidneys (Fig. 4.40). The adrenal medulla is innervated by preganglionic sympathetic fibres. Premotor sympathetic cells in the medulla (e.g. rostral ventrolateral medulla) send fibres down the spinal cord to the preganglionic cells and modulate their activity. The efferent preganglionic parasympathetic cells are in the nucleus ambiguus and the dorsal motor nucleus of the medulla, and the fibres run in the vagus nerve, mainly to the atria. Afferent fibres from the

cardiovascular somatic and visceral receptors reach the medulla in the glossopharyngeal and vagal nerves and by spinal afferents, and synapse in the nucleus tractus solitarius. Interneurons link the afferent fibres with higher centres in the brain and with the sympathetic premotor and vagal preganglionic fibres (Fig. 4.40).

The most important sympathetic premotor cells in the medulla are in the rostral ventrolateral medulla (RVLM). The parasympathetic preganglionic cells are in the nucleus ambiguus and the dorsal motor nucleus of the medulla. The medullary RVLM sympathetic cells have inherent activity and produce a basal sympathetic nervous output (or tone) to the heart and the vessels, which is held in check by reflex baroreceptor inhibition. In contrast, the medullary parasympathetic cells are inherently quiescent and are stimulated by the baroreceptor reflex: with each heartbeat the systolic

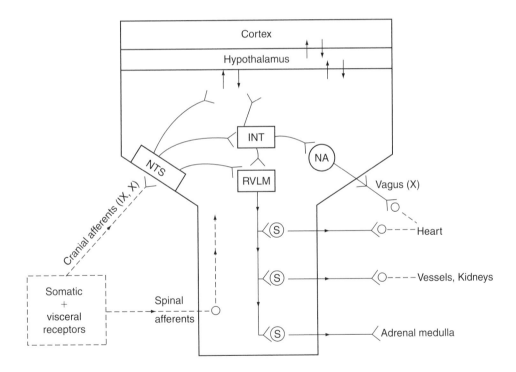

RVLM – Premotor sympathetic nerve cells (rostral ventrolateral medulla)
NA   – Preganglionic parasympathetic nerve cells (nucleus ambiguus)
INT   – Interneurons
NTS   – Nucleus tractus solitarius
S   – Preganglionic sympathetic nerve cells in the intermediolateral
        column of the spinal cord

Figure 4.40 *The sympathetic and parasympathetic nerves in the brain and spinal cord involved in cardiovascular control.*

rise in blood pressure reflexly produces a burst of inhibitory vagal nerve activity to the atria.

CENTRAL NERVOUS SYSTEM CONTROL AND INTEGRATION OF THE CARDIOVASCULAR SYSTEM

Areas of the central nervous system involved in the integrated control of the cardiovascular system include:

- Central sympathetic nerve cells
- Central parasympathetic nerve cells
- The nucleus tractus solitarius
- The cerebellum
- The midbrain periaqueductal grey (PAG)
- The hypothalamus
- The limbic system
- The cerebral cortex

Central sympathetic system nerve cells

The sympathetic premotor neurons in the brain and the preganglionic cells in the intermediolateral column of the spinal cord are the central nervous system cells responsible for the sympathetic nervous outflow to the cardiovascular system.

CARDIOVASCULAR SYMPATHETIC PREMOTOR NEURONS

Five specific sympathetic premotor cell groups innervate preganglionic outflow to all the sympathetic ganglia and the adrenal medulla: RVLM; rostral ventromedial medulla; caudal raphe nuclei; paraventricular nucleus in the hypothalamus; and the A5 noradrenergic cell group in the caudal ventrolateral pons. Of these, the RVLM cells have a crucial role in the control of arterial blood pressure, and their destruction produces dramatic hypotension. The RVLM premotor neurons send excitatory bulbospinal fibres to the sympathetic preganglionic cells in the intermediolateral column of the spinal cord (glutamate may be the excitatory transmitter released).

The RVLM neurons are tonically active, and generate most of the resting sympathetic nervous output to the cardiovascular system, increasing cardiac output and total peripheral resistance. As well as this generalized effect, RVLM cells may have specific vasomotor effects on individual tissue circulations. Afferent fibres subserving cardiovascular reflexes arriving at the RVLM originate from other sites in the medulla, the nucleus tractus solitarius (NTS), the hypothalamus and higher centres in the brain. RVLM cells are excited by excitatory amino acids (e.g. glutamate), ADH (vasopressin), angiotensin and acetylcholine, and inhibited by GABA, enkephalins and catecholamines. The RVLM cells are influenced by and integrate inputs from many sources: arterial baroreceptors, chemoreceptors, cardiac receptors, somatic receptors, and higher brain centres. The tonic output of the RVLM sympathetic premotor neurons is inhibited by the arterial baroreceptor reflex, and GABA may be the transmitter.

Other regions of the medulla modulate the output of the RVLM premotor neurons. One example is the caudal ventrolateral medulla (CVLM), which has a depressor effect on the cardiovascular system, and reduces peripheral resistance and cardiac contractility. It is thought that the CVLM, under the influence of various excitatory and inhibitory inputs, tonically inhibits RVLM cells by GABA release. Another example is the area postrema (a vascular area on the dorsum of the medulla that lacks a blood–brain barrier) where circulating angiotensin gains access to and stimulates the neurons, which then raise arterial blood pressure by excitation of RVLM cells.

SYMPATHETIC PREGANGLIONIC CELLS IN THE SPINAL CORD

In the spinal cord most of the sympathetic preganglionic nerve cells are in the intermediolateral columns of the thoracic and upper lumbar segments. Sympathetic preganglionic nerve cells have functional specificity for different organ circulations, although the sympathetic ganglia and the adrenal medulla are innervated from several spinal cord segments. The sympathetic preganglionic nerve cells release acetylcholine as a transmitter in the ganglia, but they also produce various potential co-transmitters including enkephalins, substance P, somatostatin and nitrous oxide (NO).

The sympathetic preganglionic nerve cells integrate the many influences that descend upon them from the medulla and from spinal afferents from skin, viscera and skeletal muscle. In general, excitatory inputs to the sympathetic preganglionic nerve cells release glutamate, and inhibitory fibres release GABA. For example, the medullary RVLM cells produce a tonic excitatory effect on the sympathetic preganglionic nerve cells by the release of glutamate. Some fibres terminating on these cells may release monoamines and neuropeptides as co-transmitters.

Central parasympathetic system nerve cells

The central parasympathetic preganglionic nerves are in the nucleus ambiguus in the ventrolateral medulla and in the dorsal motor nucleus of the vagus. Due to a stimulatory baroreceptor input carried via the NTS, the cells discharge synchronously with the cardiac cycle. A direct inhibitory input from medullary inspiratory neurons reduces preganglionic parasympathetic nerve discharge, and produces the tachycardia of inspiration (sinus arrhythmia).

The nucleus tractus solitarius

The NTS in the dorsomedial medulla is the principal site of termination of primary cardiovascular afferents (glossopharyngeal and vagus nerves) and of second-order afferents from other visceral and somatic receptors. The NTS sends fibres directly and via interneurons to cardiovascular neurons in the spinal cord, medulla, hypothalamus and the cerebral cortex. The NTS has an integral role in cardiovascular control, and ablation of these nerve cells produces sustained hypertension. The NTS relays information about any rise in afferent arterial baroreceptor activity to stimulate the nucleus ambiguus and increase parasympathetic output to the heart. In contrast, the NTS relays an increase in baroreceptor activity to the cells of the CVLM that inhibit the RVLM and decrease sympathetic output to the heart, vessels, kidney and adrenal medulla. The excitatory transmitter released by the primary cardiovascular afferents at their termination in the NTS may be glutamate. Afferent inputs arrive at the NTS from

cardiovascular nerve groups throughout the brain, including the cortex, hypothalamus and medulla (RVLM). It is thought that these inputs modulate the transmission of afferent baroreceptor information in the NTS and change the sensitivity of the arterial baroreceptor reflex. For example, stimulation of the hypothalamic 'defence area' depresses the baroreceptor reflex by the release of the inhibitory transmitter GABA in the NTS, and increases arterial blood pressure.

The cerebellum

The cerebellum is responsible for the control of posture and coordination of movement, and is also involved in the regulation of the cardiovascular responses to the integrated muscle and joint activities of exercise. Input to the cerebellum is from the cerebral cortex, the brainstem via the extrapyramidal tracts and the vestibular system, and the ascending spinal pathways via the dorsal and ventral spinocerebellar tracts. The spinocerebellar pathways form the major afferent pathway to the cerebellum and transmit proprioceptive information from the joints, muscles and skin. Neural impulses from the cerebellar nuclei are transmitted directly to the brainstem nuclei and then to the cerebral cortex via the thalamus, or to the spinal cord. In particular, the fastigial nucleus and the uvula in the cerebellum have important cardiovascular effects. Electrical stimulation of the fastigial nucleus increases sympathetic nerve activity and arterial blood pressure, and destruction of the nucleus impairs the pressor response of animals to exercise. Electrical stimulation of different areas of the uvula can produce cardiovascular pressor or depressor effects, possibly by excitation or inhibition of RVLM sympathetic premotor cells. The uvula has afferent inputs from brainstem nuclei for balance, sight, hearing, somatosensation and pain, and may mediate the cardiovascular responses to alerting stimuli, as are present at the onset of exercise.

The midbrain periaqueductal grey

The midbrain periaqueductal grey (PAG) has important roles in cardiovascular control, antinociception and reactions to threat. For example, stimulation of areas of the PAG produces features of the 'defence reaction' with increased arterial blood pressure, vasodilatation of skeletal muscle arterioles, and renal vessel vasoconstriction. Within the PAG, lateral areas produce pressor effects and vasoconstriction, and ventrolateral regions cause depressor effects and vasodilatation. PAG neurons control specific vascular beds: rostral cells in the lateral and ventrolateral areas affect skeletal muscle vessels, and caudal cells influence renal vessels. Vasomotor cells are arranged in the PAG according to the vascular beds they supply, and make specific connections with the RVLM sympathetic premotor neurons to those circulations.

The hypothalamus

The hypothalamus provides both neural and endocrine control of the internal organs of the body, and discrete cell groups have important cardiovascular effects. The 'defence area' of the hypothalamus lies in the anterior perifornical region. In animals, electrical stimulation of this discrete area can produce a rise in pulse rate, increased cardiac output, hypertension, dilatation of skeletal muscle vessels, constriction of gastrointestinal and renal vessels, and features of fear or rage behaviour. The defence area inhibits the baroreceptor reflex at the NTS, stimulates the sympathetic premotor RVLM cells, and inhibits vagal output to the heart. The hypothalamic defence area is activated by the limbic system. The hypothalamic 'depressor area' is found in the anterior hypothalamus, and stimulation produces effects similar to the arterial baroreceptor reflex: reduced sympathetic nerve activity and increased cardiac vagal output. In the anterior hypothalamus the supraoptic and paraventricular nuclei produce ADH (vasopressin), and release of this hormone is stimulated by local osmoreceptors and by an input from the arterial baroreceptor reflex via the NTS. With a rise in body temperature, the 'temperature regulating' area in the anterior hypothalamus promotes heat loss from the body by reducing the vasoconstrictor outflow to the skin and increasing sweating.

The limbic system

The limbic system is formed by parts of both frontal lobes and consists of the anterior cingulate, the posterior orbital gyrus, the hippocampus and the amygdala. The amygdala subserves fear and rage behaviour by activating the hypothalamic defence area. In animals, stimulation of the amygdala produces a cardiovascular response similar to that provoked by danger: a rise in heart rate, increased arterial blood pressure, skeletal muscle arteriolar vasodilatation and renal vasoconstriction. The limbic system may also produce the 'playing dead' reaction seen when some young animals are in danger, in addition to severe bradycardia and hypotension.

The cerebral cortex

The cerebral cortex can affect cardiovascular function, and this may be important for the rapid changes present at the onset of exercise. Neuronal projections have been found from one cortical area, the insular cortex, to the amygdala, hypothalamus, RVLM and the NTS.

Summary of the integrated control of the cardiovascular system

A central neuronal axis of groups of cells in the medulla, cerebellum, PAG, hypothalamus, limbic system and cortex controls the autonomic outflow to the cardiovascular system. Afferent information from within the heart and the vessels is fed back via the NTS to influence the activity of this neuronal axis. This system delivers appropriate and specific cardiovascular responses to exercise, pain, emotions and temperature change, and to disruptions, such as haemorrhage, that could cause inappropriate alterations in arterial blood pressure.

EFFERENT PATHWAYS AND EFFECTORS

The efferent pathways for the control of the cardiovascular system are the parasympathetic (vagus) and sympathetic nerves, and the hormones epinephrine, norepinephrine, ADH, renin, angiotensin, aldosterone and atrial natriuretic factor (ANF). The effectors involved in cardiovascular control are the heart, the vessels, the kidneys, and thirst and water intake.

The sympathetic nerves and the hormones epinephrine and norepinephrine have widespread effects on the heart (increased force and rate of contraction), the arteriolar resistance vessels (vasoconstriction), and the venous capacitance vessels (reduced capacitance). Vagal influence on the cardiovascular system is limited to the heart, especially the atria and the AV node (reduced rate of SA node discharge and AV node conduction). ADH from the posterior pituitary increases water reabsorption in the renal collecting ducts and produces arteriolar vasoconstriction. Sympathetic nerve activity releases renin from the granular cells of the juxtaglomerular apparatus, and activates the renin–angiotensin–aldosterone system, with conservation of water and electrolytes in the body. Angiotensin II is a potent vasoconstrictor that also increases sympathetic nervous system activity by both central and peripheral mechanisms. In addition, angiotensin II increases body water content (and therefore arterial blood pressure) by stimulating both thirst and ADH secretion. ANF from distended and stretched cardiac atrial cells increases renal sodium and water excretion.

SENSORS AND MEASURED VARIABLES

The arterial blood pressure is the primary measured variable involved in the reflex control of the cardiovascular system, and is monitored by the carotid sinus and aortic arch baroreceptors. Information is also sent to the cardiovascular centres in the brain from large veins, atria, ventricles, chemoreceptors, lungs, and receptors for pain, temperature and other sensations.

The arterial baroreceptors

The nerve fibres of the arterial baroreceptors are stimulated when the aortic and carotid walls are stretched by a rise in blood pressure. Firing of these receptors inhibits medullary sympathetic

outflow and stimulates parasympathetic nerve activity. A rise in arterial blood pressure increases the discharge rate of the carotid and aortic baroreceptors, and reflexly decreases sympathetic and increases parasympathetic activity from the medulla. A rise in arterial blood pressure therefore reflexly decreases the rate and force of cardiac contraction and dilates the resistance vessels. A fall in arterial blood pressure reduces the discharge rate of the carotid and aortic baroreceptors, and reflexly increases sympathetic and decreases parasympathetic activity from the medulla. A fall in arterial blood pressure therefore reflexly increases the rate and force of cardiac contraction and constricts the resistance vessels.

The carotid sinus is a thin-walled dilatation at the origin of the internal carotid arteries, and the aortic baroreceptors lie in the transverse arch of the aorta. Both sets of baroreceptors send afferent impulses to the NTS in the medulla: the carotid sinus via the glossopharyngeal nerve, and the aortic arch receptor via the vagus nerve. The baroreceptors are spray-like free nerve endings with many associated mitochondria and are stimulated by stretching of the vessel wall, not by pressure directly. (In experiments, a plaster cast placed around the arterial baroreceptors prevents the response to increased perfusion pressure.) The thin tunica media of the carotid sinus enhances baroreceptor stretching during systole.

Baroreceptors contain many unmyelinated C fibres and fewer larger myelinated A fibres. The threshold mean arterial blood pressure at which baroreceptors begin to fire is around 60 mmHg. C fibres tend have to have higher thresholds, and A fibres have lower thresholds and are more sensitive at lower pressures. Each individual baroreceptor neuron fires over only a narrow pressure range, but the combination of many fibres gives the carotid sinus and aortic baroreceptors a wide effective range. In an experimental isolated carotid sinus preparation, the recorded sinus nerve firing rate varies with the applied perfusion pressure. In addition, the firing rate at any given perfusion pressure is greater with pulsatile than with constant pressure. This is due to both the dynamic sensitivity of the baroreceptor fibres (increased response to changes in pressure), and the recruitment of higher threshold fibres with the high point of the pulse pressure. The baroreceptor firing

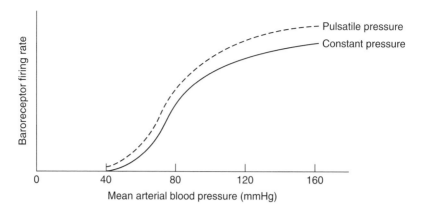

Figure 4.41   *The relationship between baroreceptor firing rate and mean arterial pressure.*

rate is therefore proportional to mean arterial and pulse pressure (Fig. 4.41).

At normal mean arterial pressure a proportion of the carotid sinus and aortic baroreceptor fibres fire and inhibit medullary tonic sympathetic output and stimulate vagal activity reflexly. Additional baroreceptors fire during systole, producing a burst of vagal discharge with more inhibition of sympathetic output. With hypotension there is reduced baroreceptor firing, the inherent sympathetic tone is freed from reflex inhibition, and there is no stimulation of parasympathetic output. With hypertension there is increased baroreceptor firing, medullary sympathetic tone is suppressed, and parasympathetic output is stimulated.

The arterial baroreceptors can be reset to higher or lower blood pressures. If the distending pressure in an isolated baroreceptor preparation is changed and then held constant, the discharge frequency increases before returning to the resting rate. The baroreceptors accommodate to the new pressure, perhaps by the opening of potassium channels that return the membrane potential to the resting value at the new distending pressure. The baroreceptors can also be reset by central mechanisms, as during exercise, and sympathetic nerve activity to the carotid sinus can increase the firing rate at a given pressure. As the response of the baroreceptors to pressure can be reset by so many influences, it is clear that they cannot be responsible for the long-term regulation of arterial blood pressure. Instead, the arterial baroreceptors regulate blood pressure in the short term, minimizing fluctuations in the face of abrupt changes in posture, cardiac output or peripheral

resistance. The long-term control of arterial blood pressure is by the balance between fluid intake and fluid excretion, which determines blood volume. Renal function is the most important long-term determinant of body fluid (and hence blood) volume and arterial blood pressure.

Cardiopulmonary receptors

There are three main groups of cardiopulmonary receptors involved in cardiovascular control: myelinated vagal veno-atrial stretch receptors; unmyelinated vagal and sympathetic cardiac mechanoreceptors; and vagal and sympathetic chemosensitive fibres. When stimulated together these cardiopulmonary receptors have an overall inhibitory effect on cardiac function, revealed by their simultaneous stimulation by veratridine injection, which causes a reflex bradycardia, vasodilatation and hypotension (Bezold–Jarisch response). However, individual groups of cardiopulmonary receptors have quite different effects on the cardiovascular system.

The veno-atrial stretch receptors are myelinated vagal fibres in the endocardium at the junction of the vena cava and the pulmonary vein with the atria. There are two types, 'A' and 'B'. Type A veno-atrial stretch receptors fire during atrial contraction, with the 'a' wave of the atrial pressure curve. The Type B veno-atrial stretch receptors fire during atrial filling (with the 'v' pressure wave), and send information to the brain about central venous pressure and cardiac distension. Stimulation of the veno-atrial stretch receptors produces a rise in heart rate, and increased urine volume and salt excretion (Bainbridge effect). The tachycardia is due to a selective increase in sympathetic nerve activity to the SA node. The increased urine output and salt excretion may be accomplished via reduced renal sympathetic nerve activity, inhibition of ADH secretion, and increased atrial muscle ANF production. The functions of the veno-atrial stretch receptors may be to regulate cardiac size when venous pressure is high, and to adjust blood volume.

Unmyelinated vagal and sympathetic cardiac mechanoreceptors form a fine network of fibres in both atria and mainly the left ventricle. There are also myelinated vagal mechanoreceptors around the coronary arteries. The left ventricular fibres fire during ventricular contraction, but only some of the atrial mechanoreceptors discharge at the height of atrial filling during inspiration. The combined effect of these atrial and ventricular mechanorecep-tors is to produce a reflex bradycardia and vasodi-latation. Ablation of afferent input from either the arterial baroreceptors or the atrial and ventricular mechanoreceptors (e.g. a heart transplant recipient) does not alter arterial blood pressure significantly, but loss of both produces sustained hypertension. Therefore, the combined input of the arterial baroreceptors and the atrial and ventricular mechanoreceptors may be important for arterial blood pressure control. Also, the mechanorecep-tors in the left ventricle may produce reflex vaso-vagal syncope when they are stimulated during orthostatic hypotension by vigorous ventricular contractions at a reduced filling volume.

Vagal and sympathetic chemosensitive fibres in the heart are stimulated by products released from ischaemic cardiac muscle cells. The sympathetic chemosensitive afferents are thought to mediate the pain of myocardial ischaemia and infarction. Convergence of these sympathetic afferent fibres with somatic pathways in the spinothalamic tracts of the spinal cord is the basis for the referred pain of myocardial ischaemia felt in the arms, neck and chest wall.

Peripheral chemoreceptors

The peripheral chemoreceptors in the carotid and aortic bodies are stimulated by hypoxia and hypercapnia. The direct cardiovascular effects are hypertension and bradycardia, but the latter is offset by the chemoreceptor stimulation of inspi-ratory neurons and lung stretch receptors, which both produce a tachycardia. The net cardiovascu-lar effect of chemoreceptor stimulation is a rise in both peripheral resistance and heart rate. The chemoreceptors are also stimulated by the stag-nant hypoxia and metabolic acidosis of a very low arterial blood pressure, and this is important for the cardiovascular response to severe hypoten-sion. The arterial baroreceptors do not fire below an arterial blood pressure of 60 mmHg, and at lower pressures the chemoreceptors drive further increases in sympathetic output (in such a situ-ation cutting the chemoreceptor afferent nerves produces a marked fall in pressure). The chemo-receptor response to low arterial blood pressure explains the tachypnoea of hypotensive shock.

Other receptors

Many other sensations have reflex cardiovascular effects: hypertension and tachycardia with somatic pain; hypotension and bradycardia with severe visceral pain; tachycardia and hypertension with bladder distension; hypertension with cold temperatures; activation of the defence response with a threatening sound or sight; the 'diving response' of bradycardia and peripheral vasocon-striction with facial nerve stimulation by cold water in some species.

EFFECTS OF THE ARTERIAL BARORECEPTOR REFLEX ON ARTERIAL BLOOD PRESSURE

The arterial baroreceptor reflex influences many factors that determine arterial blood pressure by modulating efferent sympathetic and parasympa-thetic activity. As discussed on pages 148–149, the mean arterial blood pressure is determined by the amount of blood in the arterial system at any point in time. The volume of blood in the arteries is determined by the amount of blood entering the aorta as the cardiac output, and the blood vol-ume leaving the arteries as peripheral run-off into the capillaries. The cardiac output is the product of the heart rate and the stroke volume, and the peripheral run-off is determined by the mean

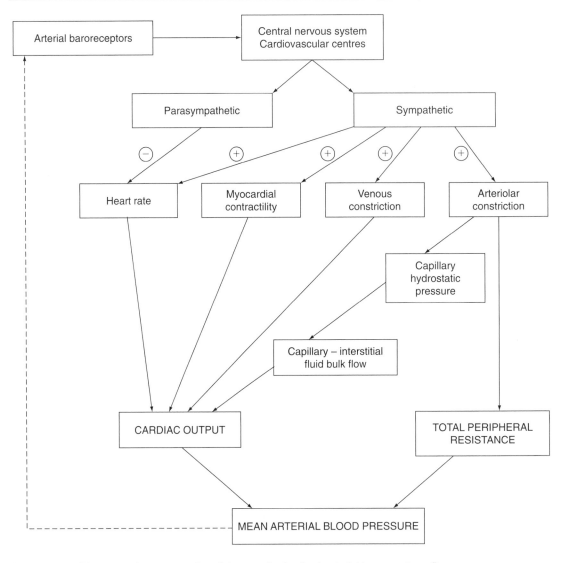

Figure 4.42   *Diagrammatic representation of the negative feedback arterial baroreceptor reflex.*

arterial blood pressure acting across the total peripheral resistance of the arterioles. The main determinants of mean arterial blood pressure are therefore the cardiac output and the total peripheral resistance, acting via changes in the arterial blood volume. The negative feedback arterial baroreceptor reflex influences many factors controlling cardiac output, peripheral resistance and arterial blood pressure (Fig. 4.42).

In addition, factors controlling longer-term body fluid balance, including renal blood flow, the renin–angiotensin–aldosterone system, ADH, and thirst and water intake, are affected by the baroreceptor reflex. The other cardiopulmonary

receptors also contribute to reflex modulation of the factors determining arterial blood pressure, and the arterial chemoreceptors stimulate sympathetic drive in severe hypotension.

CONTROL OF SPECIAL CIRCULATIONS

The resting cardiac output is 5–6 L/min, and individual organ blood flows at rest are: brain 750 mL/min; coronary 250 mL/min; kidneys 1100 mL/min; skeletal muscle 1200 mL/min; abdominal organs 1400 mL/min; skin 500 mL/min. The circulation to different organs is determined by the perfusion pressure and the resistances of their vascular beds. That is:

$$Flow = \frac{pressure}{resistance}$$

and

$$Organ\ blood\ flow = \frac{mean\ arterial\ pressure}{organ\ vascular\ resistance}$$

As all the systemic vascular beds are exposed to the same mean arterial pressure, the distribution of the cardiac output to individual organs is determined by the state of contraction or relaxation of the smooth muscle in their resistance vessels, the arterioles.

Control of arteriolar smooth muscle tone

Local factors that influence arteriolar smooth muscle tone and radius include:

- *Local myogenic control.* Myogenic control is dominant in the circulations of the brain and kidneys, and produces autoregulation of blood flow over a range of arterial blood pressures. Myogenic control is much less important in skeletal muscle and the skin.
- *Local metabolic control.* The metabolic control of arteriolar tone is especially important in the coronary, skeletal muscle and brain circulations.
- *Local tissue vasoactive chemical control.* The salivary, sweat and intestinal glands increase blood flow during activity by kallikrein release and activation of bradykinin.

Systemic factors that influence arteriolar smooth muscletone and radius include:

- *Extrinsic sympathetic nerve control.* The extrinsic sympathetic nervous control of arteriolar tone is important in the skin, kidneys and gut, but less important in the brain and coronary circulations. Skeletal muscle has a second 'sympathetic cholinergic' vasodilator supply, under cortical control.
- *Extrinsic parasympathetic nerve control.* The vessels of the external genitalia have a dual autonomic nerve supply, with parasympathetic dilator nerves as well as a sympathetic constrictor supply. The coronary, brain and pulmonary circulations each have a parasympathetic nerve supply, but the role is unclear.
- *Extrinsic hormonal control.* In the heart and skeletal muscle circulations epinephrine produces vasodilatation because of the relative preponderance of β_2 receptors. In the gut and the skin circulations, where there are relatively more α receptors, epinephrine produces vasoconstriction.

Heart (*see* page 121)

The high oxygen extraction of cardiac muscle means that coronary blood flow must increase when myocardial oxygen consumption rises. The tone in the coronary arterioles is high at rest, and the coronary circulation is controlled primarily by local metabolic factors, although some myogenic control is also present. Compression of the coronary vessels (especially to the left ventricle) during systole means that blood flow takes place mainly during diastole.

The driving pressure in the coronary circulation is the aortic pressure, but this is affected by extravascular compression of vessels during ventricular contraction. This is especially important in the left ventricle: in early systole, blood flow in the vessels is reversed, and most flow to the left ventricle takes place during diastole. This effect is less important in the right ventricle, as the pressure developed by contraction (25 mmHg) is much lower (Fig. 4.2). In diastole, there is no compression of the coronary vessels. Coronary blood flow to the left ventricle is intermittent, being maximal in diastole but stopping in early systole. In contrast, right ventricular coronary blood flow is pulsatile and is slightly higher during systole. Some 80 per cent of total coronary blood flow takes place during diastole, and the aortic diastolic pressure is therefore an important determinant of coronary perfusion. With tachycardia, the proportion of time in diastole in each cardiac cycle diminishes, but the expected effect on coronary perfusion is compensated for by a metabolic arteriolar dilatation secondary to the increased myocardial oxygen consumption. With bradycardia, the time spent in diastole increases, but the reduced myocardial oxygen consumption leads to a compensatory arteriolar constriction. The systolic compression of vessels to the left ventricle contributes to the high resistance of the coronary circulation.

At rest, myocardial oxygen delivery is 8–10 mL/min per 100 g (20 times greater than resting skeletal muscle). Even at rest the oxygen extraction ratio of cardiac muscle is high (around 75 per cent), and coronary venous blood oxygen content is only 5 mL/100 mL. The major determinants of myocardial oxygen demands are wall tension/stress (30–40 per cent), heart rate (15–25 per cent), myocardial contractility (10–15 per cent), basal metabolism (25 per cent) and external work (10–15 per cent). Oxygen requirements for electrical or activation work is about 0.7 mL/min per 100 g myocardial tissue. During heavy exercise the increased cardiac oxygen demand is met mainly by an increase in coronary blood flow (from the resting value of 80 mL/min per 100 g up to 300–400 mL/min per 100 g), although there is also a rise in coronary oxygen extraction to 90 per cent. Coronary blood flow rises in proportion to cardiac metabolic activity, through the release of local vasodilator substances from the working myocardial cells. Many factors have been proposed as being responsible for this metabolic control of arteriolar tone in the coronary circulation, including reduced oxygen tension, increased carbon dioxide tension, hydrogen ions, potassium, lactic acid, pyruvate, inorganic phosphate, interstitial fluid osmolarity, NO and adenosine. Local myogenic control is also important in the heart, and coronary blood flow is held constant between aortic diastolic pressures of 60 to 180 mmHg.

Coronary arteries are supplied by sympathetic vasoconstrictor (epinephrine transmitter, α receptor) and parasympathetic vasodilator (muscarinic) fibres, but their roles are unclear as the direct effects on the vessels are secondary to the metabolic consequences on the heart. Sympathetic nerve output to the heart raises the myocardial oxygen requirement by increasing the force and rate of contraction, and produces coronary arteriolar vasodilatation via the release of local metabolites. In this way, the direct sympathetic vasoconstrictor effect is ablated. Parasympathetic nerve activity to the heart reduces myocardial oxygen requirement by slowing the heart rate, and local metabolic control leads to vasoconstriction, despite the direct vagal vasodilatatory effect. Circulating epinephrine produces coronary vasodilatation via β_2 receptors.

Skeletal muscle

Skeletal muscle blood flow at rest is 1200 mL/min, and can rise with exercise to 20 000 mL/min. The activity of skeletal muscle determines whether extrinsic nerve or local factors predominate in blood flow control. Extrinsic sympathetic vasoconstrictor nerve control (α effect) of skeletal muscle arteriolar resistance is an integral component of the arterial baroreceptor reflex control of blood pressure. During exercise, local metabolic control of arteriolar tone predominates. Skeletal muscle also has a 'sympathetic cholinergic' vasodilator nerve supply that may be involved in the defence response and at the onset of exercise. Circulating epinephrine causes vasodilatation in skeletal muscle at low concentrations (β effect), but vasoconstriction at higher concentrations (α effect).

At rest, there is significant tone in the skeletal muscle arterioles, and a basal sympathetic nerve output contributes to this (only one-third of skeletal muscle capillaries are perfused at rest). The large bulk of body muscle means that the skeletal arterioles represent significant peripheral resistance that can be raised or lowered by the arterial baroreceptor reflex increasing or decreasing the basal sympathetic nerve activity. In severe blood loss, skeletal muscle perfusion is reduced to 20 per cent of the normal value. In contrast to the skin and gastrointestinal circulations, skeletal muscle veins have a sparse sympathetic nerve supply and cannot alter their venous capacity significantly.

Skeletal muscle vessels are compressed by muscular activity, and blood inflow is reduced and outflow increased during intermittent contractions. During exercise the skeletal muscle pump increases the venous return: contraction and relaxation of limb muscles squeeze blood from the veins towards the heart. The veins are compressed and blood is expelled during a contraction, and the veins fill again when the muscles relax (the venous valves ensure one-way flow). Skeletal muscle blood flow may be stopped for a time with strong prolonged contractions.

Splanchnic

The splanchnic circulation features two large capillary beds partially in series with one another. Blood from the capillaries of the gastrointestinal tract, spleen and pancreas perfuses the liver via the portal vein. The hepatic artery also supplies blood to the liver. Extrinsic sympathetic nerve control causes arteriolar vasoconstriction and venoconstriction, and moves large volumes of blood out of the liver into the systemic circulation. Gastrointestinal blood flow is increased after food ingestion, by the production of local metabolic factors and hormones by the gut. Local myogenic and metabolic mechanisms modulate the high hepatic arteriolar tone to compensate for changes in portal venous flow.

The arterial supply to the gastrointestinal tract is by the coeliac, and superior and inferior mesenteric arteries. In the intestinal villi the direction of blood flow in the capillaries and small veins is opposite to that in the arterioles, forming a counter-current exchange system that promotes the absorption of nutrients. However, this arrangement also allows oxygen to diffuse from the arterioles directly to the venules, promoting villous necrosis when intestinal blood flow is compromised. Sympathetic nerve activity leads to vasoconstriction (α effect) and moves blood from the gut into the systemic circulation during haemorrhage or as part of the defence reaction. Ingestion of food increases gastrointestinal blood flow by the local secretion of gastrin and cholecystokinin, and by the action of products of digestion, including glucose and fatty acids.

The liver blood flow is one-quarter of the cardiac output, and is from the portal vein and the hepatic artery. The portal vein normally accounts for three-quarters of the blood supply, but the hepatic artery provides three-quarters of the oxygen consumed by the liver. The hepatic lobule – the basic histological unit – consists of a central hepatic efferent venule with cords of hepatocytes and sinusoids converging onto the efferent venule (see Chapter 6, page 207). The acinus is the functional unit and consists of a parenchymal mass between two centrilobular veins, and is supplied by terminal branches of the hepatic artery and portal veins, which drain into the sinusoids and then into the hepatic venules. The sinusoids form a low-pressure microcirculatory system of the acinus with sphincters at the hepatic arteriole, hepatic venous sinusoid and arteriolar–portal shunts. Thus, the sinusoids act as a significant blood reservoir depending on the sphincters' tone, which is determined by sympathetic nerve activity. The mean blood pressure is 10 mmHg in the portal vein, 90 mmHg in the hepatic artery, and 5 mmHg in the hepatic veins. Blood reaches the sinusoids at a pressure of less than 10 mmHg because the hepatic arterioles have a high resting tone, controlled by local myogenic and metabolic factors and by extrinsic sympathetic nerve control. If blood flow in the portal vein falls (or liver metabolic activity increases), local metabolic control increases hepatic artery flow (up to 50 per cent of total liver blood flow). If blood flow in the portal vein rises, local myogenic control reduces hepatic artery flow. Sympathetic nerve activity constricts the sinusoidal capacitance vessels, and can move half of the liver blood volume into the systemic circulation during haemorrhage.

Kidneys (see Chapter 7, pages 227–228)

Renal blood flow amounts to 1100 mL/min, or one-fifth of the cardiac output. The main resistance vessels in the renal circulation are the afferent and efferent arterioles, and contraction of either reduces blood flow. Renal blood flow remains relatively constant over a mean arterial blood pressure range of 75 to 170 mmHg, and this is produced by altered afferent arteriolar tone in response to changes in perfusion pressure. A rise or fall in perfusion pressure leads to a corresponding rise or fall in afferent arteriolar resistance by myogenic control and by tubuloglomerular feedback. Renal blood flow is reduced by sympathetic nerve activity as part of the arterial baroreceptor response to decreased blood pressure.

Tubuloglomerular feedback involves the macula densa, which releases more adenosine if the renal perfusion pressure rises, and reduces production if the pressure falls. Adenosine constricts the afferent arterioles in the kidney. The vasodilator NO may be produced by the macula densa when renal perfusion pressure falls. The kidneys are supplied by noradrenergic sympathetic nerves that constrict the afferent and efferent arterioles and reduce blood flow. Renal sympathetic nerve activity also stimulates renin secretion from the juxtaglomerular apparatus. Angiotensin II constricts the afferent and efferent arterioles. Prostacyclin, and prostaglandins PGI_2, and PGE_2 are locally active renal vasodilators produced in clinical states associated with high circulating vasoconstrictor concentrations.

Brain (see Chapter 2, pages 50–51)

Cerebral function is dependent on continuous oxidative phosphorylation of glucose to provide ATP, and the brain is very sensitive to hypoxia because it has a high metabolic rate but no substrate stores. Within a mean arterial pressure range of 50–150 mmHg, cerebral blood flow is kept constant at 750 mL/min by local myogenic control. Distribution of blood flow within the brain is controlled by local metabolic factors including H^+, K^+, adenosine, phospholipid and glycolytic metabolites, and NO. Arterial blood carbon dioxide tension has important physiological effects on cerebral blood flow: at normal arterial blood pressure cerebral blood flow rises by 2–4 per cent for every mmHg increase in carbon dioxide tension (P_{CO_2} range of 20–80 mmHg (2.7–10.7 kPa)). CO_2 diffuses rapidly across the blood–brain barrier, increases the extracellular fluid H^+ concentration, and vasodilates the cerebral arterioles. Severe hypotension can abolish the cerebral circulatory response to Pa_{CO_2}. Cerebral blood flow is increased by low arterial blood oxygen tensions that produce tissue hypoxia and lactic acidosis. Cerebral blood flow is doubled when the Pa_{O_2} falls to 30 mmHg (4 kPa). Extrinsic nerve and hormonal control have little effect on cerebral blood flow.

The rigid cranium forms a fixed volume containing the parenchyma, cerebrospinal fluid and the blood. Within the cranial vault, changes in the volume of any one component will alter the other components of the cranial contents (Monro–Kellie hypothesis). The Cushing reflex describes the rise in arterial blood pressure that tends to maintain cerebral blood flow in the presence of a raised intracranial pressure. The mechanism of this reflex is stimulation of sympathetic neurons in the medulla by brainstem compression (the arterial baroreceptor reflex then also produces a bradycardia).

Skin

The main function of the cutaneous circulation is the maintenance of a steady body temperature, and blood flow is controlled by sympathetic vasoconstrictor output from the hypothalamus that increases with cold and decreases with heat. The normal skin blood flow is 500 mL/min, and this can rise by a factor of 30 with heat and fall by a factor of 10 with cold temperatures (or with arterial hypotension). Venous plexuses contain considerable volumes of blood (up to 1500 mL) and contribute to the skin colour. A counter-current arrangement of skin arteries and veins permits direct heat exchange between them, so that heat can be conserved or lost in cold or warm ambient temperatures, respectively.

There are two types of skin resistance vessels: the arterioles and the arteriovenous anastomoses that connect arterioles, and venules in the ears, nose, lips, fingers and toes, palms and soles. When dilated, the arteriovenous anastomoses are a low-resistance shunt pathway that increases skin blood flow and delivers more heat to the skin. With heat, the arteriovenous anastomoses dilate, blood flow increases, and skin heat loss rises; with cold, they constrict and loss diminishes. The thick smooth muscle walls of the arteriovenous anastomoses are constricted by sympathetic nerve activity, epinephrine and norepinephrine (α effect), and they have no local myogenic or metabolic control. The temperature regulation centre in the hypothalamus regulates extrinsic sympathetic vasoconstrictor activity to the arteriovenous anastomoses and to the cutaneous arterioles. In addition, the hypothalamus activates sympathetic cholinergic fibres to the sweat glands and dilates cutaneous arterioles by bradykinin release. The hypothalamus also mediates blushing and pallor of the skin with the emotions of embarrassment or fear. However, local metabolic control is also involved in the cutaneous arterioles: with prolonged cold exposure the initial (sympathetic) vasoconstriction can be overcome by local metabolic factors so that the skin vessels vasodilate.

Lungs (see Chapter 3, pages 102–103)

The pulmonary and systemic circulations are in series, and have almost identical blood flows. The pulmonary circulation operates at low pressures with very distensible low-resistance vessels distributing the venous return over the large alveolar wall surface area. The right ventricle works at lower pressures, and hence the muscle is thinner than the left ventricle. The low pressures of the pulmonary circulation also minimize the transudation of fluid into the lung interstitial spaces. As pressure in the pulmonary vessels is low, the effect of hydrostatic pressure is significant with the lung in the upright position, and perfusion pressure decreases from the base to the apex so that mismatch of perfusion with alveolar ventilation can occur.

Regional pulmonary blood flow is controlled by local metabolic factors. Blood is diverted from poorly ventilated areas of the lung by hypoxic vasoconstriction of small pulmonary arteries in the presence of a low alveolar P_{O_2}. Nitric oxide is continuously synthesized by pulmonary artery endothelium in the presence of a normal alveolar P_{O_2}. If the alveolar P_{O_2} falls below 70 mmHg, then endothelial NO synthesis is reduced, producing vasoconstriction. This is important for matching local lung perfusion to ventilation. This mechanism is important at birth when generalized hypoxic pulmonary vasoconstriction is diminished by the first breath. The role of extrinsic nerve control in the lungs is unclear as the vessels are normally maximally dilated.

INTEGRATED CARDIOVASCULAR RESPONSES

HAEMORRHAGE

Haemorrhage decreases the mean systemic filling pressure of the circulation and consequently reduces venous return such that cardiac output falls. The physiological effects of haemorrhage depend on the rate and degree of blood loss. Multiple compensatory mechanisms are activated, and these are important for modulating vascular resistance in the various tissues, bringing about a redistribution of the cardiac output. The blood flow to the brain and the myocardium is preserved as long as the compensatory processes are adequate.

Immediate responses

As a result of the decrease in blood volume during haemorrhage there is a drop in arterial blood pressure caused by a decrease in cardiac output (Fig. 4.43). The fall in arterial pressure will initiate powerful sympathetic reflexes by activating the baroreceptors and low-pressure vascular stretch

receptors in the thorax. This increased sympathetic vasoconstriction results in the following important effects:

- Constriction of arterioles in most parts of the body, producing an increased total peripheral resistance
- Constriction of the venous reservoirs, thereby maintaining venous return despite reduced blood volume
- Increased heart rate and myocardial activity

When the blood loss is less than 10 per cent, the pulse pressure is reduced but the mean arterial blood pressure may be normal because of the increase in heart rate and high systemic vascular resistance. Blood flow to tissues that are highly innervated by the sympathetic nervous system will be reduced. Venous compliance is reduced predominantly by splanchnic and cutaneous venoconstriction. Renal blood flow is autoregulated (myogenic control maintains a constant renal blood flow in the arterial blood pressure range 75 to 170 mmHg) and the decrease in renal blood flow will depend on the severity of blood loss. Sympathetic stimulation does not cause significant cerebral or coronary vasoconstriction. In addition, both these regional circulations are well autoregulated so that blood flow through the brain and heart is maintained as long as the mean arterial pressure does not fall below 70 mmHg.

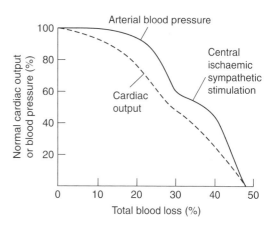

Figure 4.43 *Effect of haemorrhage on cardiac output and blood pressure.*

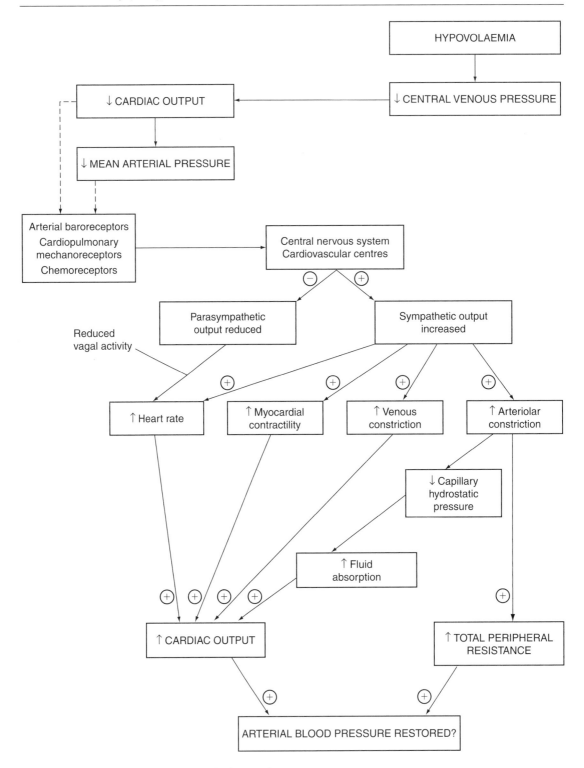

Figure 4.44 *Cardiovascular changes during haemorrhage.*

When the blood loss is greater than 20 per cent, both the arterial blood pressure and cardiac output decrease rapidly because the compensatory mechanisms become inadequate. Various mechanisms that are important for returning the blood pressure to normal may be activated. When the arterial blood pressure falls below 50 mmHg, a central nervous system ischaemic response is elicited, causing a powerful sympathetic stimulation throughout the body (Fig. 4.44). Sympathetic stimulation is maximal within 30 s after a haemorrhage. Irreversible hypotension can occur with a blood loss of greater than 30 per cent of the blood volume. Inadequate tissue perfusion leads to increased anaerobic glycolysis, with the production of large amounts of lactic acid. The resulting lactic acidosis depresses the myocardium and

reduces the peripheral vascular responses to catecholamines.

Stimulation of the sympathetic nervous system also causes precapillary vasoconstriction, and this lowers the capillary hydrostatic pressure and promotes fluid absorption from the interstitial compartment into the vascular compartment. As much as 1 L of fluid can be transferred into the vascular compartment by this mechanism. Depending on the extent of blood loss this reabsorption of fluid may take as long as 12–24 h to reach completion. A loss of blood volume causes the venous return curve to be shifted to the left because of a fall in the mean systemic filling pressure, but this is partially restored by the increased sympathetic activity (Fig. 4.45). The increased sympathetic activity also shifts the cardiac output curve

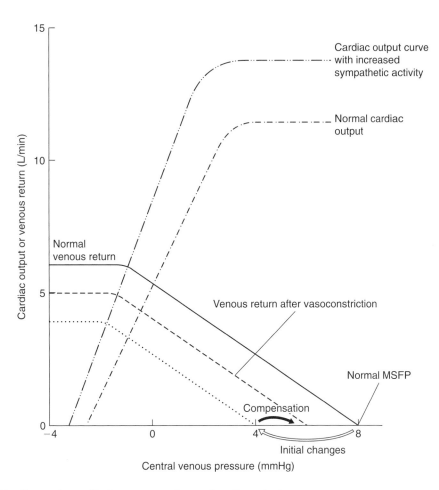

Figure 4.45 *Haemorrhage: effects on venous return and cardiac output curves.*

upwards, and consequently a new equilibrium point is established.

Hormonal responses

As a result of the decreased venous return, stretch of the right atrium is diminished. This results in a decrease in ANF release and stimulates the release of ADH. Renal vasoconstriction causes renin secretion from the macula densa, which activates the renin–angiotensin pathway, enhancing the release of aldosterone from the adrenal cortex. ADH and aldosterone promote water and sodium reabsorption in the kidneys. The angiotensin and vasopressin mechanisms take between 10 min and 1 h to respond completely, but nevertheless help to restore arterial pressure and the mean systemic filling pressure and thereby increase venous return to the heart (Fig. 4.46). Increased sympathetic activity also causes the release of cortisol and catecholamines from the adrenal gland.

Haematological responses

The reabsorption of the interstitial fluid results in a dilutional anaemia, and the plasma proteins derived from the liver are replaced in 3–6 days. Increased erythropoietin production by the kidneys

occurs, and this stimulates the bone marrow to produce red blood cells.

THE VALSALVA MANOEUVRE

Forced expiration against a closed airway is termed a Valsalva manoeuvre. Clinically, this can be performed by a person blowing into a mercury column to produce a pressure of 40 mmHg and holding it for 10–15 s. This results in a rise in the intrathoracic, intra-abdominal and cerebrospinal fluid pressures. The central venous pressure increases by about 7 mmHg for a 10 mmHg rise in the mouth pressure (Fig. 4.47).

The normal cardiovascular changes associated with the Valsalva manoeuvre may be divided into four phases:

- In Phase I, at the onset of the manoeuvre, there is a transient small rise in blood pressure with a brief fall in heart rate. This is due to the transmission of the increased intrathoracic pressure onto the aorta.

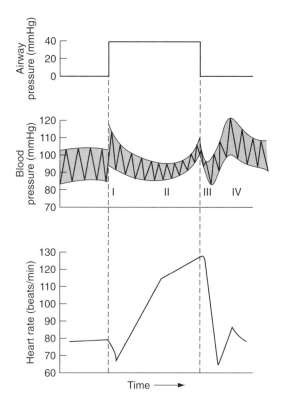

Figure 4.47 *Blood pressure and heart rate changes during the Valsalva manoeuvre.*

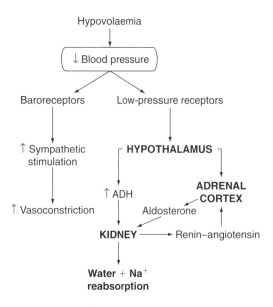

Figure 4.46 *Endocrine responses in haemorrhage.*

- In Phase II, the raised intrathoracic pressure causes a decrease in venous return to the right heart that reduces cardiac output and causes a fall in blood pressure. This fall in blood pressure stimulates the baroreceptors, and reflex compensatory mechanisms are activated. Sympathetic stimulation causes an increase in heart rate and peripheral vasoconstriction. These changes tend restore the blood pressure.
- In Phase III, immediately after the release of the positive airway pressure, there is a transient fall in blood pressure with a further rise in the heart rate. This is brought about by the loss of the transmitted raised intrathoracic pressure on the aorta.
- In Phase IV, with the intrathoracic pressure returning to baseline, venous return is restored and a normal cardiac output results. The delivery of a normal cardiac output into a constricted peripheral vascular bed causes an overshoot of the blood pressure. This rise in blood pressure is sensed by the baroreceptors, resulting in a reflex bradycardia caused by vagal activation. Peripheral vascular relaxation restores the blood pressure to normal.

Abnormal responses to the Valsalva manoeuvre may occur in patients with diminished baroreceptor reflex, e.g. with quadriplegia and diabetic autonomic neuropathy. In these patients, there is an excessive fall in blood pressure in Phase II, and an absence of overshoot and bradycardia in Phase IV (Fig. 4.48).

In congestive cardiac failure (Fig. 4.49), a square-wave response is observed. The blood pressure is elevated throughout Phase II, and there is no overshoot in Phase IV and little change in heart rate. The increased blood volume and raised peripheral venous pressure maintains venous return to the heart and the cardiac output. The raised intrathoracic pressure is transmitted on to the aorta, resulting in a raised blood pressure.

The Valsalva manoeuvre may be used clinically to assess autonomic function and also to slow supraventricular tachycardia. The autonomic function can be assessed by determining the Valsalva ratio. The Valsalva ratio is equal to the minimum heart rate (longest R–R interval) in Phase IV divided by the maximum heart rate (shortest R–R interval) in Phase IV. The Valsalva ratio is normally

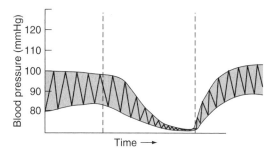

Figure 4.48   *The Valsalva response in autonomic dysfunction: excessive fall in blood pressure in Phase II and absence of overshoot and bradycardia in Phase IV.*

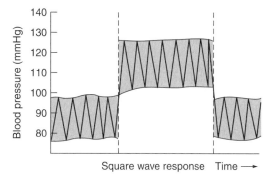

Figure 4.49   *The Valsalva manoeuvre in congestive heart failure.*

greater than 1.5, but in patients with impaired autonomic function it is less than 1.5. A supraventricular tachycardia can be terminated by the reflex increase in vagal tone in Phase IV.

EXERCISE

Exercise is associated with extensive changes in the cardiovascular and respiratory systems in order to meet the increased needs of oxygen supply, and for the removal of carbon dioxide, heat and metabolites. There are two types of exercise: static ('isometric') and dynamic ('isotonic'). Although similar muscle groups may be active in both types of exercise, there are differences in the patterns of contraction, energy consumption and blood flow.

Energy sources and production in exercise

Energy for muscle work is derived from the breakdown of ATP and creatine phosphate. The most readily available source of ATP stored within the sarcoplasm of muscle cells is small, and can supply ATP for only 1–2 s during vigorous exercise. Creatine phosphate is the next available energy source. Although the amount of creatine phosphate in the muscle is about four times that of ATP, it is still a small source of energy and can supply ATP for only a few seconds. As stores of ATP are limited, exercise lasting more than a few seconds requires an increased supply from aerobic and anaerobic metabolism. In mild exercise ATP is produced by aerobic oxidation of fatty acids and ketones, whereas in more vigorous exercise glycogen is oxidized (Fig. 4.50). At higher exercise levels

additional ATP is produced by the anaerobic conversion of glycogen to lactic acid. Brief bursts of energy are largely supplied from reserves within the muscles and produced by anaerobic processes. Anaerobic energy sources can supply a limited quantity of energy at a rapid rate. On the other hand, sustained energy can only be produced by aerobic processes in unlimited quantities at a slower rate. During vigorous exercise lasting a few seconds, aerobic metabolism contributes little to the total energy utilized. ATP derived from aerobic pathways increases to approximately 50 per cent for exercise lasting under 1 min, and exceeds 90 per cent in exercise lasting 10 min or more.

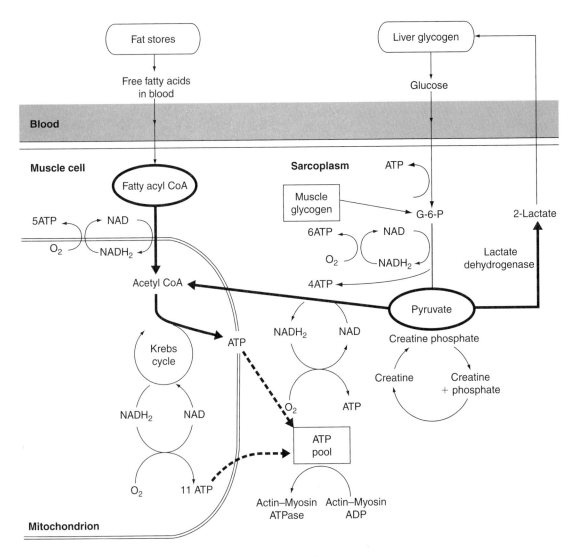

Figure 4.50 *Metabolic pathways for energy production in muscle.*

The most rapidly available supply of ATP is derived from glycolysis – an anaerobic process in which glucose is converted to pyruvate to produce two ATP molecules per molecule of glucose. Pyruvate accumulates under anaerobic conditions and is reduced to lactate. The lactic acid produced diffuses into the blood and is carried to the liver where it is converted back to pyruvate.

Under aerobic conditions, the overall yield is 16 ATP molecules per 2-carbon unit of a fatty acid, and 38 ATP per molecule of glucose or 39 ATP per glucose-6-phosphate molecule derived from glycogen. About 80 per cent of the ATP produced by aerobic metabolism is derived from fat in the liver, adipose tissue and muscle. Slow and sustained production of ATP occurs through the consumption of acetyl coenzyme A in the Krebs (citric acid) cycle. Acetyl coenzyme A is derived either from the production of pyruvate from glucose by glycolysis, or from the breakdown of fat to free fatty acids.

When fat is oxidized to provide energy substrates, glycolysis is inhibited because phosphorylase is inhibited by glucose-6-phosphate and by ATP. Energy from carbohydrate sources supplements the supply from fat as required. Intramuscular stores of glycogen can be used for a limited quantity of anaerobic energy production at short notice. Under anaerobic conditions, glycolysis is inhibited by intracellular acidosis caused by accumulation of lactate. The release of epinephrine during exercise activates lipoprotein lipase (which mobilizes free fatty acids from fat depots), and liver and muscle phosphorylase (which promotes glycogenolysis). In long-term exercise, glycogen stores are depleted and the ability of muscles to use fat becomes important.

Anaerobic energy sources are called upon when the energy demands for exercise exceed the aerobic supply. This occurs at the onset of any exercise until the oxygen supply catches up with the demand. In static exercise, the oxygen supply may be impaired whereas in severe dynamic exercise the demand may exceed the maximal oxygen supply, because there is inadequate mitochondrial capacity or insufficient oxygen. Lactate accumulates with anaerobic metabolism and can rise from a resting value of $1–1.5\,mmol/L$ to a peak of $10–15\,mmol/L$. Lactate still present in the muscle at the end of exercise may be converted to pyruvate and oxidized via the citric acid cycle. However, most of the muscle lactate diffuses into the blood. About 20 per cent of the blood lactate is utilized by cardiac muscle, while 80 per cent is synthesized to glycogen in the liver.

Muscle fatigue reduces the peak tetanic tension and velocity of contraction, and prolongs the relaxation time of the muscle. It is caused by diminished excitability of the sarcolemma and T-tubules, inhibition of Ca^{++} release from the sarcoplasmic reticulum, impaired Ca^{++} reuptake by the sarcoplasmic reticulum, impaired Ca^{++} binding to troponin, and changes to ATP hydrolysis and cross-bridge cycling in actin–myosin binding. Fatigue appears more rapidly as the intensity of exercise increases because increased anaerobic ATP production leads to lactic acid generation. The resultant intracellular acidosis changes the kinetics of many cellular processes.

Oxygen consumption in muscles

Metabolic changes in exercising muscle increase both oxygen extraction and blood flow. The oxyhaemoglobin saturation in venous blood from exercising muscles falls to less than 15 per cent in maximal exercise. The oxyhaemoglobin dissociation curve is displaced to the right by rises of CO_2, H^+ and temperature.

The muscle oxygen uptake reaches a steady state only after several minutes if exercise commences abruptly, followed by a constant work output. The lower oxygen consumption at the beginning of an abrupt exercise compared with the muscle oxygen demands is due to anaerobic metabolism that causes a partial depletion of ATP and creatine phosphate, and accumulation of lactate. This difference between oxygen demands and oxygen consumption of the muscle is known as 'oxygen deficit'. When exercise ceases, a period of increased oxygen consumption called 'oxygen debt'. provides oxygen used to 'repay' the oxygen deficit (Fig. 4.51). Following vigorous or prolonged exercise, the oxygen debt may be greater than the oxygen deficit because glycogen synthesis from lactate in the liver requires more energy. This increased oxygen consumption is due to the higher metabolic rate caused by an increased body temperature and elevated circulating catecholamine

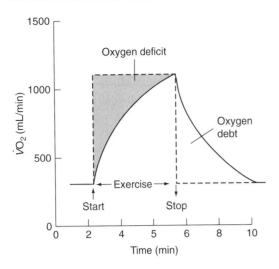

Figure 4.51 *Oxygen consumption during moderate exercise.*

and thyroxine levels, and energy required to restore intracellular electrolytes to normal.

Muscles convert about 20 per cent of the energy of ATP to external work, and the remaining energy appears as heat. At rest the basal metabolic processes in the muscle produce resting heat. Heat produced during muscle contraction in excess of resting heat is called initial heat, and this is made up of activation heat (heat produced by the muscle whenever it is contracting) and shortening heat (heat produced that is proportionate to the distance the muscle shortens). Shortening heat is due to structural changes in the muscle. After contraction, heat production in excess of resting heat continues for about 30 min. This is called recovery heat and is caused by metabolic processes that restore the muscle to its resting state. During isotonic contraction of muscle, extra heat in addition to recovery heat is produced to restore the muscle length to its previous length. This is called relaxation.

Muscle blood flow during exercise

In resting muscles, sympathetic nervous activity maintains blood flow at 20–30 mL/min per kg by constricting the arterioles. At rest the precapillary sphincters are closed and most of the blood flow in the microcirculation of the muscle is in the main channels. During exercise the partial pressure of oxygen falls, whereas the partial pressure of carbon dioxide, temperature, and the concentrations of H^+, K^+ and ADP in interstitial fluid rise. These relax the precapillary sphincters. Total muscle blood flow increases to a maximum of 500 mL/min per kg with the opening of closed capillaries. The blood flow in muscles increases 20-fold and the number of patent capillaries five-fold. The diffusion of oxygen into the muscle is more rapid and the total oxygen uptake by the muscle can increase 40-fold. During muscle contraction, intramuscular pressure rises and this can impede muscle blood flow. In static muscle contraction, muscle blood flow can be significantly reduced but in rhythmic dynamic muscle contraction, blood flow occurs during relaxation between muscle contractions. Skeletal muscle constitutes 40 per cent of the total lean body mass and receives less than 20 per cent of the cardiac output at rest, but at maximal muscle contraction it can increase to 80–90 per cent of the total cardiac output.

Blood flow to other organs during exercise

Coronary blood flow increases to meet the extra oxygen consumption of increased cardiac work, mediated by local control mechanisms, although circulating catecholamines may contribute to coronary vasodilatation via activation of β_2 receptors. However, sympathetic activity reduces blood flow to the gastrointestinal tract and the kidney, redistributing blood to the exercising muscles. Skin blood flow increases to dissipate heat produced during exercise. Blood flow to the brain remains constant at all levels of exercise.

Cardiac output

During exercise, cardiac output can increase five-fold as a result of venoconstriction, vasodilatation, increase in venous return by the 'muscle pump' mechanism, and increased myocardial contractility and heart rate (Table 4.2). When the limb muscles contract, the deep veins are intermittently compressed and venous return of blood to the

Table 4.2   *Changes in the cardiac output during exercise*

	Resting (mL/min)	Exercise		
		Light (mL/min)	Medium (mL/min)	Heavy (mL/min)
Cerebral	750 (13.0)	750 (8.0)	750 (4.0)	750 (3.0)
Coronary	250 (4.5)	350 (3.7)	650 (4.2)	1000 (4.0)
Renal	1100 (19.0)	900 (9.5)	600 (3.9)	250 (1.0)
Splanchnic	1400 (24.0)	1100 (11.6)	600 (3.9)	300 (1.2)
Skeletal muscle	1200 (20.5)	4500 (47.0)	10 800 (70.0)	22 000 (88.0)
Total cardiac output (mL/min)	5800	9500	15 500	25 000

Values in parentheses are percentages of total cardiac output.

heart is enhanced, provided that the venous valves are competent. The venous return to the heart is further increased by the 'thoracic pump' mechanism, an effect produced by changes in differential pressure between the abdomen and thorax during breathing. During inspiration, the intrathoracic pressure falls whilst the intra-abdominal pressure rises, compressing the abdominal veins and enhancing venous return to the heart. The increased depth and frequency of breathing in exercise further enhance this effect. Vasoconstriction of the splanchnic and renal circulations diverts blood to the muscles. In addition, generalized venoconstriction caused by sympathetic stimulation reduces venous capacitance to increase venous return. Arteriolar dilatation produced by local metabolites in the muscle causes a fall in total peripheral vascular resistance, and this sustains the increase in cardiac output. The resulting increase in muscle blood flow also produces a greater venous return to enhance the cardiac output.

The heart rate increases in a linear manner with the severity of the exercise, towards a maximal heart rate of 200 beats per minute in a young adult (Fig. 4.52). The progressive rise in heart rate is due to a decrease in vagal activity initially and, later, to an increasing sympathetic drive.

There is also a non-linear increase in stroke volume during exercise. The increase in stroke volume occurs mainly in light to moderate exercise, with only a small further rise in maximal exercise. Most of this increase in stroke volume is from an increase in end-diastolic volume (due to increased venous return to the heart) and from a decrease in

end-systolic volume (caused by increased emptying due to increased sympathetic activity).

Systolic arterial blood pressure can rise to 190–225 mmHg during exercise. Diastolic blood pressure shows only a small rise, and in some subjects may fall slightly. Consequently, the pulse pressure can increase two- to three-fold.

The baroreceptor reflex set point at which arterial pressure is regulated is reset to higher levels in severe exercise. The sympathetic nervous system is activated by commands from the motor cortex and sensory nerves detecting movement and metabolites in the active muscles.

There is a sudden increase in cardiac output at the start of exercise, followed by a gradual rise to a steady state. When exercise is stopped there is an abrupt decrease in cardiac output followed by an exponential fall. The abrupt changes at the onset and the cessation of exercise are caused by the effect of the muscle pump resulting in an increased venous return, as well as motor cortical activity and sensory nerve activity associated with movement. The slower changes in cardiac output reflect the time course of vasodilatation in the muscles and stimulation of the cardiovascular system.

Respiratory responses in exercise

The tidal volume and the frequency of breathing increase in proportion to the increased demand for oxygen and excretion of carbon dioxide – rapid and gradual changes similar to cardiac output. The sudden increase in ventilation at the onset, and

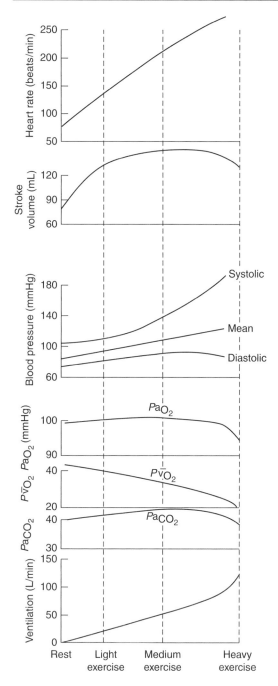

Figure 4.52 *Cardiorespiratory changes during exercise.*

exercise is thought to be by increased sensitivity of the carotid bodies. At higher levels of exercise, there is a disproportionate increase in minute ventilation and removal of carbon dioxide. This is caused by lactic acidosis causing additional stimulation of the carotid bodies. At very intense levels of exercise, oxygen consumption reaches a plateau, the 'maximal oxygen consumption'.

The two useful physiological indices of exercise capacity are maximum oxygen consumption and the anaerobic threshold. The maximum oxygen consumption is determined by exercising a subject gradually to a maximum over 15 min. The anaerobic threshold – the workload above which blood lactic acid levels rise rapidly – is a better index because it represents the highest level of exercise that can be performed without fatigue occurring from acid accumulation in the muscles. It is the point when muscle metabolism shifts from aerobic to anaerobic metabolism. The anaerobic threshold is approximately 60 per cent of the maximal exercise level.

Cardiorespiratory control

Initially, ventilation rises in close proportion to oxygen consumption and carbon dioxide production and Pa_{O_2} and Pa_{CO_2} are close to normal (*see* Fig. 4.52). Towards maximal exercise intensity, ventilation rises faster than oxygen consumption, and Pa_{CO_2} falls. Within the first 5–10 s of starting exercise, heart rate rises by 10–15 beats per minute due to a slight decrease in vagal tone. In moderate exercise, after the initial abrupt rise, there is a progressive heart rate rise over 5–10 min to a steady level. At the end of exercise, both heart rate and ventilation fall sharply at first, and then more slowly. During exercise, the baroreceptor reflex is reset to operate over a higher range of blood pressures: heart rate, cardiac output and blood pressure rise. In moderate exercise this compensates for the fall in peripheral vascular resistance caused by metabolic vasodilatation in the exercising muscles. In more severe exercise there is an increase in sympathetic activity. The respiratory chemoreceptor reflexes appear to be reset during exercise, and there is a greater response to changes of Pa_{O_2}. In severe exercise, the rise in blood lactate levels is an additional stimulus.

In moderate exercise, the core temperature rises by about 1°C. Skin blood flow increases, and

abrupt decrease at the end, are from neural input to the inspiratory centre from the motor cortex and also from proprioceptors in the exercising limbs. The sustained increase in ventilation during this enhances heat loss from the peripheral tissues. Sweating and increased heat loss by evaporation from the respiratory tract also occur. In sustained heavy exercise, the core temperature rises progressively with more intense visceral vasoconstriction.

During heavy exercise about 15 per cent of the plasma volume can shift into the interstitial space of the muscle as a result of the increased hydrostatic pressure in the vasodilated capillaries and accumulation of osmotically active substances in the interstitium. There is a loss of water and sodium in sweat, and a loss of K^+ to the interstitium.

Training

Anaerobic training for short-term power leads to an increased blood pressure during anaerobic training, and this can produce left ventricular hypertrophy. Aerobic or endurance training can increase the maximal oxygen consumption as a result of an increase in cardiac output is brought about by an increased stroke volume, as a result of an increase in the ventricular volume and in the circulating blood volume. With training, cardiac output at rest is normal, but the heart rate falls because of the increased stroke volume.

Isometric and dynamic exercise

Isometric exercise causes sustained compression of the blood vessels in the muscles. This reduces blood flow through the muscles, and total peripheral vascular resistance may rise. There is an increase in sympathetic activity, leading to rises in arterial blood pressure, heart rate and cardiac output. With isotonic or dynamic exercise there is peripheral vasodilatation, resulting in a fall in diastolic blood pressure, and a greater increase in heart rate.

Reflections

1. The circulation is organized so that the right heart pumps blood through the lungs (pulmonary circulation) and the left heart pumps blood to the rest of the body (the systemic circulation). The two circulations are in series. The principal types of blood vessels are the arteries, the arterioles, the capillaries, the venules and the veins. The larger arteries have a high proportion of elastic tissue in their walls. The greatest resistance to blood flow, and hence the greatest pressure drop, in the arterial system occurs at the level of the arterioles. Pulsatile flow and pressure is dampened by the elasticity of the arteriolar walls and the frictional resistance of the small arteries and arterioles so that capillary blood flow is largely non-pulsatile. Velocity of blood flow is inversely related to the cross-sectional area of the blood vessel. About 65 per cent of the blood volume is located in the venous system which forms a capacitance system.
2. The heart has an inherent rhythmicity that is independent of any extrinsic nerve supply. Excitation is initiated by the sinoatrial node.

The slow action potential from normal sinoatrial node is characterized by a less negative resting potential, a smaller amplitude, a less steep upstroke and a shorter plateau. The upstroke in the slow-response fibres is mediated by activation of L-type Ca^{++} channels. Slow-depolarization during Phase 4 is a hallmark of the automaticity of the sinoatrial node. Slow-response fibres are absolute refractory at the beginning of the upstroke, and partial excitability occurs very late in Phase 3.

Fast-response action potentials occur in atrial and ventricular myocardial muscle fibres. The action potential is characterized by a steep upstroke, a large amplitude and a relatively long plateau. There are five phases in the fast-response action potentials: Phase 0 – a rapid depolarization phase due to activation of fast Na^+ channels; Phase 1 – a notch due to early re-polarization due to efflux of potassium via transmembrane channels which conduct the transient outward current, i_{to}; Phase 2 – the plateau mediated by an influx of Ca^{++} via the L-type channels, and to a lesser extent the efflux of potassium through several types of K^+ channels; Phase 3 – final repolarization initiated when the efflux of K^+ exceeds Ca^{++} influx, with the resultant partial repolarization rapidly increasing K^+ conductance and restoring full repolarization; and Phase 4 – the resting potential of the fully repolarized cell determined by K^+ conductance via i_{K1} channels. The absolute refractory period of the fast-response fibres begins at the upstroke of the action potential and persists until midway through Phase 3. The fibre is relatively refractory during the remainder of Phase 3 and regains full excitability at Phase 4 when it is fully repolarized.

3. The electrocardiogram records small potential differences (about 1 mV) arising from sequential electrical depolarization and repolarization of the heart by placing electrodes at different points on the body surface, and measuring voltage differences between them. The P wave of the ECG is due to atrial depolarization, the QRS complex to ventricular depolarization, and the T wave to ventricular repolarization. Atrial repolarization is hidden within the QRS complex. The PR interval is due to the delay in transmission through the AV node.

4. Repeated alternating contraction and relaxation of the chambers of the heart enables the heart to pump blood from the venous to the arterial circulations and is called the cardiac cycle. Ventricular contraction begins at the peak of the R wave of the ECG. The atrioventricular valves close to give rise to the first heart sound. Isovolumetric contraction precedes the systolic ejection phase. At the end of systole, ventricular relaxation begins, ventricular pressure falls below that in the aorta and pulmonary artery, and the semilunar valves close to give rise to the second sound. Isometric relaxation of the ventricle precedes a rapid ventricular filling following opening of the AV valves. Atrial systole adds the final 20 per cent to the ventricular volume.

The myocardium functions as a syncytium with a all-or-none response to excitation. Cell to cell conduction occurs via gap junctions that connect adjacent cells. On excitation, voltage-gated channels open and extracellular Ca^{++} enters the cell. The Ca^{++} influx triggers the release of Ca^{++} from the sarcoplasmic reticulum and the elevated intracellular Ca^{++} elicits contraction of the myofilaments. Relaxation occurs when the cytosolic Ca^{++} is pumped back into the sarcoplasmic reticulum and exchanged for extracellular Na^+ across the sarcolemma. Velocity and force of contraction are functions of intracellular free Ca^{++}. Force and velocity are inversely related so that velocity is maximal when there is no load. An increase in diastolic fibre length increases the force of ventricular contraction and this relationship is known as the Frank–Starling relationship or Starling's law of the heart. The preload of the ventricle is the stretch of the myocardial fibres during ventricular filling (end-diastolic volume). The factors that control venous return include the mean systemic filling pressure, the right atrial pressure, the total blood volume, the venous tone, the inotropic state of the heart, the muscle pump, the respiratory pump, the drawing of blood into the atria during ventricular systole due to the descent of the AV ring and the suction of the ventricles as they relax during diastole. The afterload is the arterial pressure against which the ventricle ejects the blood. Contractility is an expression of cardiac performance at a given preload and afterload.

5. Cardiac output, the product of heart rate and stroke volume, is the volume of blood pumped each minute by the ventricle and varies according to the metabolic demands of the body. The amount of blood filling the ventricles at the end of diastole is called the end-diastolic volume, and about two thirds of this is ejected during systole – the stroke volume (70 mL). Cardiac function is regulated by a number of intrinsic and extrinsic mechanisms. Heart rate is regulated by the autonomic nervous system; vagal effects are dominant. Other reflexes that regulate heart rate include the baroreceptor, chemoreceptor, atrial receptor (Bainbridge) and ventricular reflexes. The principal intrinsic mechanisms that regulate myocardial contraction are the Frank–Starling mechanism and rate-induced regulation. The Frank–Starling mechanism states that the energy of myocardial

contraction is proportional to the initial fibre length of the muscle fibres. The resting myocardial fibre length influences subsequent contraction by altering the affinity of the myofilaments for calcium and by altering the number of interacting cross bridges between the thick and thin filaments. Rate-induced regulation refers to a process by which a sustained increase in the frequency of contraction increases the strength of contraction by enhancing the rate of influx of Ca^{++} into the cell. The autonomic nervous system regulates myocardial performance via changes in Ca^{++} conductance of the cell membrane mediated by adenyl cyclase. Hormones such as epinephrine, adrenocortical steroids, thyroid hormones, insulin, and anterior pituitary hormones regulate myocardial performance. Changes in the arterial blood concentrations of O_2, CO_2, and H^+ can directly or indirectly (via chemoreceptors) alter cardiac function.

6. The arterial pressure is determined by both the cardiac output and the total peripheral resistance which is determined by the total cross-sectional area offered by the arterioles to blood flow. Arteriolar resistance is determined by the calibre of the vessel, its length and the viscosity of blood. The mean arterial blood pressure is a time-weighted average plus one third of the pulse pressure. The contour of the systemic arterial blood pressure is distorted as it travels from the aorta to the periphery. The high-frequency components of the wave during ventricular systole are damped, the systolic pressure is elevated, and the dicrotic notch appears later in the early diastolic component of the wave. Arterial blood pressure is closely regulated by autonomic nerves, hormones and changes in blood volume on a long-term basis. A rise in blood pressure results in increased firing of baroreceptor afferents, which leads to reflex slowing of the heart, peripheral vasodilatation, and a fall in blood pressure. Long-term regulation is achieved by maintenance of normal extracellular volume via the renin–angiotensin–aldosterone system and atrial natriuretic peptide (ANP).

7. Capillaries offer little resistance to blood flow. Blood flow in the capillaries is steady. The chief determinant of capillary blood flow is the calibre of the arterioles supplying the capillary bed. The capillary pressure is about 32 mmHg at the arteriolar end and declines to 12 mmHg at the venous end.

8. Veins are capacitance vessels that contain about two-thirds of the total blood volume. The pressure in the venules is about 10 mmHg and falls to nearly zero in the right atrium. Gravity, breathing, and the pumping action of the skeletal muscles influence venous return and central venous pressure.

9. The cardiac function curve expresses the Frank–Starling mechanism; the cardiac output varies directly with the preload over a wide range of central venous pressures. The vascular function curve describes the inverse relationship between cardiac output and central venous pressure; a rise in cardiac output decreases central venous pressure. The principal factors that govern the vascular function curve are the arterial and venous compliances, the peripheral vascular resistance, and the total blood volume. The intersection of the cardiac and vascular function curves represents the equilibrium conditions that operate in an individual.

10. The diameter of a blood vessel is determined by the degree of contraction of the smooth muscle in its wall. Intrinsic and extrinsic mechanisms may superimpose on the resting tone of a vessel. Intrinsic factors include myogenic contraction in response to stretch of the vessel wall, dilatation in response to tissue metabolites, and local vasoactive substances. The myogenic response contributes to the resting tone and autoregulation. Nitric oxide is a potent vasodilator released from endothelial cells in response to shear stress, acetylcholine, and bradykinin. Nerves and hormones exert extrinsic influences on the blood vessels. Sympathetic vasoconstrictor fibres are the most widespread and important nerves that alter the calibre of blood vessels. Some arterioles such as those of the muscle, liver and heart also possess β receptors. Hormonal mechanisms such as epinephrine, ADH, ANP and the renin–angiotensin–aldosterone system provide

extrinsic control of plasma volume and vascular tone.

11. The microcirculation blood flow is regulated by contraction of the arterioles. The capillaries have thin walls consisting of a single layer of endothelial cells and provide a large surface area for exchange of solutes between blood and the tissues. Most exchange between blood and tissues occurs by diffusion. Oxygen and carbon dioxide diffuse through the endothelial cells and equilibrate rapidly by transcellular exchange. Water-soluble substances diffuse through gaps in the capillary wall (paracellular exchange). The capillary wall is almost impermeable to proteins. Capillary filtration and absorption are described by the Starling equation. The bulk flow of fluid between plasma and the interstitial fluid is determined by the net filtration pressure. The rate of fluid movement depends on the permeability of the capillary wall. In general filtration exceeds absorption and excess interstitial fluid is returned to the circulation via the lymphatics.

5

Gastrointestinal physiology

LEARNING OBJECTIVES

After studying this chapter the reader should be able to:

1. Describe the nervous and hormonal regulatory mechanisms operating within the gut
2. Describe the physiology of swallowing, salivary secretion and the functions of saliva
3. Describe the physiology of the lower oesophageal sphincter
4. Outline the physiology of gastric emptying and the factors that influence gastric emptying time
5. Describe gastric secretion, motility and its control by hormones and neural mechanisms
6. Outline the functions of the small intestine: its secretions, motility and absorption of nutrients
7. Describe the exocrine function of the pancreas
8. Outline the function of the gall bladder and biliary secretions
9. Describe the digestion of food and absorption of nutrients
10. Describe the role of the large intestine in the absorption of water and electrolytes and the importance of intestinal flora

ORAL CAVITY

The oral cavity forms the first part of the alimentary system, with the lips forming its entrance. The oral cavity is lined by stratified squamous epithelium, and numerous salivary glands open onto the mucosal surface.

The oral cavity is functionally responsible for mastication, which is the initial phase of digestion. Food is broken into smaller particles and mixed with saliva so as to soften and form a bolus for swallowing. The tongue and the cheek muscles keep the bolus between the masticatory surfaces of the teeth. The process of mastication is controlled by a reflex involving the trigeminal mesencephalic nucleus. The presence of a bolus of food in the mouth reflexly inhibits the masticatory muscles, which allows the lower jaw to drop suddenly. This in turn initiates a stretch reflex of the jaw muscles and a rebound contraction which raises the jaw to compress the teeth. Usually, mastication occurs subconsciously, but it may be subject to cortical control.

Salivary glands

The three major paired salivary glands (parotid, sublingual and submandibular) and numerous small glands in the mucosa of the oral cavity secrete saliva into the mouth. Each salivary gland consists of acini that open into intercalated and striated ducts that empty into the excretory ducts. There are three types of acini: serous, mucous (which produce mucin), and mixed. The parotid gland is entirely serous, and the sublingual and the small glands are entirely mucus-secreting. The submandibular gland, although mixed in nature, is predominantly serous.

Each salivary gland is innervated by both parasympathetic and sympathetic fibres. Preganglionic parasympathetic fibres to the parotid gland first synapse in the otic ganglion. The preganglionic parasympathetic fibres to the submandibular and sublingual glands synapse at the submandibular ganglion, and postganglionic fibres are distributed along the lingual nerve.

Saliva

Approximately 0.5–1 L of saliva is secreted daily. The secretion contains electrolytes, water and protein; the electrolyte composition and tonicity of saliva vary with secretory flow rates. Generally, saliva is slightly hypotonic and contains higher K^+ (15 mmol/L) and HCO_3^- (50 mmol/L) and lower Na^+ (50 mmol/L) and Cl^- (15 mmol/L) concentrations compared with plasma. The pH of saliva varies with salivary flow and is between 6.2 and 7.4. Proteins are found in low concentration in saliva, and include an α-amylase enzyme (ptyalin) and mucin, a glycoprotein. Antigens of AO and Lewis blood groups can be found in saliva.

Saliva is important in lubricating food to enable swallowing and also moistens the mouth to aid speech. It is important in dissolving food to stimulate taste, and it contains bactericidal substances, e.g. thiocyanate and lysozyme, which maintain oral hygiene. The high bicarbonate content of saliva buffers sudden pH changes. Ptyalin breaks down starch to maltose.

The secretory activity of the salivary glands is controlled by the autonomic nervous system. Parasympathetic stimulation causes the release of large volumes of serous juice from the parotid gland and a mixture of serous and mucous secretions by the submandibular glands, mediated by acetylcholine.

Sympathetic stimulation of the salivary glands produces saliva with a high concentration of mucus. The secretion of saliva is stimulated by the thought, smell or taste of food, or by the presence of food within the alimentary system. Thus, salivary secretions are controlled by cephalic, oral and intestinal factors. In the awake state, the basal rate of salivary production is about 0.5 mL/min, but this may increase to 5 mL/min with intense stimulation.

PHARYNX AND OESOPHAGUS

The pharynx is an incomplete tube enclosed by three constrictor muscles. The cricopharyngeus muscle is an important component of the inferior constrictor and acts as an upper oesophageal sphincter that keeps the oesophageal inlet closed, except during swallowing. The cricopharyngeus muscle is supplied by the recurrent and external laryngeal nerves. The other striated muscles are supplied by the pharyngeal branch of the vagus nerve, and the mucosa by the glossopharyngeal nerve.

The oesophagus is a muscular tube which connects the pharynx to the stomach. The muscular coat of the upper third of the oesophagus, which consists of an outer longitudinal and an inner circular layer of striated muscle, contracts rapidly so that the bolus of food passes down the oesophagus. The muscles of the lower two-thirds of the oesophagus are smooth muscles with intrinsic peristaltic activity. The mucous membrane of the oesophagus comprises stratified squamous epithelium, with mucous glands opening into the oesophageal lumen.

There are two oesophageal sphincters:

- The upper cricopharyngeal sphincter is an anatomical sphincter consisting of the cricopharyngeus and the circular smooth muscle. It has a high resting intraluminal pressure (50–100 mmHg or 6.7–13.3 kPa), and opens on swallowing.
- The lower or gastro-oesophageal sphincter is a physiological sphincter formed by the lowest 2–4 cm segment of the oesophagus. It maintains tonic contraction of the circular muscle fibres, with a resting pressure 15–25 mmHg (2–3.3 kPa) above the gastric pressure.

Swallowing or deglutition

Swallowing is a complex reflex, controlled by the swallowing centre in the medulla oblongata, which transfers food from the oral cavity to the stomach. Swallowing occurs in three phases (Fig. 5.1): an initial voluntary oral phase, followed by involuntary pharyngeal and oesophageal phases

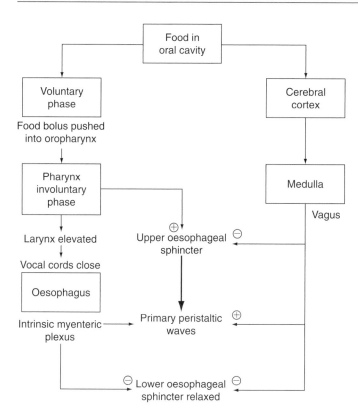

Figure 5.1 *Regulation of swallowing.*

coordinated by the swallowing centre in the medulla and pons.

ORAL PHASE

The tongue forms a bolus of food, which it pushes into the oropharynx by pushing up and against the hard palate. The stimulation of receptors in the posterior wall and soft palate results in activation of the swallowing reflex which is coordinated within a swallowing centre in the medulla oblongata and produces the involuntary movements of the next two phases of swallowing.

PHARYNGEAL PHASE

In this stage, respiration is inhibited for 1–2 s and food passes into the upper oesophagus. The nasopharynx is closed by the soft palate, and the laryngeal inlet is closed by the adduction of the vocal cords and the aryepiglottic muscle. The larynx is raised and the epiglottis swings down to close the larynx. The bolus of food is pushed into the oesophagus by contraction of the pharynx and opening of the upper oesophageal sphincter.

OESOPHAGEAL PHASE

After reaching the upper oesophagus, food is propelled into the stomach by peristaltic contractions. As soon as food enters the oesophagus, the upper oesophageal sphincter contracts and the lower oesophageal sphincter relaxes. The primary, slow, peristalic oesophageal waves with a velocity of 2–4 cm/s are initiated by the swallowing centre via the vagus with pressures between 20 and 60 mmHg. Gravity promotes the flow of fluids at a more rapid rate compared with solids. The presence of food within the oesophagus initiates the secondary peristaltic waves mediated by the enteric nervous system of the oesophagus. Stretch receptors within the wall of the oesophagus are stimulated by distension and activate the intrinsic nervous system.

Lower oesophageal sphincter

The lower oesophageal sphincter (LOS), formed by the lowest 2–4 cm segment of the oesophagus, acts as a physiological sphincter by the tonic contraction of the circular muscle fibres. The sphincteric action is enhanced by the oblique gastro-oesophageal angle, forming a flap–valve mechanism, and the crura of the diaphragm producing a pinch-cock mechanism.

The resting pressure of the LOS is 15–25 mmHg (2–3.3 kPa) above gastric pressure which prevents gastro-oesophageal reflux. The sphincter relaxes 1–2 s after swallowing is initiated and remains relaxed for 8–9 s and then contracts. This relaxation is mediated by nitric oxide and vasoactive intestinal peptide. After the passage of the food bolus, the LOS actively contracts to 1–15 mmHg (0–2 kPa) above the resting tone for 10–15 s before returning to its resting level. The resting tone of the LOS is primarily due to intrinsic or myogenic activity of the muscle.

The precise mechanism of the LOS tone has not been defined, but there is evidence to suggest that the relaxation–contraction cycle of the LOS is controlled by both the medullary centres and afferent stimuli within the oesophagus. A rise in intragastric pressure causes an increased LOS tone, which is abolished by bilateral vagotomy, suggesting a vagal mediated or local reflex. Hormones also modify LOS tone; gastrin, motilin and α-adrenergic stimulation increase the tone, whereas secretin, glucagon, vasoactive intestinal peptide (VIP) and gastric inhibitory peptide (GIP) decrease it. Drugs that increase LOS include metoclopramide, anticholinesterases, α-adrenergic agents, histamine and suxamethonium. Drugs such as the antimuscarinic agents (atropine), dopamine, ethanol, opioids, ganglion blockers and β-adrenergic agents relax the LOS. LOS pressure is usually measured by intraluminal manometry, using continuously infused catheter pressure probes that are withdrawn from the stomach in small increments, with pressure recordings made until all the transducers are in the oesophagus.

STOMACH

The stomach serves three basic functions: (1) storage of food, (2) mixing and digestion of food, and (3) emptying of food. The stomach has three smooth muscle layers: the longitudinal, circular and oblique layers. Each muscle layer of the stomach forms a functional syncytium acting as a unit. Functionally, the stomach can be divided into two

regions: the proximal stomach, comprised of the fundus and the body (or corpus); and the distal stomach or antrum. The junction between the stomach and the duodenum is formed by a thickened circular smooth muscle called the pyloric sphincter.

The extrinsic nerve supply to the stomach is from the sympathetic (via the coeliac plexus) and the parasympathetic (via the vagus) nervous systems. The sympathetic innervation inhibits motility, whereas parasympathetic activity stimulates motility.

The intrinsic nervous innervation is formed by the submucosal (Meissner's) plexus and the myenteric (Auerbach's) plexus located between the circular and longitudinal muscles of the stomach. The intrinsic innervation of the stomach is directly responsible for peristalsis and other contractions.

Receptive relaxation and accommodation

When food enters the stomach, the proximal stomach relaxes. This is a consequence of both receptive relaxation, a vago-vagal reflex initiated by the passage of food along the oesophagus, and adaptive relaxation induced by the presence of food in the proximal stomach. Both these processes involve the activation of postganglionic non-adrenergic non-cholinergic myenteric inhibitory neurons. Vasoactive intestinal peptide and nitric oxide are thought to be released from the postganglionic fibres of the enteric nervous system.

Peristalsis

Each muscle layer of the stomach forms a syncytium and acts as a unit. In the fundus – where the muscle layers are relatively thin – the strength of contraction is weak, whereas in the antrum contractions may be strong as the muscle layers are thick.

Peristaltic contractions are initiated near the border of the fundus and the body and progress to the pylorus, thus producing a peristaltic wave that propels food towards the pylorus. The peristaltic contractions are produced by changes in the membrane potential, called slow waves. The gastric slow waves, arising from pacemaker cells within the longitudinal muscle, consist of depolarizing waves (from −60 mV to −30 mV) and originate from specialized cells within the muscle layers called the interstitial cells of Cajal. An area along the greater curvature of the proximal stomach generates the greatest frequency (about 3 per minute) and acts as the dominant pacemaker which drives the more distal areas at this rhythm.

The contractile activity of the stomach is regulated by myogenic, neural and hormonal mechanisms. Vagal stimulation increases the frequency of the slow waves. Gastrin and acetylcholine increase the force of peristaltic contractions by increasing the amplitude of the slow wave plateau potential via activation of second messengers which release Ca^{++} from the sarcoplasmic reticulum.

When food reaches the pylorus, a forceful contraction of the antrum causes a retrograde movement of food through the antral ring, a process called retropulsion. The back-and-forth movement of food produced by peristalsis and retropulsion breaks the food into smaller fragments and enhances its mixing with gastric secretions.

Gastric emptying

Gastric emptying is brought about by a coordinated emptying of chyme in the stomach into the duodenum and is determined by stomach contents and its motility. The rate of gastric emptying depends on the pressure generated by the antrum against pyloric resistance. Liquids empty much faster than solids.

The rate of gastric emptying is chiefly influenced by the regulation of antral pump activity, as pyloric resistance is normally of minor importance. The antral activity is influenced by gastric volume, gastrin and the composition/volume of chyme entering the duodenum.

Increased gastric volume produces distension which provokes vago-vagal excitatory reflexes leading to increased antral pump activity and hence gastric emptying (Fig. 5.2). Both increased antral distension and high protein content of food stimulate gastrin secretion which enhances gastric emptying.

The composition of gastric contents entering the duodenum has a major influence on gastric emptying rate by influencing the release of gut hormones from the stomach and the intestine. Carbohydrate-rich meals empty most rapidly, protein-rich meals more slowly and fatty meals slowest of all.

A variety of stimuli on the duodenum provoke the inhibitory enterogastric reflexes which retard gastric emptying. These include high acidity, the presence of fat and protein breakdown products in the chyme, and distension of the duodenum

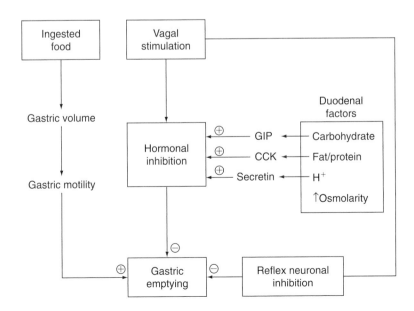

Figure 5.2 *Regulation of gastric emptying.*

(Fig. 5.2). Isosmotic gastric contents empty faster than hypo- or hyperosmotic contents because of feedback inhibition produced by duodenal osmoreceptors. Cholecystokinin (CCK) is released from the duodenum in response to breakdown products of fat and protein digestion and blocks the stimulatory effects of gastrin on the antral smooth muscle. Secretin, released from the duodenum in response to acid, has a direct inhibitory effect on the gastric smooth muscles. However, gastrin released into the circulation by antral distension increases gastric contractions.

During the interdigestive period, food remaining in the stomach is removed by a peristaltic wave called the migrating motor complex (MMC); this begins within the oesophagus and travels through the whole gastrointestinal tract every 60–90 min. Motilin released by the epithelium of the small gut can enhance the strength of the MMC.

Gastric secretions

Gastric secretions (approximately 2 L per day) include pepsin, gastric lipase, hydrochloric acid (0.15 M), mucus, gastrin and intrinsic factor. The enzymes continue the digestion of food initiated by the salivary enzymes. The secreted mucus forms a gel on the surface of the epithelial cells that creates an unstirred layer of HCO_3^- ions (secreted by these cells) and protects them from acid and pepsin in the stomach lumen.

Gastric secretion occurs in three phases: the cephalic, the gastric and the intestinal phases. The cephalic phase is initiated by the thought, sight, taste and smell of food mediated via the vagus. About 30 per cent of gastric secretions (HCl, gastrin and pepsinogen) released during a meal may occur as a result of the cephalic phase.

The gastric phase is initiated by the entry of food into the stomach. Local and vasovagal reflexes caused by distension of the body of the stomach result in HCl secretion. Antral distension results in gastrin release from antral G cells. The acidic pH enhances pepsinogen secretion mediated by local reflexes. The gastric phase results in a prolonged secretion but at a lower rate, contributing to nearly 50 per cent of gastric juice production. The intestinal phase is initiated by nervous stimuli associated with distention of the small intestine,

but overall little gastric secretion (less than 10 per cent) occurs at this phase. Of more importance is the fact that chyme entering the intestine normally decreases gastric secretion and motility. This negative feedback matches the delivery of chyme to the handling capacity of the small intestine and is of great physiological importance. Decreased pH, fat and hyperosmolarity in the duodenum suppresses gastric secretion and gastric emptying. Decreased gastric emptying is mediated by CCK whereas the inhibition of gastric secretion is mediated by secretin and gastric inhibitory peptide (GIP) and the enterogastric reflex which is transmitted the enteric nervous system and through the extrinsic sympathetic and vagus nerves.

Gastric juice

About 2 L of gastric juice are produced per day. This secretion is slightly hyperosmotic (325 mOsm/L), rich in K^+ (10 mmol/L), H^+ (150–170 mmol/L), Cl^- (180 mmol/L) and poor in Na^+ (2–4 mmol/L). It has a pH of 1–1.5 and also contains enzymes and intrinsic factor, which is essential for the absorption of vitamin B_{12} at the ileum. The high acidity may provide some antimicrobial properties. Tubular glands in the proximal stomach secrete gastric juice containing pepsinogens, acid and mucus. The glands contain mucus-secreting neck cells, chief cells that secrete pepsinogens, parietal cells that secrete acid and intrinsic factor, small numbers of enteroendocrine cells such as enterochromaffin-like cells that secrete histamine and D cells that secrete somatostatin. The antrum is free of parietal cells but have numerous mucus-secreting cells, some pepsinogen-secreting cells and G cells that release gastrin into the interstitial fluid.

Acid secretion

The parietal (oxyntic) cells in the body and fundus of the stomach secrete HCl and intrinsic factor. Hydrochloric acid facilitates the breakdown of protein in addition to providing an optimal pH for pepsin activity.

The pyramidal parietal cell contains numerous mitochondria and large quantities of cytoskeletal proteins such as actin, tubulin and cytokeratins. Within the cytoplasm there are numerous intracellular canaliculi, and the apical membrane has microvilli. The apical or canalicular membranes contain proteins that utilize ATP for HCl production and are the sites of gastric acid production.

GASTRIC ACID SECRETION RECEPTORS

Specific receptors for histamine, acetylcholine and gastrin stimulate acid secretion. However, somatostatin, epidermal growth factor, β-adrenergic agonists and enteroglucagon inhibit acid secretion, presumably by an indirect mechanism as specific receptors for these factors have not been identified. Histamine binds to H_2 receptors at the basolateral membrane of the parietal cell to stimulate acid secretion. The parietal cells also contain a muscarinic receptor which increases acid secretion when vagal stimulation releases acetylcholine. Some species possess gastrin receptors in the gastric parietal cells, but in humans gastrin receptors may be absent. However, gastrin is a potent stimulant of acid secretion in humans, and its effects appear to be mediated by the release of histamine from other cells in the gastric glands binding to H_2 receptors in the parietal cell to stimulate acid production.

The H_2, muscarinic and gastrin receptors are all located in the basolateral membrane of the parietal cell, and release second messengers (Fig. 5.3). The H_2 receptors on the parietal cell activate adenyl cyclase, which results in an increase in cyclic AMP. The cyclic AMP activates protein kinases to phosphorylate proteins that rearrange the intracellular canaliculi; cAMP also activates enzymes for the production of metabolic energy. Gastrin and muscarinic agonists release inositol triphosphate (IP_3) and diacylglycerol from the hydrolysis of membrane phospholipids by phospholipase. Diacylglycerol activates protein kinase, whereas IP_3 releases intracellular stores of Ca^{++}. The intracellular Ca^{++} ions released activate Ca^{++}-dependent protein kinase C in the parietal cells.

MECHANISM OF HYDROCHLORIC ACID SECRETION

The H^+ secreted is believed to be derived from the dissociation of H_2O: $H_2O \rightarrow H^+ + OH^-$. The

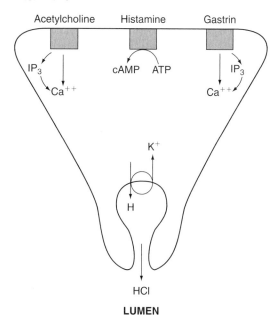

Figure 5.3 *Parietal cell receptors.*

OH^- combines with H^+ derived from the dissociation of carbonic acid, which is formed from CO_2 and H_2O, catalysed by carbonic anhydrase. CO_2 is derived from intracellular metabolism and by diffusion from blood. A Cl^-/HCO_3^- exchanger at the basolateral membrane rapidly moves HCO_3^- out of the cell into the blood, creating the alkaline tide associated with gastric acid secretion.

The apical cell membrane of the parietal cell is dependent on an H^+/K^+-ATPase (gastric proton pump), which actively pumps H^+ ions out of the cell in exchange for K^+ entering (Fig. 5.4). At the basolateral membrane, the Cl^-/HCO_3^- pump which balances the entry of Cl^- into the cell with HCO_3^- ions entering the blood, is stimulated by H^+ secretion. The Cl^- ions inside the cell leak out to the lumen down an electrochemical gradient, so that HCl is secreted into the lumen. This creates a negative potential inside the canaliculi, and K^+ flows into the lumen, associated with an increase in K^+ conductance through the K^+ channels in the apical membrane of the parietal cells.

The electrolyte composition of gastric juice varies with the secretory rates. The gastric juice contains small amounts of K^+ and H^+ and a higher concentration of NaCl at lower secretion rates. As

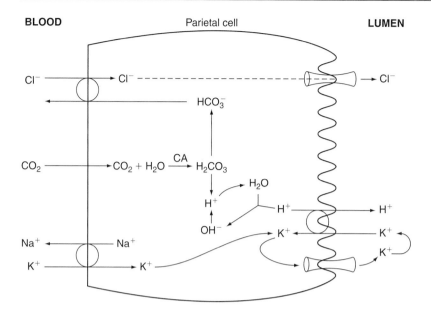

Figure 5.4 *Hydrochloric acid production by parietal cells. CA, carbonic anhydrase.*

the gastric secretion rate increases, the concentration of H^+ increases and that of NaCl decreases.

REGULATION OF GASTRIC ACID SECRETION

The cephalic phase enhances the secretion of HCl as the vagus releases acetylcholine and inhibits the release of somatostatin by the interneurons within the enteric nervous system.

During the gastric phase, amino acids and peptides directly stimulate the parietal cells to secrete acid. High gastric acidity directly inhibits acid secretion, which is also reduced by the release of somatostatin.

During the intestinal phase, the presence of protein digestion products within the duodenum and amino acids in the blood stimulate HCl secretion. Intestinal contents, which have a high content of fatty acids and H^+ ions, inhibit acid secretion due to the release of gastric inhibitory peptide (GIP) and secretin. GIP promotes the release of somatostatin which in turn directly inhibits both the parietal and G cells.

Pepsinogen secretion

The chief (peptic) cells at the base of the gastric gland produce pepsinogen I, which is stored as intracellular storage granules. Pepsinogen I is produced by the mucous glandular cells. Pepsinogen is released by exocytosis following stimulation by cholinergic, muscarinic and β-adrenergic influences. In the acidic pH of the stomach, autocatalytic cleavage of pepsinogen produces pepsin, a proteolytic enzyme which facilitates protein digestion.

During all three phases, pepsinogen is released from the chief cells via vagal stimulation. During the cephalic phase, the cholinergic enteric neurons, through vagal stimulation, directly stimulate the chief cells to secrete pepsinogen. At the gastric phase, low pH activates local reflexes to increase pepsinogen secretion. The presence of H^+ in the duodenum during the intestinal phase releases secretin, which promotes the secretion of pepsinogen.

Intrinsic factor

Intrinsic factor is a glycoprotein required for the absorption of vitamin B_{12}. It is produced by the parietal cells, mainly those in the fundus, and forms a complex with vitamin B_{12} that is absorbed in the terminal ileum.

Gastrin

CHEMISTRY

Gastrin consists of a family of peptides, the biological activity of which is produced by the four

C-terminal amino acids. G17 or 'little gastrin' (a heptadecapeptide) accounts for 90 per cent of gastrin produced by the antral G cells. A 'big' gastrin called G34 is also produced, usually during the interdigestive or basal state. After a large meal, large amounts of antral gastrin, primarily G17, are secreted. Although G17 and G34 are equipotent, their half-lives of 7 and 38 min, respectively, are clearly different.

RELEASING FACTORS

Gastrin appears to be the only gut hormone that is released directly by neural stimulation. During the cephalic phase, vagal activity increases the release of bombesin (gastrin-releasing peptide; GRP) which stimulates the G cells to secrete gastrin. The vagus also inhibits the release of somatostatin, which directly inhibits gastrin secretion. Hydrogen ions (acidity) inhibit gastrin secretion. Peptides and amino acids, alcohol and caffeine directly stimulate the release of gastrin, as may gastric distension. Both secretin and glucagon inhibit gastrin release.

ACTIONS OF GASTRIN

The primary physiological action of gastrin is the stimulation of gastric acid secretion, gastrin being approximately 1500 times more potent than histamine. Gastrin has a trophic effect on the mucosa of the small intestine and colon and the parietal cell mass of the stomach. It also increases gastric and intestinal motility, pancreatic secretions, and gall bladder contraction.

Vomiting

Vomiting may be defined as the involuntary, forceful and rapid expulsion of gastric contents through the mouth. Vomiting begins with a deep inspiration. The glottis and nasopharynx are closed, and preceded by a large retrograde intestinal contraction (initiated by central, peripheral and enteric nervous systems) that forces intestinal contents into the stomach. The oesophagus, LOS and the body of the stomach then relax, followed by the contraction of the abdominal and thoracic muscles. The diaphragm descends at the same time, and together these markedly raise the intra-abdominal pressure, forcing the gastric contents into the oesophagus and out through the mouth.

The sequence of vomiting is integrated by the vomiting centre (Fig. 5.5) which is associated with the respiratory and vasomotor centres, and the salivary nuclei. The vomiting centre, located bilaterally in the dorsal part of the lateral reticular formation in the medulla oblongata, coordinates the interrelated activity of the smooth and striated muscles involved. The chemoreceptor trigger zone (CTZ) is located bilaterally on the floor of the fourth ventricle in the area postrema near the vagal nuclei. The CTZ lies outside the blood–brain barrier and is sensitive to chemical stimuli such as drugs. It has a high density of D_2 dopamine and serotonin receptors, and impulses from the CTZ pass to the vomiting centre. Impulses arising from endolymph movements in the utricle and saccule of the vestibular apparatus are relayed via the vestibular nucleus and VIIIth nerve to the CTZ. Histamine (H_1) and muscarinic receptors are activated by vestibular stimuli such as motion sickness. The main receptors involved in the control of vomiting are dopamine D_2, histamine H_1, serotonin $5\text{-}HT_3$ and muscarinic receptors. Afferent pathways to the vomiting centre include stretch and chemoreceptors throughout the gastrointestinal tract via the vagal and sympathetic nerves, pharyngeal touch receptors via the glossopharyngeal nerve, and the cerebral cortex. The efferent pathways involve vagus, hypoglossal, glossopharyngeal, trigeminal and facial nerves to the upper gut and the spinal nerves to the diaphragm and abdominal muscles (Fig. 5.5).

Gastric mucosal barrier

The gastric mucosal barrier, is a barrier in the surface layer of gastric mucosa which keeps H^+ out of the mucosa and Na^+ ions in, while maintaining a potential difference across the surface of the mucosa. This is an important defence against autodigestion (damage by intraluminal HCl) and ulceration.

This protective mechanism is poorly understood and depends on good mucosal blood supply and the cytoprotective effects of prostaglandins. Mucus produced by the mucous glands consists of

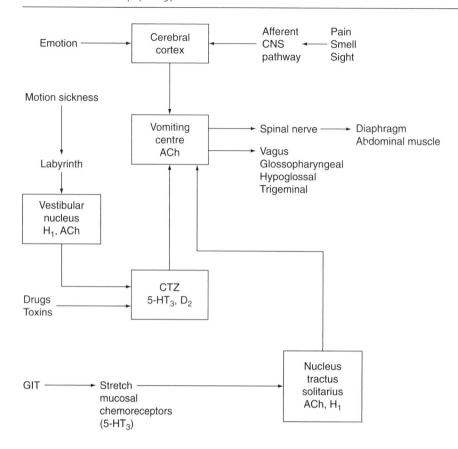

Figure 5.5 *Neural pathways of vomiting.*

mucopolysaccharide and glycoprotein. The functions of gastric mucus are to protect the mucosa, lubricate food and perhaps trap bacteria.

Gastric digestion and absorption

A small degree of digestion occurs in the stomach. Digestion of carbohydrate depends largely on the actions of salivary α-amylase (ptyalin), which remains active until it is inactivated by gastric acidity. Oligosaccharides are split from amylopectin and glycogen. Starch is converted to maltotriose, maltose and isomaltose. The hydrolytic actions continue for about 30 min until the α-amylase is inactivated at a pH of 4.

About 10 per cent of ingested protein is digested by pepsin, an endopeptidase with specificity for peptide bonds of aromatic L-amino acids, to form polypeptides.

Fat digestion is minimal, as gastric lipase activity is restricted to triglycerides with short-chain fatty acids. Minimal absorption of nutrients occurs in the stomach. Only highly lipid-soluble substances (e.g. ethanol) are rapidly absorbed in the stomach.

SMALL INTESTINE

The small intestine may be divided into three parts: duodenum, jejunum and ileum. It has several physiological roles, including the absorption of nutrients and electrolytes, secretion of various hormones, motility for mixing and propelling food, immunological and endocrine functions.

Anatomically, the small intestine is adapted for its various functions. The luminal surface is increased by approximately 600-fold through the presence of: (1) valvulae conniventes – prominent mucosal folds in the duodenum and jejunum; (2) villi which line the entire mucosal surface; and (3) microvilli along the luminal or apical surface of the intestinal cells.

The organization of the blood supply of the villus promotes the absorption of nutrients. Each villus has an arteriole which gives rise to capillary systems at the tip of the villus, which subsequently drain into venules that flow into the portal vein. Lymphatic vessels called lacteals extend to the tip of the villus and carry absorbed fats to the thoracic duct.

Intestinal motility

Functionally, intestinal motility is important for (1) mixing of food and digestive secretions; (2) ensuring maximal contact of luminal contents with the mucosa; and (3) propagation of the luminal contents from the duodenum to the colon, which usually takes 2–4 h.

Segmentation and peristalsis are the two basic movements. In the interdigestive period, basal contractions due to the migrating motor complex occur every 60–90 min, each contraction lasting for 10 min. These basal contractions generate intraluminal pressures of 4–12 mmHg.

ELECTRICAL ACTIVITY

The smooth muscle cells of the small intestine produce slow waves (basic electrical rhythm) at 12–18 cycles/min in the distal ileum. The spontaneous slow waves do not produce contraction. Spike potentials (or action potentials) are superimposed on the slow waves if the membrane potential is sufficiently depolarized, and these produce contractions. During digestion, gastrin, CCK and motilin increase the slow wave amplitude, whereas secretin and glucagon reduce it. As the amplitude of slow waves increases, the frequency of spike potentials generated also increases, leading to enhanced strength of contraction.

SEGMENTATION

Segmentation produces to-and-fro movement of intestinal luminal contents which serves to mix chyme with digestive enzymes and allows contact with the absorptive mucosal surface, thus enhancing absorption. These contractions of the intestinal wall occur about 12 times each minute in the duodenum, and 8 times each minute in the ileum; each contraction lasts 5–6 s. Segmentation is regulated by intrinsic smooth muscle excitability, the enteric nervous plexus and gut hormones.

PERISTALSIS

Propulsive motility in the small intestine is partially provided by peristalsis. Vagal innervation is not important for intestinal peristalsis. However, the myenteric plexus, intrinsic smooth muscle excitability, and gut hormones modulate peristaltic activity. Muscular activity is stimulated by gastrin, CCK and motilin, but inhibited by glucagon and somatostatin.

Although peristalsis aids the movement of chyme distally down the gut, segmentation is considered to contribute more to the propulsion of chyme. This results from the higher frequency of segmentation in the proximal intestine compared with that in the ileum.

Pancreatic secretions

The exocrine secretions are produced by acinar cells and the ductal cells of the pancreas. Approximately 1.5 L of pancreatic juice are produced per day. At rest, the secretion is plasma-like but with higher flow rates, an alkaline fluid rich in HCO_3 is secreted. The organic components of the pancreatic juice consist of digestive enzymes produced by the acinar cells. HCO_3^-, the chief inorganic component, and water are secreted by the duct cells.

ENZYMES

The pancreatic digestive enzymes are synthesized as proenzymes in the ribosomes lining the rough endoplasmic reticulum of the acinar cells. These proenzymes migrate to the Golgi complex, forming zymogen granules. The proenzymes are secreted by exocytosis. Distinct receptors which bind acetylcholine, gastrin and CCK release the enzymes by activation of the phosphatidyl inositol system. Secretin receptors are also found on the basolateral membrane, and these promote the release of enzymes via the adenyl cyclase–cyclic AMP pathway.

Three major types of hydrolytic enzymes are secreted by the pancreas:

- Proteolytic enzymes (trypsinogen and chymotrypsinogen) are secreted as inactive proenzymes (zymogens). In the gut lumen, trypsinogen is converted to trypsin by enterokinase, an enzyme found on the apical plasma membrane of the duodenal epithelial cells, or by trypsin itself (autocatalysis).

- Pancreatic α-amylase is secreted in its active form and hydrolyses glycogen, starch and other complex carbohydrates (except cellulose) to disaccharides.
- Pancreatic lipases (lipase, phospholipase) hydrolyse dietary triglycerides to glycerol and fatty acids for absorption. However, bile salts are required to render the triglyceride esters water-soluble before the lipases can break down the triglycerides.

INORGANIC COMPONENTS

The major inorganic components of pancreatic juice are water and electrolytes produced mainly by the duct cells. The duct cells possess specific transport mechanisms in their apical and basolateral membranes that facilitate the secretion of electrolytes, especially HCO_3^-.

As the pancreatic juice leaves the acinar cells, Na^+, K^+, HCO_3^- and Cl^- concentrations reflect those of plasma. However, stimulation of the duct cells after a meal results in a significant rise in HCO_3^- and a fall in Cl^-, with Na^+ and K^+ remaining constant. The pH of pancreatic juice entering the duodenum is approximately 8.

The alkalinity of the pancreatic juice buffers gastric acid, thus protecting the duodenal mucosa. Another function of pancreatic HCO_3^- is to provide the optimal pH for pancreatic enzyme activity.

The secretion of HCO_3^- and Cl^- by the duct cells is brought about by coupling of transport systems (Fig. 5.6). Plasma HCO_3^- diffuses down a concentration gradient through the basolateral membrane of the duct cells into the cytoplasm. Another source of HCO_3^- in the duct cells is from the hydration of CO_2 from plasma, as well as that produced by cellular metabolism. The HCO_3^- within the duct cell is secreted into the lumen by two mechanisms. First, HCO_3^- can diffuse freely across the apical membrane down its concentration gradient. Second, an apical membrane Cl^-/HCO_3^- transport system exchanges HCO_3^- for luminal Cl^- in a one-to-one exchanger system. This results in a high HCO_3^- and low Cl^- content of the pancreatic juice. H^+ produced by the hydration reaction is transported out of the duct cell by a Na^+/K^+ exchange system with energy provided by the Na^+/K^+-ATPase pump at the basolateral membrane. As Cl^- accumulates in the duct cell, it diffuses through the basolateral membrane into the blood. Water diffuses freely through the paracellular pathways to maintain osmotic balance.

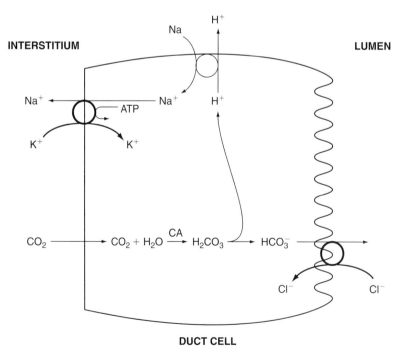

Figure 5.6 *Bicarbonate secretion by the pancreatic duct. CA, carbonic anhydrase.*

CONTROL OF PANCREATIC EXOCRINE SECRETIONS

Pancreatic juice is secreted under the influence of the neural and hormonal influences. Pancreatic secretion is stimulated by cholecystokinin, secretin and acetylcholine.

- *Cephalic phase*. This phase of pancreatic secretion is mediated by acetylcholine released by the vagus and accounts for 20 per cent of pancreatic enzymes after a meal. The sight, smell, thought or taste of food increases pancreatic enzyme and HCO_3^- secretion. The efferent cholinergic fibres of the vagus innervate the acinar cells, and indirectly influence the duct cells by modulating the peptidergic nerves that innervate these cells. The vagus nerve stimulates HCO_3^- secretion to a lesser extent.
- *Gastric phase*. This is mediated by the vagus and antral gastrin release and accounts for 5–10 per cent of enzymes secreted after a meal. Gastric distension increases pancreatic secretion by a vasovagal reflex mediated by acetylcholine. Amino acids and peptides enhance pancreatic secretions by causing the G cells to release gastrin. In turn, gastrin produces pancreatic juice that is rich in enzymes.
- *Intestinal phase*. Acidification of the proximal duodenum is a major factor that increases pancreatic exocrine secretion, producing a 'watery' secretion. Special cells in the duodenum – the amine precursor uptake and decarboxylation cells – synthesize and release secretin. Secretin increases the secretion of bicarbonate by the duct cells of the pancreas and biliary tract, thus producing a watery and alkaline pancreatic juice. CCK increases pancreatic enzyme secretion. Amino acids (phenylalanine), fatty acids and monoglycerides are the major stimulants for CCK release. The effects of secretin and CCK are also potentiated by acetylcholine.

Intestinal secretions

Various intestinal secretions are produced by the intestinal epithelium. Alkaline mucus is secreted by Brunner's glands and goblet cells in the crypts of Lieberkuhn and this is mediated by vagal stimulation, gastrointestinal hormones (especially secretin), and tactile stimuli of the overlying mucosa. Secretin-mediated secretion by the glands contain large amounts of HCO_3 ions. Enzymes are associated with the microvilli of the epithelial cells and they break down small peptides and disaccharides during absorption. Water is secreted by all the epithelial cells of the intestine.

Immunological function of the small intestine

The small gut is an immunologically active organ. T cells are present in the peripheral zone of Peyer's patches, and B cells in the germinal centres of Peyer's patches. Large antigens may penetrate the intestinal lining and promote production of immunoglobulins. The plasma cells in the lamina propria of the small intestine contain IgA, IgM and IgG. IgA is an important secretory immunoglobulin that has a protective function.

Endocrine function

The small intestine is a rich source of peptides such as secretin, CCK, gastrin, substance P and GIP. These peptides may have a regulatory role in the stomach (gastrin), pancreas (CCK, secretin), small intestine (motilin, gastrin, GIP, vasoactive intestinal peptide).

Digestion

CARBOHYDRATE

Although partial starch digestion by salivary α-amylase (ptyalin) occurs in the stomach, almost all carbohydrates are digested in the small intestine. Pancreatic α-amylase hydrolyses the 1:4 glycosidic bonds of dietary carbohydrates to maltose (disaccharide), maltriose (trisaccharide) and α dextrans. These are then further broken down by specific brush-border enzymes such as maltase, lactase and sucrase, which produce fructose, glucose and galactose for absorption. Cellulose, a plant polysaccharide, is not digested in the human digestive tract.

PROTEIN DIGESTION

Proteins found in the intestines are derived from endogenous sources (secretory proteins and desquamated cells) and exogenous proteins (dietary protein). Although 10–15 per cent of protein in the gastrointestinal tract is digested by gastric pepsin, protein digestion products in the stomach are important because they stimulate the secretion of proteases by the pancreas. Most protein digestion results from the actions of pancreatic proteolytic enzymes. Trypsinogen and chymotrypsinogen are activated by enterokinase or by autocatalysis with trypsin. Tryspin and chymotrypsin, which are endopeptidases, cleave internal peptide linkages to produce dipeptides, tripeptides and other small peptide chains that can be absorbed by intestinal cells. Carboxypeptidase (exopeptidase) is also produced by the pancreas and intestinal epithelial cells. The carboxypeptidases cleave the ends of a peptide chain producing free amino acids, which are readily absorbed by the intestine. The last step in the digestion of protein is achieved by enterocytes that line the villi. At the brush border, aminopeptidases split larger polypeptides into tripeptides, dipeptides and some aminoacids. These are transported into the enterocytes where multiple peptidases digest the dipeptides and tripeptides to aminoacids which then enter the blood.

FAT DIGESTION

The first step in fat digestion is emulsification by bile acids and lecithin. Emulsification is a process by which fat globules are broken into smaller particles by the detergent actions of bile salts and lecithin, and this increases the total surface area of the fats. The lipases are water-soluble enzymes and attack fat globules only on their surfaces. Pancreatic lipase cleaves 1–1' ester linkages of the triglycerides to yield two monoglyceride and two free fatty acid molecules. Cholesterol esterase cleaves fatty acids from cholesterol esters, while phospholipase A cleaves fatty acids from phospholipids.

The fatty acids and the monoglycerides form micelles with the bile salts; this enables them to be transported to the absorbing enterocytes. Bile salts form micelles that remove monoglycerides and free fatty acids from the vicinity of the fat globules. A micelle consists of a central fat globule containing monoglycerides and free fatty acids with bile salt molecules projecting outward to cover the surface of the micelle. The fat-soluble interior of the micelle also contains fat-soluble vitamins and cholesterol.

Absorption

CARBOHYDRATES

Carbohydrates are absorbed in the form of monosaccharides such as glucose, galactose and fructose. Glucose is absorbed by passive diffusion and carrier-mediated transport. Some 80 per cent of luminal glucose is absorbed by diffusion through the intestinal cells and also paracellularly. The remaining 20 per cent is absorbed by an active Na^+/glucose carrier-mediated co-transport system. When the luminal glucose concentration falls to 5 mmol/L, active glucose transport becomes dominant. Galactose also shares the active Na^+/glucose co-transport system. Fructose is absorbed by a carrier-mediated facilitated transport system.

After being absorbed into the enterocytes, the monosaccharides are transported across the basolateral membrane by facilitated diffusion.

PROTEIN ABSORPTION

In the intestinal lumen, proteins are absorbed through the luminal membrane of the intestinal epithelial cells in the form of dipeptides, tripeptides, and aminoacids by an active Na-dependent carrier-mediated system. Separate carrier systems specific for neutral, dibasic, acidic and methionine–phenylalanine groups exist. The basic amino acids are absorbed mainly by facilitated diffusion from lumen to blood. The amino acids are transported into the basolateral membrane by simple or facilitated diffusion, and then enter the capillaries of the villus.

FAT ABSORPTION

The mixing action of segmental contractions enables the micelles to come into contact with, and adhere to, the absorbing cell surface. The lipids, cholesterol and fat-soluble vitamins diffuse freely out of the core of the micelle into the apical membrane of the enterocyte. The bile salts are recycled, although a small proportion may be reabsorbed by a Na^+-dependent active transport system at the terminal ileum.

After entering the enterocyte, the monoglycerides and fatty acids diffuse passively through the enterocyte cell membrane and enter the smooth endoplasmic reticulum (ER). Within the smooth ER, the monoglycerides recombine with fatty acids to form triglycerides, phosphatides combine with fatty acids to form phospholipids, and cholesterol is re-esterified.

These re-formed triglycerides aggregate within the Golgi apparatus into globules that contain cholesterol and phospholipids. The phospholipids are arranged so that the fatty portions are oriented towards the centre and the polar portions on the surface, providing an electrically charged surface that makes the globule miscible with water. The globules are released from the Golgi apparatus and excreted by exocytosis into the basolateral spaces. From there the globules pass into the lymph in the central lacteal of the villi. The globules are then called chylomicrons. The chylomicrons are transported in the lymph via the thoracic duct to be emptied into the great veins of the neck.

SODIUM ABSORPTION

The main mode of Na^+ reabsorption in the small intestine is a coupled Na^+/Cl^- co-transport process where a luminal membrane carrier binds Na^+ and Cl^-. Other pathways for Na^+ absorption across the intestinal mucosa are Na^+ channels at the luminal membrane and Na^+/glucose or amino acid-coupled co-transport. The important driving force for Na^+ entry into the enterocyte is the basolateral Na^+-ATPase pump, which keeps the intracellular Na^+ concentration low. Cl^- flows passively through the enterocyte down the electrochemical gradient generated by the active transport of Na^+. Some Cl^- secretion occurs in the crypt cells, thought to be maintained by a $Na^+/K^+/Cl^-$ co-transport system at the basolateral membrane.

WATER REABSORPTION

The small intestine reabsorbs about 7–8 L of water each day. The movement of water occurs by a passive isosmotic movement process secondary to the active reabsorption of electrolytes and nutrients, which creates an osmotic gradient that favours water reabsorption.

CALCIUM ABSORPTION

Calcium is mainly absorbed from the proximal small intestine by a carrier-mediated mechanism. Vitamin D is important for Ca^{++} absorption.

VITAMIN ABSORPTION

Vitamins are absorbed in different ways according to their water- or fat-solubility:

- *Water-soluble vitamins.* Vitamin B_{12} is absorbed in the terminal ileum. Intrinsic factor, which is a glycoprotein secreted by the gastric parietal cells, complexes with vitamin B_{12}. At the ileum, the complex binds to specific receptors and becomes internalized through pinocytosis. Receptor binding requires Ca^{++} or Mg^{++} and an alkaline pH. Thiamine, vitamin C and folic acid are absorbed mainly by passive diffusion.
- *Fat-soluble vitamins.* Vitamins A, D, E and K are absorbed with other lipids, and require the formation of bile salt micelles for absorption.

LARGE INTESTINE

The large intestine or colon absorbs most of the fluids and some nutrients that are passed into it from the small intestine.

Motility

There are two forms of electrical activity in the smooth muscles of the colon. Slow waves coordinate gut motility and occur either as fast rhythms (6–12 cycles/min) or slower rhythms (2–4 cycles/min). Spike activity indicates bursts of electrical activity that occur at the time of muscle contraction. Intraluminal pressure waves occur in the colon at a frequency corresponding to that of the slow-wave activity.

NON-PROPULSIVE SEGMENTAL MOVEMENTS

This occurs commonly in fasting and consists of bands of muscle dividing the large intestine into segments called haustrations. Haustral shuttling squeezes and moves the chyme back and forth along the colon similar to segmentation contractions in the small intestine. These movements aid mixing and enhance the absorptive capacity of the colon. Haustral shuttling is found most commonly in the proximal half of the colon.

PROPULSIVE MOVEMENTS

Several types of propulsive movements are present in the colon. Haustral propulsive movements can slowly move the colonic contents. Multihaustral movements are often seen when a fresh ileal effluent passes over more solid contents. This softens hard faeces and facilitates further movement.

Mass movements occur when a peristaltic wave carries chyme along the colon, usually about three to four times a day. These mass movements usually force material into the rectum.

INTRAHAUSTRAL MOVEMENT

The annular intrahaustral constrictions occur in the distal colon and are efficient in loosening more solid faecal material.

Absorption, secretion and degradation

Approximately 1500 mL of isotonic chyme enters the colon from the ileum. The colon absorbs 1300–1400 mL of water, leaving 100–150 mL in the faeces. The proximal colon is important for absorption of electrolytes and water. Water absorption in the colon is passive, secondary to Na^+ reabsorption regulated by aldosterone. K^+ absorption is passive, related to electrochemical gradients. Active chloride absorption coupled to bicarbonate secretion also occurs in the colon. The large intestine can absorb a maximum of 5–7 L of fluid and electrolytes each day. Thus, the semi-fluid stool becomes semi-solid.

Goblet cells in the colon secrete mucus which facilitates faecal movement. Bacterial putrefaction also occurs in the colon.

Gas formation

About 100 mL of gas, derived from swallowing, bacterial action and diffusion, is normally present in the colon. Flatus consists of nitrogen (12–60 per cent), CO_2 (40 per cent) and methane (up to 20 per cent). The composition depends on diet and bacterial population in the gut. CO_2 and methane are derived from bacterial action in the gut.

The faeces

The faeces normally consists of three-quarters water and one-quarter solid matter. The solid matter is composed of dead bacteria (30 per cent), fat (10–20 per cent), inorganic matter (10–20 per cent) protein (2–3 per cent) and undigested roughage of food, sloughed epithelial cells and bile pigment (30 per cent). The brown colour of faeces is due to stercobilin and urobilin. The odour of faeces is due to indole, skatole, mercaptan and hydrogen sulphide.

Reflections

1. The major functions of the gastrointestinal tract include: food ingestion and transport along the tract; secretion of fluids, electrolytes, and digestive enzymes; the absorption of nutrients and the elimination of indigestible remains. Structural features common to all regions of the gut are the outer serosa, a layer of longitudinal muscle, a layer of circular muscle, the submucosa and the mucosa. The gastrointestinal tract is regulated by a system of intramural plexuses (enteric system) which mediate intrinsic reflexes that control secretory and contractile activity. Afferent and efferent extrinsic nerves, endocrine and paracrine hormones play an important role in regulating gastrointestinal activity. Contractile activity of the smooth muscle promotes mixing and propulsion of food. The splanchnic circulation, which supplies blood to the stomach, liver, pancreas, intestine and spleen, accounts for 20–25 per cent of the cardiac output at rest.

2. In the mouth, food is mixed with saliva. About 1500 mL of saliva (which is slightly hypotonic and contains mucus and α-amylase) is produced

per day by the acinar cells of the salivary glands. Parasympathetic stimulation promotes a watery salivary secretion that is rich in amylase and mucus. Sympathetic stimulation decreases the rate of salivary secretion and blood flow to the glands and increases the output of amylase.

3. Swallowing occurs in three phases. The first oral phase is voluntary but the subsequent phases are involuntary under autonomic control. As swallowing occurs peristalsis is initiated and this propels the food bolus through the upper oesophageal sphincter into the oesophagus and moves the food down to the lower oesophageal sphincter which relaxes to allow food to enter the stomach.

4. The functions of the stomach include storage of food, mixing of food to produce chyme and the secretion of acid, enzymes, mucus and intrinsic factor. The stomach wall has a third oblique layer of smooth muscle in addition to the circular and longitudinal smooth muscle layers and this promotes churning movements. The surface epithelium of the gastric mucosa is composed of goblets cells that secrete alkaline mucus which provides a protective coating for the mucosa. Prostaglandin E increases the thickness of this layer and stimulates bicarbonate production.

Gastric glands empty into gastric pits in the epithelium. They contain mucus cells, chief cells which secrete pepsinogens, parietal cells which secrete gastric acid, and intrinsic factor and G cells which secrete gastrin. Gastric juice contains salts, water, hydrochloric acid, pepsinogen, intrinsic factor and mucus. Gastric acid is secreted by the parietal cells of the gastric glands in response to food. Hydrogen ions are actively transported out of the parietal cells against a large concentration gradient. Chloride ions move out of the parietal cell against an electrical and chemical gradient. When gastric acid is produced, bicarbonate ions are added to the plasma creating an 'alkaline tide' in the venous blood draining the stomach. The chief cells of the gastric glands secrete proteolytic enzymes that are released as inactive pepsinogens which are activated by the acidic environment of the gastric lumen and hydrolyse peptide bonds within protein molecules to liberate polypeptides. The

parietal cells also secrete intrinsic factor, a glycoprotein that is essential for the absorption of vitamin B_{12} in the terminal ileum. Gastric secretion occurs in three phases: cephalic, gastric and intestinal. The cephalic phase of secretion occurs in response to sight, smell and anticipation of food. The vagus (parasympathetic) stimulates secretion via acetylcholine. The arrival of food in the stomach initiates the gastric phase of secretion and distention and the presence of acid and peptides stimulates HCl and pepsinogen secretion. Gastrin is an important mediator of the gastric phase. Gastric secretion is inhibited when gastric pH falls to about 2 or 3. As partially digested food enters the duodenum, a small quantity of gastrin is secreted by G cells in the intestinal mucosa and this stimulates further gastrin secretion. Secretin, CCK and GIP inhibit gastric secretion.

The stomach can store large amounts of food because the intragastric pressure rises little despite significant distention. After a meal peristaltic contractions in the stomach begin and these increase in strength as they approach the antrum where mixing is most vigorous. Peristalsis results from the slow-wave rhythm and the basic electrical rhythm (BER) of gastric smooth muscle and is enhanced by mechanical distention and gastrin. Gastric emptying occurs at a rate compatible with full digestion and absorption by the small intestine. Distention of the stomach increases the rate of emptying whilst the presence in the chyme of fats, proteins, acids and hypertonicity delays gastric emptying.

5. In the small intestine, which is a major site for both digestion and absorption, chyme is mixed with bile, pancreatic juice, and intestinal secretions. The folded mucosal surface and the villi of the small intestine provide a large surface area for nutrient absorption. The brush-border membranes of the mucosal epithelial cells house enzymes. Simple tubular glands called the crypts of Lieberkuhn lie between the villi. The epithelia of both the villi and crypts of Lieberkuhn contain mucus-secreting goblet cells, phagocytes and endocrine cells. Loss of small intestinal epithelial cells (which are replaced and renewed every 6 days) at the tips of the villi release enzymes such as enterokinase from the brush border of the

enterocytes into the lumen. Enterokinase activates pancreatic trypsin which then activates other proteolytic enzymes. The crypts of Lieberkuhn secrete 2–3 L of isotonic fluid per day. Chloride is transported out of the cell, and sodium and water follow passively via paracellular spaces. Brunner's glands in the duodenum secret alkaline fluid which neutralizes the acidic chyme arriving form the stomach. Secretion of the small intestine is stimulated by vagal activity, and by CCK, secretin, gastrin and prostaglandins.

6. The acinar cells of the pancreas secrete enzymes and fluid into a system of ducts that produce an alkaline fluid which modifies the composition of the acinar secretion. At high rates of secretion the bicarbonate content of pancreatic juice is higher than at lower rates. Precursors of proteolytic enzymes (trypsins) are stored as inactive zymogen granules to prevent autodigestion and these are activated in the duodenum. Pancreatic α-amylase is secreted in its active form and digests starch to oligosaccharides in the duodenum. Several lipases present in pancreatic juice hydrolyse water-insoluble triglycerides to release free fatty acids and monoglycerides. Bile salts are necessary for this process as they form an emulsion on which the lipases can act. The control of exocrine pancreatic secretion is chiefly hormonal via secretin and CCK. The initial cephalic phase (via the vagus) and gastric phase (mediated by gastrin) make a small contribution to the secretion of pancreatic juice. About 70 per cent of pancreatic juice secretion occurs during the intestinal phase. Secretin and CCK are released by the upper intestinal mucosa in response to products of fat and protein digestion.

7. Almost all of the absorption of water, electrolytes and nutrients occurs in the small intestine. Monosaccharides are absorbed in the duodenum and upper jejunum via sodium-dependent co-transport systems driven by the sodium–potassium pump. Amino acids utilize similar mechanisms but four separate transporters exist for four types of amino acids. Bile salts, which are essential for the digestion and absorption of fat and fat-soluble vitamins, emulsify fats in the small intestine so that they become more accessible to pancreatic lipases, which break them down to fatty acids and monoglycerides. The products of fat digestion are incorporated into micelles along with bile salts, lecithin, cholesterol and fat-soluble vitamins. This enables them to come in close contact with the enterocyte membrane so that the fat components can diffuse into the enterocytes. The fats are processed in the smooth endoplasmic reticulum to form chylomicrons which are exocytosed across the basolateral membrane and enter the lacteals of the villi. Fat-soluble vitamins are absorbed along with the products of fat digestion. Water-soluble vitamins are absorbed by facilitated transport. Vitamin B_{12} absorption occurs via a specific uptake process involving gastric intrinsic factor. The gastrointestinal tract absorbs about 10 L of water and electrolytes each day. Active transport of sodium and nutrients is followed by anion movement and water via an osmotic gradient. The main function of the large intestine is to store food residues, absorb residual water and electrolytes and secrete mucus. The colon absorbs about 1 L of water per day. Sodium is actively transported from the lumen to the blood; chloride is exchanged for bicarbonate and water is absorbed passively. Fermentation reactions caused by intestinal flora produce short-chain fatty acids (which are absorbed by the enterocytes) and flatus. Vitamin K is also synthesized by intestinal flora. Intestinal motility is an inherent property of intestinal smooth muscle that is controlled by both intrinsic and extrinsic neurons and neurotransmitters of the enteric plexuses. Parasympathetic activity enhances intestinal motility. Segmentation in the small intestine mixes chyme with enzymes and also exposes the chyme to the absorptive mucosal surfaces. Peristalsis propels chyme towards the ileocaecal valve. Segmentation is characterized by closely spaced contractions of the circular smooth muscle at a frequency that is determined by the slow-wave activity of the gut. Swaying pendular movements of the villi enhance mixing of the chyme whereas piston-like contractions of the villi facilitate the removal of fat digestion products from the lacteals of the villi. Haustrations (mixing movements) and sluggish propulsive movements occur in the large intestine.

6

Liver physiology

After reading this chapter the reader should be able to:

1. Outline the functions of the liver
2. Describe the mechanism and regulation of the formation of glycogen from glucose in the liver
3. Outline the function of glycolysis, gluconeogenesis, and the pentose phosphate pathways
4. Describe the role of the liver in the synthesis and interconversion of amino acids
5. Describe the synthesis and metabolism of fatty acids by the liver
6. Describe the liver as a storage organ for iron, and fat soluble vitamins
7. Describe the synthesis of plasma proteins in the liver and their functions
8. Describe the portal circulation and its physiological role

ANATOMICAL ASPECTS

The liver is the largest visceral organ in the body, weighing 1.2–1.5 kg in adults, and is relatively larger in children. The hepatic lobule, the basic histological unit, consists of a central hepatic efferent venule with cords of hepatocytes and sinusoids converging onto the efferent venule. The acinus is the functional unit and consists of a parenchymal mass between two centrilobular veins. The centre of the acinus is formed by the portal triads consisting of the portal vein, hepatic artery and bile duct. The acinus is supplied by terminal branches of the hepatic artery and portal veins, which drain into the sinusoids. The blood in the sinusoids drains into the hepatic venules. The hepatic acinus may be divided into three functional zones: (1) periportal, (2) mediolobular, and (3) centrilobular zones (Fig. 6.1). The periportal hepatocytes (zone 1) receive blood with the highest oxygen content and have the highest metabolic rate, and are especially involved in protein synthesis. The

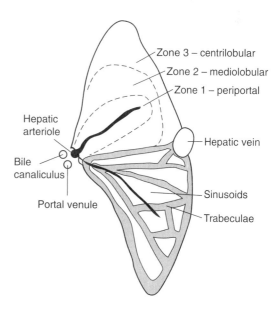

Figure 6.1 *The blood supply of the acinus.*

centrilobular hepatocytes (zone 3) receive the least oxygen, but contain high concentrations of cytochrome P-450 and therefore are important sites for drug biotransformation. These centrilobular hepatocytes are also predominantly involved in utilizing glucose, whereas the periportal cells secrete glucose into the sinusoidal blood. The mediolobular (zone 2) hepatocytes receive blood with an oxygen content intermediate between zones 1 and 3, and have intermediate enzyme activities. The sinusoids form a low-pressure microcirculatory system of the acinus with sphincters at the hepatic arteriole, hepatic venous sinusoid and arteriolar–portal shunts. Thus, the sinusoids act as a significant reservoir for blood, depending on the tone of the sphincters.

There are two distinct types of cells: the hepatocytes and the Kupffer cells (Fig. 6.2). The hepatocytes form 60 per cent of the liver cell mass and 80 per cent of liver volume. The hepatocytes are arranged in laminae through which the sinusoids interconnect.

The cells are polygonal in shape and have three surface types: one facing the space of Disse and sinusoid, a second facing the bile canaliculi and a third facing adjacent hepatocytes. Microvilli project from the surfaces in contact with the sinusoids and the bile canaliculi. These microvilli increase the surface area of the cell for active secretory and absorption functions.

The hepatocytes contain a large variety of organelles. The endoplasmic reticulum is a complex of intracellular membranes comprising high-density fatty acid complexes forming a large surface area. The smooth endoplasmic reticulum is associated with drug biotransformation. The rough endoplasmic reticulum is characterized by the presence of aggregates of ribosomes (RNA) on the tubules, and is responsible for protein synthesis. In addition, peroxisomes are contiguous with both types of endoplasmic reticulum and are sites of β-oxidation of fatty acids and storage of catalase.

The hepatocyte has numerous mitochondria, which are important for intermediary metabolism and production of adenosine triphosphate (ATP). The mitochondria in the hepatocytes are also important for steroid and nucleic acid metabolism and deamination of catecholamines.

The Golgi complex consists of a series of cytoplasmic vesicles which store albumin, lipoproteins

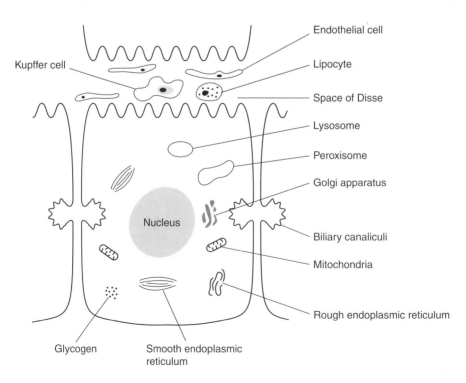

Figure 6.2   *Hepatocyte structure.*

and bile, and synthesize glycoproteins. Microtubules present in the cytoplasm may be involved in bile secretion.

Lysozymes are also present and contain proteolytic enzymes that cause autolysis of the hepatocytes when the lysosymes rupture. Lysozymes also act as sites of pigment deposition such as ferritin, lipofuscin, copper and bile pigment.

Microtubules are also abundant in the cytoplasm and form the cytoskeleton of the hepatocytes. They may be involved in bile secretion.

Kupffer cells

The Kupffer cells are macrophages that form part of the reticuloendothelial system and line the sinusoids. They are important in the phagocytosis of bacteria, destruction of endotoxin, protein denaturation and accumulation of ferritin and haemosiderin. In the fetus, the Kupffer cells have a haemopoietic function, which ceases within a few weeks of birth.

Other specialized cells are also found within the sinusoids. These include the endothelial cells, Pitt and Ito cells. Endothelial cells line the vasculature and are fenestrated, allowing molecular exchange between the hepatocyte and the space of Disse. Pitt cells are mobile lymphocytes attached to the endothelium and play a defensive role against infection and tumour cells. Ito cells contain fat, and also store vitamin A and other retinoids. The space of Disse lies between the endothelial cells of the sinusoid and the hepatocyte membrane. Collagen, fibronectin and proteoglycans are found within this space, which is also important for lymphatic transport.

FUNCTIONS OF THE LIVER

The functions of the liver are summarized in Table 6.1.

CARBOHYDRATE METABOLISM

The liver is important for the maintenance of blood glucose concentrations within narrow limits. The healthy adult liver contains about 100 g of glycogen, which can be released as free glucose into the circulation.

Table 6.1   *Functions of the liver*

- Metabolic, involving:
  a. Carbohydrate metabolism
  b. Protein and lipoprotein metabolism
  c. Metabolism of fatty acids
  d. Biotransformation of drugs
- Storage of vitamins A, D, E and K, iron, copper and glycogen
- Excretion of bilirubin, and urea formation
- Immunological functions associated with the synthesis of immunoglobulins, and the phagocytic action of Kupffer cells
- Filtration of bacteria and the degradation of endotoxins
- Haematological functions associated with haemopoiesis in the fetus, and as a blood reservoir

After dietary digestion, carbohydrates are broken down into monosaccharides and disaccharides. Enzymes in the intestinal mucosal cells (such as maltase, lactase, sucrase) cleave the disaccharides into hexoses. Uptake into portal venous systems is an active, energy-dependent process, whilst the uptake of glucose into hepatic cells is not energy-dependent. Glucokinase in the hepatocytes converts glucose to glucose-6-phosphate so that a low intracellular glucose is maintained, allowing continued diffusion of glucose into the cell (Fig. 6.3).

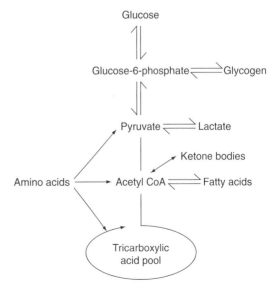

Figure 6.3   *Outline pathway of glucose utilization in the liver.*

Glycogen formation from glucose-6-phosphate proceeds via glucose-1-phosphate and uridine diphosphate glucose (UDP-glucose). The amount of glycogen in the liver is controlled by two enzymes: an anabolic enzyme, glycogen synthetase; and a catabolic enzyme, glycogen phosphorylase. In humans, glycogen arises mainly from lactate and to a lesser extent from pyruvate, glycerol and gluconeogenic amino acids. Glycogen may also arise from fructose, via triose phosphates. Overall, about 10 per cent of dietary glucose is converted to glycogen.

Gluconeogenesis in the liver is facilitated by glucagon, which enhances the transport of alanine across the hepatocyte membrane and pyruvate across the mitochondrial membrane. Cortisol increases both peripheral tissue proteolysis and plasma concentrations of amino acids, thus promoting gluconeogenesis. With feeding, insulin is secreted in response to an increase in portal blood sugar and this enhances glucose phosphorylation, activates glycogen synthetase, increases glycolysis, stimulates pyruvate dehydrogenase activity with increased acetyl co-enzyme A (CoA) formation and inhibits glycogenolysis and gluconeogenesis. Recent reports have suggested that the perivenous hepatocytes are primarily responsible for glycolysis and the periportal cells for gluconeogenesis.

The breakdown of glucose to carbon dioxide and water with the production of energy is called glycolysis. Glucose catabolism proceeds by two pathways, either by cleavage to trioses producing pyruvic acid and lactic acid (Embden–Meyerhorf pathway) or via oxidation and decarboxylation to pentose (hexose monophosphate shunt). The net energy gain from glycolysis is three molecules of ATP. Pyruvic acid enters the citric acid cycle by conversion to acetic acid with the loss of one molecule of CO_2. The citric acid cycle generates 12 molecules of ATP for every molecule of acetic acid. In total, 38 molecules of ATP are produced by the aerobic breakdown of glucose to pyruvate and its incorporation into the citric acid cycle. Pyruvic acid can be formed from the metabolism of amino acids and fat.

Glycolysis produces acetyl CoA, which is used as a substrate for lipogenesis and subsequently the production of triglycerides.

Another important property of the liver is the formation of reduced nicotinamide adenine dinucleotide phosphate (NADPH) via the pentose phosphate pathway. Two NADPH molecules and ribose-5-phosphate are produced from one glucose molecule. NADPH is required for microsomal and mitochondrial hydroxylation of steroid hormones and biotransformation of many drugs.

LIPID METABOLISM

The liver has two main roles in lipid metabolism: (1) synthesis of fatty acids which are converted to triacylglycerol and very low-density lipoproteins (VLDL) and (2) partial oxidation of fatty acids to ketone bodies (Fig. 6.4).

After absorption fat is either metabolized to yield energy or stored as triglyceride in fat deposits. Some 50 per cent of triglycerides derived from the diet are hydrolysed to glycerol and fatty acids, whilst 40 per cent are partially hydrolysed to monoglycerides. Short-chain fatty acids (fewer than 12 carbon atoms) are transported directly to the liver via the portal vein without re-esterification. Longer-chain fatty acids are re-esterified after absorption and then covered with a phospholipid and protein layer to form chylomicrons. Lipoprotein lipases hydrolyse the chylomicrons, producing free fatty acids that may be taken up by adipocytes for storage or metabolized within body tissues as an energy source.

β-Oxidation of fats to acetyl coenzyme A occurs rapidly in the liver. Excess acetyl CoA is converted to acetoacetic acid, a highly soluble molecule that can be transported to other tissues where it can be reconverted to acetyl CoA and used for energy.

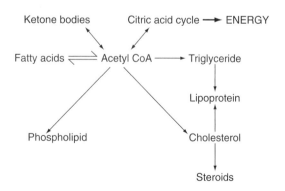

Figure 6.4 *Outline pathway of fat utilization in the liver.*

The liver is also important for cholesterol metabolism, which is controlled by the enzyme hydroxymethylglutaryl CoA (HMG-CoA). About 80 per cent of the cholesterol synthesized in the liver is converted to bile and the remainder is transported in the blood by lipoproteins. Phospholipids are also transported in the blood by lipoproteins. Both cholesterol and phospholipids are used by cells to form membranes and intracellular structures.

BILE PRODUCTION

The liver produces about 1 L of bile per day, and this passes into the gall bladder where it is concentrated to about one-fifth of its volume. Bile consists of electrolytes, protein, bilirubin, bile salts and lipids. Bile acids (cholic acid and chenodeoxycholic acid) are produced in the liver from cholesterol. In the gut, bacterial action on cholic and chenodeoxycholic acids produces secondary bile acids such as deoxycholic acid and lithocholic acid. The bile acids conjugate with glycine or taurine to form bile salts (Fig. 6.5). Bile salts are more

water soluble and less lipid soluble which limits the passive absorption in the small intestine so that the bile salts remain within the gut. The main function of the bile salts is the emulsification of dietary fat, this being essential for fat absorption. In addition, bile salts are also important for the absorption of the fat-soluble vitamins, especially vitamins A, D, E and K. At the terminal ileum bile salts are reabsorbed by the apical sodium-dependent bile transporter. The reabsorbed bile salts are carried to the liver in the portal circulation, mostly bound to plasma proteins. The recirculation of bile salts is referred to as enterohepatic circulation.

BILIRUBIN METABOLISM

Haemoglobin is broken down in the reticuloendothelial system, particularly in the spleen. Haem is broken down by haem oxygenase and nicotinamide adenine dinucleotide phosphate (NADPH)-cytochrome P-450 to biliverdin. Biliverdin is then converted to bilirubin by a reductase enzyme. About 85 per cent of bilirubin is derived from the haem moiety of the red-cell haemoglobin, the

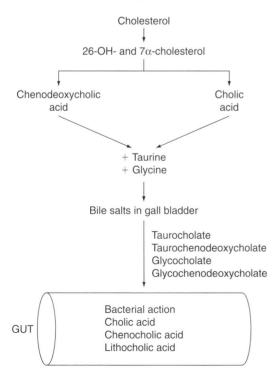

Figure 6.5   *Pathways of bile acid metabolism.*

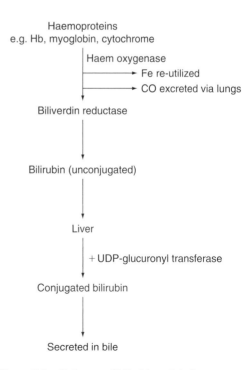

Figure 6.6   *Pathways of bilirubin metabolism.*

remainder is derived from the breakdown of other haem-containing compounds. Bilirubin is bound to serum albumin and is transported to the liver. In the liver, the unbound bilirubin enters the hepatocyte where rapid uptake occurs with two binding proteins being involved. The bilirubin is conjugated with glucuronides, rendering it water-soluble, and the conjugate is then secreted in the bile (Fig. 6.6). In the gut, conjugated bilirubin is broken down by bacteria to form urobilinogen, which then undergoes enterohepatic circulation and is excreted in the urine.

PROTEIN METABOLISM

The liver plays a central role in protein catabolism and anabolism. It also plays a key role in amino acid metabolism, by removing amino acids from the blood for gluconeogenesis and protein synthesis. The liver also releases amino acids into blood for utilization by the peripheral tissues and plays a major role in the breakdown of amino acids, removing the nitrogen as urea.

Anabolic processes

Amino acids and short peptide sequences are delivered to the liver after active uptake by enterocytes of the small intestine. The liver synthesizes a variety of proteins including albumin and clotting proteins. The liver is also responsible for the oxidative deamination of amino acids that are no longer required, a process that liberates energy and generates urea.

Albumin is synthesized in the liver at a rate of 120–300 mg/kg body weight per day. This synthesis is regulated by a number of factors, including nutritional status, endocrine balance and plasma oncotic pressure. However, as plasma albumin has a long half-life (20 days) it is a poor marker of acute liver damage.

The liver also synthesizes α_1, α_2 and β globulins, all of which are important transport or binding proteins and also form the complement proteins (50–80 per cent). The liver is an important site for the synthesis of haptoglobin (which binds to free haemoglobin), α_1 anti-trypsin, α_2 macroglobulin, antithrombin III, α_1 acid glycoprotein and C-reactive protein.

Vitamin K-dependent clotting factors (II, VII, IX, X) and some vitamin K-independent factors (V, VIII, XI, XII and XIII) are synthesized in the liver. Other nitrogen-containing compounds such as the purine and pyrimidine bases are formed from D-ribose-5-phosphate and carbamyl phosphate, respectively.

Protein catabolism

The liver has an important role in protein catabolism. The rate of protein turnover in the liver is 10 days, which contrasts sharply with the rate of 180 days for muscle proteins. Amino acid degradation is by transamination, deamination and decarboxylation. Oxidative deamination breaks down surplus amino acids and releases energy. Deamination may be coupled with the transfer of an amino group from one amino acid to another (transamination). These reactions produce acetyl CoA, oxoglutarate, succinyl CoA, oxaloacetate and fumarate, all of which enter the citric acid cycle. Amino acids (such as arginine, histidine, lysine, methionine, threonine, phenylalanine and tryptophan) are degraded mainly in the liver, whilst aspartic acid, glutamic acid, glycine, proline and alanine are metabolized in both hepatic and muscle tissue.

UREA SYNTHESIS

The nitrogenous end product of amino acid degradation is ammonia (Fig. 6.7). Surplus ammonia is toxic in concentrations greater than $1 \mu g/mL$, and it is converted to urea for excretion by the kidneys. Urea is synthesized from ammonia in the liver by the ornithine cycle, an energy-dependent process utilizing three ATP molecules per molecule synthesized. Ornithine initially combines with CO_2 and NH_3 to form citrulline. Citrulline then combines with a second molecule of NH_3 to produce arginine which is hydrolysed to yield water, ornithine and urea. About 30 g of urea is produced daily from 100 g of protein contained in the diet. As two hydrogen ions are produced for each molecule of urea synthesized, about 1000 mmol of hydrogen ions are formed. This H^+ production may be important in neutralizing the alkali load resulting from the metabolism of neutral amino acids.

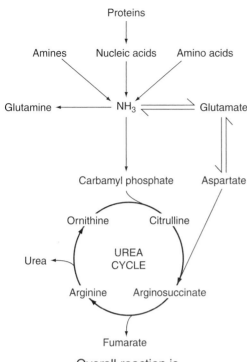

Overall reaction is
$$2NH_4^+ + CO_2 = NH_4CONH_2 + 2H^+ + H_2O$$

Figure 6.7 *Outline pathway of nitrogen metabolism in the liver.*

CREATINE SYNTHESIS

Creatine is synthesized in the liver from methionine, glycine and arginine. In muscle, creatine is phosphorylated to form phosphocreatine, which forms a back-up energy store for ATP production. Creatinine is formed from phosphocreatine and is excreted at a relatively constant rate in the urine.

PHAGOCYTIC FUNCTIONS

Kupffer cells are able to phagocytose many substances that mediate infection, inflammation and tissue injury. Substances phagocytosed by Kupffer cells include bacteria, viruses, endotoxins, immune complexes, denatured albumin, thrombin, fibrin–fibrinogen complexes and even tumour cells. Opsonins or recognition factors may be required. The particulate matter is phagocytosed and then fuses with lysosomes. The phagocytosed materials are then degraded by lysosomal enzymes.

Endotoxins derived from enteric bacteria are usually pinocytosed by the Kupffer cells.

STORAGE FUNCTIONS

The liver stores glycogen, triglycerides, vitamins (A, D, E, K, riboflavin, nicotinamide, pyridoxine, folic acid, B_{12}) iron and copper. It stores a sufficient amount of vitamin D to prevent vitamin D deficiency for about 4 months, sufficient vitamin A to prevent vitamin A deficiency for 10 months, and sufficient vitamin B_{12} to prevent vitamin B_{12} deficiency for 1 year. Excess iron is taken up by liver cells where it combines with apoferritin to from ferritin.

DRUG METABOLISM

The liver has an important role in eliminating both endogenous and exogenous compounds. As most compounds are lipophilic or partially ionized, biotransformation in the liver converts these compounds to hydrophilic substances that may be readily eliminated either by the kidneys, or in bile. The biotransformation reactions are divided into two distinct phases: Phase I reactions (oxidation, reduction, hydrolysis), and Phase II reactions (glucuronidation, sulphation, acetylation).

Phase I reactions increase the hydrophilicity of drugs, the majority being oxidative reactions catalysed by cytochrome P-450, a group of isoenzymes of at least 11 families located in the smooth endoplasmic reticulum. Reductase and hydrolase enzymes are mainly located in the cytoplasm. Some of the products of Phase I reactions may be pharmacologically active.

Phase II reactions consist of conjugation reactions that occur primarily in the cytoplasm and produce more polar compounds. Glucuronidation, the most common conjugation reaction in the liver, is mediated by glucuronosyl transferases, present mainly in the periportal hepatocytes. The majority of Phase II reactions produce inactive compounds, although there are some exceptions.

Overall, the clearance of a drug by the liver is influenced by a number of factors, including hepatic blood flow, plasma protein binding of the drug and hepatic enzyme activity.

LIVER BLOOD FLOW

Liver blood supply is unique in that it is derived from both the hepatic artery and hepatic portal vein. Total liver blood flow is approximately 1.5 L/min, or 25 per cent of cardiac output in the average adult. Both hepatic arterial and portal venous blood flows contribute to hepatic oxygenation.

Hepatic artery

The hepatic artery, a high-pressure and high-resistance system, delivers 30 per cent of the total hepatic blood flow. It contributes about 40–50 per cent of the total hepatic oxygen supply, the blood having an oxygen saturation of 98 per cent. The pressure in the hepatic artery is similar to the systemic arterial blood pressure. However, as a result of the high resistance in the hepatic arterioles, the hepatic arteriole pressure is about 35 mmHg. The ratio of presinusoidal to postsinusoidal (i.e. precapillary/postcapillary) resistance results in a low hepatic sinusoidal pressure (2 mmHg). The hepatic artery has an innervated muscular coat and therefore can constrict and dilate, albeit to a lesser extent compared with other arterial systems. The average flow velocity of blood is 16–18 cm/s.

Portal vein

The portal vein is a valveless vein and drains blood from the large and small intestines, spleen, stomach, pancreas and gall bladder to the liver. The hepatic portal vein contributes 70 per cent of the total liver blood flow, and 50–60 per cent of basal oxygen supply. In the fasting state, the oxygen saturation of portal venous blood is approximately 85 per cent, but this decreases with increased gut activity. The higher O_2 saturation in portal venous blood at resting conditions, compared with the mixed venous O_2 saturation, is due to the high mesenteric arterial shunting through the intestinal capillaries draining into the portal system. The velocity of blood flow in the portal system is 9 cm/s – approximately half that in the hepatic artery. Thus, the hepatic portal system is a low-pressure (5–10 mmHg), low-resistance and low-velocity system. The portal venous pressure depends on the state of constriction/dilatation of the mesenteric arterioles and on intrahepatic resistance. The resistance in the portal system is approximately 6–12 per cent of that in the hepatic artery.

Hepatic veins

Venous blood from the liver returns to the inferior vena cava via the right and left hepatic veins. A separate set of veins drain the caudate lobe of the liver. Various stimuli may cause hepatic venoconstriction, and these include norepinephrine, angiotensin, hepatic nerve stimulation and histamine. Hepatic venous pressure is also influenced by a number of external factors such as intermittent positive pressure ventilation, intra-abdominal pressure, gravity and gut-wall activity.

Hepatic microvasculature

The small vessels of the portal vein and hepatic artery run parallel to each other in the substance of liver, accompanied by a bile canaliculi, forming the hepatic triad. The hepatic triad vessels (i.e. arteriole and venule) eventually form anastomoses to form the sinusoids. Thus, the sinusoids form a specialized capillary system which facilitates exchange with the hepatocytes. The mixed portal venous and hepatic arterial blood flow from the periphery of the acinus to the central veins.

The sinusoids are therefore the terminals of the blood inflow into the liver, forming the microvasculature of the liver (Fig. 6.8). The pressure within

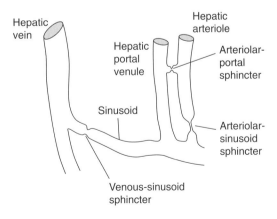

Figure 6.8   *The microcirculation system of the liver.*

the liver sinusoids is low (2 mmHg) due to the high resistance of the presinusoidal sphincters. The unique anatomy of the hepatic microcirculation does not permit vascular shunts away from the acini. Oxygen extraction is also more efficient compared with that of other organs, so that oxygen consumption of the liver is maintained by increased O_2 extraction at the sinusoids.

Capacitance function

The liver also forms a reservoir of blood with a volume of 450 mL (30 mL/100 g liver tissue), half of which may be mobilized if hypovolaemia occurs. The portal blood can bypass the sinusoids as blood is shunted from portal venules to hepatic venules by relaxation of the hepatic venule sphincters. Catecholamines can mobilize blood from the sinusoids. The liver can also buffer against an increase in blood volume. Hepatic compliance (i.e. distensibility) is higher at high venous pressures than at low venous pressures.

Regulation of hepatic blood flow

Given the high blood flow to the liver it is not surprising that any increase in oxygen demand is met by an increase in oxygen extraction rather than an increase in blood flow. However, hepatic artery blood flow does change in response to changes in flow in the portal vein. A reduction in portal vein flow is associated with an increase in hepatic arterial flow of between 22 and 100 per cent.

INTRINSIC CONTROL

Some degree of autoregulation can be demonstrated in the hepatic artery. When the hepatic arterial pressure is reduced, flow is maintained by a decrease in hepatic arterial resistance until systolic pressure is below 80 mmHg. Thus, the basal tone of the hepatic artery is minimal at pressures below 80 mmHg systolic.

The portal venous system has no autoregulation, and flow is related linearly to pressure.

The second intrinsic mechanism regulating hepatic blood flow is a semireciprocal interrelationship between portal venous and hepatic arterial flows. A reduction in portal venous blood flow causes a decrease in hepatic arterial resistance, hence increasing arterial flow. As portal blood flow is not autoregulated, there is little change in the portal venous blood flow when blood flow in the hepatic artery is reduced. This semireciprocal relationship is also termed the 'hepatic arterial buffer response'. It is suggested that this buffer response is due to intrahepatic levels of adenosine. With reduced portal blood flow, a build-up of adenosine occurs that vasodilates the hepatic artery.

An increase in hepatic venous pressure also causes an increased hepatic arterial resistance, possibly due to a myogenic mechanism. This is seen in congestive heart failure where an elevated hepatic venous pressure induces a decreased hepatic arterial blood flow.

EXTRINSIC CONTROL

Neural and blood-borne factors can modify hepatic arterial flow and resistance. The hepatic artery has α and β adrenergic receptors and dopamine receptors, but the portal vein has only α adrenergic and dopamine receptors. Epinephrine causes portal venous constriction, and initial vasoconstriction (α effect) followed by vasodilatation (β effect) of the hepatic artery. Dopamine has minimal effects on the hepatic vasculature at physiological concentrations. Glucagon also increases hepatic blood flow by vasodilation. Vasoactive intestinal peptide (VIP) and secretin vasodilate the hepatic artery, but have minimal effects on the portal vein. Angiotensin II vasoconstricts both the hepatic artery and portal vein. Vasopressin vasoconstricts the hepatic vasculature and thus reduces portal blood flow.

Feeding increases intestinal blood flow, and this dramatically increases hepatic blood flow. During normal spontaneous breathing, hepatic venous outflow is reduced with inspiration and increases during expiration. Vigorous exercise causes splanchnic vasoconstriction, leading to reduced hepatic blood flow. Positive-pressure ventilation decreases hepatic blood flow, most likely due to a decrease in cardiac output. Hypocapnia can reduce hepatic blood flow by 30 per cent, mainly due to a reduction in portal venous blood flow caused by increasing resistance in the portal system. Hypercapnia generally increases hepatic blood flow due to an increase in portal venous blood flow. Hyperoxia has little effects on both hepatic arterial and portal venous flows. Graded hypoxia initially decreases hepatic arterial flow, which returns to baseline within 20 min and has minimal effects on hepatic portal venous flow.

In acute haemorrhage, there is a greater reduction in portal venous blood than arterial flow. The oxygen supply to the liver is maintained by increased extraction. Further, sympathetic stimulation associated with hypovolaemia can mobilize about 50 per cent of the reservoir blood (30 mL/100 g tissue) of the liver into the systemic circulation.

Spinal and epidural anaesthesia reduce total hepatic blood flow – an effect which appears to be largely due to a reduced portal venous blood flow, and reduced mean arterial pressure.

Inhalational agents generally reduce total hepatic blood flow. Among the volatile agents, halothane causes the greatest reduction in total hepatic blood flow. Halothane appears to reduce hepatic arterial blood flow to a greater extent compared with other agents due to an increased hepatic arterial resistance, and despite reduced portal venous flow. This suggests a markedly reduced hepatic arterial buffer response. Enflurane reduces portal venous blood, and also hepatic arterial flow to a lesser extent, so that liver perfusion is reduced less compared with halothane. With isoflurane, hepatic arterial flow is unchanged or increased, with reduced portal venous flow indicating that the hepatic arterial buffer is better preserved.

Overall, halothane produces the greatest decrease in hepatic perfusion. Isoflurane, desflurane and sevoflurane appear to maintain hepatic oxygenation due to unchanged or even increased hepatic arterial flow.

Intravenous agents such as thiopentone, etomidate, althesin and propofol have been shown to cause a dose-dependent reduction in hepatic blood flow, presumably as a result of reduced cardiac output and obtundation of the hepatic arterial buffer mechanism.

Measurement of hepatic blood flow

Hepatic arterial, portal venous blood and total hepatic blood flow can be measured by direct (invasive) and indirect methods.

DIRECT METHODS

Hepatic artery and portal blood flows can be measured directly by the application of electromagnetic flowmeters around the respective vessels at laparotomy. However, this has limited application as surgery and anaesthesia are required. Further, anaesthesia and implantation of the devices may decrease total hepatic blood flow.

INDIRECT METHODS

Clearance techniques

Liver blood flow can be estimated by the clearance of markers such as indocyanine green (ICG), sulphobromophthalein (SBP) and iodine-131-labelled albumin. SBP undergoes enterohepatic circulation, and most clinical studies use ICG (extraction ratio 0.74) which is eliminated by the liver without any recirculation.

Single bolus technique

A single bolus of ICG (0.5 mg/kg) is injected intravenously, and venous blood samples are collected every 2 min for 14 min. The concentration–time delay curves are analyzed by non-linear regression analysis. Clearance is calculated from the formula (where AUC is area under the curve):

$$\text{Clearance} = \frac{\text{dose}}{\substack{\text{area under concentration} \\ \text{versus time curve}}}$$

As the extraction ratio of ICG is 0.74, hepatic blood flow is calculated using the formula:

$$\text{Hepatic blood flow} = \frac{\text{clearance}}{\text{extraction ratio}}$$

This technique assumes that there is adequate mixing of the dye in blood, and exclusive hepatic extraction.

Continuous infusion technique

After a loading dose of ICG (0.5 mg/kg body weight of subject), a constant infusion of ICG is administered for 20 min to achieve equilibration. Samples are taken simultaneously from a peripheral artery and the hepatic vein. Hepatic blood flow is calculated using the formula:

$$\text{Hepatic blood flow} = \frac{\text{clearance}}{\text{extraction ratio}}$$

where

$$\text{Clearance} = \frac{\text{infusion rate}}{\text{art. conc. ICG}}$$

and

$$\text{Extraction ratio} = \frac{(\text{art. conc. ICG} - \text{ven. conc. ICG})}{\text{art. conc. ICG}}$$

where 'art. conc. ICG' is the arterial concentration of ICG; and 'ven. conc. ICG' is the venous concentration of ICG.

Both the single bolus and continuous infusion clearance techniques assume that ICG is extracted exclusively by the liver, and that their hepatic venous samples reflect liver efflux – which may not be so in liver disease due to intra- and extra-hepatic shunting.

Reticuloendothelial cell uptake

As the Kupffer cells remove radiolabelled colloidal substances such as iodine-131-albumin, colloidal gold-198, technetium-99-sulphur colloid, the rates of clearance of these colloidal particles from the circulation can be used to estimate liver blood flow. A gamma camera is used to determine isotope uptake as a function of time. The area under the initial exponential phase is used as a measure of liver blood flow.

Microspheres techniques

Radiolabelled microspheres are embolized through the hepatic artery or portal vein. Samples of liver are excised and the radioactivity counted. Thus, this technique is only applicable in animal studies.

Reflections

1. The liver is essential in regulating metabolism, synthesizing proteins and other molecules, storing glycogen iron and vitamins, degrading hormones, inactivating and excreting drugs and toxins, in immune defense and as a reservoir of blood.

2. The liver regulates the metabolism of carbohydrates, lipids and proteins. The liver and skeletal muscle are two major sites of glycogen storage in the body. When the concentration of glucose is high in the blood, glycogen synthesis (mediated by insulin) occurs in the liver cells. When blood glucose is low, liver glycogen is broken down by glycogenolysis which is mediated by glucagons. The liver therefore helps to maintain a relatively constant blood glucose concentration. The liver is a major site of gluconeogenesis, the conversion of amino acids, lipids, or simple carbohydrates (e.g. lactate) into glucose.

3. The liver is centrally involved in lipid metabolism. Chylomicron by-products, rich in cholesterol, are taken up by liver cells and degraded. Hepatocytes also synthesize and secrete very-low-density lipoproteins (VLDLs) which are converted to lipoproteins. β-Oxidation of fatty acids provides a major source of energy for the body. In the liver, oxidation of fatty acids produces acetoacetate, β-hydroxybutyrate, and acetone. These compounds are called ketone bodies. Ketone bodies are released from hepatocytes into the circulation and utilized in the tissues.

4. The liver is involved in protein metabolism. Proteins are broken down to amino acids and are deaminated to form ammonia. Ammonia, which is toxic to the body, is converted to urea in the liver via the urea cycle. The liver also synthesizes all the non-essential amino acids, and all the major plasma proteins including lipoproteins, albumin, globulin, fibrinogen and coagulation factors.

5. The liver is an important storage site for iron, and some vitamins such as vitamin A, D, K and B_{12}. Hepatic storage protects the body from transient deficiencies of these vitamins.

6. The liver transforms and excretes many hormones, drugs and toxins. These substances are converted to inactive metabolites in the hepatocytes. The smooth endoplasmic reticulum of hepatocytes contain cytochrome P-450 isoenzymes that are responsible for chemical transformation of many substances. Other enzymes catalyse conjugation reactions with glucuronic

acid, glycine or gluthatione, which renders the compounds more water soluble so that they can be excreted by the kidneys. Some metabolites are secreted into bile.

7. The liver produces about 500–1000 mL of bile per day. Bile is stored and concentrated in the gallbladder which contracts to deliver bile into the duodenum following a meal. Bile acids are conjugated with amino acids to form bile salts. Bile salts have hydrophobic and hydrophilic regions (amphipathic) and aggregate at high concentrations to form micelles. The formation of bile is enhanced by bile salts, secretin, glucagons and gastrin. The release of bile stored in the gall bladder is stimulated by CCK in response to the presence of chime in the duodenum. Bile pigments (bilirubin) are excreted in the bile.

8. The Kupffer cells are macrophages that are important innate defence cells. In addition the liver synthesizes complement proteins which are involved in both innate and acquired immunity.

9. The liver stores about 600 mL of blood and this can be redistributed during hypovolaemia or sympathetic nervous system activity.

7

Renal physiology

After studying this chapter the reader should be able to:

1. Understand the general functions of the kidneys
   Describe:
   a. Water and electrolyte homeostasis
   b. Excretion in the urine of waste products of metabolism
   c. Excretion of chemicals
   d. Hormone production
   e. Gluconeogenesis
   f. Acid–base balance
2. Understand fluid and electrolyte balance and dietary requirements
   Describe:
   a. Body water compartments
   b. Water balance
   c. The obligatory urine loss
   d. The normal daily urine output
   e. Sodium balance
   f. The average minimum dietary requirements of water, sodium, potassium and chloride
3. Understand the functional anatomy of the kidneys
   Describe:
   a. The nephron
   b. The renal corpuscle
   c. The proximal tubule
   d. The loop of Henle
   e. Superficial cortical, midcortical and juxtamedullary nephrons
   f. The distal tubule
   g. The collecting ducts
   h. The renal blood vessels

   i. The juxtaglomerular apparatus
4. Understand the physiology of micturition
5. Understand glomerular filtration
   Describe:
   a. Factors determining glomerular filtration including the net filtration pressure (glomerular capillary hydrostatic pressure, Bowman's capsule hydrostatic pressure and glomerular capillary oncotic pressure), the filtration coefficient, constituents of glomerular filtrate and the filtration fraction
6. Understand the control of renal blood flow
   Describe:
   a. Renal blood flow
   b. Autoregulation of the renal blood flow and glomerular filtration rate (the myogenic mechanism, tubuloglomerular feedback)
   c. The sympathetic nerve supply to the kidney, renin release, angiotensin II, eicosanoids and atrial natriuretic factor
7. Understand tubular reabsorption and secretion
   Describe:
   a. Reabsorption of the glomerular filtrate, transport mechanisms in the renal tubule and reabsorption of sodium chloride, water, glucose, urea and protein
   b. Mechanisms of tubular secretion
   c. Proximal tubular secretion of organic ions and cations and hydrogen ion secretion (with bicarbonate reabsorption)

8. Understand renal clearance
   Describe:
   a. The definition of renal clearance
   b. The estimation of renal clearance using the renal clearance equation
   c. The renal clearance of glucose, inulin, *para*-aminohippuric acid, creatinine and urea

9. Understand the loop of Henle and the production of concentrated urine
   Describe:
   a. The general function of the loop of Henle including the role of the descending and ascending limbs
   b. The function of the counter-current multiplier of the loop of Henle
   c. The role of the vasa recta (the counter-current exchanger)
   d. The role of urea
   e. The production of dilute urine and of concentrated urine by the loop of Henle

10. Understand the overall tubular handling of the glomerular filtrate
    Describe:
    a. The role of the proximal tubule, the loop of Henle, the distal convoluted tubule and the collecting ducts

11. Understand the hormonal control of tubular function
    Describe:
    a. The renin–angiotensin system: renin and the control of renin secretion
    b. Angiotensin II
    c. Aldosterone
    d. Antidiuretic hormone
    e. Atrial natriuretic factor

12. Understand the control of renal sodium, water and potassium excretion
    Describe:
    a. The control of renal sodium excretion and the amount of sodium filtered per day
    b. Direct and indirect effects of body sodium on the glomerular filtration rate
    c. Body sodium, extracellular fluid volume and control of tubular sodium reabsorption
    d. Control of renal water excretion
    d. The control of renal potassium excretion, the effect of primary changes in body sodium balance on potassium excretion, the effect of primary changes in body water content on potassium excretion and the effect of alkalosis on potassium excretion

13. Understand the renal control of acid–base balance
    Describe:
    a. The carbon dioxide–bicarbonate buffering system
    b. Renal bicarbonate handling
    c. Hydrogen ion balance
    d. The glomerular filtration of hydrogen ions (the amount of hydrogen ions filtered per day)
    e. The glomerular filtration of bicarbonate ions (the amount of bicarbonate filtered per day)
    f. The reabsorption of bicarbonate by the renal tubule
    g. The renal extraction of hydrogen ions and the addition of new bicarbonate to blood (phosphate, ammonium)
    h. The renal response to respiratory and metabolic acidosis and alkalosis

14. Understand the mechanisms of action of diuretic drugs and the effects of diuretics on potassium excretion

FUNCTIONS OF THE KIDNEYS

Water and electrolyte homeostasis

The primary function of the kidneys is the regulation of the fluid and electrolyte composition of the body. The kidneys have a high blood flow, and from this a very large volume (180 L/day) of ultrafiltrate of plasma is produced in the renal corpuscles (glomerular capillaries and Bowman's capsule). This large glomerular filtrate is necessary for the excretion of waste products of metabolism in the urine. The filtrate passes along the nephron, where the specific processes of tubular reabsorption and secretion occur. Most of the filtered fluid is reabsorbed. The proximal tubule alone reabsorbs 60 per cent of the water and sodium filtered into Bowman's capsule, and the normal urine volume is only 1.5 L/day. Substances can also be removed from the peritubular capillary blood into the nephron lumen by specific tubular secretory mechanisms; many drugs are handled in this way. The final volume and composition of the urine are modulated to maintain normal body fluid and electrolyte balance by factors governing the processes of glomerular filtration, tubular reabsorption and tubular secretion. The kidneys have an integral role in the long-term regulation of body water and electrolyte composition, and therefore renal function is an important determinant of the long-term regulation of blood volume and arterial blood pressure.

Excretion in the urine of waste products of metabolism

Urea (from protein metabolism), creatinine (from muscle), uric acid (from nucleic acids) and bilirubin (from haemoglobin) are all excreted from the body in the urine.

Other kidney functions

These include:

- *Excretion of chemicals.* Many ingested chemicals, including drugs, are excreted from the body in the urine.
- *Hormone production.* The kidney produces renin, erythropoietin and the active form of vitamin D, 1,25-dihydroxyvitamin D_3.
- *Gluconeogenesis.* During starvation, the kidneys can produce glucose from amino acids.
- *Acid–base balance.* By varying the urinary excretion of bicarbonate and hydrogen ions, the kidneys have an important role in acid–base balance.

FLUID AND ELECTROLYTE BALANCE AND DIETARY REQUIREMENTS

Body water

Approximately two-thirds of the body weight is water. A 70-kg male has a total body water content of 42 L, divided between the intravascular (3 L plasma), interstitial (11 L), and intracellular (28 L) fluid compartments (the extracellular fluid is the sum of the intravascular and interstitial volumes). Total body water can be estimated using dilution techniques with markers that diffuse throughout

the total body water compartment, such as isotopically labelled water, using deuterium (^{2}H) or tritium (^{3}H). Markers used to determine the extracellular fluid must cross capillaries but not cell membranes; these include inulin, mannitol, radiosodium, radiochloride and thiosulphate (the latter is the most widely used). The intravascular fluid volume can be determined with a marker that remains within vessels, such as radiolabelled albumin, or the dye Evans blue which binds to plasma albumin. The interstitial volume cannot be measured directly, and is calculated by subtraction of the plasma volume from the extracellular volume.

Water balance

Each day, a total of 2550 mL of water is ingested (70 kg male). Of this water, 1000 mL is present in food, and 350 mL is produced by metabolic processes, including the electron transport chain. Drinking accounts for 1200 mL.

Each day, 900 mL of water leaves the body as insensible loss from the skin and lungs; in normal climates, 50 mL is lost as sweat. The faeces contain 100 mL of water. In order to achieve a balance between intake and loss, the urine output is 1500 mL.

The obligatory urine loss

During water deprivation, the urine volume can be reduced. However, the solute load of 600 mOsmol/day (urea, sulphate, phosphate and other waste products) that the kidney has to excrete means that there is a daily obligatory minimum urine output to avoid accumulation of waste products in the blood. The maximum urinary concentration of the kidney is 1400 mOsmol/kg H_2O. To excrete the solute load, a minimum of 430 mL (600 mOsmol/1400 mOsmol/kg H_2O) of urine has to be lost per day.

The normal daily urine output

The average daily water intake of 2550 mL is in excess of the 1480 mL required to cover the 430 mL obligatory urine loss, the 900 mL insensible loss, the 100 mL in faeces, and the 50 mL in sweat. The normal situation is therefore of surplus water intake, and the kidney produces a larger volume of less-concentrated urine than the obligatory loss.

Sodium balance

The average daily food intake of sodium is 457 mmol (10.5 g). As 11 mmol is lost in the sweat and 11 mmol in the faeces, the kidneys must excrete 435 mmol (10 g).

Average minimum dietary requirements

The daily requirement of adults for water is of the order of 25–35 mL/kg, sodium 1–1.4 mmol/kg, potassium 0.7–0.9 mmol/kg, and chloride 1.3–1.9 mmol/kg. As an approximation, adults require an intake of 100 mmol of sodium and 60 mmol of potassium each day.

FUNCTIONAL ANATOMY OF THE KIDNEYS

The nephron

Within each kidney there are one million nephrons. A protein-free filtrate of plasma is formed at the beginning of the nephron by the renal corpuscle, and the fluid then passes along the lumen, through the proximal convoluted tubule, the loop of Henle, the distal convoluted tubule and to the collecting ducts. The inner medullary collecting ducts join with others, forming large papillary collecting ducts that empty into a calyx of the renal pelvis and into the ureter (Fig. 7.1).

The nephron is made up of a single layer of epithelial cells separated from the peritubular capillaries by a basement membrane. In the distal convoluted tubule and the collecting ducts there are more than one type of cell (the principal cells and the type A and B intercalated cells).

The renal corpuscle

The renal corpuscles, in the cortex of the kidney, comprise tufts of glomerular capillaries in distinct loops that invaginate the Bowman's capsules. Fluid is filtered from the glomerular capillaries into Bowman's space under the action of opposing hydrostatic and oncotic pressures. The glomerular capillaries are supplied by an afferent arteriole and are drained by a second, efferent, arteriole. The filtration barrier to the movement of fluid and solutes from the glomerular capillary to Bowman's space comprises the capillary endothelium, a layer of basement membrane, and the capsular epithelial cells, the podocytes. The capillary endothelium has many perforations, or fenestrae. The podocytes have many lengthened foot processes, which overlap and are embedded in the basement membrane layer. There are slits between the foot processes, and these are covered by very thin diaphragms. Mesangial cells are present within the loops of the glomerular capillaries and these may influence the diffusion barrier.

The proximal tubule

The proximal tubule collects the large volume of filtrate from Bowman's capsule, and reabsorbs some 60 per cent of it back into the bloodstream. The proximal tubule reabsorbs water, sodium, chloride, potassium, bicarbonate, calcium, glucose, urea, phosphate and any filtered proteins. Substances secreted from the blood into the lumen by the proximal tubule include hydrogen ions, ammonium, urate and organic anions and cations.

The loop of Henle

The loop of Henle consists of a thin limb which descends into the medulla, followed by a hairpin bend and an ascending limb which returns to the cortex (Fig. 7.1). In nephrons with short loops, the ascending limb is thick. In nephrons with long loops that course through the medulla to the tips of the papillae, the ascending limb is at first thin and becomes thickened as it passes through the outer medulla back to the cortex. The purpose of the loop of Henle is to create an increasing interstitial osmotic gradient in the medulla, permitting reabsorption of water from the collecting ducts and the production of a concentrated urine

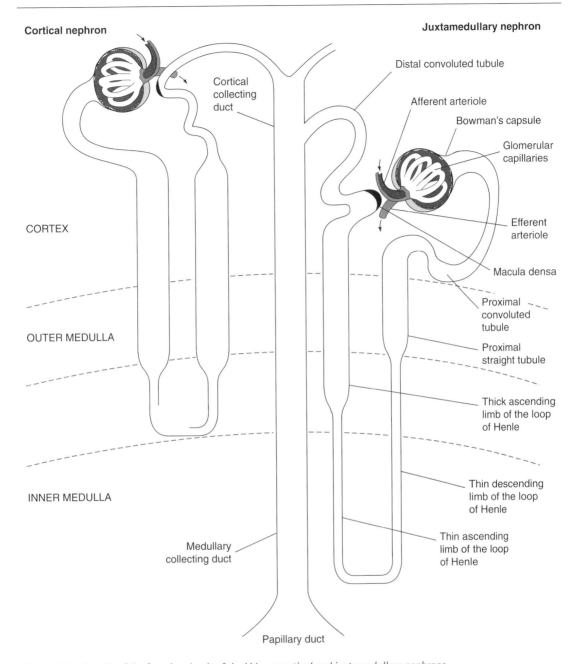

Figure 7.1   *Details of the functional unit of the kidney: cortical and juxtamedullary nephrons.*

(up to 1400 mOsmol/kg) in the presence of anti-diuretic hormone (ADH). The descending limb of the loop of Henle reabsorbs water. The thick ascending limb reabsorbs sodium, potassium, chloride and bicarbonate, and secretes hydrogen ions.

Superficial cortical, midcortical and juxtamedullary nephrons

The position of the renal corpuscle in the cortex determines the length of the loop of Henle. Superficial cortical nephrons have short loops with efferent arteriole branches into peritubular capillaries that surround the nephron segments, while juxtamedullary nephrons, with corpuscles close to the medulla, have long loops with the efferent arteriole forming not only the peritubular capillaries but also the vasa recta. Midcortical nephrons may have either short or long loops.

The distal tubule

Reabsorption of sodium chloride, bicarbonate and calcium takes place in the distal tube, and potassium and hydrogen ions are secreted into the lumen. Water is not reabsorbed by this section of the nephron.

The collecting ducts

The collecting duct is composed of two types of cells: principal cells and intercalated cells. The principal cells have a moderately invaginated basolateral membrane and contain few mitochondria. Intercalated cells have a high density of mitochondria. The later parts of the tubules fine-tune the composition of the urine. The collecting ducts course down through the medulla and empty into the papillary ducts. Aldosterone stimulates sodium reabsorption and potassium secretion by the principal cells of the cortical collecting ducts. ADH increases the permeability of the cortical and medullary collecting ducts to water.

The renal blood vessels

The afferent arterioles to the high-pressure glomerular capillaries arise from small branches of the renal artery. The glomerular capillaries form distinct loops before joining together, and are drained by the efferent arteriole which then supplies the low-pressure peritubular capillaries which receive the large amount of fluid and electrolytes reabsorbed by the tubules. The efferent arterioles of some juxtamedullary glomeruli supply the descending vasa recta, which follow the long loops of Henle into the inner medulla and are then drained by the venous ascending vasa recta. The vessel loop formed by the descending

vasa recta and the ascending vasa recta forms the counter-current exchange mechanism.

The juxtaglomerular apparatus

The juxtaglomerular apparatus (Fig. 7.2) is formed where the ascending thick limb of the loop of Henle passes between the afferent and efferent arterioles, close to the glomerulus. The three components of the juxtaglomerular apparatus are the granular cells, the macula densa and mesangial cells. The granular cells produce and store renin and are found in the walls of the afferent arteriole. The macula densa is a morphologically distinct region of the thick ascending limb of the loop of Henle which passes through the angle formed by the afferent and efferent arterioles of the same nephron. The cells of the macula densa contact the extraglomerular mesangial cells and the granular cells of the afferent arteriole and are involved in the control of renin production and renal blood flow.

Micturition

The smooth muscle of the bladder wall, the detrusor, is under the control of a spinal stretch reflex with descending inputs from the central nervous system. Afferents from bladder stretch receptors excite efferent parasympathetic motor fibres to the detrusor muscle. The smooth muscle of the internal urethral sphincter, at the base of the urethra, is pulled open by contraction of the detrusor. The external urethral sphincter is composed of striated muscle and is under voluntary control. As the bladder fills, afferents from the stretch receptors elicit a spinal reflex stimulating parasympathetic motor output to the detrusor and inhibiting activity in the motor nerves to the external sphincter, emptying the bladder. The stretch receptors also signal the sensation of bladder fullness. Descending inputs from the central nervous system modulate this spinal reflex, permitting voluntary delay or initiation of micturition.

GLOMERULAR FILTRATION

In the renal corpuscle, fluid filters from the glomerular capillaries into Bowman's space by the

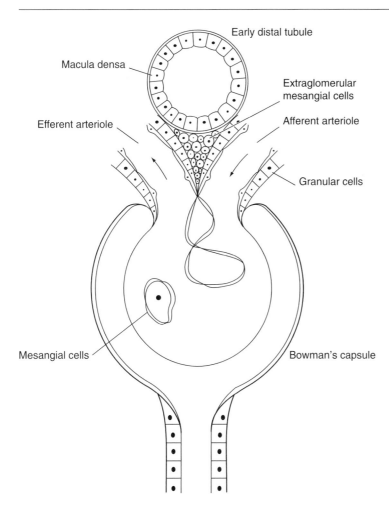

Figure 7.2 *Juxtaglomerular apparatus.*

Labels: Early distal tubule, Macula densa, Extraglomerular mesangial cells, Efferent arteriole, Afferent arteriole, Granular cells, Mesangial cells, Bowman's capsule

balance of hydrostatic and osmotic pressures acting across the thin diffusion barrier of the capillary endothelium fenestrae, the basement membrane and the slit diaphragms between the podocytes. The glomerular capillary hydrostatic pressure is determined by the renal blood flow, which is autoregulated for a range of arterial blood pressure, and the resistance of the afferent and efferent arterioles. The fluid in Bowman's space is a protein-free filtrate of plasma.

Factors determining glomerular filtration in the renal corpuscle

The rate of filtration across the glomerular capillary membrane is related to the surface area, permeability, and the net filtration pressure (hydrostatic and osmotic) acting across it. The factors of surface area and membrane permeability are included in the term the filtration coefficient (K_f):

Glomerular filtration =
  K_f × net filtration pressure

The net filtration pressure (NFP) is the balance of the capillary hydrostatic pressure moving fluid out of the capillary, the plasma oncotic pressure in the capillary which tends to retain fluid in the vessel, and the hydrostatic pressure in Bowman's capsule which tends to oppose fluid movement out of the capillary. There is no need to include a factor for the oncotic pressure in Bowman's space as the protein content of the filtrate is low (Fig. 7.3):

NFP (mmHg) =
  glomerular capillary hydrostatic pressure
  − Bowman's capsule hydrostatic pressure
  − glomerular capillary oncotic pressure

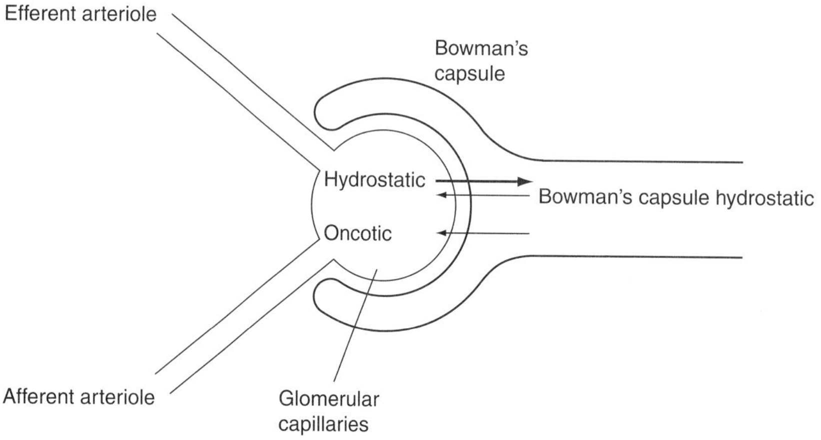

Figure 7.3 *Pressures in the renal corpuscle.*

THE NET FILTRATION PRESSURE

The filtration pressure drops from the beginning to the end of the glomerular capillary as the hydrostatic pressure falls due to vessel resistance and the oncotic pressure rises as protein-free fluid is filtered off into Bowman's capsule. The NFP at the afferent end of the capillary is thought to be of the order of 24 mmHg (glomerular capillary hydrostatic pressure of 60 mmHg − Bowman's capsule hydrostatic pressure 15 mmHg − glomerular capillary oncotic pressure 21 mmHg = 24 mmHg). At the efferent end of the capillary, the NFP falls to 10 mmHg (58 − 15 − 33 = 10 mmHg). The average NFP acting across the surface of the glomerular capillaries is thought to be 17 mmHg, which is sufficient to produce 180 L/day of glomerular filtrate.

GLOMERULAR CAPILLARY HYDROSTATIC PRESSURE

The pressure in the glomerular capillaries tends to rise with arterial blood pressure, but the effects of this are minimized by the factors involved in the autoregulation of renal blood flow. The capillary pressure is affected by the relative resistances of the afferent and efferent arterioles. Afferent arteriolar dilatation and efferent arteriolar constriction would increase the capillary hydrostatic pressure and increase glomerular filtration (one effect of the peptide hormone, atrial natriuretic factor). Afferent arteriolar constriction and efferent arteriolar dilatation would decrease the capillary hydrostatic pressure and decrease glomerular filtration.

BOWMAN'S CAPSULE HYDROSTATIC PRESSURE

If there is any obstruction to the flow of urine, the pressure in the Bowman's capsule will rise, and glomerular filtration will be reduced.

GLOMERULAR CAPILLARY ONCOTIC PRESSURE

If the renal blood flow is reduced, as by renal sympathetic nerve activity, the filtration of fluid from the reduced volume of plasma increases the oncotic pressure within the glomerular capillary and reduces the NFP.

THE FILTRATION COEFFICIENT (K_F)

The K_f, the permeability of the glomerular capillaries to water and solutes, is very high and is the product of the intrinsic permeability of the glomerular capillary and the glomerular surface area available for filtration, so the NFP can produce 180 L/day of glomerular filtrate. K_f can be altered by factors affecting the dimensions of the contractile mesangial cells between the glomerular capillaries. For example, angiotensin II may decrease the filtration rate by producing contraction of the mesangial cells with a reduction of the glomerular surface area and K_f.

THE GLOMERULAR FILTRATE

The glomerular filtrate contains electrolytes, glucose and amino acids in the same concentration as plasma. Cells and large molecular weight molecules, including proteins, are not filtered by the renal corpuscle.

Molecular weight is an important determinant of whether a substance is filtered. Below molecular weights of 7000 Da, substances are freely filtered by the glomerulus. With increasing molecular weights above 7000 Da, filtration decreases, and albumin, with a molecular weight of 70 000 Da, is only filtered in very small quantities.

Electrical charge also affects the ability of a molecule to pass across the glomerulus. The surface of the glomerular filtration barrier is covered with a layer of negatively charged substances. Negatively charged molecules, including plasma proteins, are less able to filter across the glomerulus.

FILTRATION FRACTION

The renal plasma flow is 600 mL/min, and the glomerular filtration rate (GFR) is 125 mL/min. The filtration fraction, the fraction of renal plasma flow filtered off into the tubule, is 125/600, or about 20 per cent.

CONTROL OF RENAL BLOOD FLOW

Renal blood flow

The renal blood flow amounts to 20 per cent of the cardiac output, whilst the kidneys are only less than 1 per cent of the body weight. Although the kidneys have a high metabolic rate, the renal blood flow is still 10 times more than is required. The very high renal blood flow is required for the production of large amounts of glomerular filtrate, which is necessary for the urinary excretion of waste products. Most of the blood perfuses the renal cortex (which contains the corpuscles), with the medullary flow being 10 times less (although the latter is still equivalent to brain blood flow). The main resistance vessels in the renal circulation are the afferent and efferent arterioles, and contraction of either reduces renal blood flow.

Autoregulation of the renal blood flow and the glomerular filtration rate

The renal blood flow remains relatively constant over a mean arterial blood pressure (MAP) range of 75 to 170 mmHg (Fig. 7.4). This is an intrinsic

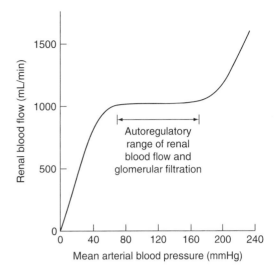

Figure 7.4   *The relationship between renal blood flow and mean arterial blood pressure.*

phenomenon and can be demonstrated in isolated kidneys. The GFR is also relatively constant over the same range of mean arterial blood pressure. This autoregulation of renal blood flow is produced by changes in the contraction of the afferent arteriole smooth muscle in response to changes in perfusion pressure (the efferent arteriole is not involved). A rise or fall in perfusion pressure leads to a corresponding rise or fall in afferent arteriolar resistance, maintaining stability of renal blood flow and GFR. The autoregulation of renal blood flow is by myogenic and tubuloglomerular feedback mechanisms.

THE MYOGENIC MECHANISM

The smooth muscle of the afferent arterioles contracts in response to the stretching produced by an increase in transmural pressure. An increase in perfusion pressure produces contraction and increased resistance (and a fall in perfusion pressure reduces resistance) by this mechanism.

TUBULOGLOMERULAR FEEDBACK

In the juxtaglomerular apparatus the macula densa lies in the wall of the ascending limb of the loop of Henle, close to the renal arterioles. The contraction of the smooth muscle of the afferent arteriole to the glomerulus is controlled by a vaso-constrictor, adenosine, from the macula densa (although the vasoconstrictor was previously thought to be renin). The macula densa releases more adenosine if the renal perfusion pressure rises, and reduces production if the pressure falls. Adenosine production by the macula densa is determined by the composition of the fluid in the ascending loop of Henle. If the perfusion pressure increases, the glomerular capillary pressure and glomerular filtration also increase. The macula densa senses the increased flow of sodium and chloride in the ascending limb of the loop of Henle, and releases more adenosine, which con-stricts the afferent arterioles, reducing the glomerular capillary pressure and the GFR. The vasodilator nitric oxide may be produced by the macula densa when the renal perfusion pressure falls.

PHENOMENON OF PRESSURE DIURESIS

The renal blood flow and GFR of the cortical nephrons are autoregulated by changes in tone of the afferent arterioles. Renal blood flow is more pre-cisely regulated than GFR over the MAP range of 75–170 mmHg. Consequently, glomerulotubular balance is disturbed and the filtration fraction increases and urine flow increases as the MAP increases. If autoregulation did not exist, as the MAP increases, the renal blood flow and GFR would increase many fold causing glomerulotubular imbalance and resulting in concomitant solute and water loss. The blood flow to the juxtamedullary nephrons is not autoregulated. Therefore when the blood pressure increases blood is diverted from the cortical nephrons to the juxtamedullary nephrons and the vasa recta capillaries. This tends to wash solutes (Na, Cl and urea) from the medulla and reduces the concentrating ability of the kidney lead-ing to an increase in urinary loss of water and solutes. This phenomenon is sometimes known as pressure diuresis.

The sympathetic nerve supply to the kidney

The kidneys are supplied by noradrenergic sympa-thetic nerves, which constrict the afferent and efferent arterioles. Renal sympathetic nerve activity reduces renal blood flow, but the relative reduction in the GFR is much less. The sympathetic nerves constrict both the afferent and efferent arterioles, increasing the glomerular capillary hydrostatic pressure, favouring filtration. However, the associated reduction in the renal blood flow increases the glomerular capillary oncotic pressure, reducing glomerular filtration. The end result is that the direct α effect of sympathetic nerve activity in the kidney is to reduce the GFR a little.

Renin release and renal sympathetic nerve activity

Renin secretion from the juxtaglomerular appar-atus granular cells is stimulated by renal sympa-thetic nerve activity, by a direct β_1 effect. Also, the fall in GFR induced by the sympathetic nerves reduces the flow of sodium and chloride to the macula densa, and this stimulates renin release. Renin release from the kidneys produces an increased blood concentration of angiotensin II.

ANGIOTENSIN II

Angiotensin II constricts the afferent and efferent arterioles, reducing renal blood flow. The effect of angiotensin II on the glomerular hydrostatic pres-sure and oncotic pressure is similar to that of the sympathetic nerves. However, by an action on the mesangial cells, angiotensin II also reduces the fil-tration coefficient, K_f, so that the net effect is a sig-nificant reduction in GFR.

EICOSANOIDS

Various products of arachidonic acid metabolism affect the renal resistance vessels. Prostacyclin and prostaglandins PGI_2 and PGE_2 are locally active renal vasodilators produced by the kidneys in clin-ical states associated with high concentrations of circulating vasoconstrictors. In such clinical states inhibition of prostaglandin production by the administration of cyclo-oxygenase inhibitors can reduce renal blood flow and impair kidney function.

ATRIAL NATRIURETIC FACTOR

Atrial natriuretic factor (ANF) increases the glomerular hydrostatic pressure and the GFR by dilating the afferent arteriole whilst constricting the efferent arteriole. ANF is produced by cardiac atrial cells, the stimulus for secretion being distension of the atria.

TUBULAR REABSORPTION AND SECRETION

The renal tubules reabsorb most of the large quantities of water and electrolytes filtered by the glomerulus. In addition, some substances are secreted by specific tubular mechanisms into the lumen. The nephron is one cell thick, and a basement membrane separates the cells from the peritubular capillaries. At the luminal membrane, there are tight junctions between the nephron cells (Fig. 7.5).

The hydrostatic pressure in the peritubular capillaries is less than the plasma oncotic pressure, so the balance of forces acting across the vessel wall favours reabsorption of fluid. There are two routes for the reabsorption of fluid and solutes in the renal tubule. The transcellular route is across the luminal and basolateral membranes of the tubule cell. The paracellular route is between the cells, across the tight junctions.

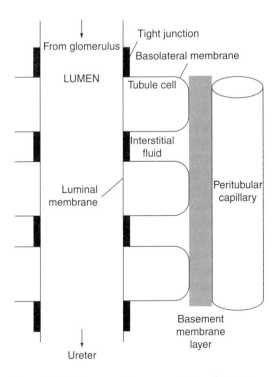

Figure 7.5 *Diagrammatic representation of tubular epithelium.*

Transport mechanisms in the renal tubule

ACTIVE TRANSPORT

Sodium is moved out of the tubular cell across the basolateral membrane, against its electrochemical gradient, by an active transport Na/K-ATPase pump (Fig. 7.6). This lowers the intracellular sodium concentration below that of the tubular fluid. Sodium ions move across the luminal membrane by a number of processes, out of the tubular lumen, and into the cell. The tubular reabsorption of sodium is therefore accomplished by the basolateral membrane Na/K-ATPase pump. The same active transport of sodium ions is also responsible for the reabsorption of glucose, amino acids, chloride, potassium and water in the nephron.

FACILITATED DIFFUSION

A molecule can cross a membrane, down its electrochemical gradient, by binding with specific membrane carrier proteins.

SECONDARY ACTIVE TRANSPORT

Two substances can move across a membrane at the same time using the same protein carrier. One substance moves down its electrochemical gradient, releasing energy that moves the other substance against its electrochemical gradient. Co-transport is the movement of the two substances in the same direction across the membrane, that is both into and out of the cell. Counter-transport is the simultaneous movement of the two substances in different directions across the membrane.

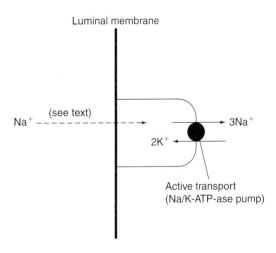

Figure 7.6 *Active transport of sodium.*

Tubular reabsorption

SODIUM REABSORPTION IN THE PROXIMAL TUBULE

In the proximal tubule sodium crosses the luminal membrane into the cell by co-transport with nutrients (including glucose and amino acids), and by counter-transport with hydrogen ions (or ammonium). These processes are dependent on the basolateral membrane Na/K-ATP-ase pump, which reduces the intracellular sodium concentration and maintains the negative intracellular potential. Luminal sodium ions flow down their electrochemical gradient into the cells, releasing energy for the co-transport and counter-transport of other substances (Fig. 7.7).

The activity of the basolateral membrane Na/K-ATP-ase pump is essential for the reabsorption of most substances in all segments of the nephron.

GLOMERULOTUBULAR BALANCE

Sodium reabsorption by the proximal tubule is adjusted to match GFR so that if the GFR increases then the amount of Na^+ reabsorbed by the proximal tubule also increases. This phenomenon is called glomerulotubular balance. Two mechanisms are responsible for glomerulotubular balance. One mechanism is mediated by an increase in the filtered load of glucose and amino acids. As GFR increases the filtered load of glucose and aminoacids also increases. The reabsorption of sodium is coupled to that of glucose and amino acids and so Na^+ and

water reabsorption increases as the filtered load of glucose and amino acids increases in association with an increase in GFR. The other mechanism responsible for glomerulotubular balance is related to the oncotic and hydrostatic pressures between the peritubular capillaries and the lateral intercellular space. An increase in GFR raises the protein concentration in the glomerular capillary plasma. This protein-rich plasma enters the peritubular capillaries and increases the oncotic pressure in the peritubular capillaries which enhances the movement of solutes and water from the lateral intercellular space into the peritubular capillaries. This increases the net solute and water reabsorption by the proximal tubule. Thus a constant fraction of the filtered Na^+ and water is reabsorbed from the proximal tubule despite variations in GFR. Consequently the net result of glomerulotubular balance is to minimize the impact of changes in GFR on the amount of Na^+ and water excreted in the urine.

SODIUM REABSORPTION IN THE ASCENDING THICK LIMB OF THE LOOP OF HENLE

In the ascending thick limb of the loop of Henle sodium ions move across the luminal membrane into the cell by a co-transport mechanism with both potassium and chloride, and by counter-transport with hydrogen ions (Fig. 7.8).

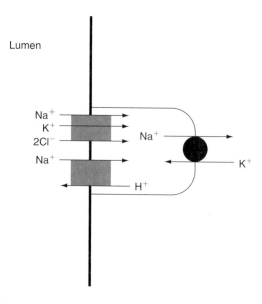

Figure 7.8 *Sodium reabsorption in the thick ascending limb of the loop of Henle.*

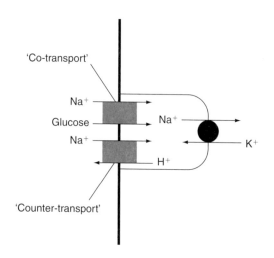

Figure 7.7 *Sodium reabsorption in the proximal tubule.*

SODIUM REABSORPTION IN THE DISTAL CONVOLUTED TUBULES

In the distal convoluted tubules sodium moves into the cell either by through specific sodium channels or by co-transport with chloride ions (Fig. 7.9).

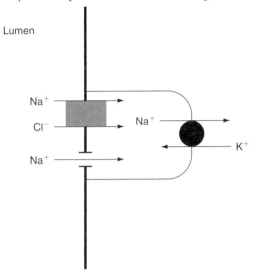

Figure 7.9   *Sodium reabsorption in the distal convoluted tubule.*

SODIUM REABSORPTION IN THE COLLECTING DUCTS

In the principal cells (the main cells in the collecting ducts), sodium enters the cell through sodium channels in the luminal membrane (Fig. 7.10).

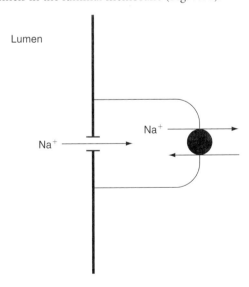

Figure 7.10   *Sodium reabsorption in the collecting duct principal cell.*

WATER REABSORPTION IN THE NEPHRON

Water reabsorption is by diffusion through the cell membranes and the tight junctions. The reabsorption of sodium and other solutes decreases the osmotic pressure of the luminal fluid, and water is reabsorbed by osmosis. The ascending limb of the loop of Henle is impermeable to water. ADH increases the water permeability of the collecting duct membrane.

CHLORIDE REABSORPTION IN THE NEPHRON

Chloride reabsorption is both paracellular and transcellular, and is coupled to a large extent to sodium reabsorption. However, in the collecting duct type B intercalated cells, chloride is reabsorbed by a process independent of the Na/K-ATP-ase pump (Fig. 7.11).

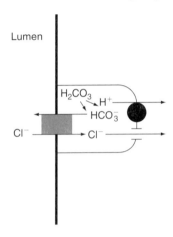

Figure 7.11   *The collecting duct type B intercalated cell.*

In the collecting duct type B cells, chloride moves across the luminal membrane by counter-transport with bicarbonate ions produced from intracellular carbonic acid. This process is dependent on the activity of the basolateral membrane hydrogen–ATP-ase pump. Chloride leaves the cell through a channel in the basolateral membrane.

GLUCOSE REABSORPTION IN THE NEPHRON: THE TRANSPORT MAXIMUM

Glucose is reabsorbed in the proximal tubule by co-transport with sodium ions, as described earlier. The proximal tubule reabsorbs all of the glucose in the tubular fluid. However, the specific carrier mechanism for glucose can be overloaded as the proximal tubule has a transport maximum for glucose (and other nutrients). If the filtered load exceeds the proximal tubule transport maximum, as in diabetes mellitus, glucose appears in the urine. In humans, at a normal GFR of 125 mL/min, glucose begins to appear in the urine at a threshold plasma glucose concentration of 10–12 mmol/L. At a plasma glucose concentration of 15 mmol/L, and with a normal GFR, the proximal tubular carrier mechanism is completely saturated and the transport maximum is reached at a filtered glucose load of 125 mL/min × 15 mmol/L = 1.88 mmol/min.

UREA REABSORPTION IN THE NEPHRON

The proximal tubule reabsorbs half of the filtered urea. This is by passive diffusion, as the luminal concentration of urea increases when water is removed from the proximal tubule, creating a gradient for reabsorption. The loop of Henle, distal convoluted tubule and the cortical collecting ducts are impermeable to urea, and the luminal fluid concentration rises as more water is reabsorbed. The inner medullary collecting duct is permeable to urea, and another tenth of the filtered load is reabsorbed. ADH increases the permeability of the medullary collecting duct to urea. In all, 60 per cent of the urea in the glomerular filtrate is reabsorbed by the nephron.

PROTEIN REABSORPTION IN THE NEPHRON

The glomerular membrane is relatively impermeable to large molecules, but some albumin is filtered and the concentration in the glomerular filtrate is 10 mg/L. Most of this is reabsorbed by the tubules, and the urine content of protein is only 100 mg/day. Large proteins molecules are taken up at the tubular luminal membrane by endocytosis and broken down by lysosomes to amino acids which diffuse into the peritubular capillaries. There is a transport maximum for protein reabsorption, and if this is exceeded large amounts of protein appear in the urine.

Mechanisms of tubular secretion

PROXIMAL TUBULAR SECRETION OF ORGANIC ANIONS

The proximal tubule secretes organic anions into the lumen by an active transport carrier. Organic anions produced in the body and secreted in this way include urate, bile salts, fatty acids and prostaglandins. Drugs and other exogenous chemical substances secreted by the proximal tubule include *para*-aminohippuric acid (PAH), penicillin, probenecid and aspirin. Some of these organic anions, such as urate, are also filtered at the glomerulus and reabsorbed by the tubule, but the secretory mechanism is involved in the regulation of the plasma concentration. Others, such as aspirin, being highly protein-bound, are not filtered and proximal tubular secretion is important for their excretion from the body.

PROXIMAL TUBULAR SECRETION OF ORGANIC CATIONS

The proximal tubule also actively secretes organic cations including creatinine, acetylcholine, the catecholamines and histamine. Drugs secreted by this mechanism include pethidine, morphine and atropine.

HYDROGEN ION SECRETION AND BICARBONATE REABSORPTION

Each day, more than 4 moles of bicarbonate ions are filtered by the renal glomerulus (24 mmol bicarbonate in each litre of the 180 L/day of filtrate), and this large amount of alkali must be reabsorbed by the tubules. The mechanism for this involves active secretion of hydrogen ions into the tubular lumen. The hydrogen ions are generated from intracellular carbonic acid, formed from water and carbon dioxide reabsorbed from the lumen. The luminal carbon dioxide and water are produced by the combination of the secreted hydrogen ion with a filtered bicarbonate ion. The bicarbonate ion produced in the cell is reabsorbed into the peritubular capillaries (Fig. 7.12).

Intracellular carbonic anhydrase catalyses the reaction between carbon dioxide and water to form carbonic acid. In the proximal tubule carbonic anhydrase is also present in the luminal cell membrane, where it catalyses the breakdown of carbonic acid to water and carbon dioxide.

Hydrogen ion secretion into the lumen is active, and by three different tubular transport mechanisms. Firstly, there is a primary active hydrogen-ATPase in the tubules, as shown in

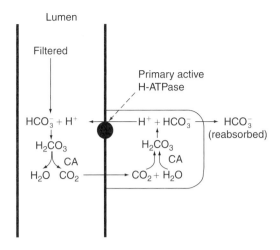

Figure 7.12   *Hydrogen secretion and reabsorption of filtered bicarbonate. CA, carbonic anhydrase.*

Fig. 7.12. Secondly, there are counter-transporters in the proximal tubule and the ascending limb of the loop of Henle that secrete hydrogen and reabsorb sodium ions. Thirdly, type A intercalated cells in the collecting ducts have a H/K-ATPase, which reabsorbs potassium and secretes hydrogen.

The tubular secretion of hydrogen ions that combine with intraluminal bicarbonate therefore prevents the loss of large amounts of filtered bicarbonate, but it does not add acidify the urine. Hydrogen ions are excreted in the urine when combined in the lumen with phosphate ions (Fig. 7.13).

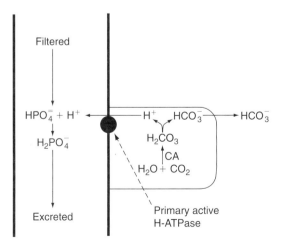

Figure 7.13   *Excretion of hydrogen ions in the urine by combination with filtered phosphate.*

RENAL CLEARANCE

Definition

For any substance, the renal clearance is the volume of plasma completely cleared of the substance by the kidneys per unit time. The units of renal clearance are therefore volume of plasma over time (e.g. mL/min or L/day).

Formula

The renal clearance can be calculated by dividing the amount of a substance in urine (collected over a given time) by the plasma concentration (P).

$$\text{Renal clearance} = \frac{\text{amount of substance in urine per unit time}}{\text{plasma concentration of substance } (P)}$$

The amount of a substance in urine over a given time is the volume of urine produced in that time (V) multiplied by the urinary concentration of the substance (U). Thus, for any substance:

$$\text{Renal clearance} = \frac{U \times V}{P}$$
(plasma volume/unit time; mL/min, L/day)

Glucose

Normally, the renal clearance of glucose is zero L/day (i.e. the filtered glucose load is lower than the proximal tubule transport maximum). Glucose is freely filtered in the renal corpuscle, but it is all reabsorbed by the proximal tubule at normal blood concentrations (at abnormal, higher, concentrations, not all of the glucose is reabsorbed and it is then found in the urine).

Inulin

Inulin is freely filtered by the glomerulus, and is neither reabsorbed, secreted, metabolized nor synthesized by the renal tubules. The amount of inulin filtered at the glomerulus is the same as the amount that appears in the urine. The renal clearance of inulin is therefore equal to the volume of fluid filtered from the glomerular capillaries into Bowman's capsule per unit time. Inulin clearance is used as a reliable estimation of the GFR, which is 125 mL/min or 180 L/day (accurate measurement of the GFR requires the use of a continuous infusion technique to establish a constant plasma inulin concentration and a steady renal excretion rate).

Para-aminohippuric acid (PAH)

PAH is filtered at the glomerulus, and any remaining in the peritubular capillaries is secreted into the lumen by the proximal tubules. When the PAH concentration is low, all the plasma-perfusing, filtering and secreting parts of the kidney (the effective renal plasma flow; 85–90 per cent of the total renal plasma flow) is completely cleared of PAH. The renal clearance of PAH is therefore equal to the effective renal plasma flow, from which the effective renal blood flow can be calculated:

$$\text{Effective renal blood flow} = \frac{\text{effective renal plasma flow}}{1 - \text{blood haematocrit}}$$

Creatinine

Creatinine is filtered by the renal tubule and is not reabsorbed. Creatinine clearance is used routinely as a method of estimating the GFR. However, a small amount of creatinine is secreted by the tubules into the lumen so that the creatinine clearance is slightly greater than the true GFR.

Urea

Urea passes freely through the renal corpuscles, and tubular reabsorption varies between 40 and 60 per cent of the filtered amount. The calculated urea clearance is about half of the GFR. However, the variable tubular reabsorption means that urea clearance is an inaccurate method of estimating the GFR.

THE LOOP OF HENLE AND PRODUCTION OF CONCENTRATED URINE

The function of the loop of Henle

The loop of Henle creates the high medullary interstitial osmolality which is essential for the production of concentrated urine (maximum osmolality of 1400 mosmol/kg H_2O) from the glomerular filtrate (that has the same osmolality as plasma, 300 mosmol/kg H_2O). In the presence of ADH, water is reabsorbed through the walls of the collecting ducts because of the high medullary interstitial fluid osmolality.

THE DESCENDING LIMB OF THE LOOP OF HENLE

The descending limb of the loop of Henle is permeable to water, but impermeable to sodium and chloride.

THE ASCENDING LIMB OF THE LOOP OF HENLE

Sodium, potassium and chloride are actively reabsorbed from the ascending thick limb by a co-transport mechanism. This active transport of sodium (with potassium and chloride) into the intersitial fluid is the prime cause of the high interstitial osmolality. The thick ascending limb of the loop of Henle is impermeable to water. There is also a significant reabsorption of sodium by paracellular diffusion in the thick ascending limb. The positive electrical potential in the lumen and high sodium permeability favour the paracellular reabsorption of sodium ions in the thick ascending limb.

THE EFFECT ON MEDULLARY INTERSTIAL FLUID OSMOLALITY

Active transport of sodium and chloride from the thick ascending limb increases the osmolality of the interstitial fluid, and dilutes the fluid within the tubular lumen. Because of the raised interstitial osmolality, water moves out of the descending limb. In the loop of Henle, water reabsorption (descending limb) is separated from sodium and chloride reabsorption (ascending limb). The net result is that the osmolality of both the interstitial fluid and the fluid within the descending limb increases (to 400 mOsmol/kg H_2O), whilst the osmolality within the ascending limb decreases (to 200 mOsmol/kg H_2O).

THE COUNTER-CURRENT MULTIPLIER

The concentrating effect described earlier on the interstitial fluid is multiplied in the kidney by the counter-current flow of tubular fluid within the two limbs of the loop of Henle. Tubular fluid flows down the descending limb, and then in the opposite, counter-current, direction back up the ascending limb. The effect of this is to multiply the maximum interstitial osmolality from 400 mOsmol/kg H_2O to 1400 mOsmol/kg H_2O in the inner medulla.

The active transport of sodium and chloride in the ascending limb and the water permeability of the descending limb produce, as described earlier, an osmolality of 400 mOsmol/kg H_2O in the descending limb and the interstitial fluid, and an osmolality of 200 mOsmol/kg H_2O in the ascending limb. The processes of fluid flow down and back up the loop, active sodium reabsorption from the ascending limb, and water reabsorption from the descending limb are continuous and move the more concentrated tubular fluid down into the hairpin of the loop. The effect is to increase the interstitial osmolality at the hairpin tip of the loop in the medulla and dilute the fluid leaving the ascending limb of the loop of Henle (Fig. 7.14).

The long loops of Henle reach from the cortex down into the tips of the papilla in the medulla and then back into the cortex. The length of the loops determines the osmolality in the inner medulla and the maximum urine concentration, which is 1400 mOsmol/kg H_2O in humans. Species with relatively longer loops of Henle can produce more concentrated urine and therefore have a relatively lower obligatory water loss than humans.

Fluid entering the loop of Henle from the proximal tubule has an osmolality of 300 mOsmol/kg H_2O. At the tip of the loop, the osmolality is 1400 mOsmol/kg H_2O. At the end of the ascending limb the osmolality is 100 mOsmol/kg H_2O. Of the fluid leaving the proximal tubule, the loop of Henle reabsorbs 25 per cent of the sodium and chloride, but only 10 per cent of the water.

THE THIN ASCENDING LIMB OF THE LOOP OF HENLE

In the long loops of Henle the ascending limb is initially thin. The process of sodium reabsorption here is not fully understood, but it is not by the active transport mechanism of the thick ascending limb.

THE VASA RECTA: THE COUNTER-CURRENT EXCHANGER

The vasa recta, the blood vessels to the loops of Henle and the collecting ducts, are also arranged in hairpin loops, with closely applied descending and ascending limbs. As they descend into the medulla, water is lost from and salt absorbed into the vessels, and at the tip of the hairpin the fluid may have an osmolality of 1200 mOsmol/kg H_2O. However, the process is reversed in the ascending vasa recta, so that fluid leaving the vessels has an osmolality of only 320 mOsmol/kg H_2O. This counter-current exchange between the descending and ascending vessels ensures that the blood flow to the medulla does not wash away the interstitial medullary gradient created by the loop of Henle. This is a passive process facilitated by the slow flow of blood in the vasa recta (Fig. 7.15).

THE ROLE OF UREA

The high medullary interstitial osmolality is not only accounted for by the high sodium and chloride concentration, but is partly due to the presence of urea (although the active transport of sodium and chloride from the ascending limb of the loop of Henle is essential). The loop of Henle, distal convoluted tubule, the cortical collecting duct and the outer medullary collecting ducts are all impermeable to urea. As a result, the urea concentration within the inner medullary collecting ducts is very high, and it is reabsorbed into the interstitial fluid by a mechanism stimulated by ADH. Consequently, the inner medullary interstitial urea concentration is high and accounts for 650 mOsmol/kg H_2O of the total osmolality of 1400 mOsmol/kg H_2O of the fluid present (with sodium and chloride accounting for most of the remainder).

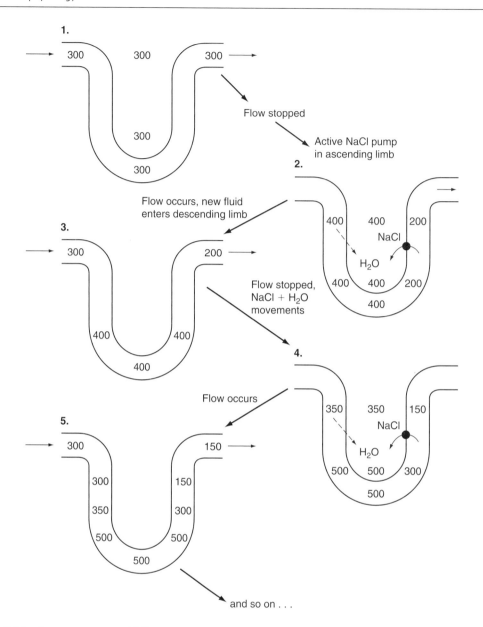

Figure 7.14   *Counter-current multiplier system in the loop of Henle. Values shown are mosmol/kg H_2O.*

THE PRODUCTION OF DILUTE URINE

During water excess, the fluid leaving the loop of Henle, osmolality 100 mOsmol/kg H_2O, is further diluted by the distal convoluted tubule, which is always impermeable to water whilst reabsorbing sodium and chloride. Similarly, the collecting ducts are relatively impermeable to water, and the fluid within them is unaffected by the high medullary interstitial fluid osmolality. The collecting ducts

reabsorb more sodium and chloride so that large volumes of dilute urine are excreted.

PRODUCTION OF CONCENTRATED URINE

During water deprivation, the distal convoluted tubule reabsorbs sodium and chloride but remains impermeable to water, and the fluid leaving it still has an osmolality of less than 100 mOsmol/kg H_2O. ADH changes the permeability of the

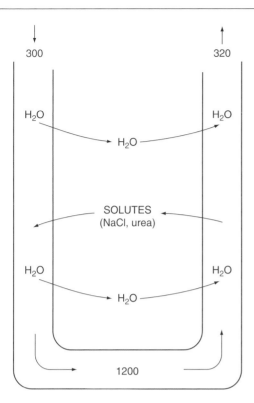

Figure 7.15 *Counter-current exchange in the vasa recta. Values shown are mosmol/kg H₂O.*

collecting duct wall so that it becomes permeable to water. In the presence of ADH, tubular fluid leaving the cortical collecting duct has an osmolality of 300 mOsmol/kg H₂O (the same as cortical interstitial fluid osmolality). With ADH, the medullary collecting duct reabsorbs much water because of the high medullary interstitial fluid osmolality produced by the loop of Henle, and a small volume of concentrated urine of osmolality 1400 mOsmol/kg H₂O is excreted (the obligatory urine loss). Urea accounts for half of the osmolality of this concentrated urine and sodium, chloride, potassium, creatinine, and other solutes account for the remainder (Fig. 7.16).

SUMMARY OF TUBULAR HANDLING OF THE GLOMERULAR FILTRATE

The proximal tubule

Each day, 180 L of the filtrate of plasma, with an osmolality of 300 mOsmol/kg H₂O, enters the proximal convoluted tubule from Bowman's capsule. The proximal tubule reabsorbs 65 per cent of the filtered sodium, chloride and water, and 55 per cent of the filtered potassium.

Active sodium reabsorption in the proximal tubule lowers the osmolality of the tubular fluid, so that water is reabsorbed in equal amounts to sodium. The movement of water increases the intraluminal concentration of other solutes, including potassium, chloride and urea, which are then reabsorbed by diffusion.

Much of the chloride reabsorption is by para-cellular diffusion, but active chloride reabsorption also occurs in the proximal tubule. Organic ions in the tubular cells dissociate into bases and hydrogen ions, and there is counter-transport of chloride into the cell and the bases out of the cell, with simultaneous sodium reabsorption by counter-transport with excreted hydrogen ions. The active reabsorption of sodium in the proximal tubule brings about the reabsorption of many nutrients, and the secretion of hydrogen ions.

At the end of the proximal tubule, only 35 per cent of the filtered water remains in the tubular lumen. The osmolality of the tubular fluid is 300 mOsmol/kg H₂O.

The loop of Henle

In the loop of Henle, 10 per cent of the filtered water is reabsorbed (descending limb), together with 25 per cent of the filtered sodium and chloride and 30 per cent of the filtered potassium. The loop of Henle creates the high medullary interstitial osmotic pressure gradient, which is necessary for water reabsorption in the collecting ducts. At the end of the loop of Henle 25 per cent of the filtered water remains in the tubule, and the tubular fluid is hypo-osmotic (100 mOsmol/kg H₂O).

The distal convoluted tubule

Sodium and chloride reabsorption continues in the distal convoluted tubule, but water is not reabsorbed, so the osmolality of the tubular fluid falls further. At the end of the distal tubule 25 per cent of the filtered water remains in the tubule.

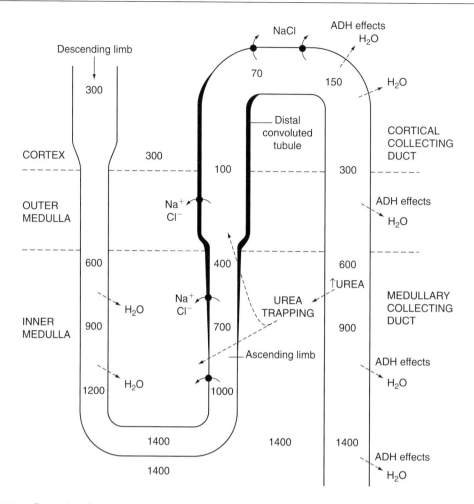

Figure 7.16   *Formation of concentrated urine.*

The collecting ducts

ADH changes the water permeability of the walls of the collecting ducts. In the presence of high concentrations of ADH, less than 1 per cent of the filtered water remains in the tubule lumen at the end of the collecting ducts, and the fluid osmolality is 1400 mOsmol/kg H_2O. In the absence of ADH, 20 per cent of the filtered water remains in the tubule at the end of the collecting ducts, and 30 L/day of urine is excreted, with an osmolality of less than 100 mOsmol/kg H_2O.

POTASSIUM HANDLING IN THE DISTAL CONVOLUTED TUBULE AND COLLECTING DUCTS

The principal cells of the distal convoluted tubule and cortical collecting duct can secrete potassium,

while the type A cells of the distal convoluted tubule and cortical collecting duct reabsorb potassium. Quantitatively, the cortical collecting duct is more important than the distal convoluted tubule. The net effect depends on the dietary intake of potassium. With a normal intake, the net effect of the distal convoluted tubule and the cortical collecting duct is potassium secretion, but during potassium depletion, the net effect is potassium reabsorption. The medullary collecting duct always reabsorbs potassium.

HORMONAL CONTROL OF TUBULAR FUNCTION

The renin–angiotensin system

Aldosterone is produced in the adrenal cortex, and stimulates both sodium reabsorption and potassium

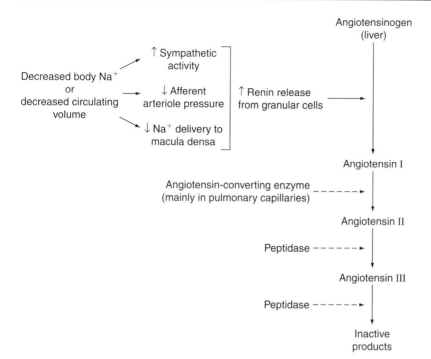

Figure 7.17 *The renin–angiotensin pathway.*

secretion by the principal cells of the cortical collecting ducts. The renin–angiotensin system is an important regulator of aldosterone secretion.

Renin

The granular cells of the juxtaglomerular apparatus secrete the proteolytic enzyme, renin, which splits off a 10-amino acid peptide, angiotensin I, from angiotensinogen, a large protein produced in the liver. Angiotensin-converting enzyme is present in capillary endothelium (especially in the lungs), and converts angiotensin I to the active angiotensin II by removing two amino acid moieties. Renin is the rate-limiting enzyme in this process (Fig. 7.17).

THE CONTROL OF RENIN SECRETION

The secretion of renin by the juxtaglomerular cells is controlled by the renal sympathetic nerves, intrarenal baroreceptors, the macula densa, and angiotensin II. The renal sympathetic nerves and circulating catecholamines each increase renin secretion via β_1 receptors. Baroreceptor reflexes, which detect low systemic cardiovascular pressures, increase renin secretion.

The granular cells of the juxtaglomerular apparatus, located in the wall of the afferent arteriole, are intrarenal baroreceptors, and increase renin secretion when intrarenal vascular pressures are low. Within the juxtaglomerular apparatus the macula densa lies in the wall of the ascending limb of the loop of Henle, close to the renal arterioles. The macula densa increases renin secretion if there is a reduction in the sodium chloride content of the flow of tubular fluid through the loop of Henle, produced by a fall in GFR, or by an increase in proximal tubular reabsorption.

Renin secretion by the granular cells is directly inhibited by angiotensin II under the effect of a negative feedback loop that controls the plasma concentration of angiotensin.

Angiotensin II

Angiotensin II has many effects which reduce sodium and water excretion and maintain circulating blood volume and blood pressure:

- It stimulates sodium reabsorption in the nephron by increasing aldosterone production and also by a direct effect on the tubules. In addition, its vasoconstrictor effect reduces the pressure in the peritubular capillaries, enhancing fluid reabsorption.

- Angiotensin II is a potent vasoconstrictor and increases sympathetic nervous system activity by both central and peripheral mechanisms. Its cardiovascular effects include increases in peripheral resistance, cardiac output and arterial blood pressure.
- Angiotensin II constricts the afferent and efferent arterioles in the kidney, reducing renal blood flow. It also reduces the filtration coefficient of the renal corpuscle, K_f, and the net effect is a significant reduction of the GFR.
- Angiotensin II increases thirst and water intake by a direct hypothalamic effect. It also stimulates ADH secretion, reducing renal water excretion.

Aldosterone

The release of the steroid hormone aldosterone from the zona glomerulosa of the adrenal cortex is stimulated by angiotensin II, increased plasma potassium concentration and adrenocorticotrophic hormone (ACTH). Aldosterone acts in the renal cortical collecting duct where it stimulates sodium reabsorption and potassium secretion by the principal cells, and hydrogen secretion by the type A intercalated cells.

Aldosterone is the main determinant of tubular sodium reabsorption, and controls the reabsorption of over 500 mmol of sodium, which is more than the normal dietary intake. Aldosterone induces the production of proteins in the cortical collecting duct cells, including the basolateral membrane Na/K-ATPase pump and sodium and potassium channels in the luminal membrane. Aldosterone similarly increases sodium reabsorption from the gut, sweat and salivary glands.

Antidiuretic hormone (ADH; vasopressin)

The peptide ADH is synthesized in the hypothalamus and secreted from the posterior pituitary. ADH increases the water permeability of the collecting duct luminal membrane so that the high inner medullary interstitial can produce water reabsorption from the collecting ducts. ADH, via the intracellular second messenger cyclic AMP,

inserts protein channels for water into the luminal membrane. ADH also increases urea reabsorption from the inner medullary collecting ducts, which maintains the contribution of urea to the maintenance of the high inner medullary osmolality.

At high blood concentrations, ADH is a vasoconstrictor ('vasopressin') and can reduce renal blood flow and GFR. ADH also stimulates sodium reabsorption and potassium secretion by the cortical collecting duct.

The release of ADH from the posterior pituitary is under the control of baroreceptor and osmoreceptor reflexes that act via the hypothalamus. Hypothalamic osmoreceptors sense changes in plasma osmotic pressure. A rise in osmotic pressure increases ADH secretion, and a fall in osmotic pressure reduces ADH secretion. The set point of the system is defined as the plasma osmolality at which ADH secretion begins to increase and the slope of this relationship is quite steep reflecting the sensitivity of the system, and below this virtually no ADH is released. The set point varies from 280 to 295 mOsm/kg H_2O. A fall in plasma volume is detected by arterial, venous and particularly cardiac atrial baroreceptors, which reduce their afferent firing rate to the hypothalamus, in turn increasing ADH release from the posterior pituitary. The sensitivity of the baro-receptor mechanism is less than that of the osmoreceptors. A 5–10 per cent decrease in blood volume is required before ADH secretion is stimulated. Changes in blood volume also influence the secretion of ADH in response to changes in plasma osmolality. When a decrease in blood volume occurs, the set point shifts to lower plasma osmolality values and the slope is steeper. This allows the kidney to conserve water, even though the water retention will reduce the osmolality of body fluids. The opposite response occurs with an increase in blood volume and the set point shifts to higher osmolality values and the slope is decreased. The same osmoreceptors and baroreceptors control the sensation of thirst via hypothalamic centres close to those producing ADH.

Atrial natriuretic factor (ANF) and renal natriuretic peptide

Distended and stretched cardiac atrial cells secrete a peptide, ANF, which increases the renal excretion of sodium and water by a number of mechanisms. ANF increases the GFR by dilating the afferent arteriole but constricting the efferent arteriole (increasing net filtration pressure), and by increasing the filtration coefficient, K_f. ANF inhibits both renin secretion and aldosterone release from the adrenal cortex, increasing sodium and water excretion. In addition, ANF directly inhibits sodium reabsorption in the collecting ducts. Renal natriuretic peptide is produced by and has effects within the kidney. Renal natriuretic peptide increases GFR and reduces Na^+ reabsorption in the collecting ducts but does not inhibit ADH action on the collecting ducts.

CONTROL OF RENAL SODIUM, WATER AND POTASSIUM EXCRETION

Control of renal sodium excretion

Each day, 180 L of glomerular filtrate is formed in the kidneys, and as this volume of fluid has the same sodium concentration as plasma (140 mmol/L), it contains a large amount of sodium:

$$\text{Sodium filtered} = 140 \text{ mmol/L} \times 180 \text{ L/day}$$
$$= 25\,200 \text{ mmol/day}$$

As described earlier, the average daily food intake of sodium is 457 mmol (10.5 g), while 11 mmol is lost in the sweat and 11 mmol in the faeces, and the kidneys excrete 435 moles (10 g) per day. It is clear that most of the sodium filtered at the glomerulus is reabsorbed by the renal tubule.

$$\text{Sodium excretion in urine} =$$
$$\text{sodium filtered} - \text{sodium reabsorbed by}$$
$$\text{renal tubule}$$

As the plasma sodium concentration is normally kept constant, the renal excretion of sodium depends on the GFR and tubular reabsorption. Both of these are varied to maintain body sodium balance.

Sodium is the main extracellular cation and, with its associated anions, determines the extracellular fluid volume of the body which comprises the interstitial and plasma volumes. A reduction in body sodium content is associated with a fall in the extracellular volume and a fall in the circulating plasma volume. A rise in body sodium content increases the plasma volume. Changes in body sodium content are reflected by changes in plasma volume and hence cardiovascular hydrostatic pressures. Renal sodium excretion is altered by these changes in cardiovascular pressures affecting the GFR and tubular sodium reabsorption. The cardiovascular pressures have direct effects on the kidney, and indirect effects via arterial, venous and cardiac baroreceptor reflex modulation of the sympathetic nervous system, the renin–angiotensin system, and other hormones.

Body sodium, extracellular fluid volume and the control of GFR

As described earlier,

$$\text{Glomerular filtration} = K_f \times \text{NFP}$$

and

$$\text{NFP (mmHg)} =$$
$$\text{glomerular capillary hydrostatic pressure}$$
$$- \text{Bowman's capsule hydrostatic pressure}$$
$$- \text{glomerular capillary oncotic pressure}$$

DIRECT EFFECTS OF BODY SODIUM ON GFR

A reduced body sodium content with a fall in extracellular fluid volume has direct effects that reduce the GFR. A fall in plasma volume and arterial blood pressure lower the glomerular capillary hydrostatic pressure. Also, with low extracellular volume, the glomerular capillary oncotic pressure is increased. The net filtration pressure is reduced, and the GFR falls.

INDIRECT EFFECTS OF BODY SODIUM ON GFR

Arterial, venous and cardiac baroreceptor reflexes indirectly reduce the GFR when the body sodium content falls. An increase in renal sympathetic outflow to the kidney reduces renal blood flow and, to a lesser extent, GFR. Activation of the renin–angiotensin system releases angiotensin II, which reduces GFR markedly (vasoconstriction and reduced K_f). ANF production falls when atrial hydrostatic pressures are low.

These direct and indirect processes reduce the GFR when the extracellular volume is low, retaining sodium in the body. The same processes work in reverse to increase sodium filtration when the extracellular volume is elevated.

Body sodium, extracellular fluid volume and control of tubular sodium reabsorption

When the body sodium content falls, many factors increase the tubular reabsorption of sodium, including aldosterone, angiotensin II, the renal sympathetic nerves, effects on renal interstitial hydrostatic pressure, ADH and a reduction in ANF secretion:

- *Aldosterone.* When body sodium, extracellular volume and cardiovascular hydrostatic pressures fall, renin secretion is stimulated by sympathetic nerve activity induced by cardiovascular baroreceptor reflexes, by intrarenal baroreceptors, and by the reduced flow of sodium at the macula densa. Aldosterone increases the reabsorption of sodium by the principal cells of the cortical collecting ducts.
- *Renal interstitial hydrostatic pressure.* When body sodium concentration is low, the reduced arterial pressure, vasoconstriction of the renal arterioles (with a fall in peritubular capillary hydrostatic pressure) and raised blood oncotic pressure reduce the hydrostatic pressure of the interstitial fluid between the renal tubular cells and the peritubular capillaries. A reduction in renal hydrostatic pressure enhances the movement of sodium and water from the tubule into the peritubular capillary.
- *Angiotensin II.* In addition to indirect effects of increased aldosterone release and arteriolar vasoconstriction, angiotensin II directly stimulates sodium reabsorption by the renal tubules.
- *Renal sympathetic nerves.* In addition to indirect effects on renin secretion and arteriolar vasoconstriction, the renal sympathetic nerves directly stimulate sodium reabsorption by the renal tubules.
- *ANF.* When total body sodium is low, ANF production is reduced, the inhibitory effects of

this hormone on sodium reabsorption are removed, and sodium is conserved in the body.
- *ADH.* A fall in extracellular volume increases ADH production by baroreceptor reflexes. ADH increases sodium reabsorption in the cortical collecting duct.

These processes increase tubular sodium reabsorption when the extracellular volume is low. The same processes work in reverse to decrease sodium reabsorption when the extracellular volume is elevated.

Control of renal water excretion

The daily urinary excretion of water is the balance of the volume filtered by the glomerulus minus the volume reabsorbed by the tubule:

Water excretion in urine =
    water filtered − water reabsorbed
    by renal tubule

The primary control of water excretion is by modulation of tubular reabsorption, although cardiovascular hydrostatic pressures changes and baroreceptor reflexes affecting GFR (as described earlier) have some effect. The main control of renal water excretion is by modulation of water reabsorption in the collecting ducts which are only permeable to water when ADH is present. As described previously, ADH secretion from the posterior pituitary is controlled by hypothalmic baroreceptor and osmoreceptor reflexes.

Normally, more water is ingested than is required to balance insensible loss, sweat, fluid in the faeces and the minimum obligatory urine volume of 430 mL (required to clear the renal solute load). The average daily urine output is 1500 mL. In the complete absence of ADH, as in diabetes insipidus, 25–30 L/day of urine is produced each day.

Control of renal potassium excretion

Potassium is the main intracellular cation. The electrical potential across resting cell membranes is determined by the ratio of intracellular and extracellular potassium concentrations, according to the Nernst equation. Small changes in the plasma potassium concentration produce significant effects on excitable tissues, including the heart.

Various factors alter the intracellular/extracellular distribution of potassium ions. Insulin and epinephrine stimulate the cell membrane Na/K-ATPase, so that potassium is shifted into the cell after a meal and during exercise, respectively. During acidosis, when the plasma hydrogen ion concentration is high, potassium ions move out of the cells and hydrogen ions move in. During alkalosis, potassium moves into cells and hydrogen ions out.

Each day, if the plasma potassium concentration is 5 mmol/L, 900 mmol of potassium ions are filtered into the renal tubules.

$$\text{Potassium filtered} = 5 \text{ mmol/L} \times 180 \text{ L/day}$$
$$= 900 \text{ mmol/day}$$

Normally most of the filtered potassium is reabsorbed, and only 30 to 100 mmol are excreted in the urine per day. However, the urinary excretion of potassium can exceed the amount of filtered potassium, indicating that potassium is also secreted by the renal tubules. Urinary potassium excretion therefore depends on the amount filtered, the amount reabsorbed, and the amount secreted by the renal tubules.

Potassium excretion in urine =
  potassium filtered
  + potassium secreted by the cortical
    collecting duct
  − potassium reabsorbed by the renal tubule

There is little control over the filtration or reabsorption of potassium. The proximal convoluted tubule constantly reabsorbs 55 per cent, probably mainly by a passive process, and the ascending limb of the loop of Henle reabsorbs 30 per cent of the filtered load of potassium. The medullary collecting duct always reabsorbs potassium.

The principal cells of the distal convoluted tubule and cortical collecting duct secrete potassium, whilst the type A intercalated cells reabsorb potassium. The contribution of the cortical collecting duct to potassium secretion is quantitatively greater than that of the distal convoluted

tubule. The net effect depends on the dietary intake of potassium. With a normal intake, the net effect of the distal convoluted tubule and the cortical collecting duct is potassium secretion. However, during potassium depletion, secretion is reduced and the net effect is potassium reabsorption. Urinary potassium excretion is therefore determined by changes in the rate of potassium secretion by the principal cells of the distal convoluted tubule and the cortical collecting duct.

CONTROL OF TUBULAR POTASSIUM SECRETION

The main factors determining the rate of potassium secretion by the principal cells of the distal convoluted tubule and the cortical collecting duct are plasma potassium concentration and aldosterone. A raised plasma potassium concentration directly stimulates the basolateral membrane Na/K-ATPase pump in the principal cells. A raised plasma potassium concentration also directly stimulates aldosterone release from the adrenal cortex. Aldosterone induces increased production of the basolateral membrane Na/K-ATPase pump and potassium channels in the luminal membrane of the principal cells (Fig. 7.18).

Potassium secretion is influenced by the flow of fluid through the distal convoluted tubule and the cortical collecting duct. The movement of

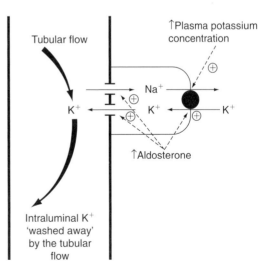

Figure 7.18 *The effect of plasma potassium concentration and aldosterone on potassium secretion by the principal cells of the cortical collecting duct.*

potassium out of the principal cells into the tubular lumen is by passive diffusion down a concentration gradient, and this depends on the potassium ions in the lumen being continuously washed away by tubular flow. The secretion of potassium varies directly with the tubular flow rate.

THE EFFECT OF PRIMARY CHANGES IN BODY SODIUM BALANCE ON POTASSIUM EXCRETION

Primary changes in body sodium content and extracellular volume do not affect potassium excretion. If total body sodium is high, aldosterone secretion is inhibited, and this tends to reduce potassium secretion. However, when body sodium and extracellular volume are high, the flow rate through the renal tubules is increased, and this increases tubular potassium secretion. The effect on potassium secretion of the reduced aldosterone is therefore balanced by the increased flow to the cortical collecting ducts. Conversely, if total body sodium is low, urinary excretion of potassium does not change as the effect on potassium secretion of increased aldosterone is balanced by the reduced tubular flow rate.

THE EFFECT OF PRIMARY CHANGES IN BODY WATER CONTENT ON POTASSIUM EXCRETION

Primary changes in total body water content do not alter urinary potassium excretion, because ADH stimulates tubular potassium secretion. When total body water content is high, ADH secretion is low and urine production high. Potassium secretion is increased by the high tubular flow rate but decreased by the low ADH, and the urinary excretion of potassium is not changed. Conversely, when total body water content is low, urinary potassium excretion is unchanged as the reduction in potassium secretion associated with the low tubular flow rate is balanced by the stimulatory effect of ADH on potassium secretion.

THE EFFECT OF ALKALOSIS ON POTASSIUM EXCRETION

The urinary excretion of potassium is increased by alkalosis as the basolateral membrane Na/K-ATPase in the principal cells is stimulated by a low plasma hydrogen ion concentration.

RENAL CONTROL OF ACID–BASE BALANCE

The importance of the carbon dioxide–bicarbonate buffering system

On a normal diet, 50–80 mmol of hydrogen ions are produced by body metabolism each day and these must be excreted by the kidneys in order to prevent acidosis (the normal plasma hydrogen ion concentration is only 36 nmol/L, i.e. 36×10^{-9} mol/L). Buffers in the body, comprising weak acids and their bases, minimize changes in plasma hydrogen ion concentration until the metabolically produced hydrogen ions can be excreted by the kidneys.

As the main extracellular buffer system, carbon dioxide and bicarbonate play a crucial role in acid–base balance. The real importance of this buffer system is that is not closed; the concentration of the acid (carbon dioxide) and the base (bicarbonate) can be changed. The plasma concentration of carbon dioxide can be changed by the lungs, and the plasma bicarbonate concentration can be altered by the kidneys. As both components of the buffer pair are under independent control, the carbon dioxide–bicarbonate system is a very efficient physiological buffer.

In aqueous solution, carbon dioxide behaves as an acid, and reacts with water to release hydrogen ions. It also releases bicarbonate ions, the corresponding buffer base:

$$CO_2 + H_2O \rightleftharpoons H^+ + HCO_3^-$$

The normal plasma carbon dioxide concentration is 1.2 mmol/L (0.03×40 mmHg), and the bicarbonate concentration is 24 mmol/L. The pK_a of the carbon dioxide–bicarbonate buffer system is 6.1, and therefore, the normal plasma pH can be calculated by the Henderson–Hasselbalch equation to be 7.4:

$$\begin{aligned} pH &= pK_a + \log \frac{[HCO_3^-]}{[CO_2]} \\ &= 6.1 + \log \frac{24}{1.2} \\ &= 7.4 \end{aligned}$$

From this equation it is clear that the kidneys can compensate for changes in the carbon dioxide tension of blood by increasing or decreasing the plasma bicarbonate concentration.

RENAL BICARBONATE HANDLING

The kidneys therefore have an important role in adjusting plasma bicarbonate concentration. Bicarbonate is freely filtered in the renal corpuscle and most is normally reabsorbed by the renal tubules. To correct an alkalosis, the kidneys reabsorb less bicarbonate and reduce the plasma bicarbonate concentration. To correct an acidosis, the kidneys reabsorb more bicarbonate by excreting hydrogen ions into the urine.

In the kidney the excretion of a hydrogen ion in the urine is equivalent to the addition of a bicarbonate ion to blood. This is because the equilibrium reaction of carbon dioxide and water, producing hydrogen ions and bicarbonate, is driven to the right if hydrogen ions are removed in the kidney. The plasma bicarbonate concentration is therefore increased by the urinary excretion of hydrogen ions (Fig. 7.19).

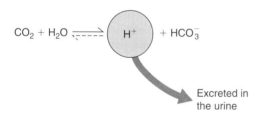

Figure 7.19 *Hydrogen excretion in the urine.*

Similarly, the loss of a bicarbonate ion in the urine is equivalent to the addition of a hydrogen ion to blood (Fig. 7.20).

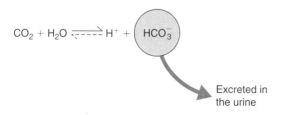

Figure 7.20 *Bicarbonate excretion in the urine.*

Hydrogen ion balance

Each day, 15 000–20 000 mmol of carbon dioxide is produced by tissue aerobic metabolism (glycolysis, the citric acid cycle and the electron transport chain). The carbon dioxide from the tissues combines with water to produce hydrogen ions and bicarbonate ions. Normally, an equal amount of carbon dioxide is lost in the lungs by the combination of hydrogen and bicarbonate ions in the pulmonary capillaries. Therefore, the tissue metabolic production of carbon dioxide does not normally produce a daily net gain or loss of hydrogen ions in the body.

Each day, non-volatile acids (which cannot be eliminated by the lungs) are produced; these include lactic acid, ketone bodies, phosphoric acid and sulphuric acid. The production of these non-volatile acids depends on dietary intake, but in general they equate to 40–80 mmol of hydrogen ions a day.

The loss of gastrointestinal fluid may produce a gain or loss of hydrogen ions as gastric fluid is acidic, but intestinal fluids are alkaline.

As discussed earlier, urinary bicarbonate loss is equivalent to the gain of hydrogen ions in plasma, and hydrogen ions excreted in urine represent a gain of bicarbonate in the blood. The urinary excretion of bicarbonate or hydrogen is altered according to the body hydrogen balance. Normally, all of the bicarbonate in the glomerular filtrate is reabsorbed and 40–80 mmol of hydrogen ions (from the metabolic production of non-volatile acids) are excreted by the kidney each day.

The glomerular filtration of hydrogen ions

As the plasma hydrogen ion concentration is only 36 nmol/L, a very small amount of hydrogen ions is filtered each day:

Hydrogen ions filtered $= 36$ nmol/L $\times 180$ L/day
$$= 6840 \text{ nmol/day}$$
$$= 6.84 \times 10^{-6} \text{ mol/day}$$
$$= 0.684 \text{ mmol/day}$$

It is clear that the contribution of filtered hydrogen ions to the daily requirement for the renal excretion of 40–80 mmol/day is negligible.

The glomerular filtration of bicarbonate ions

Bicarbonate ions filtered =
    24 mmol/L $\times$ 180 L/day = 4320 mmol/day

Each day, more than 4 mol of bicarbonate are filtered by the renal corpuscle. It is clear that this very large amount of alkali cannot be lost in the urine and that it must be reabsorbed by the renal tubules.

Bicarbonate reabsorption by the renal tubule

Of the filtered load of bicarbonate, 85 per cent is reabsorbed by the proximal tubule, 10 per cent by the thick ascending limb of the loop of Henle, and the remainder by the distal convoluted tubule and the cortical collecting duct.

The process by which filtered bicarbonate is reabsorbed involves the active transport of hydrogen ions across the luminal membrane into the tubular lumen. In the lumen, a filtered bicarbonate ion combines with the secreted hydrogen, forming carbon dioxide and water, both of which diffuse into the cell. Inside the cell, under the catalytic activity of carbonic anhydrase (CA), carbon dioxide and water react to form a bicarbonate ion (reabsorbed into the peritubular capillaries) and a hydrogen ion (available for active transport out of the cell again) (Fig. 7.12).

The reaction of carbon dioxide and water is catalysed inside tubular cells by carbonic anhydrase. This enzyme is also present in the lumen of the proximal tubule where it catalyses the breakdown of carbonic acid to carbon dioxide and water. The sum effect of this process is that a filtered bicarbonate is moved out of the tubular lumen into the cell, and a bicarbonate is moved into the peritubular capillaries. This may be thought of as reabsorption of the filtered bicarbonate ion.

This process does not excrete hydrogen ions in the urine, as secreted hydrogen ions combine with bicarbonate and diffuse back into the cell as part of a molecule of water.

MECHANISMS FOR HYDROGEN ION SECRETION ACROSS THE LUMINAL MEMBRANE

All tubular segments have a luminal membrane active transport H-ATP-ase pump. In the proximal tubule and the thick ascending limb of the loop of Henle, hydrogen ion secretion is also driven by counter-transport, with reabsorption of sodium. The type A intercalated cells of the collecting duct also have a primary active H/K-ATP-ase pump, which secretes hydrogen into the lumen and reabsorbs potassium.

BICARBONATE SECRETION BY TYPE B INTERCALATED CELLS IN THE CORTICAL COLLECTING DUCT

These cells can secrete bicarbonate ions into the urine, but the significance of this is unclear as the renal tubule always reabsorbs most of the bicarbonate filtered.

Renal excretion of hydrogen ions and the addition of new bicarbonate to blood

PHOSPHATE

If a hydrogen ion is secreted into the tubular lumen where it combines with a buffer which is not reabsorbed, then it is excreted in the urine and a new bicarbonate ion is added to the peritubular capillary blood. This occurs when the secreted hydrogen ion combines with filtered phosphate ions. Each day, 36 mmol of hydrogen ions can be excreted in the urine by combining with filtered phosphate (*see* Fig. 7.13).

AMMONIUM IONS

The proximal tubular cells take up glutamine and metabolize it to ammonium ions, and thence to glutamate, further metabolism of which produces bicarbonate. The ammonium ions are secreted into the tubular lumen by counter-transport with sodium ions, and bicarbonate diffuses into the peritubular capillaries (Fig. 7.21).

Again, hydrogen ions are excreted in the urine and new bicarbonate is added to the peritubular capillary blood. The capacity of this system to excrete hydrogen ions in the urine exceeds that of phosphate. The proximal tubule is the main site of glutamine metabolism and ammonium secretion into the urine.

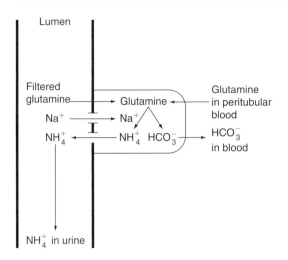

Figure 7.21 *Ammonium formation from glutamine.*

MINIMUM URINARY pH

The minimum urinary pH is 4.4, because the mechanism of active transport of hydrogen secretion is inhibited at higher hydrogen ion concentrations.

THE FATE OF HYDROGEN IONS SECRETED BY THE TUBULES

In general, hydrogen secreted into the lumen will combine with bicarbonate. When bicarbonate reabsorption is complete, the secreted hydrogen ions combine with phosphate. In the proximal tubule and the thick ascending limb of the loop of Henle, most of the secreted hydrogen ions combine with bicarbonate. In the distal convoluted tubule and the collecting ducts, little bicarbonate remains, and any secreted hydrogen ions combine with phosphate.

CONTROL OF RENAL HYDROGEN ION SECRETION

Tubular hydrogen ion secretion is increased by a raised arterial blood carbon dioxide tension and reduced by a fall in arterial blood carbon dioxide tension. The tubular secretion of hydrogen ions is increased by a high extracellular hydrogen ion concentration, and falls when the hydrogen concentration is low.

CONTROL OF RENAL AMMONIUM ION EXCRETION

Renal tubular cell glutamine metabolism and ammonium ion secretion are governed by the extracellular hydrogen ion concentration, stimulated by an acidosis and inhibited by an alkalosis.

RENAL RESPONSE TO RESPIRATORY ACIDOSIS AND ALKALOSIS

In respiratory acidosis, the kidneys increase the plasma bicarbonate concentration because the elevated carbon dioxide tension and raised extracellular hydrogen ion concentration stimulate renal tubular hydrogen and ammonium ion secretion. In respiratory alkalosis, renal secretion of hydrogen ions and ammonium ions is inhibited, less bicarbonate is reabsorbed, and the plasma bicarbonate concentration is reduced.

RENAL RESPONSE TO METABOLIC ACIDOSIS

In a metabolic acidosis the amount of filtered bicarbonate is reduced, because of the low plasma concentration. Tubular hydrogen secretion is enough to reabsorb all the filtered bicarbonate and for the urinary excretion of hydrogen ions as phosphate, even when the arterial carbon dioxide tension has been lowered by the ventilatory response to the metabolic acidosis. The tubular secretion of ammonium ions is increased by the high extracellular hydrogen ion concentration. The kidneys therefore increase the plasma bicarbonate concentration by reabsorbing all the filtered bicarbonate and by creating new bicarbonate by excreting hydrogen ions in the urine as phosphate and as ammonium ions.

RENAL RESPONSE TO METABOLIC ALKALOSIS

In a metabolic alkalosis the amount of filtered bicarbonate is increased, because of the high plasma concentration. Tubular hydrogen secretion is not enough to reabsorb all the filtered bicarbonate or for the urinary excretion of hydrogen ions as phosphate, even when the arterial carbon dioxide tension has been elevated by the ventilatory response to the metabolic alkalosis. The tubular secretion of ammonium ions is decreased by the low extracellular hydrogen ion concentration. The kidneys therefore decrease the plasma bicarbonate concentration by not reabsorbing all the bicarbonate filtered at the glomerulus.

MECHANISMS OF ACTION OF DIURETIC DRUGS

Mechanisms of action of diuretics vary according to drug type:

- *Thiazides.* Bendroflumethazide and chlortalidone reduce sodium reabsorption at the beginning of the distal convoluted tubule by inhibiting the luminal membrane sodium and chloride co-transporter.
- *Loop diuretics.* Furosemide, bumetanide and ethacrynic acid reduce sodium reabsorption in the thick ascending limb of the loop of Henle by inhibiting the luminal membrane sodium, potassium and chloride co-transporter.
- *Potassium-sparing diuretics.* These include two subgroups:
  a. Aldosterone antagonists. Spironolactone and potassium canrenoate antagonize the effect of aldosterone on the renal tubule. The main site of action is the cortical collecting duct
  b. Amiloride and triamterene. These are weak diuretics that cause potassium retention and block luminal sodium channels in the collecting ducts
- *Osmotic diuretics.* Mannitol is filtered by the glomerulus, but is not reabsorbed by the proximal tubule. As sodium reabsorption proceeds, the osmotic effect of mannitol in the tubular fluid impedes the reabsorption of water.

- *Carbonic anhydrase inhibitors.* Acetazolamide and dichlorphenamide inhibit the tubular secretion of hydrogen ions, resulting in less reabsorption of bicarbonate and sodium ions. The proximal tubule is the main site of action.
- *Mercurial diuretics.* Mersalyl poisons the active transport pumps responsible for the reabsorption of ions in the renal tubule, including the Na/K-ATP-ase pump.
- *Non-specific agents.* Drugs which increase cardiac output increase renal blood flow and urine excretion, including digoxin and plasma volume expanders. Dopamine is a renal vasodilator at low doses and increases renal blood flow. Theophylline may increase the GFR.

The effect of diuretics on potassium excretion

Potassium secretion by the principal cells is increased by increased tubular fluid flow to the cortical collecting duct. Most diuretics increase the flow of tubular fluid to the cortical collecting duct and increase potassium excretion. Thus, the thiazides and loop diuretics can produce severe potassium depletion. The aldosterone antagonists and triamterene and amiloride have specific effects on sodium excretion in the collecting ducts and do not increase urinary potassium loss.

Reflections

1. The kidneys have both excretory and regulatory functions such as regulation of body fluid osmolality and volumes, regulation of electrolyte balance, regulation of acid–base balance, excretion of metabolic products and exogenous substances, and production and secretion of hormones.
2. The nephron is the functional unit of the kidney. Each nephron consists of a renal corpuscle, proximal tubule, loop of Henle, distal tubule, and collecting duct. The renal corpuscle is formed by the Bowman's capsule and glomerular capillaries. The juxtaglomerular apparatus, an important component of the feedback mechanism that regulates renal blood

flow and glomerular filtration, consists of the macula densa, the extraglomerular mesangial cells, and renin-producing granular cells in the afferent arteriole.

3. Starling forces across the glomerular capillaries provide the driving force for ultrafiltration of plasma from the glomerular capillaries into Bowman's space. The glomerular filtration barrier is formed by the capillary endothelium, basement membrane, and filtration slits of the podocytes. The presence of negatively charged glycoproteins on the surface of all components of the filtration barrier restricts the filtration of proteins that have a molecular weight of 7000 to 70 000. Proteins with a molecular weight

greater than 70 000 are not filtered. The rate of glomerular filtration is calculated by measuring the clearance of inulin or creatinine.

4. Renal blood flow is about 20 per cent of the total cardiac output and is autoregulated. Renal blood flow determines the GFR, delivers oxygen, nutrients and hormones to cells of the nephrons, delivers substrates for urinary excretion, participates in the concentrating function of the kidneys, and modifies proximal reabsorption of water and solutes. Autoregulation is achieved by changes in renal vascular resistance mediated by tubuloglomerular feedback and the myogenic reflex, and maintains a constant renal blood flow and glomerular filtration rate despite chages in mean arterial pressure of 75–170 mmHg. Sympathetic stimulation, angiotensin II, prostaglandins, nitric oxide, endothelin, bradykinin and adenosine can override the autoregulatory mechanisms. A constant fraction of filtered sodium and water is reabsorbed from the proximal tubule despite changes in GFR; this is called glomerulotubular balance.

5. Tubular reabsorption allows the kidney to regulate the plasma concentrations of electrolytes and organic solutes. The proximal tubule reabsorbs 60–70 per cent of the glomerular ultrafiltrate, and the loop of Henle cells reabsorb about 25 per cent of the filtered NaCl and 15 per cent of the filtered water. Although the distal tubules and the collecting ducts have a limited reabsorptive capacity, the final adjustments in the composition and volume of the urine, and the regulation by hormones and other autocoids occur in the distal tubules. Excretion of various by-products of metabolism and exogenous organic anions and bases occurs by secretion into the tubular fluid.

6. The loop of Henle is central to the process of concentrating and diluting the urine. The reabsorption of NaCl by Henle's loop renders the medullary interstitial fluid hyperosmotic. This hyperosmotic medullary interstitium provides the osmotic driving force for the reabsorption of water in the counter-current multipler system of the nephron.

7. The kidneys regulate acid–base balance by excreting the daily net acid load and this is the route of excretion of the non-volatile acids. The kidneys also reabsorb nearly all the HCO_3 that is filtered at the glomerulus. The reabsorption of filtered HCO_3 and the excretion of acid is achieved by secretion of H^+ by the nephron. Phosphate is the primary urinary buffer and this is necessary for effective excretion of acid. Ammonium excretion leads to new HCO_3^- formation. Renal NH_4^+ production and excretion are regulated in response to acid–base disturbances.

8. Regulation of body fluid osmolality is achieved by the integrated interaction of ADH secretion and the hypothalamic thirst centres, and the ability of the kidney to concentrate or dilute the urine. When body fluid osmolality increases, ADH secretion and thirst are stimulated. ADH increases the permeability of the collecting ducts to water causing water reabsorption by the collecting ducts. The renal conservation of water and increased water intake restores body fluid osmolality to normal.

9. In normovolaemic states Na^+ excretion by the kidneys is matched to the NaCl intake. The kidney absorbs virtually all the filtered Na^+ and the collecting ducts adjust Na^+ excretion to achieve Na^+ balance. Aldosterone, which stimulates Na^+ reabsorption, is the major hormone that regulates Na^+ absorption by the collecting ducts. The volume of the extracellular fluid is determined by Na^+ balance. The regulation of Na^+ intake and excretion and thus the maintenance of extracellular fluid volume is integrated by the kidneys, the cardiovascular system and the sympathetic nervous system. Sensors throughout the body, especially the low- and high-pressure vascular volume sensors, monitor the effective circulating volume and then neural and hormonal factors modulate NaCl excretion to match its intake.

Acid–base physiology

The concentration of hydrogen (H^+) ions in body fluids is precisely regulated. Derangements of H^+ ion regulation can produce direct intracellular disturbances, including changes of enzyme activity, membrane excitability and energy production, and indirect systemic effects altering reflexes in the central nervous system and the release of hormones by the endocrine system. Although the body produces large quantities of volatile (carbonic acid) and fixed (non-carbonic) acids as a result of metabolism, the H^+ ion concentration of body fluids is maintained at a low concentration (40 nmol/L). Acids are removed from the body by the lungs, kidneys and the gastrointestinal tract. When an imbalance between the production and removal of H^+ ions occurs, the H^+ ion concentration deviates over a narrow range of 20–160 nmol/L (pH 6.8–7.7).

DEFINITIONS

The hydrogen ion is a hydrogen atom without its orbital electron, and therefore is a proton. In an aqueous solution, it exists as a hydrated proton called the hydronium ion (H_3O^+).

- An *acid* is a substance that donates a proton.
- A *base* is a substance that accepts protons in solution.

In solution, an acid (HA) will dissociate to a H^+ ion and a base (A^-), as shown in the equation:

$$HA \underset{k_2}{\overset{k_1}{\rightleftharpoons}} H^+ + A^-$$

The proportions of the relative reactions are determined by the dissociation constants, k_1 and k_2. If k_1 is greater than k_2 then the reaction moves towards the production of H^+ and A^-.

Henderson applied the law of mass action and described the relationship as:

$$[H^+] = K\frac{[HA]}{[A^-]} \quad \text{where } K = \frac{k_1}{k_2}$$

This shows that the concentration ([ ]) of H^+ ions in solution ($[H^+]$) depends on the ratio of the buffer pairs, A^- and HA, and the dissociation constant.

Hasselbach modified the Henderson equation using logarithmic transformation resulting in the equation:

$$pH = pK_a + \log\frac{[A^-]}{[HA]}$$

where pH is the negative logarithm of the hydrogen ion concentration, pK_a is the negative logarithm of the dissociation constant of the substance and is the pH at which the substance is 50 per cent dissociated. A substance with a lower pK_a is a stronger acid than a substance with a higher pK_a. The ability of a substance to donate or accept a proton (that is, to act as an acid or base) depends on the concentration of H^+ ions in solution (pH of the solution) and the degree of dissociation (pK) of the substance.

THE pH SYSTEM

H^+ ion concentration may be measured in two ways: directly as concentrations in nmol/L, or indirectly as pH. pH is defined as the negative logarithm (to the base 10) of the concentration of hydrogen ions. The pH is related to the concentration of H^+ as follows:

(a) $pH = \log_{10}\dfrac{1}{[H^+]}$

(b) $pH = -\log_{10}[H^+]$

(c) $H^+ = 10^{-pH}$

(d) $pH = pK + \log \text{base/acid}$

Table 8.1 *Relationship between pH and hydrogen ion concentration*

pH	Hydrogen ion concentration (nmol/L)
7.7	20
7.4	40
7.3	50
7.1	80

It is important to note that pH and hydrogen ion concentration $[H^+]$ are inversely related such that an increase in pH describes a decrease in $[H^+]$ (Table 8.1). However, the logarithmic scale is non-linear and therefore a change of one pH unit reflects a 10-fold change in $[H^+]$, and equal changes in pH are not correlated with equal changes in $[H^+]$. For example, a change of pH from 7.4 to 7.0 (40 nmol/L $[H^+]$ to 100 nmol/L $[H^+]$) represents a change of 60 nmol/L $[H^+]$, although the same pH change of 0.4, but from 7.4 to 7.8 (40 nmol/L to 16 nmol/L $[H^+]$), represents a change of only 24 nmol/L $[H^+]$.

BUFFERS

A buffer is a solution consisting of a weak acid and its conjugate base, which resists a change in pH when a stronger acid or base is added, thereby minimizing a change in pH. The most important buffer pair in the extracellular fluid is carbonic acid (H_2CO_3) and bicarbonate (HCO_3^-). The interaction between this buffer pair forms the basis of the measurement of acid–base balance.

HYDROGEN ION BALANCE

The cellular hydrogen ion turnover can be described in terms of processes that produce or consume H^+ ions in the body (Table 8.2). The total daily H^+ ion turnover in a normal adult is approximately 15 moles. Only 50–100 mmol of acid produced by the body is excreted by the kidney per day.

Carbon dioxide, produced by the oxidation of carbohydrates and triglycerides, forms the main acid load. Each day, about 15 moles of CO_2 are produced via decarboxylation reactions in the

Table 8.2 *Hydrogen ion balance*

Process	H⁺ balance (mmol/day)
I Production	
CO$_2$	15 000
Lactate	1500
'Fixed acids'	
Sulphuric acid	45
Phosphoric acid	13
Others	12
II Output	
Lungs	15 000
Liver	1500
Titratable acid	30
NH$_4^+$	40

tricarboxylic acid cycle, and about three-quarters of this is converted to 11 moles of carbonic acid daily. This carbonic acid is eliminated as CO$_2$ via the lungs, and CO$_2$ is referred to as a volatile or respiratory acid.

The majority of the non-volatile or metabolic acids are derived from protein metabolism, primarily metabolism of exogenous protein in the form of methionine and phosphoproteins. Sulphuric acid is formed from sulphur-containing amino acids such as cysteine and methionine. Hydrochloric acid is formed from the degradation of lysine, arginine and histidine. Phosphoric acid is formed by the hydrolysis of phosphoproteins. A person consuming 100 g of protein in a day produces about 1.1 moles of hydrogen ions during the conversion of protein nitrogen to urea. About 1500 mmol per day of lactic acid is produced by normal anaerobic metabolism of glucose and glycogen processes in the red blood cell, skin and skeletal muscle . The lactate is oxidized in the liver to regenerate bicarbonate. Excess lactic acid in the plasma indicates a diminished supply of oxygen to the tissues. Acetoacetic acid and β-hydroxybutyric acid are produced by the metabolism of triglycerides during fasting. Acetoacetic acid and hydroxybutyric acids, in excess of normal amounts (e.g. in diabetic ketoacidosis), are excreted by the kidneys. Acetoacetic acid can be decarboxylated to acetone which is excreted via the lungs and the kidneys. About 30 mmol of bicarbonate is lost in the faeces via the gastrointestinal tract, and this is equivalent to an acid load to the body.

Intracellular pH is finely regulated to enable optimal function of the enzyme systems. This achieved by the extrusion of protons through the membrane by the Na$^+$/H$^+$ antiport (or counter-transport) system (H$^+$ ions move in the opposite direction to the Na$^+$ ion gradient) and intracellular buffering by intracellular proteins.

ACID–BASE HOMEOSTASIS

Acid–base homeostasis maintains the H$^+$ ion concentration of plasma between 35 and 45 nmol/L (pH 7.35–7.44). This is achieved by three mechanisms: buffering, compensation and correction. Buffer systems of the body minimize the change in pH by the addition of an acid or alkali almost immediately. Compensation refers to the physiological processes that restore the HCO$_3^-$/PCO$_2$ ratio to normal. The ultimate mechanism to correct the acid–base derangement is through correction of the primary disorder.

Buffer systems

Buffers can reversibly bind H$^+$ ions to minimize or resist the change in pH. The general reaction of a buffer system is

$$\text{Buffer} + \text{H}^+ \rightleftharpoons \text{H.Buffer}$$

There are several buffers present in the extracellular fluid (ECF) and intracellular fluid (ICF). The effectiveness and capacity of a buffer system are determined by the amount of buffer present, its pK_a, the pH of the carrying solution, and whether the buffer functions as an open (physiological) system, or as a closed (completely chemical) system. Approximately 80 per cent of buffering occurs within ±1 pH unit of the pK_a of the buffer system. The major buffer systems in the body are bicarbonate, haemoglobin, protein and phosphate.

The Henderson equation provides a simple non-logarithmic (arithmetic) method of determining [H$^+$] or [HCO$_3^-$], as indicated in the following equations:

$$[\text{H}^+] = K \cdot \frac{P\text{CO}_2}{[\text{HCO}_3^-]}$$

or

$$[HCO_3^-] = K \cdot \frac{P_{CO_2}}{[H^+]}$$

This equation can be rewritten as:

$$[H^+] = 24 \cdot \frac{P_{CO_2} \text{ (mmHg)}}{[HCO_3^-] \text{ (mmol/L)}}$$

or

$$[HCO_3^-] = 24 \cdot \frac{P_{CO_2} \text{ (mmHg)}}{[H^+] \text{ (nmol/L)}}$$

where $24 = K \times sP_{CO_2}$ (K is the dissociation constant of carbonic acid and s is the solubility coefficient of CO_2 in plasma $= 0.03$ mmol/L per mmHg at $37°C$).

The Hasselbach equation describes acid–base relationships as:

$$pH = pK + \log\left(\frac{\text{kidneys}}{\text{lungs}}\right)$$

or

$$pH = pK + \log\frac{[HCO_3^-]}{sP_{CO_2}}$$

Substituting values for pK and s gives:

$$pH = 6.1 + \log\frac{[HCO_3^-]}{0.03P_{CO_2}}$$

and at physiological pH, the ratio $[HCO_3^-]/sP_{CO_2}$ is $24/(0.03 \times 40) = 24/1.2$. Therefore, pH $= 6.1 + \log(24/1.2) = 7.4$.

BICARBONATE–CARBONIC ACID

The bicarbonate buffer system consists of a weak acid, H_2CO_3 or carbonic acid, and a bicarbonate salt such as sodium bicarbonate in the extracellular fluid, and potassium bicarbonate and magnesium bicarbonate in the intracellular fluid. H_2CO_3 is formed by the reaction of CO_2 and H_2O, catalysed by the enzyme carbonic anhydrase, which is present

in large amounts in the renal tubules, red blood cells and the lung alveolar cells.

When a strong acid is added to the bicarbonate buffer, the H^+ ions released from the acid are buffered by HCO_3^- ions to form H_2CO_3. The H_2CO_3 rapidly dissociates to form CO_2 and H_2O. Pulmonary ventilation is stimulated by the consequent increase in arterial CO_2 tension, and more CO_2 from the ECF is excreted by the lungs. A strong base is buffered by carbonic acid to form sodium bicarbonate.

The bicarbonate buffer system has a low pK_a of 6.1 compared with its physiological pH of 7.4, and the concentration of its components (CO_2 and HCO_3^-) are not high. However, the bicarbonate system is the most important extracellular buffer because plasma CO_2 can be quickly regulated by changes in pulmonary ventilation, and HCO_3^- is regulated by reabsorption or excretion in the kidneys. The renal compensatory mechanisms occur slowly over several days.

HAEMOGLOBIN AS A BUFFER

Haemoglobin is present within the red blood cells (cellular component of the ECF) but is readily available for buffering extracellular acids. It is the primary non-bicarbonate buffer of blood for both respiratory and metabolic acids. Haemoglobin exists within the red cell as a weak acid (HHb) and its potassium salt (KHb). It is a weaker acid (pK_a of 6.8) than carbonic acid (pK_a of 6.1). As a result, H^+ is buffered by haemoglobin, and HCO_3^- is increased proportionately as shown in the following reactions:

$$HCl + KHb \rightleftharpoons HHb + KCl$$
$$\Updownarrow$$
$$H^+ + Hb^-$$
$$\Updownarrow$$
$$H_2CO_3 + KHb \rightleftharpoons HHb + KHCO_3$$

As a result of these reactions, red cell bicarbonate concentration increases and the bicarbonate diffuses out of the red blood cell to the plasma to maintain electrical neutrality. Therefore, although the buffering of H^+ ions occurs in the red blood cell, the increase in HCO_3^- is observed in the plasma.

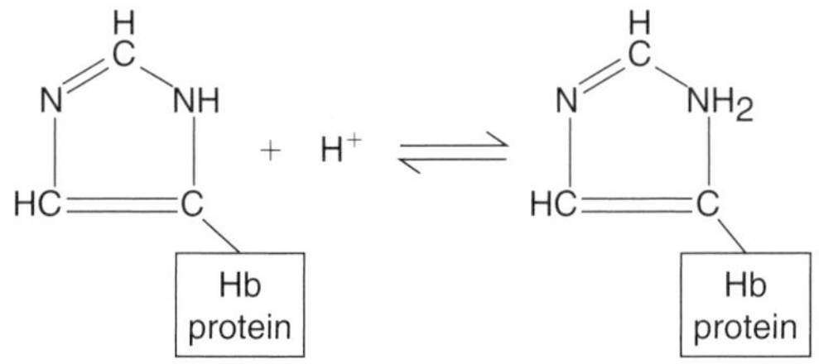

Figure 8.1 *Buffering action of imidazole group of haemoglobin.*

At the physiological pH of 7.4, the buffering action of haemoglobin is due mainly to the anionic sites of the imidazole groups of the histidine residues located in the globin chains Fig. 8.1). Haemoglobin is a very powerful buffer system because the haemoglobin molecule has 38 histidine residues that contain the imidazole side chains (pK_a 6.8) responsible for its buffering activity. In addition, haemoglobin is present within the erythrocytes at a relatively high concentration.

Another feature of haemoglobin as a buffer is that the deoxygenated form, deoxyhaemoglobin (pK_a 8.2), is a better buffer than oxyhaemoglobin (pK_a 6.6). This is because deoxyhaemoglobin dissociates to a greater extent than oxyhaemoglobin. Therefore, in the tissue capillaries, oxyhaemoglobin releases oxygen and is reduced to deoxyhaemoglobin, facilitating the uptake and buffering of H^+ ions generated by the hydration of CO_2 and dissociation of H_2CO_3. For each mmol of oxyhaemoglobin that is reduced, about 0.7 mmol of H^+ can be taken up by haemoglobin and 0.7 mmol of CO_2 can enter the venous blood without a change in pH. As this phenomenon results in little or no change in pH, it is called isohydric buffering. This explains why the pH of venous blood is only slightly more acidic than arterial blood in spite of the large amounts (15 mol/day) of CO_2 produced.

Carbon dioxide and carbonic acid can combine directly with the terminal amino groups of the amino acids of haemoglobin to form carbamino compounds as shown:

$$HbNH_2 + CO_2 \rightarrow HbNHCOO^- + H^+$$
$$HbNH_2 + H_2CO_3 \rightarrow HbNH_3^+ + HCO_3^-$$

The formation of carbamino compounds can account for about 15 per cent of the total CO_2 in the blood. In fact, 1 g of haemoglobin has three times the buffering capacity of 1 g of plasma proteins. As the concentration of haemoglobin is approximately twice that of plasma proteins, haemoglobin has six times the capacity of plasma proteins to buffer H^+ ions.

PROTEINS AS BUFFERS

All proteins contain amino and carboxyl groups in their side chains, and these can buffer H^+ ions. However, the pK_a of the amino groups is 9 and that of the carboxyl groups is 2, and these therefore contribute little to buffering at physiological pH. The imidazole groups of the histidine residues of the proteins are the only buffer groups in proteins to be of any physiological importance for buffering in the extracellular fluid as their pK_a is nearer the physiological pH.

In contrast, proteins are important intracellular buffers. This is because intracellular pH is more acidic and closer to the pK_a of the buffering moieties of proteins. Also, the intracellular protein concentration is higher than in plasma.

PHOSPHATE BUFFER

Phosphoric acid is a tribasic acid and can dissociate as:

$$H_3PO_4 \rightleftharpoons H^+ + H_2PO_4^- \quad pK_a = 2$$
$$H_2PO_4 \rightleftharpoons H^+ + HPO_4^{--} \quad pK_a = 6.8$$
$$HPO_4^{--} \rightleftharpoons H^+ + PO_4^{---} \quad pK_a = 11.7$$

At physiological pH, the biphosphate (H_2PO_4) buffer is the main phosphate buffer. On a purely chemical basis, it is potentially the most efficient buffer in the ECF as its pK_a is 6.8. As the plasma phosphate concentration is low (0.8 to 1.4 mmol/L) and phosphate operates as a closed buffer system, it is, however, of minor importance in the ECF. As intracellular concentrations of phosphate are higher than that of plasma and intracellular pH is more acidic and closer to the pK_a of phosphate buffers, phosphate does contribute substantially to intracellular buffering. Phosphate is also an important buffer in urine.

Calcium phosphate in the bone becomes more soluble with prolonged acidosis and plasma phosphate concentrations then increase as shown in these reactions:

$$CaPO_4^- \rightleftharpoons Ca^{++} + PO_4^{---}$$
$$PO_4^{---} + H^+ \rightleftharpoons HPO_4^{--}$$
$$HPO_4^{--} + H^+ \rightleftharpoons H_2PO_4^-$$

Calcium phosphate can be regarded as an 'alkali reserve', in response to a prolonged acidosis.

Distribution of buffers in body compartments

Functionally, buffers can be classified as either 'bicarbonate' or 'non-bicarbonate'. The bicarbonate system can only buffer metabolic acids, as it cannot buffer carbonic acid itself. The non-bicarbonate system comprises haemoglobin, phosphates and proteins and can buffer both carbonic acid and the fixed or metabolic acids (Fig. 8.2).

BLOOD

The two principal buffers in blood are bicarbonate and haemoglobin. The plasma proteins and phosphate contribute only a very small extent to the buffering capacity of blood (Table 8.3).

The chief buffer in plasma is bicarbonate, the plasma bicarbonate concentration being 24 mmol/L. Approximately 70 per cent of non-carbonic acids produced in the body are buffered by plasma bicarbonate. Plasma proteins and phosphates have a minor role in buffering. The buffers within the red blood cell are haemoglobin, bicarbonate and phosphate. More than 90 per cent of the capacity of blood to buffer carbonic acid is

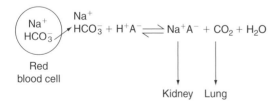

Figure 8.2 Buffering of metabolic acids.

Table 8.3 *Buffers in whole blood*

Buffer		Buffering capacity (%)
Red blood cells		
	Haemoglobin	35
	Bicarbonate	18
	Organic phosphates	3
Plasma		
	Bicarbonate	35
	Plasma protein	7
	Inorganic phosphates	2

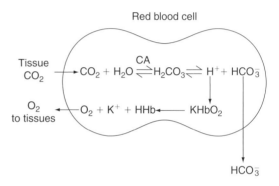

Figure 8.3 Buffering of carbonic acid in red blood cells. CA, carbonic anhydrase.

contributed by haemoglobin. Bicarbonate present in blood is formed within the red blood cell due to the presence of carbonic anhydrase, and then diffuses into the plasma. The concentration of bicarbonate within the red cell is approximately 15 mmol/L of red cells. The inorganic and organic phosphates within the red blood cell contribute less than 5 per cent of the total buffering capacity (Fig. 8.3).

INTERSTITIAL FLUID

The concentration of bicarbonate in the interstitial fluid is 27 mmol/L, and this makes it the main interstitial fluid buffer. As the interstitial fluid volume is approximately three times that of plasma, the total capacity of interstitial fluid to buffer non-carbonic acid is greater than that of blood. The concentration of phosphate in the interstitial fluid is 0.7 mmol/L.

INTRACELLULAR COMPARTMENT

Proteins and phosphates are important intracellular buffers because they are present in significant quantities within the cell. About 6 mmol/L of protein and 6 mmol/L of phosphate are present in most cells. These buffers can effectively buffer non-carbonic and carbonic acid as the lower intracellular pH (6.8–7.1) enhances the buffering actions of these two buffers with pK_a values around 6.8.

WHOLE–BODY (*IN VIVO*) TITRATION CURVES

When CO_2 is added to blood in a test tube (*in vitro*), the bicarbonate produced by the buffers in blood is confined within the test tube. However, when an acute increase in CO_2 occurs in the human body (*in vivo*), plasma bicarbonate increases to a smaller extent associated with a greater decrease in arterial pH. This is because in the whole-body (*in vivo*) situation, some of the bicarbonate produced by the buffering of CO_2 in blood diffuses from the plasma across the capillary to the interstitial space. Equilibration of the acid–base composition of blood and that of the interstitial fluid is reached within 10–20 min and consequently less bicarbonate remains in the plasma *in vivo* compared to that seen in the test tube (*in vitro*). The CO_2 is primarily buffered by haemoglobin, producing bicarbonate. As the blood volume is about one-third of the total volume of the ECF, haemoglobin may be regarded as being distributed throughout the ECF. Hence the whole ECF compartment can be regarded as having a haemoglobin concentration of 5 g per 100 mL of ECF, and a buffering capacity of about one-third that of blood (approximately 10 mmol/L/pH unit). In summary, in the *in vivo* or whole-body situation, when CO_2 is added, the change in pH is greater and the change in $[HCO_3^-]$ is less than that of blood *in vitro*, once equilibrium between plasma and interstitial fluid is achieved. The *in vivo* base excess and buffer base are both reduced compared to the *in vitro* situation because of the diffusion of bicarbonate out of the vascular compartment. The difference between the *in vitro* and *in vivo* buffer lines is only important at very high or very low

Pa_{CO_2}, or when there is severe metabolic acidosis. The main problem of the *in vitro* approach is that it wrongly suggests that a metabolic acidosis is present when the Pa_{CO_2} is acutely raised, and that a metabolic alkalosis is present when the CO_2 is acutely lowered. However, in most clinical situations, the difference between the two methods is minimal.

COMPENSATORY MECHANISMS

Acid–base derangements are minimized by physiological responses that tend to restore the $[HCO_3^-]/Pa_{CO_2}$ ratio in an attempt to normalize the pH. This is achieved by compensatory regulation of the elimination of acids and bases by ventilatory changes and renal excretion. Generally, compensation for the primary acid–base disturbance is usually not complete, and the pH is not restored fully to 7.4.

Respiratory compensation

The plasma bicarbonate ion concentration is increased in metabolic alkalosis and decreased in metabolic acidosis. pH control of ventilation is determined by chemoreceptors in the medulla that monitor the pH of the brain ECF and by the pH sensed by the carotid body chemoreceptors. In metabolic acidosis, the low plasma bicarbonate concentration and increase in blood $[H^+]$ induces hyperventilation, lowering the Pa_{CO_2} and restoring the normal $[HCO_3]/Pa_{CO_2}$ ratio. It is estimated that the ventilatory rate increases about two-fold for a change in pH of 0.1 pH unit.

In metabolic alkalosis, there is an increase in plasma $[HCO_3^-]$ concentration and the respiratory drive is decreased, resulting in a rise in Pa_{CO_2}. Thus, respiratory compensation to acid–base disturbances controls the excretion of CO_2 to restore the $[HCO_3^-]/Pa_{CO_2}$ ratio closer to normal. The capacity of the respiratory system to buffer acid–base disturbances is approximately twice the buffering capacity of the chemical buffers in the ECF. The other important feature of the respiratory control of acid–base balance is that it responds rapidly and hence prevents large acute changes in plasma $[H^+]$.

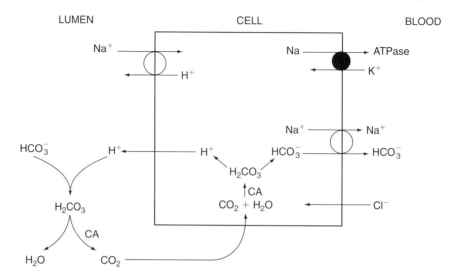

Figure 8.4 *Bicarbonate reabsorption in the proximal tubule of the kidney. CA, carbonic anhydrase.*

Renal compensation

Renal compensatory mechanisms regulate acid–base balance by altering the plasma bicarbonate ion concentration to restore the $[HCO_3^-]/Pa_{CO_2}$ ratio. This response is slow, and reaches its maximal capacity (approximately 300 mmol H^+ per day) after about 7–10 days.

The body produces 70 mmol of non-volatile acids per day. All the HCO_3^- filtered at the glomerulus (4320 mmol/day) is reabsorbed at the tubules in association with $[H^+]$ ion secretion. Thus, about 4320 mmol of $[H^+]$ is secreted each day to facilitate reabsorption of any filtered bicarbonate. An additional 70 mmol of $[H^+]$ has to be secreted to excrete the non-volatile acids each day. Thus a total of 4390 mmol of $[H^+]$ is secreted each day in the renal tubules.

Renal regulation of H^+ balance is brought about by: (1) secretion of hydrogen ions associated with reabsorption of bicarbonate ion by the renal tubules; (2) excretion of titratable acidity; and (3) excretion of ammonia.

REABSORPTION OF FILTERED HCO_3^- AND SECRETION OF HYDROGEN IONS

About 80 per cent of bicarbonate reabsorption and hydrogen ion secretion occurs in the proximal tubule (Fig. 8.4). The secretion of H^+ ions is a low-gradient and high-capacity system, and the lowest urine pH that can be achieved in the proximal tubule is around 7. In the thick ascending limb of the loop of Henle, 10 per cent of the filtered bicarbonate is reabsorbed, with the remaining bicarbonate being reabsorbed in the distal tubule and the collecting duct.

The epithelial cells of the proximal tubule, the thick ascending limb of the loop of Henle and the early distal tubule secrete H^+ ions into the tubular fluid by a secondary active Na^+/H^+ counter-transport mechanism at the luminal membrane. The energy for H^+ secretion against a concentration gradient is derived from the Na^+ gradient, which favours the influx of Na^+ ions. This gradient is established by the Na^+/K^+-ATPase pump at the basolateral membrane. Cellular carbonic anhydrase catalyses the combination of CO_2 and H_2O to form H_2CO_3, which dissociates to H^+ ion and HCO_3^-. The H^+ ion is secreted into the tubular fluid by the Na^+/H^+ counter-transport, and the HCO_3^- ion moves down the concentration gradient across the basolateral membrane into the peritubular fluid and peritubular capillary blood. For every H^+ ion secreted into the tubular fluid, one bicarbonate ion enters the blood.

In the late distal tubule and the collecting ducts the tubular epithelium secretes H^+ ions by primary active secretion by the intercalated cells and by a H^+-ATP-ase transport system which utilizes energy derived from the breakdown of ATP to ADP. The H^+ ion in the intercalated cell is derived from the reaction between CO_2 and water. One bicarbonate ion is absorbed for each H^+ ion secreted, and a chloride ion is passively secreted with each H^+ ion. Although the intercalated cells of the late distal tubule only account for 5 per cent of the total H^+ ions secreted, they are important for the formation of maximally acidic urine as they can concentrate the H^+ ions by 900-fold and reduce the urine pH to 4.5. Distal tubule H^+ secretion is described as a high-gradient, low-capacity system. The secretion of hydrogen ions in the distal tubule is controlled by aldosterone (Fig. 8.5).

FORMATION OF TITRATABLE ACIDITY

Titratable acidity refers to the hydrogen ions bound to filtered buffers in the urine, and is equal to the amount of alkali (NaOH) required to titrate the urine to a pH of 7.4. Urinary titratable acidity is due

to the conversion of monohydrogen phosphate to dihydrogen phosphate in the tubule. At the maximal urine acidity of 4.5, all the urinary phosphate is in the form of dihydrogen phosphate. After all the bicarbonate ions in the tubular fluid are reabsorbed, excess H^+ ions in the tubular fluid combine with monohydrogen phosphate to form dihydrogen phosphate. A bicarbonate ion is added to the peritubular capillary blood for each dihydrogen phosphate ion produced. Other filtered buffers in the tubular fluid, including creatinine, β-hydroxybutyrate and sulphates, contribute only a minor extent to titratable acidity. The proximal tubule is the chief site for the formation of titratable acidity (Fig. 8.6).

AMMONIA SECRETION

About 75 per cent of the metabolic acids produced in the body (approximately 50 mmol/day) are

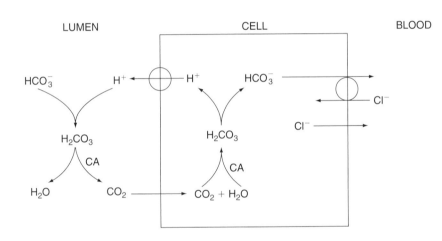

Figure 8.5 *Hydrogen secretion by the intercalated A cells of the distal tubule of the kidney.*

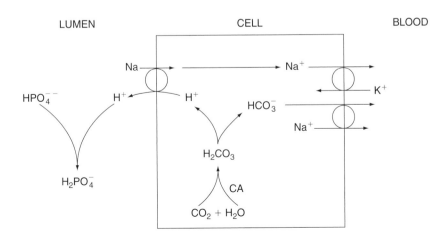

Figure 8.6 *Formation of titratable acidity in the kidney.*

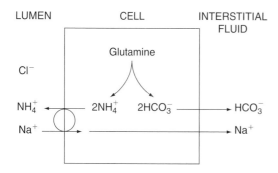

Figure 8.7 *Ammonia production in the proximal tubule of the kidney.*

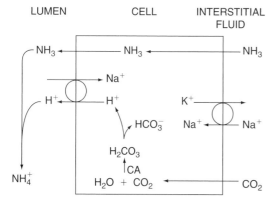

Figure 8.8 *Ammonia production in the distal tubule of the kidney. CA, carbonic anhydrase.*

excreted as the ammonium ion, NH_4^+. Glutamine is deaminated in the proximal tubule, the thick ascending limb of the loop of Henle, and distal tubules to form two NH_4^+ ions and two HCO_3^- ions. The NH_4^+ ions are secreted into the tubular lumen by a Na^+/NH_4^+ counter-transport pump. The HCO_3^- ions are reabsorbed into the blood (Fig. 8.7). In the collecting ducts, ammonia (NH_3) diffuses from the renal interstitial fluid into the collecting duct cells. Within the collecting duct cells, H^+ ions are formed from CO_2 and water, catalysed by carbonic anhydrase. The H^+ ions are secreted into the tubular fluid by a Na^+/H^+ exchange system. NH_3 is lipid soluble, and diffuses across the luminal membrane into the lumen where it combines with the secreted H^+ ions to form NH_4^+. As NH_4^+ is ionized, it cannot diffuse through to the luminal membrane and hence it is excreted in the urine. Strictly, NH_4^+ acts as a 'sink'

for H^+ ions as ammonia (NH_3) combines irreversibly with H^+. The high pK_a (9.2) of the NH_3/NH_4^+ system is valuable as it enables the kidney to excrete H^+ ions into the tubular fluid even when the tubular fluid has reached its maximal acidity (Fig. 8.8). In severe acidosis, about 300 mmol of H^+ can be excreted as the NH_4^+ ion in a day.

CLINICAL EFFECTS OF ACID–BASE CHANGES

Acid–base disturbances have widespread physiological and biochemical effects, either by direct action on various organs or indirectly through changes in the autonomic nervous or endocrine systems.

Cardiovascular system

Acidosis has direct negative inotropic effects on the myocardium by inhibition of the slow inward calcium current and diminished calcium release from the sarcoplasmic reticulum. Acute respiratory acidosis depresses myocardial contractility more than acute metabolic acidosis. The myocardial depressant effect is opposed by the increased release of catecholamines by acidosis, but if the pH falls below 7.2 the negative inotropic effect predominates.

Alkalosis is associated with increased coronary and systemic vascular resistance and a shift of the oxyhaemoglobin dissociation curve to the left, which impairs oxygen delivery to the tissues.

Acidosis increases catecholamine release from the adrenal medulla, and initially causes a tachycardia. Atrial and ventricular arrhythmias are increased. It is suggested that acidosis reduces intracellular K^+ concentrations, leading to depolarization of the resting membrane potential in the cardiac conductive pathways and that this predisposes to arrhythmias.

Acidosis causes direct arterial vasodilatation in the skin, skeletal muscles, uterus and heart, but pulmonary vasoconstriction. With mild acidosis (pH 7.2–7.3), indirect effects due to elevated circulating catecholamines produce systemic, renal and splanchnic vasoconstriction.

Respiratory system

The minute volume of ventilation increases with metabolic acidosis via stimulation of the medullary central chemoreceptors. In respiratory acidosis minute ventilation increases 2–3 L/min for every mmHg (0.13 kPa) rise in Pa_{CO_2}. The hyperventilatory response to respiratory acidosis occurs more rapidly than with a metabolic acidosis, as the blood–brain barrier is more permeable to carbon dioxide than to H^+ ions. The hyperventilation peaks at a Pa_{CO_2} of 100 mmHg (13.3 kPa), and further increases in Pa_{CO_2} cause respiratory depression. Hypercapnia causes direct bronchodilatation, although bronchoconstriction may occur as a result of vagal stimulation.

Acidosis shifts the oxyhaemoglobin dissociation curve to the right, increasing oxygen delivery to the tissues. Conversely, alkalosis decreases oxygen delivery as it shifts the oxyhaemoglobin dissociation curve to the left.

Electrolyte changes

In acidosis, H^+ ions compete for the negatively charged binding sites on albumin and displace calcium from these sites and increase ionized serum calcium concentration. The converse occurs in alkalosis. Such changes in serum ionized calcium concentrations may lead to various physiological effects.

In acidosis, H^+ ions move intracellularly whilst potassium ions move out into the ECF. It is estimated there is a 0.6 mmol/L change in serum $[K^+]$ for every 0.1 unit change in pH. Chronic metabolic acidosis can mobilize calcium from bone, leading to a rise in serum calcium concentrations.

Central nervous system

Severe acidosis can lead to impaired consciousness. However, the overall effects of acid–base changes on the central nervous system are due to changes in cerebral blood flow and intracranial pressure. Epilepsy may be precipitated by alkalosis.

Gastrointestinal tract

In addition to splanchnic vasoconstriction, acidosis may decrease gastrointestinal motility.

TEMPERATURE AND ACID–BASE CONTROL

The pH of neutral water ($[OH^-] = [H^+]$) increases by 0.017 unit for every 1°C decrease in temperature. The pH of blood increases by about 0.015 unit for every 1°C fall in temperature, and Pa_{CO_2} decreases by about 4.5 per cent (Fig. 8.9).

The subject of the optimal acid–base management during hypothermia remains controversial. The two specific forms of management that may be used are the pH stat and α stat systems. In the pH stat system, pH is maintained constant at 7.4 over varying temperatures. When blood is cooled (e.g. during hypothermic cardiopulmonary bypass), CO_2 becomes more soluble and Pa_{CO_2} decreases, although a constant CO_2 content is maintained. Because of this, during cooling, CO_2 must be added to maintain a Pa_{CO_2} of 40 mmHg and a pH of 7.40. Extracellular and intracellular $[OH^-]/[H^+]$ ratios are altered, and total CO_2 stores are elevated.

In the α stat system, the actual pH varies with body temperature. The blood pH is increased by 0.015 unit per 1°C below 37°C. The slope of this relationship is remarkably similar to the change in the pH of water in relation to temperature changes

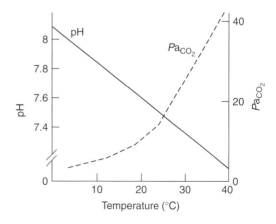

Figure 8.9 *Changes in pH and Pa_{CO_2} with temperature.*

(pH of water increases by 0.017 unit per °C decrease in temperature). The imidazole moieties of histidine residues in protein are responsible for this constant relationship of blood pH changes to that of water. This maintains a constant ratio of $[OH^-]/[H^+]$ in blood, averaging at 16:1 over a wide range of temperatures. As temperature changes, the pK_a of the imidazole buffer changes in parallel with the pH of water. The fraction of unprotonated histidine imidazole groups (known as 'a' groups) remains constant. The term 'α stat' refers to the maintenance of this constant net charge on proteins with temperature change by keeping total CO_2 stores constant. Plasma bicarbonate concentration remains constant. Theoretically, α stat management is preferable. A constant intracellular electrochemical neutrality is essential for normal enzyme function and for maintaining the Donnan equilibrium across cellular membranes, allowing normal intracellular anion concentrations and water content.

Stewart's physico-chemical approach

The 'traditional' concept of acid–base balance using the Henderson–Hasselbach equation assumes that HCO_3^- behaves as an independent variable whose concentration determines the metabolic component of pH balance. However, HCO_3^- varies with CO_2 and this can lead to confusion when measuring the metabolic component of acid–base balance. In 1983, Stewart showed that the carbon dioxide/bicarbonate system could not be viewed in isolation. By applying physico-chemical laws (law of electrochemical neutrality, law of conservation of mass, and law of mass action) he showed that biological fluids have their hydrogen ion concentrations set by multiple chemical equilibria reactions occurring simultaneously. According to Stewart, the concentration of H^+ in an aqueous solution depends on the extent of dissociation of water into H^+ and OH^-. Three independent variables (P_{CO_2}, total concentration of weak acids, and strong ion difference) and several dependent variables (concentrations of H^+, OH^-, HA, A^-, HCO_3^-, CO_3^{--}) determine the dissociation of water into H^+ and OH^-. Only the independent variables account for pH changes in a biological solution. The dependent variables cannot directly

influence any other variable. Therefore, bicarbonate, being a dependent variable, cannot exert direct control over H^+ concentration or pH. The important principle of this theory is that the dependent variables only change in response to changes in one or more of the independent variables.

Changes in P_{CO_2} mediated by altered alveolar ventilation alterations cause rapid $[H^+]$ changes in aqueous solutions due to reversible dissociation of carbonic acid. The high solubility of CO_2 allows it to pass easily between compartments and alter $[H^+]$ in all body fluids.

Total weak acids such as plasma proteins and phosphates are the next independent variable. Plasma protein concentration is controlled by the liver and changes occur over several days. Phosphates are significant in hypoalbuminaemia.

Strong ions such as sodium, potassium, magnesium, calcium, chloride and lactate dissociate completely in aqueous solutions. Strong ion difference (SID; measured in milliequivalents per litre) is the amount by which strong cations exceed strong anions. Strong ion difference (SID) is therefore obtained by adding together the concentrations (in MEq/L) of all the main cations in solution (Na^+, K^+, Ca^{++}, Mg^{++}) and subtracting the concentrations of the main cations (Cl^-, lactate) in solution. In plasma:

$$SID =$$

$$(Na^+ + K^+ + Ca^{++} + Mg^{++}) - (Cl^- + lactate)$$

The normal SID is approximately 42–46 mEq/L. The bicarbonate concentration will decrease when the SID is narrow because the Cl^- is increasing. As SID increases, there is less dissociation of water and hydrogen ion concentration is reduced (that is, pH increases) so that electrical neutrality is maintained. Conversely, as SID decreases the hydrogen ion concentration increases (pH decreases).

Stewart's hypothesis clarifies the role of the lungs, kidneys, liver and gut in acid–base control. The lungs regulate acid–base by altering P_{CO_2}. The kidneys control plasma electrolytes, especially chloride, and this allows manipulation of SID and therefore plasma pH. Liver and gut can have a major effect on acid–base balance by altering the level of total weak acid. Prolonged vomiting decreases plasma chloride concentration relative to sodium as a result of loss of gastric hydrochloric acid. Consequently, the SID decreases and metabolic alkalosis occurs. According to Stewart's hypothesis, the alkalosis is not caused by loss of H^+ ions because there is an inexhaustible supply of H^+ ions from the dissociation of water. Large-volume infusions of normal saline cause metabolic acidosis. Hyperchloraemia develops as a result of the relatively high chloride concentration of normal saline (150 mmol/L) compared to the plasma chloride concentration (100 mmol/L). This reduces the SID and increases water dissociation and hydrogen ion concentration. The administration of sodium bicarbonate to treat acidosis can be explained by the Stewart's hypothesis. The sodium load of the sodium bicarbonate infusion increases plasma sodium and this increases SID. The dissociation of plasma water decreases to maintain electroneutrality and the concentration of free H^+ decreases. The bicarbonate appears to act as a buffer but the HCO_3 ion in sodium bicarbonate cannot influence plasma pH as $[HCO_3^-]$ is a dependant variable. Metabolic alkalosis associated with chronic hypoalbuminaemia in critically ill patients occurs because the total weak acid concentration is decreased.

Therefore, the Stewart's hypothesis is useful for the analysis of metabolic acid–base disturbances caused by changes in either SID or total weak acid concentration. Measurement of chloride concentration is important: if the chloride is low or normal and there is metabolic acidosis, unmeasured ions such as lactic acid and keto-acids may be present in the blood. The classification of acid–base disturbances according to derangements of independent variables provides a better understanding of the primary clinical problem, and helps direct treatment.

CLINICAL ASPECTS OF ACID–BASE CONTROL

Terminology

- *Acidaemia* indicates that arterial blood pH is lower than the normal range (<7.36) and the hydrogen ion concentration is more than 44 nmol/L.

- *Alkalaemia* indicates that arterial blood pH is higher than 7.44 and the hydrogen ion concentration is less than 36 nmol/L.
- *Acidosis* describes the abnormal condition or process that tends to decrease the blood pH if there are no secondary changes in the response to the primary disease. The primary process is called respiratory if the changes are in Pa_{CO_2}, or metabolic if the changes are due to fixed or non-volatile acids that lead to a decrease in $[HCO_3^-]$.
- *Alkalosis* describes the abnormal condition or process that tends to increase arterial pH if there are no secondary changes in the response to the primary disease.
- *Compensation* refers to the physiological process (respiratory or renal) that tends to return or correct the arterial pH to the normal range.

Assessment of acid–base disorders

The measurements of pH, Pa_{CO_2} and calculated plasma HCO_3^- concentration are required to determine the primary disorder and the degree of compensation. Historically, a crude method involved the titration of a plasma sample to a Pa_{CO_2} of 40 mmHg (5.3 kPa) and the total content of CO_2 (called alkali reserve) was then determined, the HCO_3^- measured and pH derived using the Henderson–Hasselbach equation. This system was flawed as it did not take into account the buffering properties of haemoglobin.

STANDARD BICARBONATE

The standard bicarbonate system measures the bicarbonate concentration in whole blood which is fully oxygenated, with a Pa_{CO_2} of 40 mmHg (5.3 kPa) and a temperature of 37°C. The normal standard bicarbonate is 22–26 mmol/L. The disadvantage of this system is that it does not take into consideration the buffers present in the interstitial and intracellular compartments.

ASTRUP METHOD

The Astrup method is an *in vitro* equilibration method used to determine the Pa_{CO_2}. The pH of the patient's blood sample is initially measured with a pH electrode. Two blood samples from the patient are equilibrated with 4 per cent and 8 per cent carbon dioxide in oxygen, in a water bath at 37°C. The Pa_{CO_2} of the blood samples are determined using the relationship:

$$Pa_{CO_2} = \frac{CO_2\%}{100} \times (\text{barometric pressure} - 47)\ \text{mmHg}$$

The pH is measured at each of the two Pa_{CO_2} tensions, plotted against log Pa_{CO_2}, and a line is drawn between the two points. This line describes all combinations of pH and Pa_{CO_2} for the particular blood sample. The Pa_{CO_2} of the initial blood sample can then be found from the intersection of the measured pH with the plotted line.

BASE EXCESS

The base excess system is based on the principle that when the pH of blood is normal, the ratios and the total concentrations of the non-carbonic buffers are normal. It assumes that blood behaves as if it is a simple bicarbonate solution (that is, by H_2CO_3/HCO_3^- buffers) so that a change in $[HCO_3^-]$ reflects a change in $[H^+]$ of the system. This is achieved by altering the Pa_{CO_2} of the blood to bring the pH to 7.4 so the non-carbonic acid buffer pairs are returned to their normal ratios and concentrations. All the buffering is achieved by the bicarbonate system, so that the HCO_3^- concentration reflects the amount of acid or alkali added to the blood. The bicarbonate buffer is not affected by respiratory acidosis, as the addition of CO_2 results in the formation of equal amounts of H^+ ions and HCO_3^- from carbonic acid.

Buffer base is the sum of the concentrations of all the buffer anions in the blood including haemoglobin, bicarbonate, protein and phosphate. Whole-blood buffer base refers to the concentrations of these buffers in fully oxygenated blood. The normal range for buffer base is 45–59 mmol/L.

Base excess or deficit is defined as the amount of acid or base required to titrate whole blood at 37°C and a Pa_{CO_2} of 40 mmHg (5.3 kPa) to a pH of 7.4. An increase in buffer base (>50 mmol/L) or base excess indicates an increase in buffering capacity, and this may be from a decrease in metabolic acids or an increase in buffers (HCO_3^-, Hb, proteins).

A decreased buffer base or base deficit may result from excess metabolic acids or a decrease in buffer content.

The base excess system is used clinically because it is simple and logical. Its main limitation is that it is an *in vitro* system and does not take into account the extravascular buffers, or the interaction of the buffers of blood with the buffer systems of the interstitial and ICF compartments.

The main limitation of the *in vitro* evaluation of buffers is that it assesses blood buffers, which represent only about one-third of the total body buffering capacity, and does not take into account the ability of H^+ ions, CO_2 and HCO_3^- to diffuse between blood, interstitial fluid and ICF. Therefore, the *in vitro* assessment of acid–base disturbances tends to overestimate the acid–base changes in the whole body.

IN VIVO EVALUATION

In vivo titration curves are derived from the collation of changes of pH, Pa_{CO_2} and HCO_3^- during acute and chronic disorders. The main advantage is that this accounts for changes in $[HCO_3^-]$ and pH due to interaction of the buffers in the different compartments and also changes secondary to normal compensatory mechanisms.

In vivo assessment of respiratory acidosis

An acute increase in Pa_{CO_2} increases carbonic acid causing a rise in both $[H^+]$ and $[HCO_3^-]$. Acutely, each 10 mmHg (1.3 kPa) increase in Pa_{CO_2} above 40 mmHg (5.3 kPa) increases $[HCO_3^-]$ by 0.08 mmol/L and decreases pH by 0.07 unit (8 nmol/L $[H^+]$).

During chronic respiratory acidosis, renal compensation results in H^+ excretion and reabsorption of HCO_3^-. Therefore, in chronic respiratory failure with hypercapnia and renal compensation, each 10 mmHg (1.3 kPa) rise in Pa_{CO_2} increases $[HCO_3^-]$ by 4 mmol/L and decreases pH by 0.03 unit (3.2 nmol/L $[H^+]$).

In acute respiratory alkalosis, hypocapnia leads to reduced $[H^+]$ and $[HCO_3^-]$. For each 10 mmHg (1.3 kPa) decrease in Pa_{CO_2} below 40 mmHg (5.3 kPa), $[HCO_3^-]$ decreases by 2 mmol/L and pH increases by 0.08 unit (decrease in $[H^+]$ by 8 nmol/L).

During chronic hypocapnia, renal compensation results in a decrease in H^+ secretion and reduced

HCO_3^- reabsorption. The HCO_3^-/Pa_{CO_2} ratio is maintained. Thus, during chronic hypocapnia, for each 10 mmHg (1.3 kPa) decrease in Pa_{CO_2} below 40 mmHg (5.3 kPa), $[HCO_3^-]$ decreases by 6 mmol/L and pH increases by 0.03 unit (3. 2 nmol/L [H^+]).

In vivo assessment of metabolic disturbances

Metabolic acidosis immediately increases ventilation, but the compensation is only partial. *In vivo* titration curves indicate that Pa_{CO_2} decreases by 0. 1 mmHg (0.01 kPa) for each mmol/L decrease in $[HCO_3^-]$ and pH decreases by 0.02 unit. In metabolic alkalosis, a rapid decrease in respiratory drive results in an increase in Pa_{CO_2} which partially restores the $[HCO_3^-]/Pa_{CO_2}$ ratio to normal. The Pa_{CO_2} can be predicted using the formula:

$$\text{Predicted } Pa_{CO_2} = 0.7[HCO_3^-] + 20(\pm 2) \text{ mmHg}$$

Anion gap

In metabolic acidosis, for every molecule of acid produced, a molecule of HCO_3^- is replaced by another anion. However, the law of electroneutrality dictates that the sum of positive charges is exactly balanced by the negative charges. Routine clinical electrolyte measurements include nearly all cations (Na^+ and K^+), but only some of the anions (Cl^- and HCO_3^-). This apparent difference between the total cation concentration and total anion concentration is called an anion gap.

$$\text{Anion gap} = [Na^+ + K^+] - [Cl^- + HCO_3^-]$$

Anions such as proteins, sulphates, phosphates and organic acids are not measured in routine clinical chemistry tests, although they are present. The normal anion gap is 8–16 mmol/L. An increased anion gap indicates an increased concentration of unmeasured anions produced by a metabolic acidosis (ketosis or lactic acidosis, uraemia, and poisoning by salicylates, methanol and ethylene glycol). A metabolic acidosis with a normal anion gap and hyperchloraemia is seen with diarrhoea, pancreatic fistula, renal tubular acidosis, and treatment with HCl, NH_4Cl or acetazolamide. A decreased anion gap may be due to hypoproteinaemic states.

Osmolar gap

Osmolar gap refers to the difference between measured serum osmolality and calculated serum osmolality. The serum osmolality can be calculated as follows:

Serum osmolality =
   2[Na^+]
   + blood urea (mmol/L)
   + blood sugar (mmol/L)

Normally, the osmolar gap is less than 15 mOsm/kg. Circulating intoxicants such as alcohol increase the measured serum osmolality without altering serum Na^+. Thus, the osmolar gap will increase.

Biochemical description of an acid–base defect

An acid–base disturbance is described by three parameters:

- pH, which measures acidity or alkalinity
- Pa_{CO_2}, which measures the respiratory component
- $[HCO_3^-]$, which measures the metabolic component

pH and Pa_{CO_2} are measured directly using the pH and CO_2 electrodes respectively. The HCO_3^- concentration cannot be measured directly, but can be calculated from the Henderson–Hasselbach equation. As $[HCO_3^-]$ may vary with changes in Pa_{CO_2}, derived parameters such as standard bicarbonate and base excess are used. However, these derived parameters do not reflect the *in vivo* changes. For most clinical situations, the acid–base disturbance can be assessed by pH, Pa_{CO_2} and $[HCO_3^-]$.

Graphical evaluation

THE pH/Pa_{CO_2} DIAGRAM

Various graphical acid–base diagrams have been used, the most useful one having pH on one axis and Pa_{CO_2} on the other, as it demonstrates the *in vivo* relationship between [H^+] and Pa_{CO_2} in primary acid–base disorders. Such a diagram, in which

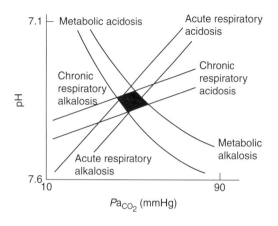

Figure 8.10   pH/Pa_{CO_2} diagram.

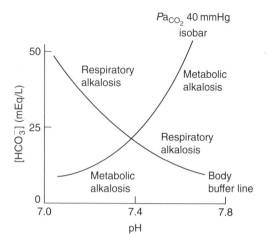

Figure 8.11   pH/HCO_3^- diagram.

arterial pH is plotted against Pa_{CO_2} is shown in Fig. 8.10. The shaded square represents the approximate limits of arterial pH and Pa_{CO_2} in normal individuals. The band labelled acute respiratory acidosis shows the 95 per cent confidence interval range of normal individuals breathing air and CO_2 mixtures for a short period. The band labelled acute respiratory alkalosis indicates the 95 per cent confidence range of values obtained in normal individuals voluntarily hyperventilating for a short period.

After a few days of CO_2 retention, renal compensation leads to an increase in plasma bicarbonate. This band is shown in Fig. 8.10 as chronic respiratory acidosis. It should be noted that the slope of the CO_2 response is less than that for acute respiratory acidosis.

The 95 per cent confidence interval band for metabolic alkalosis is located to the right, and below the shaded area for the normal range. The compensation for metabolic alkalosis is hypoventilation but is limited by hypoxia. The band for metabolic acidosis is located to the left and above the band for the normal range.

THE pH/HCO_3^- DIAGRAM

Acid–base disturbances can be defined by the P_{CO_2} 40 mmHg (5.3 kPa) isobar and the body buffer line. The P_{CO_2} isobar line represents the line obtained on titration of the blood sample with hydrochloric acid with the P_{CO_2} maintained at 40 mmHg (5.3 kPa) by a constant flow of CO_2. The body buffer line represents the line relating the plasma bicarbonate concentration to pH when an oxygenated blood sample is equilibrated with gas mixtures having different CO_2 tensions. Respiratory disorders are classified as acidosis or alkalosis, depending on whether they lie to the left or right of the P_{CO_2} 40 mmHg (5.3 kPa) isobar. Metabolic disorders are classified as alkalosis or acidosis when they lie above or below the body buffer line (Fig. 8.11).

Metabolic disturbances can be classified according to the standard bicarbonate. Metabolic acidosis is associated with a standard bicarbonate less than 24 mmol/L, and metabolic alkalosis is associated with a standard bicarbonate of greater than 24 mmol/L.

Reflections

1. The acidity of a solution is determined by its hydrogen ion concentration; the greater the hydrogen ion concentration, the more acid the solution. The degree of acidity is expressed using the pH scale. Pure water has a pH of 7 at 25°C and is neutral in acid–base terms. An acid is a proton donor, generating hydrogen ions in solution. A base is a proton acceptor and absorbs hydrogen ions. Strong acids and bases in aqueous solutions dissociate completely into

their constituent ions. Weak acids and bases are only partially dissociated and the degree of ionization depends on the hydrogen concentration. The ratio of dissociated to undissociated weak acid can be calculated from the Henderson–Hasselbach equation.

2. At constant temperature the concentration of hydrogen ions in the plasma is determined by three factors: (i) the difference between the total concentration of fully dissociated cations and that of fully dissociated anions (strong ion difference, SID); (ii) the quantity and the pK_a values of the weak acids that are present; and (iii) the partial pressure of carbon dioxide. A change in any one of these will cause a change in plasma hydrogen ion concentration.

3. Strong ions such as sodium, potassium, magnesium, calcium, chloride and lactate dissociate completely in aqueous solutions. Strong ion difference (or SID; measured in milliequivalents per litre) is the amount by which strong cations exceed strong anions. In plasma, $SID = (Na^+ + K^+ + Ca^{++} + Mg^{++}) - (Cl^- + lactate)$. The normal SID is approximately 42–46 mEq/L. The bicarbonate concentration will decrease when the SID is narrow because the Cl^- is increasing. As SID increases, there is less dissociation of water and hydrogen ion concentration is reduced (that is pH increases) so that electrical neutrality is maintained. Conversely, as SID decreases the hydrogen ion concentration increases (pH decreases).

Stewart's hypothesis clarifies the role of the lungs, kidneys, liver and gut in acid–base control. The lungs regulate acid–base by altering P_{CO_2}. The kidneys control plasma electrolytes, especially chloride, and this allows manipulation of SID and therefore plasma pH. Liver and gut can have a major effect on acid–base balance by altering the level of total weak acid. Prolonged vomiting decreases plasma chloride concentration relative to sodium as a result of loss of gastric hydrochloric acid. Consequently, the SID decreases and metabolic alkalosis occurs. Large-volume infusions of normal saline cause metabolic acidosis. Hyperchloraemia develops as a result of the relatively high chloride concentration of normal saline (150 mEq/L) compared to the plasma chloride concentration (100 mEq/L). This reduces the SID and increases water dissociation and hydrogen ion concentration. The administration of sodium bicarbonate to treat acidosis increases plasma sodium and this increases SID. The dissociation of plasma water decreases to maintain electroneutrality and the concentration of free H^+ decreases. The bicarbonate appears to act as a buffer but the HCO_3 ion in sodium bicarbonate cannot influence plasma pH as $[HCO_3^-]$ is a dependant variable. Stewart's hypothesis is useful for the analysis of metabolic acid–base disturbances caused by changes in either SID or total weak acid concentration. Measurement of chloride concentration is important: if the chloride is low or normal and there is metabolic acidosis, unmeasured ions such as lactic acid and ketoacids may be present in the blood. The weak acids and bases of the plasma can absorb some of the hydrogen ions that are formed as a result of metabolism. This is called buffering. The carbon dioxide/bicarbonate buffer system is quantitatively the most important buffer in the body, partly because it is the most abundant buffer and partly because it is an open system because the P_{CO_2} can be regulated by the respiratory system.

4. A normal person produces about 15 moles of carbon dioxide and 70 mmol of non-volatile acid (mainly sulphuric acid). To maintain plasma hydrogen ion concentration within normal limits these acids have to be removed from the body. This is achieved by two processes; CO_2 is excreted via the lungs and the non-volatile acid is excreted by the kidney. The frequency and depth of respiration is stimulated by an increase in plasma CO_2 and hydrogen ion concentration. An increase in either plasma CO_2 or plasma hydrogen ion concentration leads to an increase in alveolar ventilation and this leads to an increased loss of carbon dioxide and this mechanism provides a rapid means of adjusting plasma hydrogen ion concentration.

5. The excretion of a non-volatile metabolic acid in the urine depends on the ability of phosphate to buffer hydrogen ions and on the

ability of the kidney to generate ammonium ions. Under normal conditions approximately half of the non-volatile acid is excreted as ammonium salts. If non-volatile acid production increases, the amount of ammonium ions in the urine increases proportionately.

6. Normal arterial blood pH is 7.4. Respiratory acid–base disorders occur if the deviation in plasma pH results from a change in alveolar ventilation. A metabolic acidosis occurs if the production of non-volatile acids exceed their rate of excretion by the kidneys or if there is a loss of non-volatile base from the gut. Conversely the loss of acid from the stomach gives rise to metabolic alkalosis.

7. Following an acid–base disturbance, compensatory mechanism are activated to bring the plasma pH within the normal range. Respiratory disorders are compensated by renal adjustments of plasma bicarbonate which occurs over a few days. Metabolic disorders are initially compensated by alterations in alveolar ventilation (respiratory compensation). However, this is always insufficient to restore plasma pH to the normal range. Full compensation and correction occurs via renal mechanisms.

9

Physiology of blood

After studying this chapter, the reader should be able to:

1. Explain the principal role of blood and its chief constituents
2. Describe the origin of blood cells
3. Describe the features of iron metabolism and its role in the body
4. Describe the structure and functions of platelets
5. Describe the carriage of oxygen and carbon dioxide by the red cells
6. Describe the physiology of blood coagulation including the cell-based theory of coagulation
7. Outline the mechanisms of fibrinolysis
8. Explain the significance of blood groups and their importance in blood transfusions
9. Describe the composition and functions of plasma
10. Outline the physiological consequences of blood transfusion

HAEMOPOIESIS AND ITS CONTROL

Blood cells originate from a common pluripotential haemopoietic stem cell (PHSC). In the first 3–4 weeks of embryonic development, primitive erythroblasts are the first cells that develop from mesenchymal stem cells in the yolk sac. From 6 weeks to 7 months, the liver – and the spleen to a lesser extent – becomes haemopoietically active and reaches its peak activity at 5 months and then declines to the time of birth. Between 6 and 7 months of fetal life, the bone marrow becomes the main source of new blood cells.

At birth and during the first 5 years of life, red blood cells are produced exclusively by the bone marrow. There is progressive fatty replacement of the marrow throughout the long bones and, by about 18–20 years of age, haemopoiesis occurs in the marrow confined to the central skeleton (vertebrae, pelvis, ribs, sternum, skull) and the proximal end of the femur and humerus.

Haemopoietic progenitor cells

The PHSCs give rise to all mature blood cells that circulate freely in the peripheral blood. The stem cells undergo cell division and maturation in the bone marrow. The erythroid, granulocytic and megakaryocytic cell lines are derived from a common pluripotential stem cell, appearing as an early myeloid precursor called CFU_{GEMM} (colony-forming unit, granulocyte, erythrocyte, monocyte, megakaryocyte). The precursor cells are stimulated by haemopoietic growth factors, resulting in considerable amplification within the system and increased production of one or more cell lines in accordance to need. The bone marrow is also the primary origin of lymphocytes, and there is some evidence for a common precursor cell for both myeloid and lymphoid cells (Fig. 9.1).

The haemopoietic growth factors are glycoprotein hormones that regulate the production, differentiation and maturation of haemopoietic

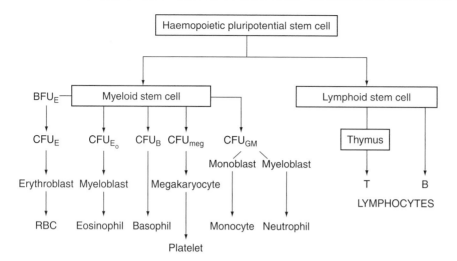

Figure 9.1 *Cell lines from pluripotential stem cells.*

precursor cells. Except for erythropoietin, which is mainly (90 per cent) synthesized in the kidney, these growth factors are produced by endothelial cells, stromal cells, T lymphocytes, monocytes and macrophages and activate specific receptors on the haemopoietic precursor cells. Interleukin-1 (IL-1) and tumour necrosis factor (TNF) are produced by lymphocytes acting on stromal cells to stimulate production of colony stimulating factor (CSF) for granulocytes, monocytes, erthryocytes and megakaryocytes. Stem cell factor (SCF) acts on pluripotential stem cells. IL-3 increases platelet production. Erythropoietin, CSF-granulocyte, CSF-monocyte and IL-5 act on late cells which are committed to one-cell lineage.

RED BLOOD CELLS

Formation

Red blood cells are formed in the bone marrow, and the most primitive cell is the proerythroblast. Over a period of 5 days, the proerythroblast gives rise to a series of progressively smaller normoblasts by undergoing cell division and maturation. The erythroblasts progressively contain more haemoglobin, while the nuclear chromatin becomes more condensed. Eventually a pyknotic nucleus is extruded from the late erythroblast to form a reticulocyte, which is the first red cell to enter the circulation (Fig. 9.2).

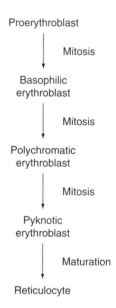

Figure 9.2 *Red blood cell development.*

The reticulocyte still contains some RNA and is capable of synthesizing haemoglobin. The reticulocyte is released from the bone marrow into the peripheral circulation and circulates for 1–2 days and matures when the RNA is completely lost.

Erythropoietic activity is regulated by erythropoietin ($T_{1/2} = 6$–9 h), which is produced by the peritubular complex of the kidney (90 per cent) and the liver (10 per cent). Erythropoietin enhances erythropoiesis by increasing the rate of differentiation of the stem cell. The final maturation of the red blood cells requires vitamin B_{12} and folic acid for the synthesis of thymidine triphosphate, which is essential for DNA formation. Deficiency of these two factors causes diminished DNA synthesis and failure of nuclear maturation and division, resulting in red blood cells that are large and fragile and have a short half-life. The mature erythrocytes survive for an average of 120 days in the circulation and are removed by phagocytosis in the reticuloendothelial system, mainly in the spleen and bone marrow.

Structure

The red blood cell is a biconcave disc, 7.5 μm wide and 2 μm thick (average dimensions). It has a large surface area relative to its volume, and this promotes gaseous diffusion into the cell. The red cell is markedly deformable and thus is able to traverse the microvasculature.

The red cell membrane is a bipolar lipid layer containing structural and contractile proteins, enzymes and surface antigens. About 50 per cent of the membrane is protein, 40 per cent is fat and 10 per cent carbohydrate. The phospho- and glycolipids are structural units with polar groups on the external and inner surfaces and the non-polar groups at the centre of the membrane. The carbohydrates are only present on the external surface. The proteins may be peripheral or integral, the latter penetrating the lipid layer. Four major proteins

(spectrin, actin, ankyrin and band 4.1) form a lattice on the inner side of the red cell membrane (Fig. 9.3) and are important in maintaining the biconcave shape of the red cell.

Haemoglobin production

The red cell contains haemoglobin, a metalloprotein of mol. wt. approximately 65 000 Da. Each molecule of haemoglobin consists of four polypeptide chains, each with its own haem group.

Haem synthesis (Fig. 9.4) occurs in the mitochondria by a series of biochemical reactions. Protoporphyrin is synthesized from the condensation of glycine and succinyl CoA under the influence of δ-aminolaevulinic acid (ALA) synthetase with pyridoxal phosphate as a coenzyme. Protoporphyrin combines with iron in the ferrous state (Fe^{++}) to form haem. Each haem combines with a globin chain, which is formed in the ribosomes. A haemoglobin molecule is formed by a tetramer of four globin chains, each with its own haem group in a hydrophobic pocket (Fig. 9.5).

Normal adult haemoglobin A (HbA) (96–98 per cent) consists of four polypeptide chains (2α and 2β), each with its own haem group. β globin production only appears after about 6 months of age. Fetal Hb (HbF) consists of 2α and 2γ chains,

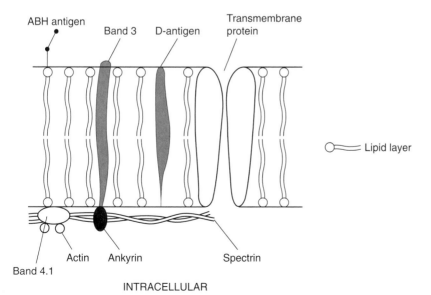

Figure 9.3 *The structure of the red cell membrane.*

Glycine + vitamin B$_6$ +   Succinyl CoA

δ-aminolaevulinic acid synthetase

Porphobilinogen

Uroporphyrinogen

Coproporphyrinogen

Protoporphyrin

— Fe^{++}

Haem

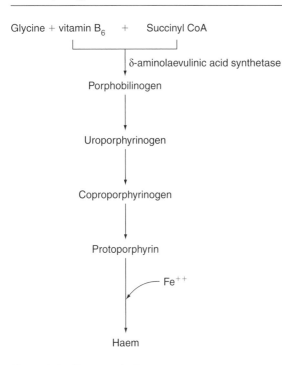

Figure 9.4 *Haem synthesis.*

and comprises only 1 per cent of adult Hb. About 1.5–3 per cent of adult haemoglobin consists of HbA$_2$, which contains 2α and 2δ chains.

Function of haemoglobin

As Hb acts as an oxygen carrier, the globin chains slide on each other. The α$_1$β$_1$ and α$_2$β$_2$ contacts stabilize the haemoglobin molecule when oxygen reacts with Hb. When O$_2$ is unloaded, the β chains are pulled apart so that 2,3-diphosphoglycerate (DPG) enters the molecule and decreases the affinity of the Hb molecule for oxygen.

Red cell metabolism

The red cell is able to generate ATP by the anaerobic glycolytic (Embden–Meyerhof) pathway for energy, and also generates NADPH as a source of reducing power to maintain reduced glutathione by the hexose monophosphate shunt (Fig. 9.6).

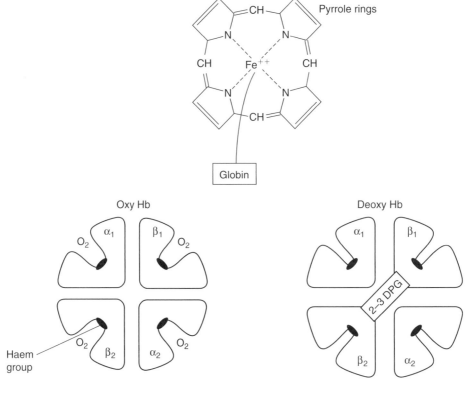

Figure 9.5 *Haemoglobin (Hb) structure.*

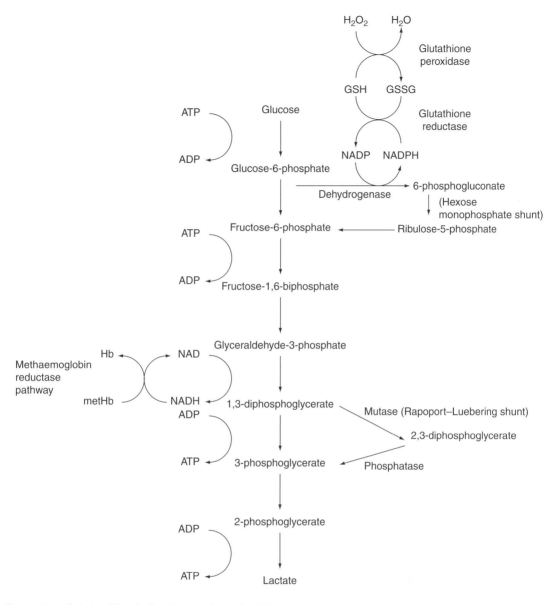

Figure 9.6   *Embden–Meyerhof pathway and associated shunts.*

The Embden–Meyerhof pathway produces two molecules of ATP for each molecule of glucose which is metabolized to lactate. The ATP is required for the maintenance of red cell shape, volume and flexibility by a Na^+/K^+-ATPase pump. NADH is also generated by the Embden–Meyerhof pathway, which is required by methaemoglobin reductase to reduce methaemoglobin to haemoglobin. Some 1,3-DPG is converted to 2,3-DPG by the Rapoport–Luebering shunt.

About 5 per cent of glycolysis undergoes oxidation by the hexose monophosphate (HMP) shunt in which glucose-6-phosphate is converted to 6-phosphogluconate and to ribulose-5-phosphate. NADPH is formed from NADP, and this is linked with glutathione which maintains sulphydryl (-SH) groups intact in the cell. The amount of glucose passing through the HMP shunt is determined by the NADPH:NADP ratio.

Destruction of red cells

With ageing, glycolysis – and hence ATP formation – decreases in the red cell, so that energy for the maintenance of cellular integrity is diminished. The old red cells are removed by the reticuloendothelial cells, especially in the spleen. The globin chains are broken down to amino acids and re-enter the amino acid pool. The iron is reutilized by the bone marrow for the synthesis of haemoglobin. The protoporphyrin ring is opened to form biliverdin. A small fraction of protoporphyrin is converted to carbon monoxide. Biliverdin is metabolized to bilirubin, which is bound to albumin and carried to the liver. In the liver, bilirubin is conjugated with glucuronic acid and excreted in the bile and hence into the small gut. In the gut, the bilirubin is converted to stercobilin, some of which is reabsorbed into the plasma and excreted by the kidney as urobilinogen in the urine (Fig. 9.7).

Small amounts of free haemoglobin may be released in the plasma and haptoglobulin, an α_2 globulin, binds to the globin moiety of the haemoglobin. A second protein, haemopexin, binds to haem when haptoglobin is saturated.

Iron metabolism

Haemoglobin contains 65–70 per cent of the total body iron, with myoglobin containing 4–5 per cent. Iron is also associated with cellular respiration through the action of iron-containing enzymes such as cytochromes, catalase and peroxidase. Iron is transported in plasma as transferrin, a β_1 globulin that binds two atoms of ferric iron per molecule. The main source of iron carried by transferrin is from the reticuloendothelial cells that destroy ageing red cells. Some iron is also stored in the reticuloendothelial cells as haemosiderin and ferritin. Ferritin is a water-soluble protein–iron complex consisting of an outer protein shell, called apoferritin, and an inner core of iron–phosphate–hydroxide.

DIETARY IRON

Iron is present in food in the form of haem–protein and ferric–protein complexes, and as ferric hydroxide. Although about 10–15 mg of iron is present in an average diet, only 10 per cent is absorbed in a normal person. However, this proportion may be increased to 20–30 per cent in pregnancy, or in iron-deficiency states (Fig. 9.8).

IRON ABSORPTION

Iron absorption occurs mainly in the duodenum. Factors favouring the absorption of iron include gastric acid and reducing agents, which maintain the soluble iron in the ferrous state. Iron absorption is reduced by alkali, chelating agents such as phytates and phosphates which form insoluble iron complexes. Soluble iron in the ferrous state enters the brush border of the mucosal cells and enters the portal blood. Excess iron combines with apoferritin to form ferritin, which is shed into the gut lumen when the mucosal cell comes to the end of its life span (3–4 days).

IRON TRANSPORT AND STORAGE

Iron is transported in plasma bound to transferrin, which is synthesized in the liver. Transferrin binds

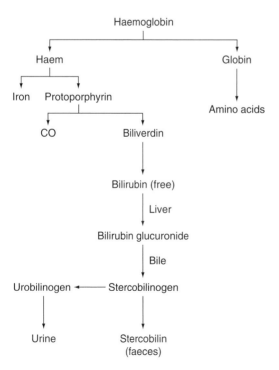

Figure 9.7 *Haemoglobin breakdown.*

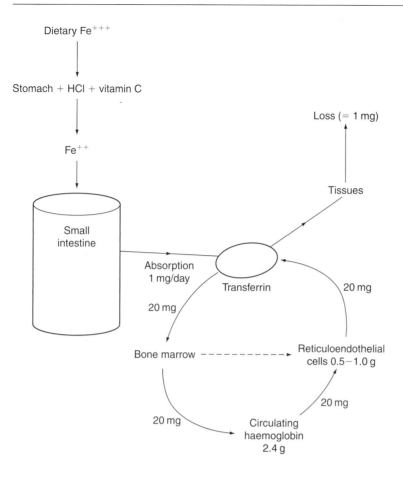

Figure 9.8 *The iron cycle.*

two atoms of iron per molecule, and has a half-life of 8–10 days and is recycled. Plasma transferrin is normally 30 per cent saturated with iron, and the total iron binding capacity of plasma is approximately 40–75 μmol/L. Transferrin gains most of its iron from the macrophages of the reticuloendothelial system. When the plasma iron concentration is raised, transferrin is saturated and iron is transferred to parenchymal cells.

The main sites of iron storage are the liver, spleen and bone marrow. Some 65 per cent of the iron is stored as ferritin (which is water-soluble) and 35 per cent as haemosiderin (which is granular and insoluble) mainly present in the liver, bone marrow and other tissues such as the pancreas and the heart.

About 0.5–1 g iron is lost each day in the faeces as desquamated epithelial cells of the gut. Small amounts of iron are lost in the urine, hair and sweat. In the female, menstruation forms an important source of iron loss. In pregnancy, considerable iron loss occur across the placenta to the fetus.

Red cell antigens and antibodies

RED CELL ANTIGENS

Over 400 red cell antigens have been described, the majority of which are inherited in a simple Mendelian fashion. Blood groups are important, as individuals who lack a particular blood group may produce antibodies reacting with that antigen. The significance of this is that such individuals may develop a transfusion reaction if red cells bearing the antigen are transfused. The biological role of most red cell antigens is not known, but they may allow the recognition of self from foreign cells by the immune system.

The most important red cell antigens are ABO and Rh. The other red cell antigens are less important because the antigens are weak and their antibodies develop only after multiple exposures. Others react only at low temperatures.

RED CELL ANTIBODIES

Naturally occurring antibodies are present in the plasma of individuals who lack the corresponding antigen, the most important being anti-A and anti-B antibodies. These are usually IgM and are reactive at 37°C, but are optimally reactive at 4°C.

Immune antibodies to red cell antigens develop upon exposure to red cells possessing antigens which the individual lacks, usually by transfusion or by transplacental passage during pregnancy. These are IgG antibodies and they react optimally at 37°C, although IgM antibodies may also develop in the early phase of the reaction. Only IgG antibodies pass transplacentally from mother to fetus; the most important is the Rh antibody, anti-D.

Natural antibodies develop in the ABO system after 3 months of age. Small amounts of group A and B antigens enter the body in bacteria and food and stimulate the formation of anti-A or anti-B antibodies.

ABO SYSTEM

The ABO system of red cell antigens was the first system of red cell antigens described, and is the most important. The ABO antigens are controlled by three allelic genes, A, B and O genes. The A and B genes control the synthesis of enzymes required for the addition of specific carbohydrate residues to the basic antigenic glycoprotein, the H antigen.

The ABO antigens are complex carbohydrates (Fig. 9.9). The core antigen is the H antigen, which is a glycoprotein precursor with L-fucose as the terminal sugar. The A antigen is formed when N-acetylgalactosamine is added to the terminal group of the H antigen. The B antigen results when D-galactose is added to the H antigen.

Natural antibodies to A and/or B antigens are found in the plasma of subjects who have red cells that lack the corresponding antigen. Two subgroups A_1 and A_2 are recognized on group A red cells. Both subgroups agglutinate with anti-A antibody, but only group A_1 agglutinates with lectin from *Dolichos biflorus*, a plant seed, and anti-A_1 antibody. This difference in agglutinability

depends on the number of antigen sites reduced in group A_2 cells.

Individuals with group O red cells lack both A and B antigen, but have the H antigen, and also anti-A and anti-B antibodies. Individuals with the AB group possess both A and B antigen, but lack both anti-A and anti-B antibodies. The frequency of blood groups in the Caucasian population is: A group (45 per cent), B (9 per cent), AB (3 per cent), O (43 per cent); with 80 per cent of group A with strong antigens (A_1) and 20 per cent with weak antigens (A_2).

Three allelic genes, A_1, B_1 and O_1, and a pair of allelic genes, H and h, determine the formation of the ABO antigens. Gene H is responsible for the enzyme, α-L-fucosyltransferase, which attaches fucose to the glycoprotein precursor to form the H antigen. Genes A and B are responsible for α-N-acetyl-D-galactosaminyl transferase and α-D-galactosyltransferase, which attach galactosamine and galactose to substance H, respectively, and thus determine the antigenic specificity of the A and B antigen.

A_1, B_1 and H antigens are present in the red cells as well as most other body cells, including white cells and platelets. These antigens are also present in the body fluids of 80 per cent of the population (who possess the secretor gene) in a water-soluble form. These water-soluble antigens are present in plasma, saliva, semen, urine, gastric juice, tears and bile, but not in the cerebrospinal fluid. As the ABO antigens are stable, their detection in dried blood and other body fluid stains are used in forensic serology.

THE RH SYSTEM

The Rhesus blood group system is next to the ABO in clinical importance. The Rh system is named after the Rhesus monkey, the red cells of which stimulated antibodies when they were injected into rabbits and guinea pigs. Although the Rhesus monkey antigen is similar to the Rh D antigen of human red cells, it is not identical. Human red cells have both the monkey antigen and a separate Rh antigen. The Rh D antigen is a protein of mol. wt. 30 000 Da. The serological activity of the Rh antigen is governed by the amino acid sequence and the presence of specific phospholipids.

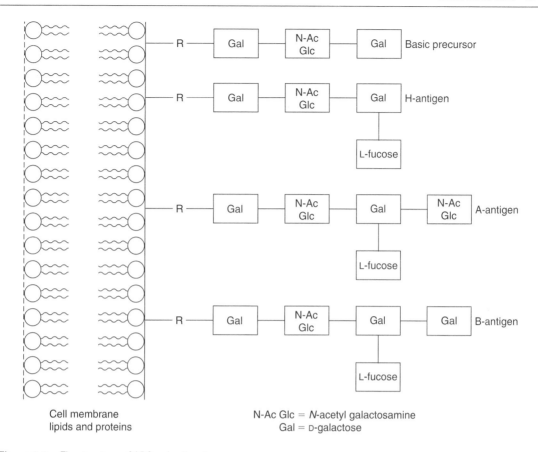

Figure 9.9   *The structure of ABO red cell antigens.*

The red cell antigens of the Rhesus system are determined by three closely linked pair of alleles. The main alleles are Dd, Ce and Ee, and the three antigenic groups are derived from one gene complex. The term Rh +ve refers to the presence of the D antigen. Each antigen is defined by a specific antibody (anti-C, anti-c, anti-E and anti-e) except for d antigen. Anti-d does not exist. Rh antigens, unlike the ABO groups, are only present on red blood cells. Rh antibodies rarely occur naturally. Most Rh antibodies are immune, 'warm' and IgG in origin. The D group is the most important, as 85 per cent of the population possess the D antigen (Rh +ve).

OTHER BLOOD GROUP SYSTEMS

Other blood group systems are clinically less important. The P, Lewis and MN systems are not uncommon, but the naturally occurring antibodies only react at low temperatures and have antigens with low antigenicity. The MN antigenicity depends on the terminal amino acid sequence of glycophorin (a red cell membrane protein). The P antigen is determined by the conversion of trihexosyl ceramide to globoside (P antigen) found on the red cell membrane. The main feature of the antigens of the Lewis system is that they are soluble substances present in saliva and plasma, which are adsorbed onto the surface of red cells from plasma.

The Kell system is the third most important blood group system after the ABO and Rh systems. The K antigen is present on red cells, leukocytes and platelets. Although the K antigen is immunogenic, it is of relatively low frequency and therefore may only cause iso-immunization in patients who have had multiple blood transfusions.

WHITE BLOOD CELLS

The white blood cells (leukocytes) consist of two groups: (1) the phagocytes, and (2) the immunocytes. The phagocytes are made up of the granulocytes (neutrophils, eosinophils and basophils) and the monocytes. The immunocytes consist of the lymphocytes, their precursor cells and plasma cells. The phagocytes and immunocytes are important for protection against infection, and are closely associated with immunoglobulins and complement.

Granulocyte formation

Granulocytes develop from the PHSCs in the bone marrow, and the earliest recognizable myeloid cell is the myeloblast. The myeloblasts, promyelocytes and myelocytes undergo mitosis to form a proliferative pool of cells. These undergo post-mitotic maturation to form metamyelocytes and segmented granulocytes. These cells undergo maturation and differentiation into polymorphonuclear neutrophils (Fig. 9.10).

Eosinophil and basophil precursors undergo similar maturation. The formation and maturation of neutrophils take 6–10 days in the bone marrow, which contains more myeloid than erythroid cells. In the normal stable state, the bone marrow stores 10–15 times the number of circulating granulocytes. Maturation in the bone marrow takes 3–4 days, after which the granulocytes are stored for 2–5 days before being released into the circulation.

Various growth factors are produced by stromal cells (endothelial cells, fibroblasts, and macrophages) and T lymphocytes. Interleukin-1, IL-3 and IL-6 act synergistically to produce progenitor cells, which produce red cells, neutrophils, eosinophils, basophils, monocytes and platelets. Colony stimulating factor (granulocyte), IL-5, and colony stimulating factor (monocyte) increase the production of neutrophils, eosinophils and monocytes, respectively.

NEUTROPHILS

The bone marrow releases neutrophils from the PHSCs into the circulation. Neutrophils in blood may be divided into circulating and marginating (attached to vascular endothelium) pools. The transit time of neutrophils in the peripheral circulation is 6–12 h, after which the neutrophils

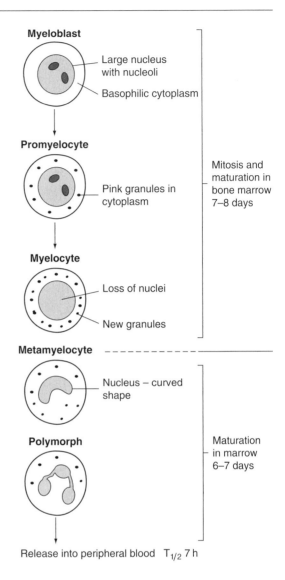

Figure 9.10 *Maturation sequence of white cells.*

migrate to the tissues, where they survive for 4–5 days before destruction as a result of senescence or defence actions.

The neutrophil cell has a dense nucleus containing primary azurophilic granules (acid phosphatase, myeloperoxidase, esterase and lysozyme) and secondary granules (aminopeptidase, lysozyme and collagenase).

EOSINOPHILS

Eosinophils develop from the PHSCs and mature in the bone marrow in 3–4 days. They circulate in the peripheral circulation for 3–8 h and enter into the tissues – primarily epithelial linings (e.g. respiratory mucosa) – where they stay for 8–12 days. Eosinophils are phagocytic and are important for allergic and parasitic diseases. Eosinophils contain granules that, when stimulated, release enzymes that inhibit the mast cell products (SRS-A) and destroy antigen–antibody complexes, thus limiting the spread of local inflammation.

BASOPHILS

Basophils are also derived from the PHSCs, and have dark cytoplasmic granules containing heparin, histamine, hyaluronic acid and serotonin. These may be present occasionally in blood, but they become mast cells in the tissues. They have IgE attachment sites and release histamine on degranulation. They are involved in allergic and parasitic diseases.

MONOCYTES

Monocytes are also derived from the common PHSCs, and circulate in blood for approximately 20–40 h. They then enter the tissues where they mature into macrophages with a life span of several months to years. They are also found in the reticuloendothelial system throughout the body in the form of Kupffer cells (liver), Langerhans' cells (skin), alveolar macrophages (lung), sinus macrophages (lymph nodes and spleen) and the follicular interdigitating cells (lymph nodes). They are phagocytic and have receptors for IgG (Fc fragment), complement, and various lymphokines. Macrophages and monocytes produce several monokines such as IL-1, prostaglandins, interferon and TNF.

Neutrophil and monocyte function

The defence functions of the neutrophils and monocytes occur in three phases: (1) chemotaxis, (2) phagocytosis, and (3) killing and digestion (Fig. 9.11).

CHEMOTAXIS

Chemotaxis is brought about by cell mobilization and migration. Neutrophils and monocytes

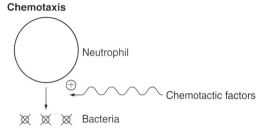

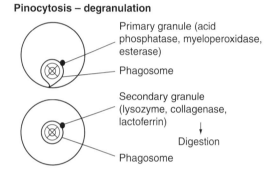

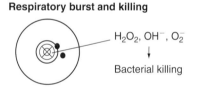

Figure 9.11 *Sequence showing phagocytosis and bacterial destruction.*

migrate through pores in the endothelium of blood vessels by diapedesis, and by amoeboid movement through tissues. These phagocytes are attracted to bacteria or inflammatory areas by chemicals released by damaged tissues, and by complement or leukocyte cohesion molecules interacting with damaged tissues.

PHAGOCYTOSIS

Phagocytosis refers to the cellular ingestion of foreign material (bacteria, fungi) or dead or damaged cells of the host. The foreign particle is opsonized with immunoglobulin or complement, and this enables the neutrophils and monocytes to recognize the foreign material as they possess Fc and C3b receptors. The binding of phagocytes to particles coated with IgG or C3 results in activation of phagocytosis (Fig. 9.11).

Macrophages also present foreign antigens to the immune system and secrete a large number of growth factors (IL-1, IL-3, IL-4, IL-6, TNF, CSF-GM).

KILLING AND DIGESTION

Foreign material may be digested by lysosomes rich in proteolytic enzymes. The macrophages also have lipases to digest the thick lipid cell membranes of bacteria. Oxidative reactions are also important for killing bacteria. Superoxide anion (O_2^-), hydrogen peroxide (H_2O_2) and hydroxyl ions are generated from O_2 and NADPH. In the neutrophils, myeloperoxidase catalyses the reaction of H_2O_2 and Cl^- to form hypochlorite, which is bactericidal. Lactoferrin is an iron-binding protein present in neutrophil granules and has a bacteriostatic action by depriving bacteria of iron.

Lymphocytes

Lymphocytes are immunologically competent cells that assist and add specificity to the defence of the body against infection. The bone marrow is the primary lymphopoietic organ where lymphocytic stem cells are produced in response to non-specific cytokines in postnatal life. In the fetus, the yolk sac, liver and spleen are the primary sites. The secondary lymphopoietic sites are the lymphoid tissue found in the lymph nodes, spleen and respiratory tract. There are two types of lymphocytes: T and B.

T LYMPHOCYTES

T lymphocyte precursors are derived from the bone marrow, and migrate to the thymus where they are conditioned to develop into helper and suppressor cells. As the T cells migrate from the thymic cortex to the medulla, they lose and gain a number of antigens. The helper cells express CD4 antigen, whilst the suppressor cells express the CD8, CD2, CD3 and other T-cell antigens.

The T cells have to recognize major histocompatibility complex antigens on the presenting monocyte.

T cells comprise 60–80 per cent of all circulating lymphocytes and 90 per cent of thoracic duct lymphocytes. They are long-lived and produce a wide range of lymphokines (IL-2, IL-3, α- and γ-interferon, lymphotoxins, TNF and β cell growth factor). The T cells are primarily involved in cell-mediated immunity.

B LYMPHOCYTES

The B lymphocytes are also derived from bone marrow stem cells, and are primarily concerned with antibody production in humoral immunity. B lymphocytes are the predominant lymphocytes in the bone marrow and the germinal centres of the lymph node follicles. The B cells comprise 5–15 per cent of the circulating lymphocytes and are short-lived. They express antigen-specific immunoglobulins on their surface (IgM, IgG, IgA, IgD) and other surface antigens.

NON-B/NON-T LYMPHOCYTES

These include the large granular lymphocytes (LGL), which may be classified into two subgroups: the natural killer (NK) cells and the antibody-dependent cytotoxic (ADC) cells. The NK cells are stimulated by IL-2 and γ-interferon, and are not MHC restricted. The NK and ADC cells are associated with graft rejection and killing of tumour and viral-infected cells.

PLATELETS

Platelets are small colourless, disc-shaped cell bodies approximately 2–3 μm in diameter and 7×10^{-15} L in volume.

Platelet production

Platelets are derived from megakaryocytes developed from pluripotent stem cells in the bone marrow (Fig. 9.12). The megakaryoblasts undergo non-mitotic nuclear replication with increasing cytoplasmic volume. As the cell enlarges, the cell membrane invaginates and platelets bud off from the surface. The time taken for the stem cell to

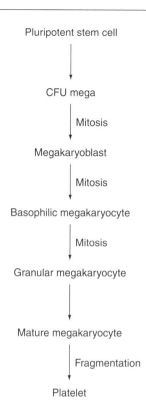

Pluripotent stem cell

↓

CFU mega

↓ Mitosis

Megakaryoblast

↓ Mitosis

Basophilic megakaryocyte

↓ Mitosis

Granular megakaryocyte

↓

Mature megakaryocyte

↓ Fragmentation

Platelet

Figure 9.12   *Platelet production.*

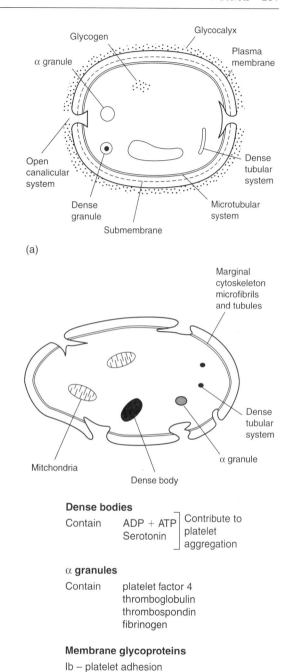

(a)

Dense bodies

Contain   ADP + ATP ⎤ Contribute to
           Serotonin   ⎦ platelet aggregation

α granules

Contain   platelet factor 4
           thromboglobulin
           thrombospondin
           fibrinogen

Membrane glycoproteins

Ib – platelet adhesion
IIb/IIIa – platelet aggregation

(b)

Figure 9.13   *Platelet ultrastructure.*

produce platelets is approximately 10 days. The production of platelets is under the control of thrombopoietin.

Young platelets are stored in the spleen for 36 h before being released into the circulation. The normal life span of a platelet is 7–10 days. Old platelets are removed by the reticuloendothelial system in the spleen and liver.

Platelet structure

Circulating platelets are fragments of megakarocyte cytoplasm enclosed within a membrane. As they lack a nucleus, they cannot synthesize protein, but they do contain a large number of structural elements. These include an external coat or glycocalyx, the platelet membrane with invaginations forming the open canalicular system, a dense tubular system, a cytoskeleton, a peripheral zone of microtubules, and numerous organelles such as α granules, dense granules, lysosomes, microperoxisomes and mitochondria (Fig. 9.13(a)).

The exterior coat (glycocalyx) contains an outer layer of membrane glycoproteins, which are important for platelet adhesion and aggregation. Adhesion to collagen is facilitated by glycoprotein

Ia. Glycoproteins Ib, IIb and IIIa are required for the attachment of platelets to von Willebrand factor (vWF) and thus to the vascular subendothelium. The binding site for glycoprotein IIb–IIIa is also an important receptor for fibrinogen, which promotes platelet–platelet aggregation.

The platelet membrane consists of a double layer of lipid and phospholipid covered by protein on either side. The membrane phospholipids serve as a source of arachidonic acid, platelet activating factor (PAF) and platelet factor 3, which is important for the activation of factor X and prothrombin.

The plasma membrane invaginates to form the open canalicular system, which provides a large surface area for the absorption of plasma coagulation proteins. In the submembranous area and throughout the cytoplasm there is the contractile protein system of microfilaments. The circumferential band of microtubules maintains the discoid shape of resting platelets.

The dense tubular system represents the residual endoplasmic reticulum, which is rich in Ca^{++}, ATP-ase, adenyl cyclase, acetyl cholinesterase, peroxidase and glucose-6-phosphatase. This may be the site of synthesis of prostaglandins and thromboxane A_2.

Within the platelet are numerous granules, the α granules and the dense granules. The α granules contain β thromboglobulin, fibronectin, fibrinogen, platelet factor 4, platelet-derived growth factor (PDGF), thrombospondin (TSP) and vWF (Fig. 9.13(b)). The platelet dense granules have an electron-dense core and store serotonin, ADP, ATP and pyrophosphate. Other organelles include lysosomes containing hydrolytic enzymes, and peroxisomes containing catalase. Anaerobic glycolysis in the mitochondria is the main source of ATP although some aerobic metabolism occurs through the Krebs citric acid cycle.

Platelet antigens

Platelets have specific surface antigens, HPA 1–5, and they also express ABO and HLA class I antigens. Thus, platelet antibodies may be found in patients after multiple transfusions. The presence of platelet antibodies may shorten the survival of transfused platelets and reduce their efficacy.

Platelet function

The main function of platelets is the formation of a haemostatic plug during the normal response to vascular injury. This is brought about by platelet adhesion, secretion, aggregation and procoagulant activity.

Platelet adhesion

Following blood vessel damage, the endothelium is disrupted and platelets are rapidly recruited from the circulating blood to adhere to the subendothelial connective tissue to form an occlusive plug. Platelet adhesion does not require metabolic activity. Adhesion to collagen is facilitated by glycoprotein Ia. Subendothelial microfibrils bind large multimers of vWF, and through these react with glycoprotein Ib–IX complex, an adhesion receptor on the platelet surface membrane (Fig. 9.14(a)). The platelet GPIIb–GPIIIa complex is then exposed and binds to vWF and fibronectin, leading to the spreading of platelets on the subendothelial matrix.

Following adhesion, metabolic processes are activated in the platelets and promote platelet release reactions, shape change and aggregation. Platelet activation can be induced by adhesion to proteins such as collagen, soluble agonists (epinephrine, ADP, serotonin and thrombin) and cell contact during platelet aggregation. The activation process involves complex biochemical events activating membrane phospholipase A_2 and phospholipase C converting inositol biphosphate to diacylglycerol and inositol triphosphate. These mediators activate protein kinase C, resulting in protein phosphorylation, and an increased cytoplasmic calcium concentration that promotes calcium-dependent and calmodulin-dependent reactions. Following platelet activation, the cytoskeleton proteins are reorganized, leading to a transformation of the platelet from a compact disc to a sphere with long pseudopods spreading onto the subendothelial matrix.

RELEASE REACTION

Collagen exposure or the action of thrombin results in the release of platelet granule contents.

(a) Adhesion

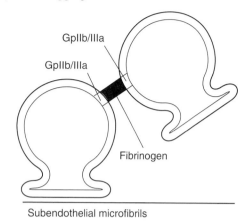

(b) Activation

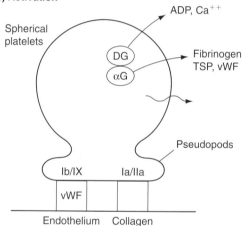

(c) Platelet aggregation

(d) Stabilization of platelet plug by fibrin

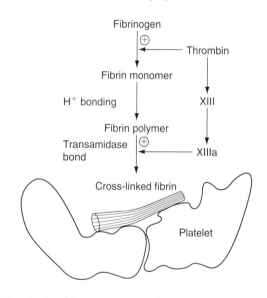

Figure 9.14   *Sequence showing platelet activity: (a) adhesion, (b) activation, (c) aggregation, and (d) stabilization of platelet plug by fibrin.*

The dense granules (DG) release ADP, epinephrine and serotonin within 30 s, and these reinforce platelet activation (Fig. 9.14(b)). After 30 s, the α granules release fibrinogen, β thromboglobulin, PAF-4, factor V, vWF, platelet-derived growth factor (PDGF) and thrombospondin (TSP). These mediate and reinforce platelet aggregation and adhesion. Thromboxane A_2, produced by the action of phospholipase A_2 on membrane phospholipids, lowers

platelet cyclic AMP, initiating the release reaction and promoting vasoconstriction and platelet aggregation.

PLATELET AGGREGATION

Platelet release reactions promote platelet aggregation in several ways (Fig. 9.14(c)). Released ADP and thromboxane A_2 cause platelet aggregation at the site of vascular injury. ADP promotes platelet–platelet adhesion, liberating more ADP and thromboxane A_2, causing secondary platelet aggregation and a positive feedback mechanism promoting the formation of a platelet plug. ADP also increases the number of fibrinogen receptors (IIb/IIIa) (IIb/IIIa) on the platelet surface (Fig. 9.14(c)). Fibrinogen and vWF released from the α granules enhance platelet adhesion and aggregation. Thrombospondin, also released from the α granules, stabilizes the platelet aggregate. Another α granule protein (GMP-140) may be involved in the interaction between platelets and white cells.

Platelet procoagulant activity

After platelet activation by thrombin and collagen, procoagulant activity increases. This procoagulant action requires Ca^{++} influx across the plasma membrane and re-orientation of phosphatidyl serine from the inner layer to the outer layer of the platelet membrane with the expression of binding sites for specific coagulation processes. Platelet factor 3, an exposed membrane phospholipid, is available for coagulation and protein complex formation. The first reaction involves factors IXa, VIII and X in the formation of factor Xa. The second (or prothrombinase) reaction results in the formation of thrombin from the interaction of factors Xa, V and II. The irreversible fusion of platelets aggregated at the site of endothelial injury is enhanced by ADP and other enzymes released during the platelet release reaction. Thrombin also promotes platelet plug formation.

PDGF, released by the α granules, stimulates vascular smooth muscle cells to multiply, and thus promotes vascular healing.

COAGULATION

Coagulation is a biological amplification system in which a few initial substances activate, by proteolysis, a cascade of circulating precursor enzymes. This results in the generation of thrombin, which converts soluble plasma fibrinogen into fibrin. The fibrin enmeshes the platelet aggregates at the site of injury so that a firm and stable haemostatic plug is formed. The coagulation cascade or sequence comprises complex interactions between inactive

and active enzymes, and can occur along two pathways (intrinsic/extrinsic) (Fig. 9.15).

The sequential activation of this enzyme cascade requires the local concentration of circulating clotting factors at the site of injury. The clotting factors are either precursors of serine proteases (enzymes) or co-factors.

The intrinsic system

The intrinsic system (Fig. 9.15) is activated by contact with negatively charged surfaces such as collagen and subendothelial connective tissue. This contact phase is unique in that the reactions do not depend on Ca^{++}. Plasma factor XII interacts with the negatively charged surfaces of collagen or subendothelial microfibrils in vivo, or glass in vitro, and activated factor XII (XIIa) is generated. XIIa activates prekallikrein, which further generates additional XIIa, amplifying the initial response. High-molecular-weight kininogen (HMWK) binds to the negatively charged surface in association with factor XII, and enhances the activation of factor XI by XIIa. HMWK also promotes the conversion of prekallikrein to kallikrein, which generates more XIIa, releasing kinins at the same time.

Activated factor XI on the subendothelial tissue or the surface of platelets next activates factor IX by limited proteolysis in the presence of Ca^{++}. Activated IX (IXa), in the presence of factor VIII, Ca^{++} and platelet phospholipid, activates factor X. This reaction is enhanced when these factors form a multimolecular complex on the phospholipid membrane of the platelets with IXa and X bound at α-carboxylglutamic acid residues via Ca^{++} bridges to the phospholipid and factor VIII bound to the lipid matrix.

It is now considered that the contact phase involving prekallikrein and HMWK only occurs in vitro. Activation of IX in vivo is considered to be mainly via factor VII activated by tissue factor. Factor VIII consists of two components, VIII C (VIII coagulant) and VIII vWF (von Willebrand factor or ristocetin co-factor). Factor VIII C is synthesized in the liver and is inactivated by protein C and protein S. Factor VIII vWF is synthesized by endothelial cells and platelets, and circulates as high-molecular-weight multimers important for platelet–endothelial cell interactions via GP Ib and IIb/IIIa receptors.

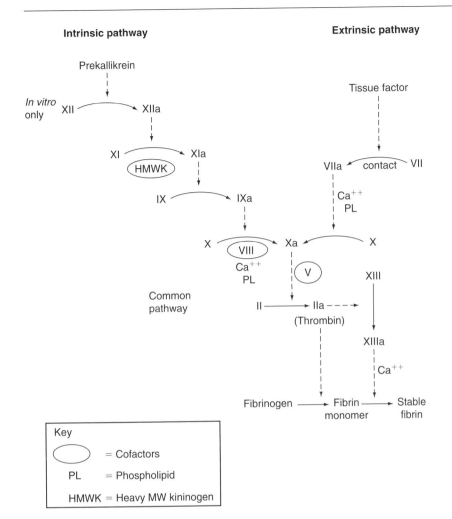

Intrinsic pathway

Extrinsic pathway

Figure 9.15 *The coagulation cascade.*

Extrinsic pathway

The extrinsic pathway (*see* Fig. 9.15) is an alternative mechanism for the activation of factor X. Tissue factor or tissue thromboplastin (III) found on the surface of perivascular tissue cells binds factor VII, which in turn activates factor X. In this process, factor VII is itself activated. *In vivo*, it now appears that the main role of factor VII is to activate factor IX rather than to activate factor X directly.

The final common pathway

In the final common pathway, activated factor X together with factor V, calcium and platelet factor 3 convert prothrombin (II) to thrombin (IIa).

Thrombin hydrolyses the arginine–glycine bonds of fibrinogen to fibrinopeptide A and B to form fibrin monomers. Hydrogen bonds link fibrin monomers to form a loose, insoluble fibrin polymer. Factor XIII activated by thrombin and calcium stabilize the fibrin polymers via covalent bond cross-links.

The extrinsic and intrinsic systems complement each other. Factor VII activated by tissue factor in the extrinsic pathway also activates factor IX in the intrinsic system. Thrombin also activates factor V, VIII and XI via positive feedback mechanisms.

The normal haemostatic response to vascular damage results in closely linked interactions between the blood vessel wall, circulating platelets and blood coagulation factors, as summarized in Fig. 9.16.

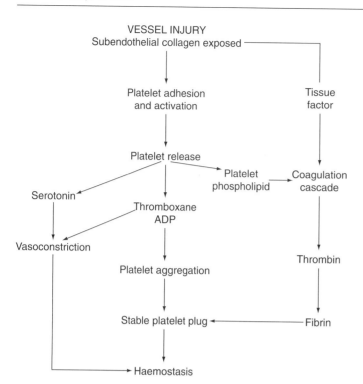

Figure 9.16 *The haemostatic response to injury.*

CELL-BASED THEORY OF COAGULATION

Several clinical and experimental observations suggest that the traditional 'cascade' theory of coagulation does not accurately reflect the events of *in-vivo* haemostasis. A cell-based model of haemostasis has been proposed (Fig. 9.17) and factor VIIa plays an important role at various stages. In this model, the process that leads to the explosive generation of thrombin that is required for the formation of a stable clot and haemostasis may be divided into three stages: (1) initiation, (2) amplification, and (3) propagation.

Small amounts of factor VII, X and prothrombin leave the vasculature to tissue spaces. The initiation phase begins when small amounts of factor Xa and thrombin are generated on tissue factor-bearing cells such as stromal fibroblast, mononuclear cells, macrophages, and endothelial cells. Activated factor VII binds to these tissue factor-bearing cells. The formation of tissue factor/ factor VII complex (TF–FVIIa) on cell surfaces activates factor X and this tissue factor VIIa–Xase complex generates small amounts of FXa and thrombin on tissue factor-bearing cells, platelets

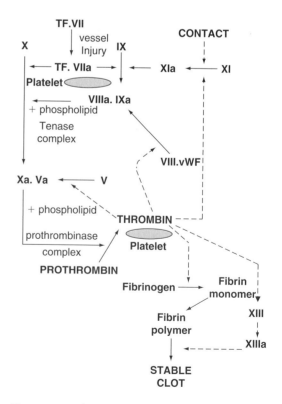

Figure 9.17 *Cell-based theory of coagulation. TF, tissue factor.*

or monocytes. This small amount of thrombin causes local activation of platelets, factor V and factor VIII. This phase may still be referred to as extrinsic because tissue factor-bearing cells are normally outside the vasculature. Tissue factor is not usually in contact with blood until injury or inflammation occurs (Fig. 9.18(a)).

(a)

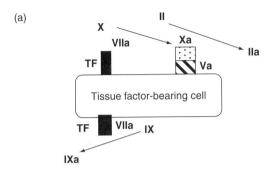

1. Factor VII, X and prothrombin leave vasculature and bind to tissue factor-bearing cells
2. Generation of small amounts of Xa and thrombin on surface of tissue factor-bearing cells by VIIa
3. Thrombin activates platelets, factor V and VIII

Figure 9.18 *(a) Initiation phase of cell theory of coagulation.*

The amplification stage begins with vascular disruption and exposure of tissue factor-bearing cells to platelets, von Willebrand factor and factor VIII. The thrombin generated in the initiation stage activates platelets forming the platelet plug. The surface of activated platelets is primed with factor Va, factor VIIIa and factor XIa. Both tissue factor–factor VIIa complex and factor XIa activate factor IX. This leads to the formation of IXa–VIIIa (tenase) complex on platelet surface generating more factor Xa and starts the propagation phase. Factor XII and other factors are not always necessary for coagulation as initially postulated by the cascade hypothesis (Fig. 9.18(b)).

The propagation phase begins with the formation with the tenase (IXa–VIIIa) complex on platelet surfaces. The tenase complex generates large amounts of factor Xa which interacts with factor Va (forming factor Va–factor Xa complex, prothrombinase) and this generates large amounts of thrombin, cleaving fibrinogen into fibrin monomers which polymerse to form a stable

(b)

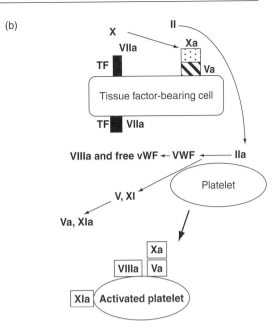

1. Thrombin generated in initiation phase activates platelets to form platelet plug and prime surface with Va, VIIIa, and XIa
2. TF–VIIa and XIa activate IX, leading to propagation phase

Figure 9.18 *(b) Amplification phase of cell theory of coagulation.*

fibrin clot. This burst of thrombin generation exerts a positive feedback by activating factor V, factor VIII, and factor XI (Fig. 9.18(c)).

In this revised hypothesis, tissue factor rather than a 'contact' factor is responsible for initiating coagulation. Tissue factor is also found in the adventitia of blood vessels, skin epithelium, intestinal mucosae, and organ capsules. Tissue factor interacts with coagulation factors when vessel wall damage occurs. Factors IX and VII are necessary for enhanced factor Xa generation and sustained coagulation.

Physiological inhibitors of coagulation

In order to prevent uncontrolled disseminated coagulation, there are a number of physiological inhibitors of coagulation. Serine protease inhibitors include antithrombin III (inhibits IIa and Xa), C_1 inhibitors (inhibit contact factors), α_2 macroglobulins (inhibit IIa and contact factors),

(c)

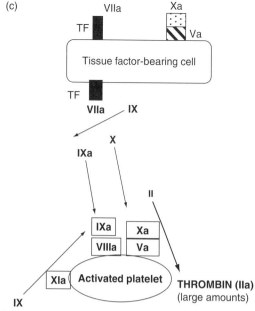

Large amounts of Xa bind to Va to form prothrombinase complex which stimulates activated platelets to generate large amounts of thrombin (thrombin burst reaction) which converts fibrinogen to fibrin monomers, and activates factor XIII to form fibrin polymers

Figure 9.18 *(c) Propagation phase of cell theory of coagulation.*

α_2 antiplasmin and α_2 antitrypsin which inhibit circulating serine proteases.

Factors Va and VIIIa are regulated by proteins C and S, which are vitamin-K-dependent serine proteases. Protein C destroys factor V and VIII, whilst protein S enhances protein C by binding it to the platelet surface. Protein C is activated by thrombomodulin formed by the binding of thrombin to the endothelial cell surface (Fig. 9.19).

Fibrinolysis

Fibrinolysis is a process where fibrin and fibrinogen are cleared by plasmin. Plasminogen (a β-globulin) in blood and tissue fluid is converted to plasmin (a serine protease) by intrinsic activation (via vessel wall activators or extrinsic activation via tissue activators). Plasminogen activators are produced by endothelial cells. Plasmin cleaves fibrin and fibrinogen to fibrin degradation products

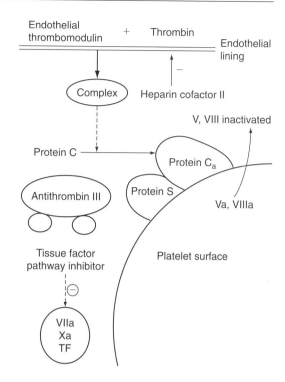

Figure 9.19 *Plasma inhibitors of activated factors.*

by hydrolysing arginine and lysine bonds. D-dimers are cleavage products of cross-linked fibrin. Various split products such fragments X, Y, D and E may be produced (Fig. 9.20).

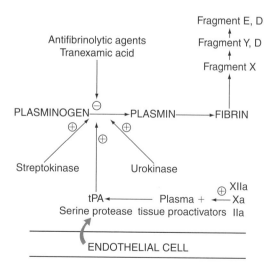

Figure 9.20 *The fibrinolytic pathway.*

BLOOD TRANSFUSION

Blood transfusion involves the infusion of safe and compatible blood (or its components) from the donor to the recipient. Compatibility between donor red cell antigens and the plasma antibodies of the recipient should be ensured to avoid fatal haemolytic reactions.

Blood group determination

The determination of the ABO and Rh(D) groups must be performed on the recipient and on the donor cells. The principles of ABO group determination involve the testing of the individual's red cells using antisera containing IgM anti-A, anti-B and anti-AB antibodies, and the serum is tested using known group A, B and O red cells (reverse blood grouping). Red cells are also tested using an antiserum containing an IgG sufficiently potent to agglutinate Rh(D)-positive cells in saline. All pregnant women and blood donors should be tested for D antigen. The potential problem in Rh D grouping is that weak agglutination due to D variants may be missed.

Compatibility testing

Compatibility testing involves the demonstration *in vitro* of serological compatibility between the recipient's serum and the donor's red cell (major cross-matching). Minor cross-matching involving the testing of the donor's serum against the recipient's red cells is seldom performed, as all donated blood is tested for irregular antibodies, and blood issued to hospitals should be free of such antibodies.

The donor's cells are prepared in several ways for presentation to the recipient's serum and the following methods are used: (1) saline agglutination test, (2) albumin technique or the papain agglutination test, (3) the low ionic strength saline (LISS) test, and (4) antiglobulin test.

SALINE AGGLUTINATION TEST

This test involves the donor's red cells being suspended in saline and tested against antibodies at room temperature. It is important for detecting IgM antibodies, which are referred to as saline or complete agglutinins. Complete antibodies such as anti-A, anti-B, anti-Lewis, anti-P, anti-M and anti-N may be detected using this test.

ALBUMIN OR PAPAIN AGGLUTINATION TEST

Agglutination occurs when red cells are close enough to allow antibodies to link to adjacent cells. When red cells are suspended in solutions containing free ions, electrical repulsion (zeta potential) between cells occurs as a result of negatively charged surfaces. This may be reduced by the presence of albumin in the medium or enzyme treatment of the red cells with papain, which removes negatively charged carbohydrates (e.g. sialic acid) from the cell surface. The albumin test at 37°C is used to enhance agglutination, usually for IgG antibodies.

LOW IONIC STRENGTH SALINE TEST

In the LISS test, suspended red cells are used to enhance the activity of some antibodies. LISS, combined with polybrene, is used to provide a rapid and sensitive method to detect most blood group antibodies.

THE ANTIGLOBULIN (COOMBS') TEST

The antiglobulin test is used for the detection of incomplete IgG and IgM antibodies (e.g. Kidd and Duffy), which may be missed by other methods. Anti-human globulin (AHG) is produced in animals (rabbits, sheep) after injection of human globulin, complement or specific immunoglobulin (IgG or IgM). When AHG is added to human red cells coated with immunoglobin or complement, agglutination of the red cells indicates a positive test. Polyspecific (anti-IgG and anti-C3d) and monospecific (anti-IgG, anti-IgM, anti-IgA, anti-C3c, anti-C3d and anti-C4) antiglobulin may be used to diagnose autoimmune haemolytic anaemias.

The direct antiglobulin test (Fig. 9.21(a)) is used for the detection of antibody or complement on the red cells that have been sensitized in the patient's body (*in vivo* sensitization). AHG is added to the patient's washed cells and agglutination indicates a positive test. A direct antiglobulin

(a)

(b)

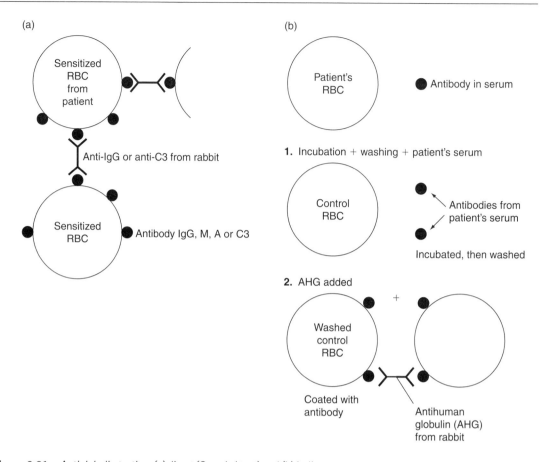

Figure 9.21   *Antiglobulin testing: (a) direct (Coombs' test) and (b) indirect.*

test is positive in haemolytic disease (Rh) of the newborn, autoimmune haemolytic anaemia, drug-induced immune haemolysis, and haemolytic transfusion reactions.

The indirect antiglobulin test (Fig. 9.21(b)) is a two-stage test used to detect antibodies that have coated the red cells *in vitro*. The 'test' red cells are first incubated at 37°C with serum, and in the second stage the red cells are washed with saline to remove the globulins. The AHG is then added to the washed red cells. Agglutination means that the original serum contained an antibody that coated the red cells *in vitro*. The indirect antiglobulin test is used for detecting antibodies in the patient's serum against donor red cells during cross-matching, atypical blood group antibodies during antibody screening, blood group antibodies in pregnant women and serum antibodies in autoimmune haemolytic anaemia.

Whole blood and red cell preparations

Normally, 400–480 mL blood is taken with 63 mL anticoagulant, which is either citrate–phosphate–dextrose (CPD; sodium citrate 1.66 g, anhydrous dextrose 1.46 g, citric acid monohydrate 206 mg, sodium acid phospate 158 g, and water to 63 mL) or CPD–adenine (sodium citrate 1.66 g, anhydrous dextrose 1.82 g, citric acid monohydrate 206 g, sodium acid phosphate 158 mg, adenine 17.3 mg, water to 63 mL) or SAG-M or ADSOL (saline, adenine, glucose and mannitol). The addition of adenine prolongs the shelf-life to 35 days. The citrate combines with calcium and anticoagulates the blood.

Blood is stored at 4–6°C. The anticoagulant solution dilutes the plasma by about 20 per cent. The properties of whole blood depend on the anti-coagulant used and on the duration of storage.

Table 9.1 *Changes in stored blood*

Change	Duration of storage (days)				
	0	7	14	21	28
Red blood cell survival (%)	100	98	85	80	75
2,3-DPG (%)	100	99	50	15	5
pH	7.2	7	6.9	6.8	6.7
Na^+ (mmol/L)	168	166	163	156	154
K^+ (mmol/L)	3.9	11.9	17.2	21	28
Glucose (mmol/L)	19.2	17.3	15.6	12.8	12.2
Free Hb (μg/L)	1.7	7.8	12.5	19	29

CHANGES IN WHOLE BLOOD DURING STORAGE (Table 9.1)

The following changes take place during storage:

- *Red cells.* As storage time increases, some red cells become spherical, due to metabolic changes, with an associated increase in cell rigidity. If red cells are transfused at the maximum recommended storage time, 10–20 per cent may be destroyed within 24 h.
- *White cells.* Granulocytes lose their phagocytic and bactericidal properties within 4–6 h after collection, but maintain their antigenic properties.
- *Platelets* become non-functional within 48 h in blood stored at 4°C.
- *Factor V and VIII levels* decrease with storage of whole blood. Factor V decreases 50 per cent by 21 days, whilst factor VIII decreases exponentially to 75 per cent by 24 h after collection and 30 per cent after 21 days of storage.
- *ATP and 2,3-DPG concentrations* fall with time, but at different rates. With CPD-A blood, 2,3-DPG decreases to 50 per cent at 14 days and to 5 per cent at 28 days. ATP decreases slowly to 75 per cent at 28 days.
- *Potassium levels.* After the first 48 h, there is a slow progressive K^+ loss from red cells into the plasma, so that the plasma K^+ concentration reaches approximately 30 mmol/L at 28 days.

Packed red cells are obtained by removing 200–250 mL plasma after centrifugation or sedimentation of 1 unit of whole blood. Packed red cells have a haematocrit of 0.75 or higher.

Heparinized blood is prepared when 500 mL blood is collected in 30 mL (2250 I.U.) of heparin. It has a shelf-life of only 24 h.

Frozen red cells are treated with glycerol (3.8 M) as the cryopreservative and stored in liquid nitrogen. The red cells must be thawed and washed extensively with electrolyte solutions to remove glycerol before transfusion.

Platelet concentrates

Platelet concentrates are available as single concentrates harvested from a single donor or as pooled platelet concentrates obtained from 4–6 units of blood. Special platelet packs made from polyolefin plastic enable better aeration of platelets, and may extend the shelf-life to 5 days if stored at 20–26°C with constant agitation.

One unit of platelet concentrates contains approximately 6×10^{10} platelets. Although platelets express only HLA class I antigens, contamination by leukocytes and red blood cells can cause alloimmunization. Thus, ABO- and Rh-compatible platelets are usually used, but HLA-matched platelets are used for patients with HLA antibodies. Nearly one-third of transfused platelets are sequestered in the normal spleen. One unit of platelet concentrate increases the blood platelet count by 10×10^9/L per m^2 body surface area.

Human plasma preparations

- *Fresh frozen plasma* is prepared from fresh blood and frozen rapidly to be stored at −30°C.

It is used for the replacement of coagulation factors, and has a shelf-life of 1 year.

- *Cryoprecipitate* is produced from freshly separated plasma by freezing at $-70°C$, followed by thawing at $4°C$. Cryoprecipitate is rich in factor VIII, fibrinogen and fibrinectin, and contains approximately 80 units of factor VIII and 250 mg fibrinogen. It is stored at $-30°C$, with a shelf-life of 12 months.
- *Freeze-dried factor VIII concentrate* is a lyophilized preparation from fresh frozen plasma. It is stored at $4°C$ and contains large amounts of factor VIII, with small amounts of fibrinogen. However, it carries a higher risk of transmitting hepatitis.
- *Freeze-dried factor IX concentrates* are used for the treatment of factor IX deficiency (Christmas disease) or haemophilia B. It also contains factors II, VII and X.

Complications of blood transfusion

Approximately 3 per cent of patients given blood transfusions have reactions mediated by immunological and non-immunological mechanisms. Fatal reactions to transfusions are rare, and are estimated to occur in 1 in 50 000 transfusions.

IMMUNOLOGICAL REACTIONS

Immunological reactions may be immediate or delayed. Immediate reactions are associated with massive intravascular haemolysis as a result of complement activating IgM or IgG antibodies (e.g. ABO antibodies). The severity of these reactions depends on the recipient's titre of antibodies. Reactions associated with the coating of red cells with IgG (e.g. Rh antibodies) result in extravascular haemolysis which is generally less severe.

Severe haemolytic transfusion reactions include urticaria, flushing, chest pain, dyspnoea, rigors, tachycardia and shock progressing to bleeding and renal shutdown. There is evidence of blood destruction with jaundice, haemoglobinuria and disseminated intravascular coagulation.

If the recipient develops antibodies to antigens present on the donor cells during or after the transfusion, a reaction can occur even if apparently compatible blood is transfused. This is caused by very low levels of antibody in the recipient which are not detectable in the cross-match procedure. Following the transfusion, there is a secondary response with a rapid rise of IgG antibodies. The antibodies most frequently involved are against antigens of Kidd (Jk), Duffy (Fy), Rhesus (Rh), Kell (K) and S blood group systems; these may have been acquired by previous exposure to the antigens during pregnancy or previous blood transfusions.

- *White cell reactions.* Febrile reactions occur in approximately 2 per cent of all transfusions and are caused by donor white cells reacting with alloantibodies induced by previous transfusions or pregnancy.
- *Graft-versus-host reactions.* Graft-versus-host disease – a rare complication of blood transfusion – is characterized by the deposition of donor lymphocytes in the recipient's skin, liver or gastrointestinal tract, leading to a rash, hepatitis or diarrhoea.
- *Post-transfusion purpura.* This is a consumptive thrombocytopenia occurring 7–10 days after transfusion of a blood product and is usually self-limiting, lasting for 2–6 weeks.
- *Anaphylaxis to plasma proteins.* Life-threatening anaphylaxis during blood transfusion usually occurs in IgA-deficient patients whose sera contain anti-IgA antibodies. Urticaria may also occur in patients when recipient antibodies react to antigens in donor plasma, especially IgA.

NON-IMMUNOLOGICAL REACTIONS

Septicaemia

Approximately 3 in 1000 units of donor blood may be contaminated with bacteria (some *Pseudomonas* strains), which can multiply in the cold. Bacteria present in platelet concentrates may produce an increased risk as platelets are stored at $22°C$. Fever, chills, hypotension and other signs of Gram-negative endotoxaemia may occur.

Disease transmission

Post-transfusion hepatitis may be due to hepatitis A, B, C and other viruses. Hepatitis A, an RNA virus, is rarely transmitted by blood transfusion. Hepatitis B, caused by a DNA virus, may lead to a chronic carrier state, although this risk is significantly reduced by screening of all donors for hepatitis B surface antigen. Transmission of cytomegalovirus (CMV) is likely to cause problems in the newborn, transplant patients and open-heart cardiac patients.

Human immunodeficiency virus (HIV) can be transmitted by both cellular and plasma components of blood. However, as all blood units are now tested for HIV by enzyme-linked immunoabsorbent assay (ELISA), the risks of acquiring HIV are reduced. Other infections such as malaria, toxoplasmosis and syphilis may be transmitted by blood transfusion.

Other complications

Air embolism, circulatory overload and iron overload may also occur. Repeated red cell tranfusions and deposition of iron in the reticuloendothelial tissues may occur in thalassemia major.

Massive blood transfusion

Massive blood transfusion is defined as transfusion of a volume of stored blood greater than the recipient's blood volume in less than 24 h.

COMPLICATIONS

- *Citrate toxicity and hypocalcaemia.* This may occur if the transfusion rate exceeds 1 L per 10 min, or when an exchange transfusion is carried out within 2 h. The clinical features of hypocalcaemia are involuntary muscle tremors, bradycardia with ST segment prolongation and prolonged QT interval. Calcium chloride should be given if clinical or ECG evidence of hypocalcaemia is present.
- *Hyperkalaemia.* As potassium leaks from red cells, the plasma K^+ concentration may reach 30 mmol/L after 30 days of storage. Usually, the potassium diffuses into the red cells after transfusion and therefore does not pose a problem unless the patient is hyperkalaemic or persistently acidotic and hypotensive.
- *Acidosis.* As stored blood progressively becomes acidotic with a pH of 6.5–6.8 after 2 weeks of storage, massive blood transfusion can aggravate any acidosis already present in the recipient. The citrate is metabolized to bicarbonate in the liver a few minutes after transfusion.

- *Hypothermia.* During rapid blood transfusion, transfusion of cold blood may cause hypothermia. The harmful effects of hypothermia include ventricular arrhythmias (ventricular fibrillation at 28°C) and may lead to cardiac arrest, reduction in oxygen delivery due to the Bohr effect, and aggravation of citrate toxicity.
- *2,3-DPG deficiency.* During storage, the 2,3-DPG concentration in the red cells decreases, reducing oxygen delivery. However, transfused blood regenerates 2,3-DPG within 24 h of infusion. The use of CPD-adenine as a preservative reduces this problem, as 2,3-DPG depletion then occurs slowly.
- *Dilutional coagulopathy.* As stored blood has low levels of factors VIII, V and XI, dilutional coagulopathy occurs especially if the total body blood volume is replaced more than twice within 24 h. In stored blood the platelets are reduced in number and are dysfunctional. Thus, dilutional thrombocytopenia may lead to coagulopathy after massive blood transfusion.
- *Microaggregates.* Microaggregates consisting of clumps of fibrin, platelets and leukocytes, over 20 μm in diameter, are formed in stored blood. Platelet fibrinogen appears to be the main determinant of microaggregate formation, which reaches a peak within 1–2 weeks of storage. When the microaggregates enter the recipient's circulation, they are trapped in the pulmonary vessels and may release lysosomes, in turn contributing to the adult respiratory distress syndrome.

PLASMA

Plasma is the fluid medium of the intravascular compartment and is important for the transport of materials between tissues and the internal environment. Plasma differs from the extravascular component of ECF in that it has a much greater content of protein. Plasma constitutes about 4 per cent of the total body weight (40–50 mL/kg).

Water and electrolytes

Approximately 93 per cent of plasma is water. The principal plasma cation is Na^+ (140 mmol/L), whilst other important cations include K^+ (4 mmol/L), Ca^{++} (1 mmol/L) and Mg^{++} (2 mmol/L). About one-third to one-half of the divalent cations are complexed with proteins (e.g. albumin) or low-molecular-weight anions, and carry negative charges as they are on the alkaline side of their isoelectric points at a pH of 7.4. Organic acids such as lactate and pyruvate make up the remaining plasma cations.

Plasma carbohydrates

Blood glucose is the main carbohydrate in plasma (3.5–6 mmol/L), with variable small amounts of fructose, galactose and mannose. Complex carbohydrates are present in small amounts in plasma; glycoproteins are formed by the covalent attachment of carbohydrate to the amino acids, asparagine, serine or threonine.

Plasma lipids

Lipids generally are complexed with plasma proteins in the circulation. A small fraction of the total fatty acids in plasma is unesterified, usually associated with albumin. The remaining fatty acids form the triglycerides and are found in plasma as lipoproteins. The lipoproteins complex with phospholipids and cholesterol to form chylomicrons, very low-density lipoproteins, low-density lipoproteins and high-density lipoproteins.

Plasma proteins

Plasma proteins are a diverse group structurally and functionally, with the total plasma protein concentration ranging from 60 to 80 g/L. All plasma proteins are globular molecules and range from simple unconjugated proteins such as albumin to complex proteins such as lipoproteins, glycoproteins and metalloproteins.

Albumin

Albumin (mol. wt. 67 000 Da) is the most abundant plasma protein, with a concentration of 40 g/L. It is synthesized in the liver, has a half-life of 20 days, and is metabolized in the liver, kidneys and gut. Approximately 13 g of albumin is synthesized and catabolized per day. Its main functional role is the transport of a wide range of substances and maintenance of plasma colloidal osmotic pressure.

α_1 GLOBULINS

α_1-Antitrypsin is a serine protease inhibitor (serpin) produced mainly by the liver. It is a potent inhibitor in plasma of trypsin, chymotrypsin, activated plasmin and other proteases.

α_1-Lipoproteins are associated with α_1-globulins, and contain 45–55 per cent lipid. The plasma lipoproteins may be divided into four classes, namely the chylomicrons, very low-density lipoproteins (VLDL), low-density lipoproteins (LDL) and high-density lipoproteins (HDL). The chylomicrons are the largest of the lipoproteins, consisting mainly of triglycerides (80–90 per cent) with only 1–2 per cent protein. They are mainly derived from dietary fat and serve to carry exogenous triglycerides from the gut to the tissues for utilization or storage.

The VLDL transport endogenous triglycerides from the liver to the peripheral tissues for storage or utilization. The LDL transport cholesterol to the tissues, while the HDL return excess cholesterol from the peripheral tissues to the liver.

α_1-Acid glycoprotein is an acute phase protein which is present in low concentrations, but its physiological role is unknown. Pharmacologically, it is important as it binds to basic drugs.

α_2-GLOBULINS

Various proteins belong to this group of globulins:

- α_2-Macroglobulin is a protease inhibitor in plasma, and is the major protein in the α_2-globulin fraction (approximately 80 per cent). It has inhibitory functions on plasma trypsin, chymotrypsin and plasmin. The primary function of α_2-macroglobulin may be to inhibit proteases produced by infectious organisms.
- Prothrombin is a clotting factor synthesized by the liver. Some 60 per cent of the extracellular pool of prothrombin is in the plasma and

40 per cent in the extravascular space. It has a rapid turnover.

- *Haptoglobulin* is a heterogeneous group of globulins that bind free haemoglobin and transport it to the liver.
- *Caeruloplasmin* is a plasma protein that carries copper, and is produced in the liver. It also functions as an oxidase enzyme and oxidizes ferrous to ferric ions before the binding of iron to transferrin. As an acute-phase protein, it may modulate inflammation by its free radical scavenging properties.

β GLOBULINS

Transferrin is the plasma protein that transports iron. Apotransferrin, its precursor, is produced in the liver. One molecule of transferrin will bind two ferric ions and is normally approximately one-third saturated with iron.

Haemopexin is a β-globulin that binds to haem and releases it to the reticuloendothelial system.

FIBRINOGEN

Fibrinogen is a large protein that is produced by the liver and has an important role in blood coagulation.

γ-GLOBULINS

The γ-globulins consist of immunoglobulins. The immunoglobulins are produced by plasma cells of the bone marrow, spleen, lymph nodes and gut. IgG accounts for 76 per cent of the total plasma immunoglobulins. IgG can bind to complement, and its main action is against soluble antigens. IgA forms 16 per cent of the circulating antibodies, and is present in seromucous secretions. It does not fix complement and the main function is protection against secretory mucosal surfaces. IgM accounts for 7 per cent of plasma immunoglobulins, is rapidly synthesized in response to particular antigens, and can fix complement to break down foreign surfaces. IgE is present in very low concentrations in normal individuals, and is involved in hypersensitivity reactions by binding to mast cells in capillaries and tissues.

Biological functions

Plasma has numerous functions. It is important for the carriage of dissolved oxygen and carbon dioxide, glucose, amino acids and excretory waste products such as urea and creatinine. The bicarbonate in plasma, derived from the red cell, is an important buffer system in blood.

The plasma proteins are a diverse group of proteins and have a wide range of biological functions.

TRANSPORT FUNCTIONS

Many plasma proteins are carriers of hormones, metals, vitamins, metabolites and excretory products in the body. Albumin transports many substances and renders them water-soluble. It transports bilirubin, free fatty acids, Ca^{++} and hormones such as thyroid hormone and cortisol, and acidic drugs (e.g. barbiturates).

The globulins transport a wide variety of substances. α- and β-lipoproteins transport triglycerides, cholesterol and fat-soluble vitamins. Iron is transported by transferrin, and copper by caeruloplasmin. Thyroxine is also transported by thyroid-binding globulin, and cortisol by transcortin. Transcobalamin is an important carrier for vitamin B_{12}.

BLOOD COAGULATION

Various plasma proteins, including prothrombin and fibrinogen, are involved in the coagulation cascade.

ENZYMES

Various enzymes are present in plasma, including plasma cholinesterase and the acute-phase proteins such as α_1-acid glycoprotein and the anti-proteolytic enzymes, α_1-antitrypsin and α_2-macroglobulin.

ONCOTIC PRESSURE

The plasma proteins exert an oncotic pressure (28 mmHg), which contributes to the total osmotic pressure (5610 mmHg) of plasma. Plasma oncotic pressure is important in the control of fluid balance between the vascular and the interstitial compartments. Quantitatively, albumin is the most important plasma protein for oncotic pressure because of its low molecular weight and high concentration compared with other plasma proteins.

BUFFERING ACTION

Plasma proteins are amphoteric and dissociate in the pH range of 7–7.8, with a net negative charge. Thus, they can accept H^+ ions, although this buffering function is minor compared with other buffering systems in blood.

Reflections

1. Blood is a body fluid that is a vehicle for communication between the tissues and serves to transport respiratory gases, nutrients and hormones around the body and transport waste materials to excretory organ systems. The formed elements of blood include erythrocytes, five types of leukocytes, and platelets suspended in plasma.

2. Mature blood cells are continuously renewed by haemopoiesis and are derived from a common population of pluripotent stem cells in the bone marrow. There are two distinct cell lines: the myeloid and lymphoid cells. The myeloid cells remain in the bone marrow and form red cells and leukocytes. Lymphoid stem cells migrate to the lymph nodes, spleen and thymus where they develop into lymphocytes. The stem cells develop into precursor cells which differentiate and mature into one of the cell types. Erythroblasts undergo successive mitosis and synthesize haemoglobin before they lose their nuclei to become reticulocytes, in which form they are released into the circulation. Erythropoiesis is controlled by erythropoietin, a hormone secreted by the peritubular cells of the kidneys. After about 120 days in the circulation the red cells are destroyed by the macrophages in the spleen and liver. Granulocytes and monocytes mature from precursor cells in a similar fashion. Platelets bud off from giant cells called megakaryocytes which are derived from pluripotent cells in the bone marrow.

3. Red cells are small non-nucleated biconcave cells that transport oxygen and carbon dioxide between the lungs and tissues. They contain haemoglobin which has a high affinity for oxygen. Antigens or agglutinogens are present on the red cell plasma membrane. In the ABO system, two types of agglutinogens, A or B agglutinogen, may be present separately, together or completely absent, giving rise four groups respectively: A, B, AB and O. In addition, human plasma may contain antibodies (agglutinins) to none or both agglutinogens (antigens). Other weaker agglutinogens found on the surface of the red cell membrane include the Rh antigen, the M, N, P and Lewis agglutinogens.

4. Blood supplies oxygen to all tissues in the body and transports carbon dioxide produced by metabolism to the lungs for removal from the body. Most of the oxygen carried in blood is loosely bound to haemoglobin within the red blood cells. The amount of oxygen carried in blood depends on the partial pressure of oxygen which is described by the sigmoidal-shaped oxyhaemoglobin dissociation curve. The curve is shifted to the right by an increase in P_{CO_2}, an increase in the level of 2,3-DPG, an increase in temperature, and by a fall in pH. This is called the Bohr effect. Carbon dioxide is carried in blood as bicarbonate ions, carbamino compounds and in physical solution. More CO_2 is carried by blood as the level of oxyhaemoglobin decreases and this is known as the Haldane effect.

5. About two-thirds of total body iron is within the haemoglobin of red cells, 4–5 per cent is within myoglobin and enzymes and the rest is stored mainly in the liver as ferritin. When red cells are phagocytosed, most of the iron is recycled and reused immediately or stored as ferritin in the liver or in the bone marrow. Iron is absorbed as ferrous ion.

6. Duodenal and jejunal epithelial cells take up iron from the intestinal lumen by a carrier-mediated process. It is stored within the enterocyte bound to an iron-binding protein called ferritin. The absorbed iron is released into the blood across the basolateral membrane where it combines with plasma transferring to be transported to tissues.

7. Plasma is about 95 per cent water and the rest consists of a variety of proteins including albumins, globulins, and fibrinogen, electrolytes, glucose and a variety of substances in transit (hormones, nutrients and excretory waste products) between tissues. Albumin carries lipids and steroid hormones in plasma. α and β globulins also transport lipids and fat-soluble substances while the γ globulins are antibodies that play an important role against infection.

8. Following damage to the vascular epithelium a cascade of events is initiated leading to the formation of a blood clot. Platelet adhesion, release and activation occurs. Platelet aggregation at the site of injury occurs within seconds. Current theories on the coagulation cascade are based on the cell-based theory involving three phases: initiation, amplification and propagation. Tissue-factor bearing cells and activated factor VII initiate the process by generating small amounts of factor Xa and IIa on platelets, monocytes, macrophages or subendothelial fibroblasts. The IIa primes the activated platelets with factor VIIIa, Va and XIa and this generates the IXa–VIIIa complex on platelets and elicits the propagation phase via Va–Xa (prothrombinase) complex, which generates large amounts of thrombin (IIa) that activates fibrinogen to fibrin. The clotting mechanism requires calcium ions and phospholipids present in the membranes of the platelets. The fibrin threads trap blood cells to form a stronger clot which retracts by shrinkage. The blood clot is then dissolved by plasmin. Undamaged vascular endothelial cells prevent clotting by releasing natural anticoagulants such as heparin and prostacyclins, and by expressing thrombomodulin, a protein which binds thrombin and activates protein C, an activator of plasmin.

Physiology of the immune system

After reading this chapter, the reader should be able to:

1. Differentiate the innate and adaptive immune system
2. Describe the passive mechanisms by which the body resists infection
3. Explain how the body recognizes invading organisms
4. Describe the natural immune system
5. Describe the complement system and explain its function
6. Describe the adaptive immune system and the role of lymphocytes
7. Outline the role of immunoglobulins
8. Differentiate and explain the clinical significance of hypersensitivity reactions

The immune system is a multicomponent defence system that recognizes and protects the host against microorganisms, toxins, mutant host cells or transplanted tissues that can potentially damage tissues or organs. It has evolved mechanisms that enable it to distinguish between 'self' and 'non-self' tissues. The immune system is functionally divided into innate and acquired components (Table 10.1).

Table 10.1 Components of the immune system

Component	Innate immunity	Acquired immunity
Physical	Skin	
	Mucous membrane	
	Cilia	
	Mucus	
Chemical	Gastric HCl	Antibodies
	Lysozyme	
	Complement	
	Acute-phase proteins	
Cells	Phagocytes	T lymphocytes
	Natural killer (NK) cells	B lymphocytes

INNATE IMMUNITY

Innate immunity refers to the natural immune mechanisms present at birth, and is present for life. It is the first line of defence against invading bacteria, fungi and helminths (e.g. worms), and has no specificity and no memory.

The innate immune system is made up of three components: physico-chemical barriers, humoral and cellular defence mechanisms.

Physico-chemical barriers

The physico-chemical barriers that prevent microorganisms from gaining access to the body include the skin and mucous membranes, mucus, cilia and hydrochloric acid produced in the stomach. The skin forms an excellent physical barrier and prevents bacterial growth by secreting antibacterial substances (e.g. lactic acid and fatty acids) in sweat and sebaceous secretions. Mucus traps bacteria and foreign particles that are then removed by ciliary motion. Gastric juice hydrochloric acid is bactericidal. The high flow rates of the urine, saliva, tears and secretions in the biliary and the lower respiratory tracts also physically remove foreign material.

Humoral components

The humoral components of the innate immune system include complement, acute-phase proteins and proteolytic enzymes (e.g. lysozyme). These are present in mucosal secretions, blood and cerebrospinal fluid.

COMPLEMENT

Complement consists of a group of at least 25 heat-labile serum proteins produced mainly by the hepatocytes and some local production by macrophages at sites of inflammation. These serum proteins are important in control of inflammation (Fig. 10.1). Complement proteins are present in the circulation in an inactive form as proenzymes. Activation of complement occurs in a sequential manner, each activated component catalysing the activation of several molecules of the next component, resulting in an amplification of the response. This is known as the 'complement cascade'.

There are three interacting pathways in the complement cascade. The first two are the classical and alternative pathways, and both activate the common or membrane-attack pathway. The classical pathway (initiated by IgG or IgM antigen–antibody complexes) and the alternative pathway (initiated by liposaccharide, endotoxin or IgA), produce complement C3, which cleaves C5 into C5a and C5b. The formed C5b combines with C6, C7, C8 and C9 to produce the membrane-attack complex that damages cell membranes. The functions of the components of complement include opsonization (coating the walls of bacteria so that they can attract and bind to phagocytic cells and be easily ingested), chemotaxis, activation of neutrophils and mononuclear phagocytes, and lysis of bacteria or foreign cells (Fig. 10.2).

IgG or IgM antigen–antibody complexes, aggregated immunoglobulins, and C-reactive protein activate the classical pathway (Fig. 10.3). The formation of an antigen–antibody complex causes a conformation change in the antibody molecule, exposing a binding site for complement C1. C1 consists of three distinct proteins, C1q, C1r and C1s,

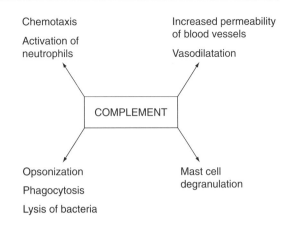

Figure 10.1 *Actions of the components of complement.*

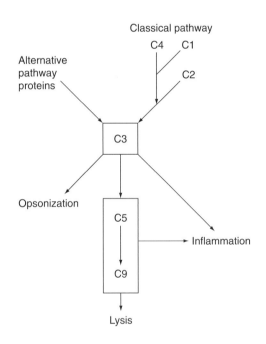

Figure 10.2 *Outline of the complement system.*

held together by calcium ions. C1q binds to the antibody and converts C1r to an enzyme that activates C1s. The C1qrs complex cleaves C2 and C4 to form C3 convertase and this cleaves C3 to C5 convertase for the membrane-attack sequence. The C3a fragments cause smooth muscle contraction, histamine release and increased vascular permeability.

The alternative pathway (Fig. 10.3) is a slow reaction and is activated by insoluble polysaccharides and non-self cells in the presence of C3b.

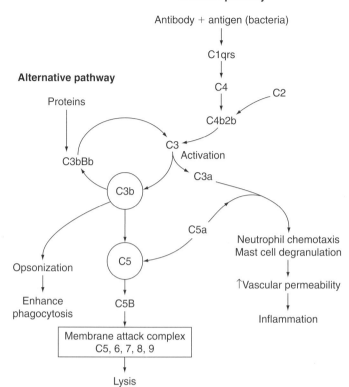

Classical pathway

Antibody + antigen (bacteria)

C1qrs

C4   C2

C4b2b

C3   Activation

C3a

C5a

Neutrophil chemotaxis
Mast cell degranulation

↑Vascular permeability

Inflammation

Alternative pathway

Proteins

C3bBb

C3b

Opsonization

Enhance
phagocytosis

C5

C5B

Membrane attack complex
C5, 6, 7, 8, 9

Lysis

Figure 10.3   *The complement cascade.*

This pathway activates the complement cascade without bound antibodies. C3 undergoes spontaneous slow hydrolysis of its thiolester bonds, to form $C3(H_2O)$. In the presence of magnesium ions, $C3(H_2O)$ binds to factor B, another circulating protein. The complex thus formed is cleaved by a third blood protein, factor D, resulting in fragments with C3 convertase activity that cleave C3 and produce highly reactive C3b. Some C3b binds factor B and the resultant complex is cleaved by protein D to C3bBb. C3b and C3bBb bind to form a larger complex that has C5 convertase activity.

The membrane attack pathway is the final common complement pathway that produces C5a and other 'killer' molecules. C5 convertase produced by the classical and alternative pathways cleaves C5 into a smaller C5a fragment and a larger C5b fragment. C5b binds to C6, and then to C7. The C5b67 complex binds to C8 and C9, forming the membrane-attack complex (MAC) that forms holes in cell membranes, resulting in cell lysis and death. Complements C1, C2 and C3 function as opsonins

(proteins coating bacteria or foreign particles which enhance phagocytosis). The formation of plasma kinins also depends on complement activation, and these promote phagocytosis. C3a and C5a are responsible for releasing anaphylatoxins that increase vascular permeability, release histamine and contract smooth muscle. C5a and the C1,4,2,3 complex have chemotactic properties.

ACUTE-PHASE PROTEINS

Acute-phase proteins are plasma proteins expressed during acute infections and include C-reactive protein, fibrinectin, α_1 antitrypsin and α_2 macroglobulin, all primarily produced by the liver. Acute-phase proteins are involved in opsonization and regulation of inflammatory mediators during the acute septic response. C-reactive protein binds components of bacterial cell walls and then activates the classical pathway of complement, independently of antibody. During an acute infection or inflammatory process, the C-reactive protein plasma concentrations rise rapidly.

Fibrinectin binds bacteria and macrophages and monocytes, enhancing the clearance of these organisms from the body.

LYSOZYME

Lysozyme is a bactericidal enzyme secreted in saliva, tears and the mucus of the respiratory and gastrointestinal tract. It is also present in neutrophils. Lysozyme breaks the bonds between N-acetylglucosamine and N-acetylmuraminic acid of bacterial cell wall proteoglycans, causing lysis.

Cellular components

The cellular elements of the innate cellular immune system present and functional at birth include the leukocytes (white blood cells), mast cells and natural killer (NK) cells (Fig. 10.4). The leukocytes are formed in the bone marrow and lymph tissue and transported in the blood to areas of inflammation to provide a rapid defence against invading infectious agents. Five types of leukocytes are normally found in the blood; polymorphonuclear neutrophils (62%), polymorphonuclear eosinophils (2.3%), polymorphonuclear basophils (0.4%), monocytes (5%) and lymphocytes (30%). The three polymorphonuclear white cells (granulocytes) and monocytes protect the body against invading organisms by ingesting via the process of phagocytosis. Neutrophils are the main cells responsible for killing and removal of bacteria and fungi, whilst eosinophils control infection with multicellular parasites such as worms. Basophils have a less well-defined role in immunity. Mast cells, characterized by the presence of abundant intracellular granules, are present in loose connective and mucosal tissues. NK cells are a subset of lymphocytes that are important for immune protection against viral infection and are possibly active against tumour cells.

Granulocytes and monocytes develop from the myelocyte pluripotential stem cells in the bone marrow. The lymphocytes are produced mainly in the lymph glands, spleen and thymus.

NEUTROPHILS

Neutrophils circulate in blood for 5–6 h and may migrate into and remain in tissues for 4–5 days. At sites of infection and inflammation, tissue damage results in the release of C5a, leukotriene B4 (LTB$_4$), and interleukin-8 (IL-8) which activate the neutrophils and the endothelium. Following this, specialized adhesion molecules facilitate neutrophil migration into the tissues; neutrophils express adhesive protein molecules called L-selectin and the endothelial cells express P-selectin (Fig. 10.5). These adhesive proteins induce the neutrophils to

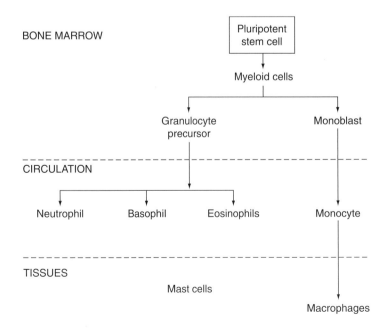

Figure 10.4 *Origin of the cells of innate immunity.*

move along the margin of the endothelium, a process called margination.

Neutrophil–endothelium adhesion is enhanced by the release of the mediators, C3b, leukotriene and IL-8. These activate the neutrophils to release adhesive proteins called integrins, and cause the endothelium to release intercellular adhesion molecule (ICAM). The integrin/ICAM proteins increases the affinity of neutrophils for the endothelial cells and enables them to migrate through the endothelium by diapedesis. The neutrophils then migrate to tissue sites in response to chemotactic factors (C5a, IL-8). This process by which directed movement of cells (neutrophils) occurs along a gradient of increasing concentration of the attracting molecule is called chemotaxis.

Neutrophils ingest bacteria or fungi by phagocytosis. Microorganisms must be coated with opsonins (complement C3b, C-reactive protein and antibody). The neutrophils form pseudopodia which engulf the microorganism (Fig. 10.6). Receptors for opsonin (complement receptors and Fc receptors that bind to the Fc portion of immunoglobulin G) are present on the neutrophil surface, and these bind the microorganism. Killing of bacteria or fungi is mediated by neutrophil oxygen-dependent and oxygen-independent mechanisms via the 'respiratory burst' reaction involving superoxide anions, hydrogen peroxide, singlet oxygen and hydroxyl radicals (Fig. 10.7). When there is tissue infection the total life span of a granulocyte is shortened to a few hours because the granulocytes migrate rapidly to the infected tissues, perform their function and in the process are destroyed.

Eosinophils

Large numbers of eosinophils are present in the tissues, especially at epithelial surfaces. Eosinophils have granules containing basic proteins toxic to worms (helminths). Eosinophils are activated by C5a, C3b and leukotrienes. The main role of eosinophils in host defence is protection against multicellular parasites, killed by the release of toxic basic (cationic) proteins onto the surface of helminths.

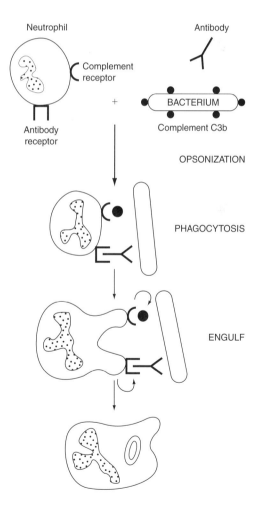

Figure 10.6 *Neutrophil function: opsonization and phagocytosis.*

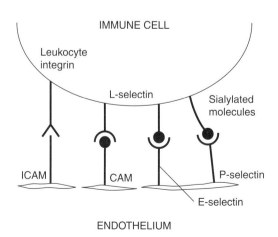

Figure 10.5 *Different types of adhesion molecules.*

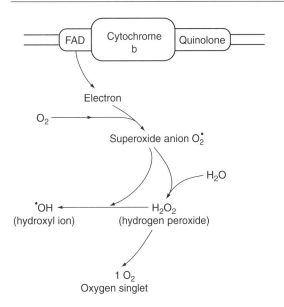

Figure 10.7 *The 'respiratory burst' in secondary granules.*

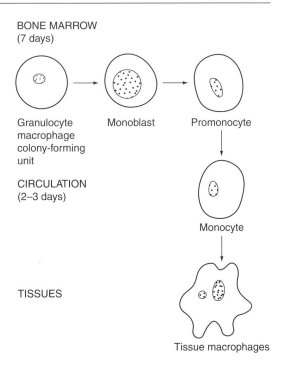

Figure 10.8 *The formation of macrophages.*

MONOCYTES AND MACROPHAGES

Monocytes, formed from the monoblasts in the bone marrow, comprise 5–10 per cent of the circulating white cells and have a half-life of approximately 24 h in blood (Fig. 10.8). They enter the tissues, where they swell to a larger size and become macrophages.

Macrophages are relatively large irregularly shaped nucleated cells with well-developed Golgi apparatus, lysosomes and a vast number of intracellular enzymes. In the central nervous system they exist as glial cells. The Kupffer cells in the liver are important macrophages in the sinusoids, where they are located strategically in the portal circulation to ingest microorganisms from the gastrointestinal tract. In the lungs, there are pulmonary alveolar macrophages and interstitial macrophages. The spleen contains a large number of macrophages that are important for phagocytosis, synthesis of complement components and antigen presentation. The mesangial cells are the macrophages of the kidney.

Macrophages have two important functions: phagocytosis (engulf and digest cellular debris and pathogens in innate immunity) and antigen presentation (presenting a protein molecule present on surface of a pathogen to a corresponding helper T cell in cell-mediated immunity). Macrophages are important for host defences against intracellular pathogens (*Listeria*), mycobacteria, parasites (trypanosomes) and fungi. The macrophages first develop an enhanced capacity for respiratory burst activity (an oxidative mechanism producing free radicals of oxygen) via stimulating factors (bacterial endotoxin, interferon-γ and proteases) by a process called priming. The primed macrophages undergo activation (Fig. 10.9), and can phagocytose microorganisms and generate a respiratory burst reaction which enhances the killing of intracellular microorganisms.

Macrophages process and present foreign antigens to helper T lymphocytes. in the presence of major histocompatibility complex (MHC) antigens on their surface that enable them to process foreign antigen. The processed antigen is bound to cell surface MHC class II receptors (Fig. 10.10).

MACROPHAGE AND NEUTROPHIL RESPONSE DURING INFLAMMATION

The tissue macrophage is the first line of defence against invading pathogens. Within minutes,

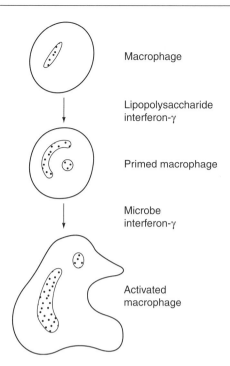

Figure 10.9   *Activation of the macrophage.*

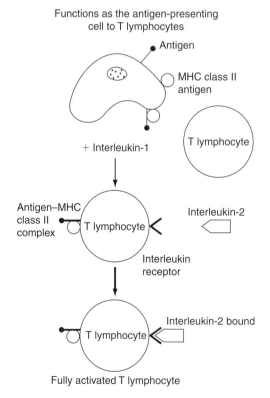

Figure 10.10   *The role of the macrophage.*

macrophages migrate to the area of inflammation in response to chemotactic factors. During the first hour, neutrophils invade the inflamed area as a result of inflammatory mediators. This is the second line of defence. As the neutrophils migrate to the inflamed area, monocytes from the blood enter the inflamed tissue and enlarge to become macrophages. This forms the third line of defence. After several days or weeks the macrophages are the dominant phagocytic cell in the inflamed area. The fourth line of defence is associated with the increased production of both macrophages and monocytes by the bone marrow mediated by tumour necrosis factor, interleukin-1, and colony-stimulating factors. The neutrophils and macrophages that engulf bacteria and necrotic tissue eventually die and form pus.

MAST CELLS AND BASOPHIL GRANULOCYTES

Mast cells and basophil granulocytes contain a large number of histamine granules, and possess high-affinity receptors for IgE. Activation of mast cells and basophils by IgE leads to degranulation. The activated complement proteins, C3a, C4a and C5a, activate basophils and mast cells in the lungs. The effect of the release of mast cell and basophil granules depends on the site of release. In the airways, histamine induces smooth muscle contraction, resulting in airway obstruction. In the mucosal membranes, histamine causes nasal discharge, conjunctivitis, mucosal oedema and itching. Mast cells therefore play a key role in the inflammatory process. Widespread systemic degranulation of mast cells and basophils can lead to anaphylaxis.

NATURAL KILLER CELLS

Natural killer (NK) cells are large granular lymphocytes that are able to kill target cells spontaneously without prior sensitization. They lack T-lymphocyte surface receptors. They are a part of cell-mediated immunity and act during innate immune response. The two major functions of the NK cells are (1) recognition of antigens on target cells (e.g. tumour cells or virus-infected cells) and (2) antibody-dependent cytotoxicity. NK cells attack host cells that have a foreign peptide on particular cell surface proteins called MHC class I molecules. The NK cells release cytotoxic granules that destroy the infected cells. NK cells do not require prior activation to perform their cytotoxic effects on the target cells by inducing apoptosis (programmed cell death) in which the nucleus of the target cell is rapidly frag- mented by endonucleases. NK cells are activated in response to interferons or macrophage-derived cytokines. They serve to contain viral infections while the adaptive immune response generates anti- gen-specific cytotoxic T cells that can clear infection.

ACQUIRED IMMUNITY

If the first line of defence provided by innate immunity is breached, the acquired or adaptive immune system is activated to produce a specific reaction against the infectious agent or provoking stimulus (Fig. 10.12). During the first exposure of the host to the agent, the host's immune system 'identifies' and 'learns' its distinctive structural or chemical features, resulting in specific recognition. Effector responses, for example killing of microbes or neutralization of toxins, are initiated to protect the host. 'Memory' of the effective components of the immune response to the particular agent is established. During a subsequent exposure to same agent, the host's defences are mobilized more quickly and powerfully. Specificity and memory are the hallmarks of acquired immunity and lympho- cytes are involved in these two processes.

In general, innate and acquired immune responses complement each other. Acquired immu- nity is brought about by the activities of lympho- cytes. Antibodies generated by acquired immunity can direct elements of the innate immune system (e.g. complement, neutrophils) to the relevant targets.

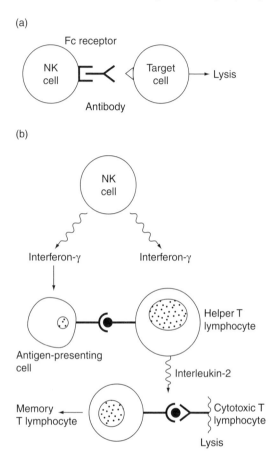

(a)

(b)

Figure 10.11   *The role of the natural killer (NK) cell: (a) antibody-dependent cellular toxicity, and (b) recruitment of T lymphocytes and macrophage activation.*

Lymphocytes

Lymphocytes are important for acquired immunity as they recognize specific molecules on the

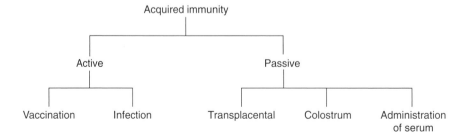

Figure 10.12   *Methods of developing acquired immunity.*

invading microbe or toxin. They are produced in the adult bone marrow and thymus, enter the systemic circulation, and finally reside in lymphoid organs such as the spleen. Lymphocytes make up about 20 per cent of the circulating white cells in blood. There are two types of small lymphocytes, T and B, present in blood in a ratio of about 1:5 (Table 10.2). T and B lymphocytes differ from NK cells in that they are the principal cells involved in

Table 10.2   Functions of macrophages and lymphocytes

Cells	Function
Macrophages	Remove and destroy antigen
	Localize antigen
	Present antigen to lymphocytes
	Regulate lymphocytes via cytokine secretion
	Bind Ag–Ab complexes
T lymphocytes	Act as helper cells for some antigens
	Secrete cytokines to regulate macrophages and B lymphocytes
	Recruit other T lymphocytes
	Present antigen to B lymphocyte cytotoxic cells
B lymphocytes	Immunoglobin production
	Bind Ag–Ab complexes
	Bind Ag to T lymphocytes

adaptive immunity; the T cells are chiefly responsible for cell-mediated immunity whereas the B cells are primarily responsible for humoral immunity (antibody production). Both T and B lymphocytes retain memory of a previous infection so that they produce a quicker and stronger response upon reinfection. T lymphocytes after exposure to an antigen are activated and secrete cytokines and cytotoxic granules. In the presence of an antigen, B lymphocytes become more active and differentiate into plasma cells.

T LYMPHOCYTES

T lymphocytes arise from the thymus and carry an antigen-specific receptor, the T-cell receptor (TCR), and constitute about 80 per cent of the circulating lymphocytes (Fig. 10.13). Lymphocytes divide rapidly and develop extreme diversity for reacting against specific antigens in the thymus. The preprocessing of the T lymphocytes occurs shortly before and after birth. The processed T cells leave the thymus and spread to lymphoid tissues throughout the body. T lymphocytes mediate cell-mediated immunity to foreign antigens, delayed hypersensitivity and allograft rejection.

About two-thirds of the T lymphocytes are called helper T lymphocytes because they possess a surface glycoprotein CD4 and release cytokines and growth factors that regulate other immune cells. The CD4 surface glycoprotein binds to MHC class II molecules on antigen-presenting cells

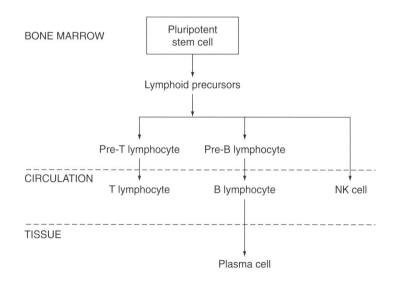

Figure 10.13   *The origin of lymphocytes.*

(Fig. 10.14). The cytokines (IL-4, IL-5, IL-6, IL-10 and interferon-γ) released by the helper T lymphocytes activate macrophages, cytotoxic and suppressor T cells, stimulate B lymphocyte differentiation to form plasma cells and antibodies, and promote B-cell maturation.

The other one-third of T lymphocytes are called suppressor/cytotoxic T lymphocytes. They express the CD8 surface glycoprotein (but not CD4) and can down-regulate immune responses and kill targets cells. The CD8 surface glycoprotein interacts with MHC class I molecules on cells, stabilizing T-cell interactions with antigen-presenting cells (Fig. 10.15). The cytotoxic T lymphocytes are important in the lysis of virally infected cells (via perforins which punch large holes in the cell membranes), tumour cells and rejection of foreign-tissue grafts. Cytotoxic T lymphocytes also secrete the cytokines, interferon-γ, IL-2, tumour necrosis factor (TNF) and lymphotoxin.

The functions of the T lymphocytes can be summarized (Fig. 10.16):

- Induction of B lymphocytes to mature into plasma cells or memory cells
- Recruitment and activation of mononuclear phagocytes
- Recruitment and activation of T cytotoxic cells
- Secretion of cytokines promoting the growth and differentiation of other T cells, macrophages and eosinophils

B LYMPHOCYTES

In humans B lymphocytes are formed and pre-processed in the liver during mid-fetal life and in the bone marrow during late fetal life and after birth. B lymphocytes comprise 5–15 per cent of the circulating lymphoid pool, and are also present in the germinal centres of lymph nodes, the spleen and mucosa-associated lymphoid tissue (MALT). They may differentiate to form plasma cells, which are non-circulating cells found in the bone marrow, medulla of lymph nodes and gut. Plasma cells are important for antibody production and secretion.

The majority of B lymphocytes express MHC class II antigens on their surface, and these are

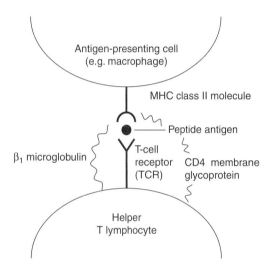

Figure 10.14 *Helper T lymphocyte interactions.*

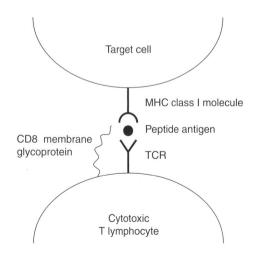

Figure 10.15 *Interaction of cytotoxic T lymphocytes.*

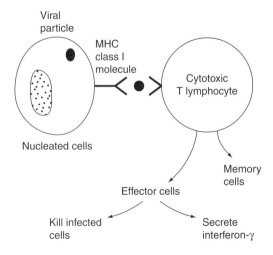

Figure 10.16 *The role of cytotoxic T lymphocytes.*

essential for cooperation with T lymphocytes. B lymphocytes also possess surface glycoproteins CD19, CD20, CD22, CD23 and CD40.

B lymphocytes have a major role in the humoral immune response. A foreign antigen initiates a specific clone of B lymphocytes to form plasma cells, usually regulated by IL-4, IL-2 and B cell differentiation factor (BCDF) released by helper T lymphocytes. Plasma cells produce antibodies that bind the antigen and generally activate complement to either destroy the antigen or opsonize it to facilitate phagocytosis by macrophages and neutrophils. Several interleukins (e.g. IL-4, IL-5, IL-6) derived from T lymphocytes activate B lymphocytes in the lymph nodes to form plasma cells and memory B lymphocytes. Activated B lymphocytes can also function as antigen-presenting cells.

The functions of B lymphocytes can be summarized:

- Production of antibody against specific antigens, with the aid of T lymphocytes
- Presentation of antigen to stimulate T lymphocyte activation

The roles of T and B lymphocytes during a microbial infection are summarized in Fig. 10.17.

ANTIGENS

An antigen is a substance capable of stimulating the immune system of the host to produce a specific response to it. An important characteristic is the specificity of the immune response for the chemical structure (antigenic determinants) of the antigen. Most antigens are proteins, polysaccharide or lipid macromolecules, but often the antigenic properties are determined by their carbohydrate moieties. Lipopolysaccharides in the cell walls of bacteria are important antigens, although the antigenic properties are determined by carbohydrate moieties. For example, *N*-acetylglucosamine is the antigenic determinant of the cell wall of *Streptococcus A*, whilst *N*-acetylgalactosamine is the main determinant for *Streptococcus C*.

An antigen must be a 'foreign substance' to a host for it to elicit an immune response. Most antigens have a high molecular weight ($\geqslant 5000\,\mathrm{Da}$). However, simple substances incapable alone of inducing an immune response, can induce a

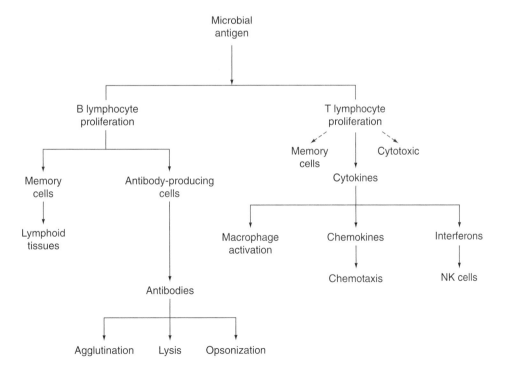

Figure 10.17 *The role of B and T lymphocytes in infection.*

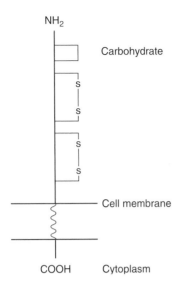

NH₂

Carbohydrate

S

S

S

S

Cell membrane

COOH   Cytoplasm

Figure 10.18   *An outline of the structure of the human MHC (HLA) transmembrane antigen.*

response when attached to a protein (e.g. serum albumin). These simple substances (called haptens) determine the antigenic specificity, and the protein to which the hapten is attached functions as a carrier molecule.

RECOGNITION AND BINDING OF ANTIGEN

The immune system has three specific elements used in the binding and recognition of antigens:

- Antibodies
- T cell receptors (TCRs)
- Major histocompatibility complex (MHC) molecules (Fig. 10.18)

These three elements are glycoproteins and belong to the immunoglobulin superfamily. Antigens interact with these by reversible forces (hydrogen bonding, electrostatic attraction, van der Waal's forces)

Antibodies are glycoproteins produced by B lymphocytes and plasma cells, and have recognition sites ('epitopes') on the antigens. Antibodies bind primarily to antigens (bacterial toxins, virus particles) and directly neutralize them. Other antibody functions include enhancement of phagocytosis by opsonization, activation of complement and interaction with mast cells and B cells.

TCRs are another group of large glycoproteins found on T lymphocytes that only interact with a small fragment of the antigen (peptide antigen) called the T-cell epitope. The TCR cannot interact with soluble (free in solution) peptide antigen. The TCR can only recognize the antigen when it is associated with MHC molecules found on the surface of cells.

MHC molecules hold the peptide antigen enclosed within a groove. The two types of MHC molecules are the class I molecules on all nucleated cells and the class II molecules on specialized antigen-presenting cells. Protein antigens are broken down into smaller peptides by proteolysis and presented in the groove of the MHC molecules. Endogenous antigens (e.g. viral proteins) interact with MHC class I molecules, whereas antigens endocytosed by macrophages or B lymphocytes interact with MHC class II molecules.

Immunoglobulins

Immunoglobulins are serum globulins with immune functions. All antibodies are immunoglobulins, but not all immunoglobulins have antigen-binding (antibody) functions.

TYPICAL STRUCTURE OF AN IMMUNOGLOBULIN MOLECULE

All immunoglobulins are composed of two identical light chains (approximately 23 kDa molecular weight) and two identical heavy chains (50–80 kDa) linked into a four-chain structure by disulphide bonds (Fig. 10.19). Each light and heavy chain has a constant portion and a variable portion. The constant portion is a segment at the C-terminal end containing a constant sequence of amino acids which mediates the effector functions of the immunoglobulin (e.g. complement activation). The variable portion lies at the N-terminal end and contains considerable variation in the amino acid sequence; this forms the antigen-binding sites.

When the immunoglobulin molecule is cleaved at the middle 'hinge' region by papain, the result is two identical fragments (Fab fragments) that retain antigen-binding capabilities, and a third larger fragment (Fc fragment) that is crystallizable

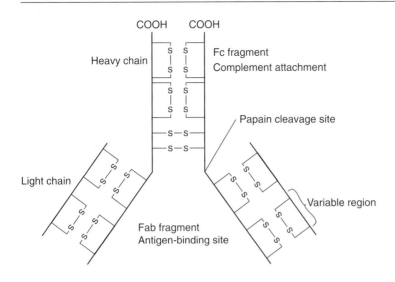

Figure 10.19   *The structure of immunoglobulins.*

Table 10.3   *The characteristics of human immunoglobulin*

Type	Mol. wt. ($\times 10^3$)	Half–life (days)	Heavy chain	Properties
IgG	160	18–23	γ	Precipitation Antitoxin Complement fixation Late antibodies
IgA	170	5–6	α	Surface antibody
IgM	960	5	μ	Agglutination Opsonin Lysis Complement fixation Early antibody
IgD	184	3	δ	(Not known)
IgE	188	2	ε	Reagin Antibody in type I hypersensitivity

and retains the effector functions (e.g. binding to cell surface receptors such as complement fixation). Pepsin cleaves the immunoglobulin molecule further towards the *C*-terminal end, leaving the heavy–heavy interchain disulphide bond intact, and produces a divalent Fab′$_2$ fragment that has two antigen-binding sites, but no effector function.

DIVERSITY OF IMMUNOGLOBULINS

Two types of light chain, kappa (κ) and lambda (λ), are present in all immunoglobulins. There are five types of heavy (H) chains, α, δ, ε, γ and μ, produced through gene variation encoding for the H chains. These immunoglobulins differ in the amino acid sequence of their H chains, in their physical characteristics and in their immunological function (Table 10.3).

IMMUNOGLOBULIN G

Immunoglobulin G (IgG) is the most abundant human immunoglobulin existing as α monomer, and makes up 75 per cent of the total serum immunoglobulin. The IgG molecule has two antigen-binding sites. IgG can be further subdivided into four subclasses (IgG1, IgG2, IgG3 and IgG4) on the basis of the four different forms of γ chains in the polypeptide heavy chains.

IgG has a high binding capacity for antigen, and is the major immunoglobulin of the secondary immune response. The half-life of IgG is 18–23 days. IgG2 production increases in response to bacterial polysaccharides, and is involved in combating encapsulated bacteria. IgG1 and IgG3 activate the classical complement pathway. Cellular receptors for the Fc region of IgG, bind and activate polymorphonucleocytes, mononuclear phagocytes and NK cells. Fc receptors in the placenta mediate the active transfer of IgG from the mother to the fetus, thereby protecting the newborn until immunocompetence has developed.

IMMUNOGLOBULIN A

Immunoglobulin A (IgA), the next most abundant immunoglobulin, is synthesized by plasma cells in the submucosal areas. It is present as a secretory IgA (dimeric form) in saliva, tears, breast milk, bronchial fluids and gastrointestinal secretions. IgA protects the mucosa against microbial invasion and growth by activation of the alternative pathway of the complement cascade, as an opsonin, and by reacting with receptors on monocytes and neutrophils. The half-life of IgA is 6 days.

IMMUNOGLOBULIN M

Immunoglobulin M (IgM) is a pentamer of five IgM monomers linked by disulphide bonds. It is the principal intravascular immunoglobulin of the primary immune response and provides an effective first-line defence mechanism against bacteraemia. IgM is a potent activator of complement via the classical pathway. The half-life of IgM is 5 days. It opsonizes and agglutinates particulate antigens. Blood group antibodies are IgM immunoglobulins.

IMMUNOGLOBULIN D

Immunoglobulin D is present in very low concentrations in the serum. Its precise function is not known and it has a half-life is 3 days. IgD is present as a cell surface receptor on B lymphocytes and may be involved in B cell activation.

IMMUNOGLOBULIN E

Immunoglobulin E (IgE) is present in low concentrations in serum and has a half-life of 2–3 days. Serum IgE concentrations increase in individuals with parasitic infections, atopy, and with immediate hypersensitivity reactions. IgE antibodies bind to mast cells and basophil surface receptors via its Fc fragment. Subsequent antigen binding to the acquired IgE produces mast cell activation and degranulation resulting in localized, and sometimes generalized, vascular effects.

IMMUNOGLOBULIN PRODUCTION: THE HUMORAL ANTIBODY RESPONSE

B lymphocytes are normally dormant in the lymphoid tissues. After exposure to an antigen there is an interval of about 2 weeks before antibodies can be found in the blood. The invading antigen (microorganism) is first localized and phagocytosed by macrophages that present it to adjacent B lymphocytes. The antigen also activates helper T lymphocytes. The B lymphocytes then proliferate and differentiate into lymphoblasts. Some lymphoblasts differentiate into plasma cells and release antibodies into the lymph to be carried to the blood. These antibodies do not reach a high concentration and do not persist. This is the primary immune response (Fig. 10.20). Some lymphoblasts form new B lymphocytes that circulate and enter lymphoid tissues where they remain dormant. Such lymphocytes are known as memory cells that produce antibodies more rapidly (within hours) and powerfully on subsequent exposure to the same antigen. This is known as the 'secondary response', and persists for months rather than weeks.

T–cell receptors for antigen

There are two types of TCR for antigen: $\alpha\beta$ TCRs and $\gamma\delta$ TCRs. $\gamma\delta$ TCRs are found on the surface of primitive T lymphocytes; their function is unknown. $\alpha\beta$ TCRs are present on over 90 per cent of peripheral T cells. TCRs recognize and bind a specific antigen with a MHC molecule that is formed when the antigenic peptide is embedded within the groove of the MHC molecule.

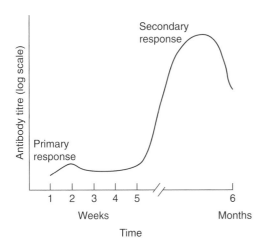

Figure 10.20 *An antibody–response curve.*

Major histocompatibility complex (MHC) proteins

The MHC molecules, expressed on the surface of a variety of cells, are proteins that are required for antigen recognition by T lymphocytes. In humans, these are known as human leukocyte antigens (HLA). T lymphocytes only respond to antigen on cell surfaces. On the presenting cells, the antigen is a short peptide embedded in a groove created by a HLA molecule. Antigen-specific cytotoxic T lymphocytes kill only target cells that possess the correct HLA molecule.

MHC is a collection of molecules that act as:

- Peptidases that cleave large protein antigens
- Transporters that load the peptides into the HLA molecules
- HLA molecules

HLA molecules are transmembrane glycoproteins. Each molecule has a peptide-binding region formed by a cleft that can bind processed antigenic fragments. Structurally, there are two classes of HLA molecules, class I and class II.

Class I HLA molecules are present on all nucleated cells and are dimers with an α heavy chain, and a light chain, β microglobulin. The main class I HLA molecules are human leukocyte antigens A, B and C (HLA-A, HLA-B and HLA-C). A virus or intracellular bacterium invades a host cell and stimulates the infected cell to synthesize proteins that are hydrolysed to form antigenic peptides that bind to class I HLA molecules. The class I HLA molecules are receptors for foreign antigens, presenting them to cytotoxic T lymphocytes.

Class II HLA molecules are also dimers (consisting of an α chain and a β chain) and are expressed on macrophages, monocytes and B lymphocytes. Class II HLA molecules are expressed by cells stimulated by interferon-γ. Exogenous antigen is taken up by an antigen-presenting cell by phagocytosis and hydrolysed into peptides which bind to class II HLA molecules. After exocytosis, the peptides are expressed on the cell surface and are presented to helper T lymphocytes. Class II HLA molecules present processed foreign antigen to helper T lymphocytes.

CYTOKINES

Cytokines are low-molecular-weight proteins released by certain cells, and regulate immunity, inflammation, cell growth and healing. Lymphokines are cytokines produced by lymphocytes, monokines are produced by monocytes, and interleukins are cytokines released by leukocytes that act on other leukocytes.

Cytokines are synthesized in response to cell surface signals (often other cytokines). They are highly potent but are released in minute quantities. Cytokine secretion is brief and usually self-limiting. Many cytokines modulate intracellular transduction such as protein phosphorylation.

Cytokines may be classified on the basis of their cell of origin, spectrum of activity, the target cells and the ligand–receptor interaction.

Interferons

Interferons (IFNs) are glycoproteins produced by virus-infected cells. There are three species of interferons: IFN-α produced by leukocytes, IFN-β, produced by fibroblasts, and IFN-γ produced by T lymphocytes in response to antigenic stimulation. The interferons prevent viral replication and have antitumour activity.

IFN-γ increases the expression of MHC class I antigens and of MHC class II antigens on antigen-presenting cells, activates macrophages to destroy intracellular pathogens and tumour cells, and enhances the cytotoxic actions of cytotoxic T lymphocytes. IFN-γ also inhibits endothelial cell growth but increases expression of adhesive molecules.

Inflammatory cytokines

The inflammatory cytokines, TNF-α, TNF-β, interleukin-α (IL-1α) and IL-1β have many similar effects, being endogenous pyrogens.

TNF-α is a proinflammatory cytokine produced mainly by macrophages and monocytes, NK cells, neutrophils and endothelial cells. It activates macrophages, granulocytes, cytotoxic T lymphocytes and endothelial cells. It also enhances expression of MHC class I molecules and has antitumour effects. TNF-β is secreted by T lymphocytes. TNF plays an important role in the acute septic response, tumour necrosis and cachexia in cancer. These actions are therefore important in the immune response to viruses and bacteria. Systemic release of TNF increases synthesis of acute reactive proteins, systemic hypotension, reduced myocardial contractility and intravascular thrombosis.

Interleukins stimulate the proliferation of T helper and cytotoxic cells and B lymphocytes. Interleukin-1 is secreted by macrophages as two polypeptides, IL-1α and IL-1β. Both IL-1α and IL-1β activate T and B lymphocytes, macrophages and endothelium and increase production of acute-phase proteins. IL-1 releases prostaglandins in the anterior hypothalamus (causing fever) and endorphins in the brain (attenuation of pain after injury).

Interleukin-6 is produced by macrophages, fibroblasts and endothelial cells. The main functions of IL-6 include enhanced production of acute-phase proteins in the liver, stimulation of immunoglobulin production by B lymphocytes, and induction of fever.

LYMPHOCYTE-DERIVED MEDIATORS

Lymphocyte-derived cytokines regulate immune function. IL-2, secreted by T lymphocytes that have been activated by antigen, stimulates the proliferation and differentiation of T lymphocytes

and enhances NK cell growth and cytotoxicity. IL-4 promotes the growth and differentiation of B lymphocytes and increases IgE production by B lymphocytes. IL-5 is produced by lymphocytes after antigen stimulation and by mast cells upon stimulation by an allergen–IgE complex, and it induces bone marrow eosinophil production. IL-9 promotes the proliferation of activated T lymphocytes, production of immunoglobulins by B lymphocytes, and production and differentiation of mast cells and haemopoietic precursor cells. IL-10 is produced by T lymphocytes, macrophages and B lymphocytes and inhibits the activity of macrophages, decreases production of proinflammatory cytokines, and increases immunoglobulin secretion and B lymphocyte proliferation. IL-13 is produced by T lymphocytes, and enhances B lymphocyte growth and differentiation, induces B lymphocyte IgE production, and inhibits monocyte and macrophage proinflammatory cytokine production.

MACROPHAGE-DERIVED CYTOKINES

The macrophage-derived cytokines are IL-12 and IL-15. IL-12, produced by macrophages and B lymphocytes, enhances cell lysis by cytotoxic T lymphocytes, NK cells and macrophages. It also induces the production of IFN-γ by T lymphocytes, promotes the proliferation of haemopoietic stem cells, and inhibits B lymphocyte IgE secretion. IL-15 stimulates the proliferation of T lymphocytes and the development of lymphokine-activated killer cells.

Chemokines

Chemokines are two related families of at least 20 small cysteine-rich peptides and include the potent chemoattractants IL-8, monocyte chemotactic peptide (MCP), and macrophage inflammatory proteins. IL-8 attracts neutrophils, whilst MCP attracts monocytes predominantly. Another chemokine, RANTES (regulated upon activation normal T-cell expressed and secreted) attracts monocytes and memory T lymphocytes, and also stimulates mast cell histamine release. Chemokines control cell migration into the extravascular compartment, and increased production may be associated with arthritis, glomerulonephritis, pulmonary diseases and skin disorders such as psoriasis.

Colony stimulating factors

Colony stimulating factors (CSFs) are soluble factors that increase the proliferation, differentiation and maturation of specific blood cells from the pluripotential haemopoietic stem cell. Macrophage-CSF (M-CSF) and granulocyte-CSF (G-CSF) are produced by monocytes, fibroblasts and endothelial cells, and promote the formation of monocytes and neutrophils, respectively. Erythropoietin is synthesized in the peritubular cells of the kidney, and stimulates and regulates erythrocyte production.

Growth factors

Platelet-derived growth factor (PDGF) and transforming growth factor-β (TGF-β) are produced by a wide variety of cells. They have other biological actions besides the promotion of cell growth. PDGF increases mitosis of cells, enhances phagocytosis and induces the secretion of proteinases (e.g. collagenase in fibroblasts and endothelial cells), thus promoting tissue repair. TGF-β promotes humoral rather than cell-mediated immune responses by inhibiting macrophage activation and T-lymphocyte function, and enhances fibrosis by stimulating the formation of extracellular matrix.

HYPERSENSITIVITY

The term 'hypersensitivity' describes exaggerated or inappropriate immune responses that cause tissue damage and even death of the host. It was originally categorized into four types (I–IV) by Gell and Coombs in 1970, based on: (1) the antibody involved; (2) the nature of the antigen; (3) the type of cell mediating the response; and (4) the duration of the reaction. Although recent advances in immunology have revealed that many diseases are more complex and do not fit into the groups proposed by Gell and Coombs, the classification is a simple and useful guide to the immunomechanisms involved.

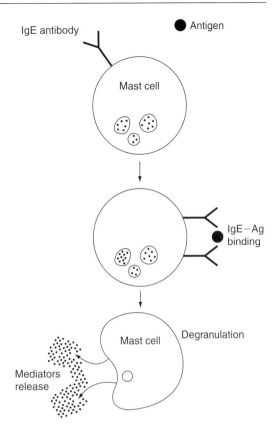

Figure 10.21 *Type I hypersensitivity.*

Type I hypersensitivity

In type I hypersensitivity (Fig. 10.21) an antigen binds to mast cell IgE, resulting in degranulation. The antigen – usually an exogenous substance such as pollen – stimulates B lymphocytes to produce specific IgE with the aid of helper T lymphocytes. The IgE then binds to mast cells via Fc receptors and sensitizes them. When the antigen reaches the sensitized mast cells, it binds to the surface-bound IgE and the mast cell degranulates, releasing mediators (histamine, 5-hydroxytryptamine, bradykinin, slow-reacting substance) that increase vascular permeability with vasodilatation, bronchoconstriction and mucus secretion. These events occur in the first 15–30 min after exposure to the antigen (allergen). About 6–12 h later, there is progressive tissue infiltration of neutrophils, followed by eosinophils and then mononuclear cells.

Type I hypersensitivity can cause local or systemic manifestations, depending on how the antigen enters the body. Hay fever and asthma are local type I reactions that appear when the antigen comes into contact with the respiratory mucous membranes in a sensitized individual. Urticaria caused by latex allergy or by food is also an example of a local type I reaction. When the antigen (e.g. drug or foreign serum) is administered parenterally to a sensitized individual, systemic manifestations of hypotension, bronchospasm, laryngeal oedema, skin rashes and sometimes death result, and this is known as anaphylaxis.

Anaphylaxis has a wide spectrum of clinical manifestations. In such a circumstance, to treat the cardiovascular collapse epinephrine (adrenaline) should be administered intravenously, with a rapid intravenous infusion of 1–2 L of colloids. The adrenaline is also beneficial for the management of bronchospasm. Endotracheal intubation should be performed to avoid airway obstruction produced by angioedema, and the inspired oxygen concentration increased to 100 per cent. Corticosteroids may be given to prevent further capillary leakage.

Type II hypersensitivity

Type II hypersensitivity (Fig. 10.22) is an antibody-mediated 'cytotoxic' reaction involving IgG or IgM antibodies binding to a cell surface antigen resulting in:

- Complement activation through the classical pathway causing cell lysis, mast cell activation, and neutrophil recruitment
- Mobilization and activation of neutrophils, eosinophils, monocytes and killer cells with antibody-dependent, cell-mediated cytotoxicity

The clinical picture of type II reactions depends on the target tissue:

- Organ-specific diseases such as myasthenia gravis and glomerulonephritis
- Autoimmune blood-cell destruction such as haemolytic anaemia or thrombocytopenia
- Transfusion reactions
- Haemolytic disease of the newborn (Rh isoimmunization)
- Hyperacute allograft rejection

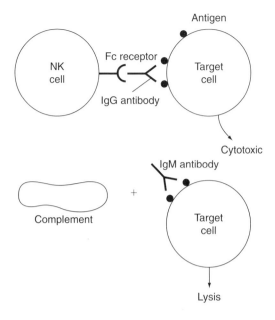

Figure 10.22　*Type II hypersensitivity.*

Type III hypersensitivity

A type III hypersensitivity (Fig. 10.23) is an immune complex-mediated reaction resulting in the deposition of antigen–antibody complexes in host tissues leading to complement activation, neutrophil infiltration and tissue damage. There are two forms of reaction: (1) complexes formed in the circulation and then deposited in the tissues causing systemic effects (e.g. serum sickness); and (2) complexes formed within the tissues resulting in localized effects (e.g. Arthus phenomenon). Normally, immune complexes deposited in small amounts in tissues are easily removed by the reticuloendothelial system, but in type III hypersensitivity these immune complexes are either too abundant or too small to be cleared effectively.

The systemic form of type III hypersensitivity appears because antigen excess leads to the formation of immune complexes in the blood that are deposited in the walls of the medium and small-sized arterioles, and these cause vasculitis. An example of this is serum sickness after immunization with serum containing antitoxins to diphtheria and tetanus (derived from horses).

The localized form of type III hypersensitivity is seen when antibody–antigen complexes form at

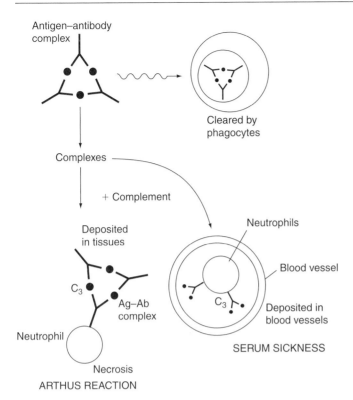

ARTHUS REACTION

Figure 10.23 *Type III hypersensitivity.*

the site of injection because an excess of antibodies localizes the offending antigen. Localized vasculitis results, and immune complexes are present in the vessel wall with perivascular infiltration of granulocytes due to chemotaxis induced by the antigen–antibody complex. This is seen in the Arthus reaction.

Type IV or delayed cell-mediated hypersensitivity

Type IV hypersensitivity (Fig. 10.24) results from an antigen presentation to T lymphocytes causing a release of cytokines (IL-2, IL-4, IFN-γ) that activate macrophages and with tissue injury. Reaction to *Mycobacterium tuberculosis* is a good example of type IV hypersensitivity. During infection, T cells recognize the antigen and proliferate, producing a population of sensitized T cells. When these cells are presented with the antigen by antigen-presenting cells, they release cytokines and the activated macrophages kill the microorganisms they contain.

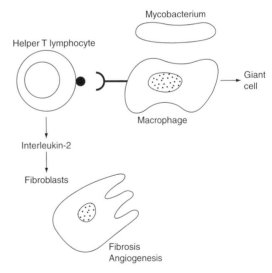

Figure 10.24 *Type IV hypersensitivity.*

The lymphocytes and macrophages arise at least 24 h after the provoking stimulus. With prolonged antigenic stimulation, the macrophages in the lesion fuse to form giant cells. A granuloma may form to wall off the infective focus, and within it extensive tissue damage leads to caseation with fibrosis and calcification.

TRANSPLANT IMMUNOLOGY

The major problem with transplantation of tissues from one person to another is that the immune system of the recipient may react with the donor's antigens, leading to rejection. The major histocompatibilty antigens (the HLAs in humans) are important determinants for this. HLAs are found on the surface of all nucleated cells and are receptors for foreign antigens, presenting them to cytotoxic T and helper T lymphocytes. Tissue rejection may be caused by cytotoxic lymphocytes, cytokines released by helper T lymphocytes, complement activation and activation of NK cells.

Allograft reaction

An allograft reaction is an immunological process initiated by the presence of transplanted cells. In host-versus-graft reaction (HVGR), the host's immune system attacks and destroys the graft. Graft-versus-host reaction (GVHR) is when immunologically competent graft cells attack the host's environment.

There are four types of HVGR:

- *Hyperacute rejection* occurs within minutes of transplantation by interaction of preformed cytotoxic antibodies in the host's circulation with HLA class I antigens expressed on the endothelium of the graft. The result is complement activation, coagulation, microvascular thrombosis and graft infarction.
- *Accelerated rejection* occurs within 4 days of transplantation by cellular and humoral mechanisms in recipients who have been sensitized previously against the donor's antigens.
- *Acute rejection* is a T-lymphocyte-mediated reaction, which occurs during the first month after transplantation.
- *Chronic rejection* is characterized by the slow loss of tissue function over a period of months or years. It may be a cellular immune response, an antibody response or a combination of the

two. It is associated with chronic immune-mediated destruction, and arteriolar narrowing with ischaemia of the graft.

The initial process in graft rejection is a sensitization phase during which the host's lymphocytes are exposed to foreign donor antigens and are primed to attack the graft. Antigens on the cells of the graft activate T lymphocytes in the host. HLA class II antigens bind to helper T lymphocytes, whilst HLA class I antigens bind to cytotoxic T lymphocytes. At the same time, macrophages release IL-1. The antigens and IL-1 activate helper T lymphocytes to release IL-2 that activates cytotoxic T lymphocytes.

The graft destruction phase follows, mediated by cellular and humoral mechanisms. Cytotoxic T lymphocytes are directly cytotoxic to graft cells. Lymphokines are released and promote macrophages to attack the graft. IL-1 is released and activates helper T lymphocytes, releases TNF and IFN-γ, and increases graft expression of HLA class II. Soluble antigens from the graft stimulate B lymphocytes to secrete antibodies that activate complement and cause further cell damage.

Graft–versus–host reaction

This is a condition resulting from an attack by the donor's immunologically active lymphocytes against the foreign antigens of the recipient, where there is an antigen difference between the donor and the recipient (as can happen after bone marrow transplantation). Acute GVHR may occur within 4 weeks of transplantation, leading to dermatitis, jaundice, hepatosplenomegaly and overwhelming infection. Chronic GVHR is seen 100 or more days after transplantation, with hepatitis, pericarditis, myositis and death from opportunistic infection.

ASSESSMENT OF IMMUNE FUNCTION

A history and physical examination will provide invaluable information in most immune disorders, and simple laboratory tests such as a full blood count, erythrocyte sedimentation rate and plasma immunoglobulin concentrations are useful.

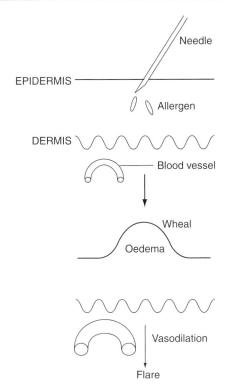

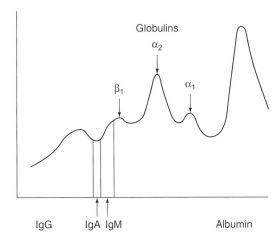

Figure 10.26   *Electrophoresis of human serum.*

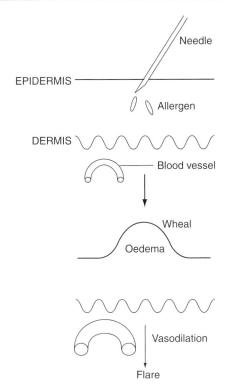

Figure 10.25   *The skin prick test.*

In vivo tests

SKIN TESTS

Skin tests detect the presence or absence of an immune response to a specific antigen (allergen) by observing a transient visible skin lesion. They are simple, convenient, inexpensive and relatively safe. Skin testing for immediate hypersensitivity reactions or anaphylaxis is the main method for evaluating allergies to neuromuscular muscle relaxants, intravenous induction agents, local anaesthetic drugs, latex and antibiotics. A positive test is characterized by a wheal (local oedema) and flare (erythema) in the skin within minutes of exposure and is due to a release of local mediators.

A cutaneous prick test (Fig. 10.25) introduces a minute quantity of allergen into the dermis to react with IgE antibodies fixed to cutaneous mast cells. In an intradermal test, a measured quantity of allergen is introduced into the skin to induce IgE-mediated, IgG-mediated or T-lymphocyte-mediated reactions. A patch test produces an allergic contact

dermatitis (cell-mediated hypersensitivity) when an antigen (allergen) reacts with sensitized T lymphocytes to release lymphokines at the site of contact with the allergen.

In vitro tests

In vitro tests (e.g. electrophoresis) measure some of the components of the complex events of hypersensitivity reactions (Fig. 10.26).

SERUM IgE CONCENTRATIONS

Total serum IgE concentrations can be measured by a radioimmunosorbent test, but this is rarely useful in diagnosing anaphylaxis in anaesthesia and surgery. Specific IgE antibodies can be measured by the radioallergosorbent test (RAST) (Fig. 10.27). This is a two-phase (solid–liquid) system using a non-soluble allergen incubated first with a specific antibody and then in radiolabelled antihuman IgE to detect specific antibodies of IgE.

Immunodiffusion is used for the qualitative and quantitative analysis of antigens and antibodies in serum or body fluids. The method depends on a precipitation reaction of an insoluble antigen–antibody complex from soluble antigen and antibodies. The enzyme-linked immunoabsorbent assay (ELISA) is an immunoassay using enzyme-linked antibody or antigen (Fig. 10.28). Either the

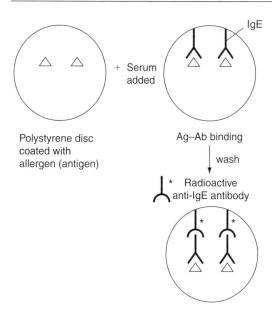

Figure 10.27   *The radioallergosorbent (RAST) test.*

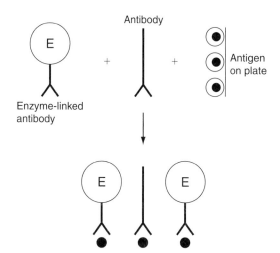

Figure 10.28   *The enzyme-linked immunosorbent assay (ELISA).*

antigen or antibody can be linked to an enzyme (e.g. horseradish peroxidase or alkaline phosphatase). In the standard indirect ELISA an antigen is adsorbed onto a polystyrene plate (solid phase). The antibody to be detected binds to this component, and a second enzyme-labelled antibody is then added. The substrate of the enzyme is added, producing a coloured reaction product that can be measured spectrophotometrically.

In a radioimmunoassay (RIA), the antibody is adsorbed or cross-linked to a solid matrix and an added unlabelled antigen in the test sample binds to the antibody. Radiolabelled antigen is then added and the amount of bound versus free labelled antigen is determined. A standard curve is constructed from data obtained by allowing varying amounts of unlabelled antigen to compete with the labelled antigen. The amount of unlabelled antigen in the test sample can be calculated from this standard curve.

SERUM CONCENTRATIONS OF COMPLEMENT

Serum concentrations of complement components can be measured by the radial immunodiffusion test. Specific antisera against individual components are incorporated into agar, and the patient's serum is placed in wells in the agar. Raised concentrations of components of complement are frequently found in acute inflammation. A reduction in concentration of complement can be due to:

- Primary genetically determined immunodeficiency disorders
- Secondary deficiencies caused by complement-consuming antibody–antigen interactions, or associated with liver or renal disease. Complement activation in allergic reactions is usually associated with decreases in C3 or C4 concentrations in serum or increases in concentrations of products of complement activation (e.g. C3a, C4a, C5a)

DETECTION OF IMMUNE COMPLEXES

Immune complexes can be detected in tissues by immunohistochemical staining with specific antisera. The detection of immune complexes in serum and biological fluids is not performed routinely. C1q has a high affinity for immune complexes, and the radiolabelled form can be used to quantitate the amount of immune complexes in tissues.

T LYMPHOCYTE ASSAYS

T lymphocytes are primarily responsible for cell-mediated immunity. Monoclonal antibodies are used to identify specific T-cell antigens. The measurement of the absolute number or percentage of helper or cytotoxic T lymphocytes is valuable in the monitoring of the treatment and progress of some diseases (e.g. acquired immunodeficiency disease). T lymphocyte function can be assessed by using the phytohaemagglutinin (PHA) and conconavalin A mitogen stimulation test. (These substances are called mitogens because they cause proliferation of T lymphocytes in response to a mitotic stimulus.) The mitogen stimulation is assessed by DNA synthesis measured by tritiated thymidine uptake.

B LYMPHOCYTE ASSAYS

Monoclonal antibodies labelled with fluorochromes or enzymes are used to detect B lymphocytes. Plasma cells have their own set of markers (PCA1, PC1), which can be detected by monoclonal antibodies. B lymphocyte function can be assessed by using staphylococcal protein A (spA) and poke-weed mitogen (PWM), which are B lymphocyte mitogens.

ASSAY OF MEDIATORS RELEASED IN A HYPERSENSITIVITY REACTION

In vitro testing may be performed using the patient's basophil cells that possess IgE and will release histamine when exposed to the allergen, thus avoiding the danger of triggering anaphylaxis in the patient. However, this test is limited to research laboratories, and non-specific, non-immune release of histamine cannot be excluded. The detection of mediators during or shortly after an allergic reaction can diagnose a hypersensitivity response that occurred during anaesthesia. However, plasma histamine concentrations increase only transiently in such reactions. Measurement of serum mast cell tryptase concentrations is more useful, as they are elevated for 1–5 h after the onset of anaphylactic reactions and indicate mast cell activation. Tryptase is a neutral protease found in mast cell granules.

EFFECTS OF ANAESTHESIA ON IMMUNE FUNCTION

There is considerable evidence that anaesthesia causes a reversible depression of immune function. The physico-chemical barriers may be impaired during and after anaesthesia, tracheal ciliary activity is decreased, and depression of phagocytosis may be demonstrated, although this appears to be proportional to the degree of surgical stress. Decreased specific immunity occurs in the postoperative period due to depressed lymphocyte function brought about by the hormonal (increased cortisol concentrations) changes associated with the stress response of surgery. T lymphocyte numbers decrease and their activity is diminished. NK cell activity changes in a biphasic manner with an initial rapid and transient increase of NK cells by recruitment from the extravascular space, lymph nodes and spleen, and then post-operatively their activity is depressed by release of suppressor monocytes.

ALLERGIC DRUG REACTIONS IN ANAESTHESIA

Allergic reactions to drugs are caused by the interaction of drugs or their metabolites with immune effector cells. Most drugs combine with host proteins to form a drug–protein complex that stimulates the host immune reaction. Anaphylactic reactions occur in between 1 in 5000 and 1 in 25 000 anaesthetic cases, with a mortality rate of 3.4 per cent.

The reported incidences of adverse reactions to induction agents, muscle relaxants and plasma volume expanders vary widely. Allergic reactions to propanidid (1 in 540 to 1 in 1700) and althesin (1 in 900 to 1 in 1100) were more common than reactions to thiopental (1 in 14 000). Non-depolarizing muscle relaxants are the most common cause of life-threatening reactions during anaesthesia, the reported incidence being as high as 1 in 1200 exposures. Most are IgE-mediated, and suxamethonium is the most common causative agent. Alcuronium and atracurium have a high incidence of reactions compared with pancuronium and vecuronium. True allergy to local anaesthetic drugs is rare, and is by IgE-mediated reactions. IgE-mediated allergies to methyl paraben, a preservative in local anaesthetic solutions, have been reported.

Protamine produces allergic reactions mediated by IgE, IgG or complement and can produce a classical anaphylactic reaction or a clinical syndrome resulting in acute pulmonary hypertension and right ventricular failure. Reactions to plasma volume expanders occur in between 1 in 1000 cases of patients given gelatin products and 1 in 10 000 cases involving plasma protein preparations. Haemaccel causes direct histamine release from mast cells without complement activation or IgE production. For intraoperative antibiotics, penicillin is the most common antibiotic causing allergic reactions, with an incidence of between 1 in 2500 and 1 in 10 000, with a fatality rate in those affected of about 10 per cent. The β-lactam ring in penicillin opens spontaneously to form the penicilloyl group, which combines with plasma proteins, and IgE antibodies are produced.

Reflections

1. The innate immune system consists of natural physical and chemical barriers and innate cellular defence mechanisms. Complement, interferons and acute-phase proteins offer innate immune protection in the blood. There are four different types of cells involved: the phagocytes, natural killer cells, mast cells, and eosinophils. Macrophages and neutrophils engulf bacteria, and kill and digest them with highly reactive oxygen free radicals.

2. Natural killer cells are large granular lymphocytes which recognize cells that become infected with a virus via cell surface markers.

3. When the body becomes infected or injured, an inflammatory response results. Local vasodilatation and increased capillary permeability occurs and this leads to local oedema and infiltration of the damaged tissues by white cells. The trigger for the inflammatory response is mast cell degranulation which releases histamine that elicits capillary dilatation and leakage and also chemotactic agents that attract neutrophils to the site of injury.

4. The adaptive immune system is specific and long-lasting (via memory) protection to the body against a range of organisms. The cells of the adaptive immune system are the lymphocytes. Two principal types of lymphocytes, B lymphocytes (which mature in the bone marrow) and T lymphocytes (which mature in the thymus) are involved. The lymphocytes leave the blood vessels, pass through the tissues, and re-enter the venous blood by way of the lymph nodes and the thoracic duct. When lymphocytes are stimulated by antigens they undergo mitosis and form a population of cells with identical specificity called a clone. Some clonal cells proliferate and produce antibodies whilst others remain in the lymphoid tissue as memory cells which can respond to a similar challenge in future.

5. After a B cell is stimulated by an antigen, it is transformed into a plasma cell which secretes antibody into the circulation. Antibodies have two main functions: (a) binding an antigen and (b) eliciting a response that removes the antigen from the body. The antibody acts together with complement to stimulate phagocytes with the result that the organism carrying the antigen is killed and digested. On initial response to an antigen the B cells secrete IgM. Plasma levels of these antibodies decline after 2–3 weeks. Subsequent re-exposure to the same organism causes a more prompt and long-lasting increase in the plasma levels of IgG.

6. Activated T lymphocytes secrete cytokines or cytotoxic molecules on to neighbouring cells. T cells may be cytotoxic or helper cells. The effects of T cell activation usually involve one target cell. Depending on the substances secreted by the T cell, the target cell may be killed or stimulated. T cells respond to cells that possess MHC molecules that have bound a foreign peptide.

7. The immune system can react powerfully to an antigen and cause a hypersensitivity reaction. There are four types of hypersensitivity reactions: (a) anaphylaxis, (b) cytotoxic hypersensitivity, (c) immune complex deposition, and (d) cell-mediated hypersensitivity.

11

Endocrine physiology

LEARNING OBJECTIVES

After studying this chapter the reader should be able to:

1. Explain the concept of hormones and the role hormones play in homeostasis and the principles of hormone action
2. Explain the second messenger model and the gene expression model of hormone action
3. Describe the role of the central nervous system in the regulation of the endocrine system via the hypothalamic–pituitary axis
4. Describe the hormones of the anterior pituitary gland, the regulation of their secretions, and their actions on target tissues
5. Explain the role of growth hormone and somatomedins in growth and lipolysis, and their effects on blood sugar
6. Describe the synthesis and release of vasopressin (antidiuretic hormone) and oxytocin and explain the main physiological effects of these hormones
7. Describe the synthesis, storage and release of thyroid hormones and explain their physiological actions on development of the central nervous system, body growth, and metabolism
8. Explain the role of the adrenal cortical steroid hormones in metabolism and fluid balance
9. Describe the role of the adrenal medullary hormones, their metabolism, and their effects on the cardiovascular system
10. Describe the role of vitamin D, parathyroid hormone and calcitonin in the handling of calcium and phosphate in the body

INTRODUCTION

The endocrine system consists of a collection of glands that are scattered throughout the body and secrete chemical substances called hormones. A hormone may be defined as a chemical messenger produced and secreted by a specific endocrine cell and circulates in trace amounts in the blood and acts on specific target cells, usually remote from the endocrine cell. The hormone binds to receptors in the target cell, initiating specific intracellular pathways, and this results in alterations of cellular actions such as uptake or output of mediators involved in the continuous, long-term regulation of physiological processes (e.g. fluid and electrolyte balance, energy metabolism, growth and development, digestion, reproduction and adaptation to physiological stresses). An important feature is signal amplification due to the self-multiplication of intracellular pathways.

There are three types of hormones: (1) peptides, e.g. produced by the hypothalamus, pituitary, pancreas, parathyroid, gut and heart; (2) amines, e.g. produced by the adrenal medulla and thyroid; and (3) steroids, e.g. produced by the adrenal cortex, ovary and testis.

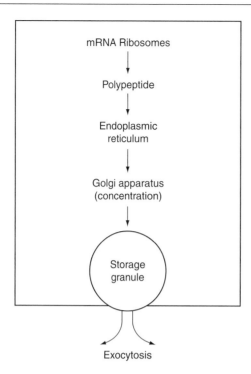

Figure 11.1 *Peptide hormone formation.*

HORMONE PRODUCTION AND SECRETION

Peptides

The peptide hormones are synthesized by mRNA transcription in the nucleus, leading to ribosomal translation with the production of the polypeptide within the endoplasmic reticulum. The polypeptide is concentrated in the Golgi apparatus and then stored in storage granules (Fig. 11.1). Large protein hormones are usually retained in the storage granules, whilst small peptide hormones are bound to specific binding proteins within the granules. The hormones are released from the endocrine cell by exocytosis following a neural, chemical, hormonal or physical stimulus. Hormones not released are usually degraded to amino acids and recycled.

Tyrosine derivatives

The tyrosine derivatives (e.g. thyroxine, cate-cholamines) are synthesized within the cytoplasm of the cells and stored either in storage granules

attached to binding proteins (e.g. catecholamines) or as part of thyroglobulin in the colloid of the follicles.

Steroids

The steroid hormones are synthesized from cholesterol, released immediately, and are not stored in the endocrine cell.

REGULATION OF HORMONE SECRETION

The immediate stimulus for the secretion of a hormone may be neural, hormonal, or a change in the level of some metabolite or electrolyte in the blood. Secretion of hormones may be regulated by a negative feedback mechanism via a short or long loop (Fig. 11.2), whereby increased levels of the hormone in the blood inhibit further hormone secretion.

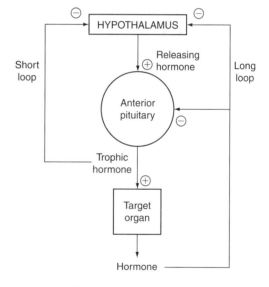

Figure 11.2 *Feedback loop systems.*

ACTIONS OF HORMONES

Hormones regulate the growth, development, metabolic activity and function of tissues. The responses are often the result of the actions of a number of hormones. Most hormones act on specific receptors that either stimulate or inhibit intracellular processes in the target cells.

Peptide hormones

The peptide hormones generally bind to specific receptors on the cell surface, and this initiates a cascade of intracellular reactions, with each step amplifying the previous reaction, so that a large final result occurs even with a small initial stimulus. Various mediators called 'second messengers' initiate intracellular biochemical reactions leading to cellular responses.

SECOND MESSENGER SYSTEMS

Cyclic AMP

This is a common second messenger system stimulated by catecholamines, glucagon, parathyroid hormone, vasopressin (antidiurectic hormone; ADH), adrenocorticotrophin (ACTH), thyrotrophin (TSH), luteinizing hormone (LH), follicule-stimulating hormone (FSH) and most hypothalamic-releasing factors. The hormone binds to a specific receptor on the cell membrane, and this activates adenyl cyclase, which converts adenosine triphosphate (ATP) to cyclic AMP (cAMP). cAMP activates intracellular protein kinase, which phosphorylates proteins that mediate the cellular reactions, such as contraction or relaxation of muscle, changes in cell permeability or cellular secretion (Fig. 11.3).

The phosphatidylinositol (PIP) system

This system results from the breakdown of the membrane phospholipid, phosphatidylinositol biphosphate (PIP_2), by phospholipase C which is activated when some hormones bind to their transmembrane receptors. Phosphatidylinositol biphosphate is broken down to two important second messengers, inositol triphosphate (IP_3) and diacylglycerol (DAG).

IP$_3$ mobilizes calcium ions from the endoplasmic reticulum and mitochondria, resulting in smooth muscle contraction, cellular secretion and ciliary activity. DAG activates protein kinase C, which is enhanced by calcium ions released by IP$_3$ (Fig. 11.4). Activated protein kinase C promotes cell division and proliferation.

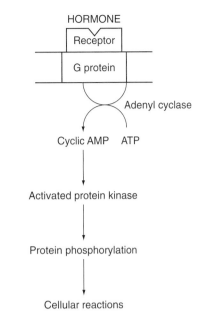

Figure 11.3 *Cyclic AMP cascade.*

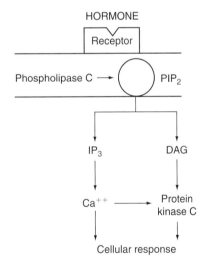

Figure 11.4 *The phosphatidylinositol system.*

Calmodulin

Calmodulin is an intracellular binding protein that acts as a second messenger. A number of peptide hormones increase intracellular calcium, either by opening calcium channels following receptor activation or secondary to the PIP system. Calcium ions bind to calmodulin, leading to a conformational change. The calmodulin–calcium complex binds to and activates various enzymes. For example, it activates myosin light-chain kinase and this results in the phosphorylation of myosin, causing smooth muscle contraction.

Tyrosine derivatives or amine hormones

The mode of action of the amine hormones depends on their chemical subtypes.

CATECHOLAMINES

The catecholamines act on cell membrane receptors by various mechanisms. An increase in cAMP results when β receptors are activated, whilst phosphatidylinositol activation occurs with activation of α receptors.

THYROID HORMONES

Thyroid hormones cross the cell membranes and bind to receptors in the nucleus, increasing gene transcription. This produces many intracellular enzymes, which increase intracellular metabolism in almost all cells. As the thyroid hormones bind to intranuclear receptors, the resultant physiological effects last for days or weeks.

Steroid hormones

Steroid hormones, being highly lipid-soluble, readily enter cells and bind to specific protein receptors in the cytoplasm. This steroid–protein complex enters the nucleus where it stimulates transcription of specific genes, leading to increased mRNA. This diffuses into the cytoplasm and promotes ribosomal translation and protein synthesis (Fig. 11.5). The proteins formed may be enzymes, transport proteins or structural proteins.

Receptor regulation

Receptor numbers in target cells alter considerably in a dynamic fashion, as they may be inactivated or destroyed; alternatively, they may be reactivated or increased. A decrease in the responsiveness of the

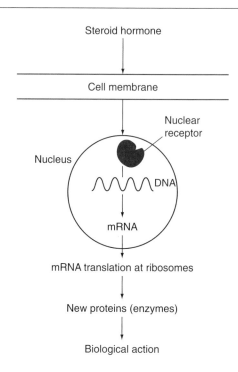

Figure 11.5   *Actions of steroid hormones.*

target tissue as a result of a decreased number of active receptors is known as 'down-regulation'. This can result from inactivation of some receptor molecules or from decreased production of receptor molecules. 'Up-regulation' may occur when increased receptors are available, usually caused by increased synthesis of receptor molecules due to the stimulating hormone. This usually results in an increased effect of the hormone.

HYPOTHALAMUS

The hypothalamus is an important regulation centre that controls autonomic functions and hormonal output from the anterior pituitary.

Hypothalamic control of autonomic function

Autonomic functions regulated by the hypothalamus include cardiovascular and temperature regulation, feeding, water balance, defence reactions and control of sexual behaviour.

REGULATION OF THE CARDIOVASCULAR SYSTEM

Stimulation of the anterior hypothalamus can lead to a decrease in blood pressure and heart rate, whilst stimulation of the posterior hypothalamus produces hypertension and tachycardia.

TEMPERATURE REGULATION

The anterior hypothalamus is stimulated by increased blood temperature and this leads to sweating and cutaneous vasodilatation. The posterior hypothalamus is stimulated by cold temperature, resulting in shivering and cutaneous vasoconstriction.

REGULATION OF FEEDING

Stimulation of the lateral hypothalamic nucleus (hunger centre) results in feeding, whilst stimulation of the ventromedial nucleus (VMN, satiety centre) inhibits feeding by inhibition of the hunger centre. The VMN is also sensitive to blood glucose levels, as hyperglycaemia inhibits feeding and hypoglycaemia promotes it.

REGULATION OF WATER BALANCE

A thirst centre is located at the paraventricular nucleus (PVN), and responds to increased electrolyte concentration within the cells of the PVN. Osmoreceptors located at the supraoptic nucleus (SON) and the PVN regulate the release of ADH, which increases water reabsorption in the renal tubules.

DEFENCE REACTIONS

The hypothalamus regulates sympathetic activity, affecting the release of catecholamines by the adrenal medulla.

SEXUAL BEHAVIOUR

The cells in the hypothalamus are sensitive to both oestrogens and androgens, thus influencing sexual behaviour.

Hypothalamic regulation of hormone output

The hypothalamus has direct neural control of the posterior pituitary gland. The anterior pituitary gland secretion is controlled by hypothalamic releasing or inhibiting factors, which are released into the hypothalamic–hypophysial portal system.

HYPOPHYSIAL PORTAL SYSTEM

The posterior pituitary derives its blood from the capillary plexus arising from the inferior hypophysial artery. The median eminence of the hypothalamus is the release centre for the hypothalamic releasing factors, and is supplied by branches of the superior hypophysial artery. The primary capillary plexus, arising from the superior hypophysial artery, forms the lesser portal veins, which give rise to a secondary capillary plexus that provides 90 per cent of the blood supply of the anterior lobe.

NEUROSECRETORY NEURONS

Two distinct populations of neurosecretory neurons are present in the hypothalamus: (1) magnocellular neurons, consisting of the SON and the PVN, synthesize and secrete ADH and oxytocin; and (2) parvocellular neurons, forming the tuberoinfundibular tract, secrete the hypophysiotrophic hormones.

NEURAL CONTROL

The function of the magnocellular system is controlled by cholinergic and noradrenergic neurotransmitters. Acetylcholine stimulates the release of ADH and oxytocin, whilst norepinephrine inhibits their secretion. Norepinephrine, dopamine and serotonin control the release of hypophysiotrophic hormones by the parvocellular system.

FEEDBACK CONTROL

Negative feedback occurs at three levels (Fig. 11.6): (1) long-loop feedback, where hormones produced by peripheral organs can exert a negative feedback on the hypothalamus and anterior pituitary, e.g. thyroid, adrenocortical and sex hormones; (2) short-loop feedback, where the anterior pituitary hormones may exert a negative feedback on the synthesis or secretion of the hypothalamic releasing or inhibiting factors; and (3) ultra-short-loop feedback, where the hypothalamic releasing or

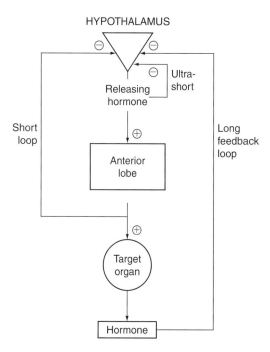

Figure 11.6 *The three levels of feedback in the hypothalamus.*

inhibiting factors can inhibit their own synthesis and secretion.

Positive feedback occurs at the mid-menstrual cycle when a high oestradiol concentration stimulates the hypothalamic release of LH, which promotes the release of ova.

ANTERIOR PITUITARY

The anterior pituitary is derived from the buccal ectoderm as an upward extension (Rathke's pouch) of the epithelium of the primitive mouth cavity (stomodeum). It consists of sinusoids and two major cell types, namely, the granular secretory cells (chromophils) and the agranular secretory cells (chromophobes). The chromophils exist in two forms: (1) acidophils (80 per cent) secrete prolactin (PRL) and somatotrophin (GH); (2) basophils (20 per cent) secrete glycoprotein trophic hormones such as TSH, ACTH, LH, FSH and β-lipotrophin (LPH).

The chromophobes are now considered to be degranulated secretory cells. Neurons in the anterior lobe are postganglionic sympathetic fibres.

Releasing factors or hormones

These releasing factors are synthesized in the cell bodies of the hypothalamus and released into the primary plexus in the median eminence. They are then transported via the portal vessels to the secondary plexus (sinusoids) in the anterior pituitary to stimulate target-cell hormone release. As these releasing and inhibiting factors are present in high concentrations in the hypophysial portal system, they exert their effects on the anterior pituitary cells, but do not have any systemic activity as their systemic concentrations are extremely low.

Thyrotrophin-releasing factor (TRF) is a tripeptide that stimulates TSH release. Gonadotrophin-releasing factor (or LH-releasing factor or FSH-releasing factor) is a 10-amino acid peptide which promotes the release of LH and FSH. Somatotrophin-releasing factor is a 43-amino acid peptide, and releases growth hormone (GH). Somatostatin (GH-inhibiting factor) is a 14-amino acid peptide which suppresses the secretion of GH, ACTH, TSH and PRL. It also inhibits the secretion of insulin and glucagon. Corticotrophin-releasing factor is a 41-amino acid peptide promoting the release of ACTH.

Growth hormone (somatotrophin)

Growth hormone (GH) is a 191-amino acid polypeptide with two disulphide bridges, and is secreted by 30–40 per cent of anterior pituitary cells. The main effect of GH is stimulation of normal growth in conjunction with the thyroid, sex and adrenocortical hormones. The effects of growth hormone on linear growth and protein metabolism are not direct and are indirectly mediated by the generation of somatomedins from the liver and locally from cartilage and muscle.

GH enhances the transport of amino acids across cell membranes into cells, and increases protein synthesis by increasing transcription of DNA in the nucleus to form more RNA and, with enhanced RNA translation in the cytoplasm, increased protein synthesis in the ribosomes.

It also has metabolic effects on most tissues in the body and overall it increases blood glucose and mobilizes free fatty acids. The effects of GH on carbohydrate metabolism are increased glycogenesis and decreased glucose uptake by the tissues, resulting in a diabetogenic (anti-insulin) effect. GH also promotes the release of fatty acids from adipose tissue and enhances the formation of acetyl CoA from fatty acids, so that fat is preferentially used as an energy source. However, excess GH can cause excessive ketone body formation from fat tissue, resulting in ketosis.

The most important effect of GH is increased growth of the skeleton by proliferation of chondrocytes and osteoblasts and increased uptake of protein by the osteogenic cells. This results in epiphyseal growth, or periosteal growth if the epiphyses have fused. Some of the effects of GH are mediated by somatomedin (insulin-like growth factor, IGF). GH stimulates the liver to produce somatomedin, which promotes cartilage growth.

The action of GH begins with its binding to specific GH receptors (which have a single transmembrane domain) on cell membranes. This induces a conformational change in the protein of the cytokine receptor that results in the binding and activation of cytosolic tyrosine kinases resulting in protein phosphorylation. The long-term effects of GH are mediated by somatomedins which induce mitosis and consequently cell proliferation.

The plasma concentrations of GH show diurnal fluctuations, with a peak occurring 1–2 h after the onset of deep sleep. Nutritional factors such as starvation, hypoglycaemia or low plasma concentrations of fatty acids can increase GH secretion. Stress factors such as exercise, excitement and trauma also stimulate GH secretion.

The release of GH is under the control of somatotrophin-releasing factor and is mediated by monoaminergic and serotoninergic pathways. The secretion of GH is stimulated by low blood glucose, high blood amino acids, and ghrelin. α Adrenergic, dopaminergic and serotoninergic agonists, opioids and amino acids (arginine, leucine, lysine, tryptophan) all stimulate release of GH. The secretion of GH is inhibited by somatostatin (from the hypothalamus) and somatomedin C (from the liver).

Overproduction of GH in adolescence produces giantism and acromegaly in adulthood when the epiphyseal plates of long bones have fused. Decreased GH secretion in adolescence leads to dwarfism. The anti-insulin effects of GH can lead to hyperglycaemia in states of chronic GH excess.

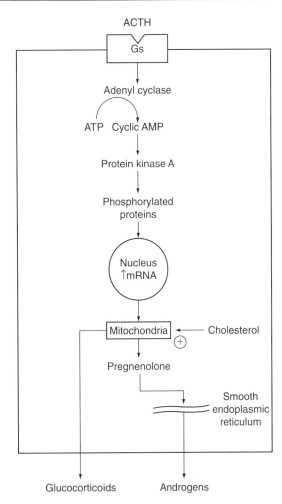

Figure 11.7   *Effects of ACTH on steroid synthesis. Gs, stimulatory G protein.*

Adrenocorticotrophin

Adrenocorticotrophin (ACTH) is a 39-amino acid polypeptide, the biological activity of which is mediated by the first 23-amino acids. The actions of adrenocorticotrophin on the adrenal cortex are the stimulation of the synthesis and release of glucocorticoids by the zona fasciculata and zona reticularis. The main effects are to cause an increase in cholesterol and steroid synthesis. ACTH has a trophic action on the adrenocortical cells.

ACTH binds to cell membrane receptors, increasing cAMP and activating protein kinase A, which phosphorylates protein enzymes responsible for steroid synthesis (Fig. 11.7).

The release of ACTH is increased by corticotrophin-releasing factor (CRF). In turn, CRF

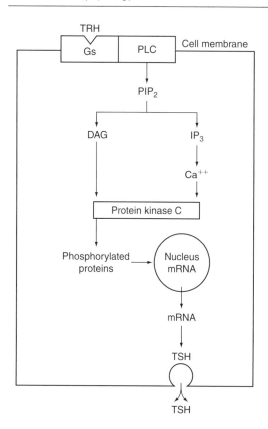

Figure 11.8 *Actions of thyroid releasing hormone (TRH). Gs, stimulatory G protein; PLC, phospholipase C.*

release is controlled by the limbic system and is increased by heat, toxins and stress. Catecholamines and vasopressin also stimulate ACTH release. The release of ACTH is inhibited by glucocorticoids.

Thyrotrophin

Thyrotrophin (TSH) is a glycoprotein with α and β subunits. Thyrotrophin-releasing factor increases TSH formation by enhanced gene expression and release through the phosphatidylinositol system (Fig. 11.8). The thyroid hormones have a direct negative feedback effect on TSH synthesis. The secretion of TSH is also promoted by cold temperature in neonates, but not in adults.

Prolactin

Human prolactin (PRL) is a 198-amino acid polypeptide synthesized by the acidophil cells of the anterior pituitary. The secretion of PRL is under tonic inhibitory control by the hypothalamus via prolactin release-inhibiting factor. The main stimulus for PRL release is suckling. PRL is important for development of mammary glands and milk production. During pregnancy, PRL promotes the development of the mammary duct. Immediately after delivery of the baby, PRL stimulates the galactotransferase enzyme, leading to synthesis of lactose, casein and fat. During lactation, increased PRL concentration suppress LH secretion, causing amenorrhoea.

Gonadotrophins

The gonadotrophins, LH and FSH, are glycoproteins with two subunits, α and β. LH stimulates ovulation and luteinization of the ovarian follicles in the female and testosterone secretion in the male. FSH stimulates the development of ovarian follicles and regulates spermatogenesis in the testes.

POSTERIOR PITUITARY

The posterior pituitary is merely an extension of the hypothalamus, and secretes two hormones, ADH and oxytocin. Both hormones are synthesized in the cell bodies of the SON and PVN of the hypothalamus. Each hormone binds to a specific transport protein, neurophysin, and the hormone–neurophysin complex is transported in tiny vesicles along the axons. These vesicles coalesce to form storage granules in the nerve terminals in the posterior pituitary, and the granules are released by exocytosis.

Antidiuretic hormone (ADH; vasopressin)

Antidiuretic hormone (ADH or vasopressin) is a nonapeptide (mol. wt. 1000 Da) with a biological half-life of 16–20 min. When the ADH–neurophysin complex is released from the posterior pituitary gland into the bloodstream, it dissociates immediately. ADH is rapidly metabolized by tissue peptidases, and about one-third is excreted by the kidneys.

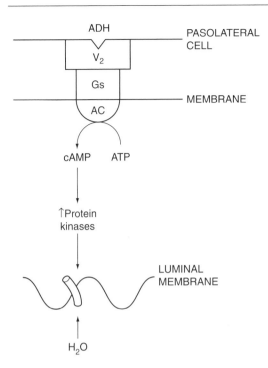

Figure 11.9 *Actions of antidiuretic hormone (ADH) on renal collecting duct cells. AC, adenyl cyclase; Gs, stimulatory G protein.*

ADH controls the reabsorption of water by the kidneys and constricts arterioles. Two vasopressin receptors have been identified: one (V_1) found in vascular smooth muscle that mediates vasoconstriction, and another (V_2) found in the distal tubules and collecting ducts of the kidney that mediates the antidiuretic effects. The major action of ADH is an increase in the water permeability of the apical membrane of the principal cells of the cortical and medullary collecting duct (Fig. 11.9). ADH binds to the V_2 receptor located at the basolateral membrane of the above cells, and increases cAMP and protein phosphorylation in the principal cells of the renal tubules. The proteins formed fuse with the apical membrane and dilate the intercellular spaces and increase the permeability of the epithelium at the lumen. As a result of the increased permeability of the apical membrane, increased water reabsorption leads to a decreased volume of urine with increased osmolarity. ADH also reduces renal medullary blood flow.

ADH is a potent vasoconstrictor. It acts on V_1 receptors in vascular smooth muscle cells to increase cytoplasmic calcium via DAG and IP_3 and stimulates vasoconstriction. The vasoconstrictor effects are important in the maintenance of arterial blood pressure during hypotensive states. ADH also stimulates platelet aggregation and degranulation by acting on V_1 receptors and the release of factor VIII by the endothelium by V_2 receptors.

In addition, ADH may have central nervous system effects, promoting memory, learning, attention and concentration, and also stimulates the release of ACTH from the anterior pituitary.

CONTROL OF ADH SECRETION

Changes in osmotic concentration and blood pressure both influence the release of ADH. The strongest stimulus for ADH secretion is hyperosmolality. Osmoreceptor cells, closely located to the PVN and SON, promote ADH secretion in response to a 1–2 per cent increase in effective plasma osmotic pressure, with a threshold starting at 280 mOsm/L. An increase of the extracellular osmotic concentration as little as 3 mOsm causes depolarization of the magnocellular neurons of the supraoptic nucleus of the hypothalamus, which triggers the release of ADH from the posterior pituitary. Sodium and mannitol are potent stimuli for ADH release, while hyperglycaemia is a less potent stimulus. However, the control of ADH secretion can be over-ridden by hypovolaemia.

Hypovolaemia (decreased effective blood volume) is a more potent stimulus to ADH release than hyperosmolality. When there is a decrease in blood volume that is greater than 10–25 per cent. The rate of ADH secretion is inversely proportional to the rate of firing of the baroreceptors. In most mammals, the low-pressure receptors in the atria are dominant sensors, but in humans and primates the baroreceptors in the carotid sinus and aortic arch are most important. ADH is also released in response to increased release of angiotensin II.

Pain, stress, emotion, exercise and drugs (morphine, nicotine, barbiturates) increase ADH secretion. Alcohol inhibits ADH secretion, leading to diuresis.

Oxytocin

Oxytocin is a nonapeptide that is synthesized in the cell bodies of peptidergic neurons of the paraventricular nuclei and stored in the posterior pituitary. Cholinergic stimulation causes oxytocin secretion.

Oxytocin stimulates the contraction of the lactating mammary gland myoepithelium during suckling, and ejection of milk results. It also stimulates contraction of the myometrium of the pregnant uterus. The sensitivity of the uterus in late pregnancy is increased. Oxytocin may have a role in labour. Ethanol, enkephalins and β-sympathetic nervous system activity inhibit oxytocin release.

PANCREATIC ISLETS

The islet of Langerhans are scattered collections of cells and form 1–2 per cent of the mass of the pancreas. There are four cell types: (1) α cells, which make up 20–25 per cent of the islet cells and secrete glucagon; (2) β cells, which constitute 75 per cent of the islet cells and secrete insulin; (3) δ cells, which make up 5 per cent of the islet cells and secrete somatostatin; and (iv) F cells, which make up 5 per cent of the islet cells and secrete pancreatic polypeptide.

Insulin

Insulin is synthesized by the β cells. The ribosomes translate mRNA at the endoplasmic reticulum to form preproinsulin (mol. wt. 11 500 Da). The preproinsulin is then cleaved in the endoplasmic reticulum to form proinsulin (mol. wt. 9000 Da). The proinsulin is cleaved in the Golgi apparatus to form insulin, which is stored in granules.

STRUCTURE OF INSULIN

Insulin is a polypeptide (mol. wt. 5734 Da) consisting of two chains. The A chain (21 amino acids) and B chain (30 amino acids) are linked by disulphide bridges (Fig. 11.10).

Insulin is stored in the β cell granules as a crystalline hexamer complexed with zinc. Insulin is secreted by exocytosis when extracellular calcium ions enter the β cell and bind to calmodulin. The plasma half-life of insulin is 5 min.

CONTROL OF INSULIN SECRETION

Control of insulin secretion is via several mechanisms and mediators:

- *Carbohydrates.* Glucose is the principal stimulus for insulin release, with a threshold of approximately 5–6 mmol/L blood concentration. Glucose enters the β cells through the glucose carrier GLUT-2 and is phosphorylated by glucokinase. Glucose metabolism in the β cells increases intracellular ATP. ATP-sensitive K^+ channels are responsible for the resting membrane potential in the β cells. ATP blocks the K^+ channels, depolarization of the cell membrane occurs, and voltage-dependent calcium channels open leading to an influx of calcium ions and exocytosis of insulin. The secretion of insulin occurs in two phases: an initial rapid phase due to release of stored insulin and a slow, sustained phase due to secretion of both stored and newly synthesized insulin.
- *Amino acids.* Arginine, leucine, lysine and phenylalanine are potent stimulators of the initial phase of insulin release.
- *Other hormones.* Other islet hormones are known to influence insulin release. Glucagon stimulates insulin release, whilst somatostatin inhibits insulin release. After feeding, gut hormones such as gastric inhibitory peptide (GIP), gastrin, secretin and cholecystokinin stimulate insulin release. Leptin inhibits both basal- and glucose-mediated insulin release.
- *Neural influences.* Unmyelinated postganglionic sympathetic and parasympathetic fibres modulate α, β and δ cell function by the secretion of neurotransmitters. Acetylcholine, which is released when blood glucose is elevated, stimulates insulin release and inhibits somatostatin secretion. Catecholamines acting on α_2 receptors inhibit insulin secretion.

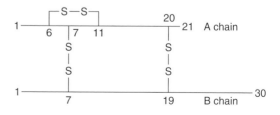

Figure 11.10 *Structure of insulin.*

METABOLISM OF INSULIN

Insulin is secreted into the portal system, where it has a half-life of 5 min. Some 80 per cent of the insulin released is degraded in the liver and kidneys. Hepatic glutathione insulin transhydrogenase splits the insulin molecule into A and B chains.

INSULIN RECEPTORS

The insulin receptor, located in the cell membrane, is made up of two α and two β chains held together by disulphide bridges. Each α chain (mol. wt. 130 000 Da) contains an extracellular insulin binding site. The β chain (mol. wt. 90 000 Da), which lies within the cell membrane, has tyrosine kinase activity.

Insulin binds to the α subunits on the cell membrane and insulin activates the tyrosine kinase in the β subunit (Fig. 11.11). Binding to the receptor results in autophosphorylation of the receptor. The active receptor phosphorylates cytosolic protein kinases and phosphatases which regulate various metabolic cascades.

The number and binding affinity of insulin receptors may be regulated by a number of factors. Receptor numbers and affinity are decreased by exposure to high concentrations of insulin, β-adrenergic agonists and glucocorticoids. They are also reduced in obesity and diabetes mellitus.

The effects of insulin are time-dependent. Within seconds of insulin binding to its receptors, increased permeability of the membrane to glucose occurs in muscle and adipose cells, but not in the brain, red blood cells and liver. This occurs because insulin stimulates the recruitment of glucose transporter protein (GLUT-4) from cytoplasmic vesicles to the plasma membrane. In addition, the cell membrane also becomes more permeable to amino acids, phosphate and potassium ions. Over the next 10–15 min, the activity of the phosphorylated intracellular enzymes is enhanced. It is suggested that the prolonged effects are from new proteins formed by mRNA translation.

PHYSIOLOGICAL ACTIONS OF INSULIN

Insulin is an anabolic hormone, which in general affects carbohydrate, fat and protein metabolism by causing enhanced cellular uptake, increased storage and decreased breakdown.

Carbohydrate metabolism

The actions of insulin on glucose metabolism are:

- Increased glucose uptake by enhanced facilitated diffusion via recruitment of glucose transport proteins (GLUT-4) in most tissues, except the renal tubules, red blood cells, intestinal mucosa and brain (excluding the hypothalamus), which are normally permeable to glucose
- Increased glycogen synthesis via enhanced glycogen synthase activity
- Reduced glycogenolysis, especially in the liver
- Reduced gluconeogenesis
- Diminished release of glucose by the liver

The enhanced uptake of glucose by the liver cells is due to the increased activity of glucokinase. Glucokinase phosphorylates glucose after it diffuses

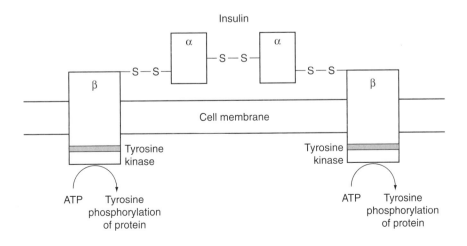

Figure 11.11
Insulin receptor action.

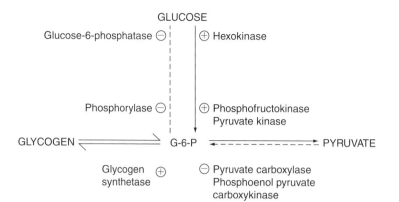

Figure 11.12 *Effects of insulin on carbohydrate metabolism.*

intracellularly and the phosphorylated glucose is held in the liver cell. Glycogen synthesis is enhanced due to the increased activity of glycogen synthase which polymerizes glucose to form glycogen.

Liver phosphorylase is inactivated so that glycogenolysis is inhibited (Fig. 11.12). Phosphoenol pyruvate carboxykinase is inhibited by insulin and glucose, and this reduces gluconeogenesis by inhibiting the reaction between pyruvate and phosphoenolpyruvate.

In muscle tissues, facilitated diffusion of glucose (via GLUT-4) into skeletal and cardiac muscle cells is enhanced by insulin. Insulin activates glycogen synthetase and phosphofructokinase so that glycogen storage and glucose utilization is promoted.

The diffusion of glucose into fat cells is enhanced by insulin by facilitating the mobilization of glucose transporters in the cell membrane. Glucose is utilized by fat cells to form fatty acids and α-glycerophosphate. This leads to esterification of the fatty acids and to the formation of triglycerides.

Protein metabolism

Insulin is an important protein anabolic hormone. Protein stores in the body are increased by the following mechanisms: (1) increased tissue uptake and decreased oxidation of amino acids; (2) increased protein synthesis; and (3) diminished protein breakdown.

The amino acids that are dependent on insulin for cellular uptake are valine, leucine, tyrosine and phenylalanine. An increased translation of mRNA in the ribosomes produced by insulin promotes protein synthesis (Fig. 11.13), which is further enhanced by increased DNA transcription in the nucleus.

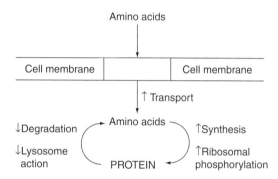

Figure 11.13 *Effects of insulin on protein metabolism.*

Fat metabolism

Insulin is a lipogenic as well as an antilipolytic hormone (Fig. 11.14). Overall, its effects on fat metabolism are: (1) enhanced glucose entry into fat cells leading to increased fat synthesis; (2) increased clearance of fat in the blood by enhanced low-density lipoprotein receptor activity and activation of lipoprotein lipase; (3) increased fatty acid and α-glycerophosphate synthesis; (4) decreased ketogenesis (especially in the liver); and (5) inhibition of hormone-sensitive lipase and thus reduced triglyceride breakdown, especially in adipose tissue.

Insulin enhances glucose transport into liver cells, where the glucose is converted to glycogen. Additional glucose is utilized to form fat when glucose is split to pyruvate in the glycolytic pathway and then converted to acetyl CoA from which fatty acids are synthesized. The liver is more important quantitatively for fat synthesis than is adipose tissue. The fatty acids thus synthesized are stored as

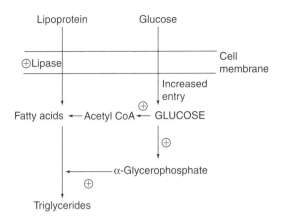

Figure 11.14 *Effects of insulin in adipose tissue.*

triglycerides in the liver. Insulin also promotes the synthesis and release of lipoprotein lipase.

The most important consequence of the enhanced glucose uptake by adipose cells is an increased synthesis of α-glycerophosphate, which is used for esterification of fatty acids. α-Glycerophosphate is also derived from glucose. Hormone-sensitive lipase is also inhibited by insulin, and this reduces the hydrolysis of triglycerides stored in fat cells.

An enhancement also occurs in the activity of lipoprotein lipase, which splits the triglyceride into fatty acids so that they can be absorbed and stored by fat cells. Lipoprotein lipase is therefore important in the formation of low- and high-density lipoprotein.

In muscle, the main effect of insulin is an increase in the facilitated transport of glucose, although it also increases glycogen synthesis and glycolysis.

Electrolyte shifts

Insulin decreases membrane permeability to Na^+, resulting in membrane hyperpolarization which causes a shift of K^+ from the extracellular to the intracellular compartment. Thus, insulin lowers serum K^+, predominantly due to K^+ uptake by muscle and hepatic tissue.

Glucagon

Glucagon is a 29-amino acid polypeptide (mol. wt. 3485 Da) which is synthesized as proglucagon

in the α cells of the islets. It has a plasma half-life of 5–10 min and is degraded by the liver.

CONTROL OF GLUCAGON SECRETION

The secretion of glucagon is regulated by metabolic by-products and gastrointestinal hormones. Blood glucose concentration is the most important factor controlling glucagon secretion, which is stimulated by hypoglycaemia and inhibited by hyperglycaemia. Amino acids, e.g. arginine, alanine, serine and glycine, also stimulate glucagon secretion, whilst fatty acids inhibit glucagon release. Cholecystokinin, gastrin and secretin all stimulate glucagon secretion, whilst glucocorticoids and β-adrenergic agonists enhance glucagon secretion. Somatostatin inhibits glucagon release.

Exercise, stress and some drugs (e.g. theophylline) enhance the release of glucagon.

PHYSIOLOGICAL ACTIONS OF GLUCAGON

Glucagon interacts with specific receptors on the cell membrane coupled to adenyl cyclase by G stimulatory proteins, with pronounced effects on metabolism.

The major site of action of glucagon is the liver, where it enhances hepatic glycogenolysis and gluconeogenesis, resulting in increased blood glucose. Glycogenolysis in the liver is enhanced by the activation of adenyl cyclase in the hepatic cell membrane. This results in increased cAMP, which then activates protein kinases leading to enhanced phosphorylase a and b activity. Glycogen is thus broken down to glucose-1-phosphate and then dephosphorylated to glucose (Fig. 11.15). It inhibits glycolysis by inhibiting phosphofructokinase and pyruvate kinase so that glucose-6-phosphate levels in the liver rise, leading to glucose release from the liver.

Glucagon also increases gluconeogenesis. The uptake of amino acids by liver cells from the plasma is enhanced by glucagon. It stimulates the actions of gluconeogenic enzymes, especially pyruvate carboxylase and phosphoenolpyruvate carboxykinase, which converts pyruvate to phosphoenol pyruvate, a rate-limiting step in gluconeogenesis.

Glucagon is also an important lipolytic hormone as it activates lipase in adipose tissue via cyclic AMP. Therefore, glucagon causes a rise in

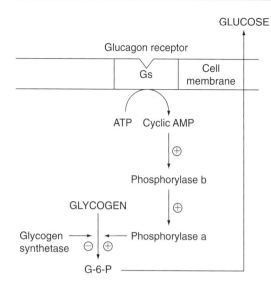

Figure 11.15   *Effects of glucagon on carbohydrate metabolism.*

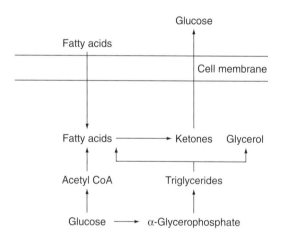

Figure 11.16   *Effects of glucagon on fat metabolism.*

plasma concentrations of fatty acids and glycerol (Fig. 11.16). Glycerol is utilized by the liver for gluconeogenesis, whilst fatty acids may be oxidized as an energy substrate. In the absence of insulin, glucagon increases ketogenesis by the oxidation of fatty acids.

Glucagon is a hormone responsible for protein catabolism because it increases amino acid oxidation and inhibits protein synthesis (Fig. 11.17).

The insulin/glucagon ratio

As insulin and glucagon produce opposing effects, the overall physiological response is determined

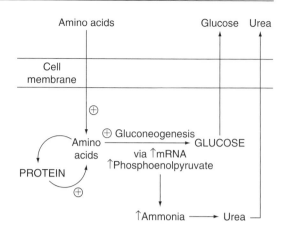

Figure 11.17   *Effects of glucagon on protein metabolism.*

by the relative levels of both hormones in the blood. In the fed state, the insulin/glucagon (I/G) ratio is approximately 30, falling to 2 after an overnight fast and to 0.5 after prolonged fasting.

CONTROL OF BLOOD SUGAR

The control of blood glucose is by a feedback system involving the pancreatic islets, liver, muscle and fat. Insulin is the main regulatory hormone responsible for fluctuations in blood glucose. Its main effect is to decrease blood glucose by increased glucose uptake into cells and enhanced glycogen synthesis, with decreased gluconeogenesis and glycogenolysis. When blood glucose is less than 3 mmol/L, glucagon, catecholamines, glucocorticoids and GH are secreted to increase the level. Glucagon increases blood glucose by enhanced glycogenolysis and gluconeogenesis. The glucocorticoids reduce glucose uptake and utilization by the cells, and increase blood glucose concentration. GH increases blood glucose by an anti-insulin effect. Severe hypoglycaemia stimulates the release of catecholamines which increase blood glucose by enhanced glycogenolysis and reduced glucose uptake.

Somatostatin

The δ cells of the islets secrete somatostatin, a polypeptide containing 14 amino acids. Somatostatin is also secreted by the hypothalamus, and has a plasma half-life of 3 min.

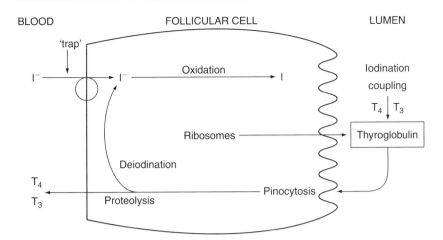

Figure 11.18 *Thyroid hormone synthesis.*

Somatostatin is released in response to increased blood glucose, amino acids and fatty acids. Its actions include the inhibition of the secretion of insulin, glucagon, pancreatic polypeptide and GH. It also reduces the motility of the stomach, duodenum and the gall bladder. Secretory processes in the gastrointestinal tract are also inhibited by somatostatin.

THYROID

The functional units of the thyroid gland are the follicles (acini). They are made up of a single layer of cuboidal epithelial cells with a central lumen filled with colloid, which is predominantly a glyco-protein, thyroglobulin. Scattered between the follicles are parafollicular cells (C cells) which secrete calcitonin.

Formation and secretion of thyroid hormone

The thyroid gland secretes two hormones: thyroxine (T_4, 93 per cent) and tri-iodothyronine (T_3, 7 per cent). Reversed T_3 (rT_3) is biologically inactive and is formed by peripheral conversion of T_4 by 5-deiodinase.

IODINE METABOLISM

A normal diet contains $500\,\mu g$ iodine per day and the minimum daily requirements of iodine are $120–150\,\mu g$. About $120\,\mu g$ iodine is taken up by

the thyroid each day, of which $80\,\mu g$ is secreted as thyroid hormones and $40\,\mu g$ is released into the plasma. Ingested iodine is converted to iodide and absorbed from the gut.

Iodide in the blood is actively taken up into the thyroid follicular cells. The basolateral membrane has a secondary active transport system of iodide (via both Na^+/I^- co-transporters and Na^+/K^+-ATPase), which actively pumps iodide into the interior of the cell, a process called iodide trapping (Fig. 11.18). The iodide pump concentrates the iodide to about 30 times its blood concentration. The transport mechanism is stimulated by TSH and inhibited by perchlorate and thiocyanate ions. A high circulating iodide concentration can also reduce the iodide trapping mechanism (Wolff–Chaikoff effect).

The oxidation of iodide to iodine is the first essential step in the formation of thyroid hormones. Thyroid peroxidase oxidizes iodide to iodine, with hydrogen peroxide accepting the electron. The peroxidase is located in the apical membrane of the follicular cell.

In the thyroid cell, the iodine rapidly binds to the 3 position of tyrosine in thyroglobulin within seconds, in the presence of iodinase enzyme, to form mono-iodotyrosine, and then di-iodotyrosine. These iodination reactions occur close to the apical membrane of the follicular cells.

Following iodination, oxidative condensation of two di-iodotyrosine residues forms T_4 and a serine residue. The condensation of mono-iodotyrosine and di-iodotyrosine leads to the formation of T_3 and a serine residue. These condensation reactions are catalysed by peroxidase enzyme.

The thyroid hormones are formed within the thyroglobulin molecule and stored in the follicular colloid. Thyroglobulin, a large glycoprotein which contains 70 tyrosine residues, is synthesized in the endoplasmic reticulum and Golgi apparatus of the follicular cell and secreted into the follicular cavity.

THYROXINE SECRETION

At their apical membrane, thyroid cells ingest colloid by endocytosis, forming vesicles. Within the vesicles, thyroglobulin is broken down by lysosomal enzymes (proteinases), which release thyroxine (T_4) and T_3. The hormones diffuse through the basolateral membrane of the thyroid cell and are released into the blood of the surrounding capillaries.

TRANSPORT OF T_4 AND T_3

The total plasma concentrations of T_4 and T_3 are 103 and 1.8 nmol/L, respectively. The plasma proteins that bind thyroid hormones are albumin, thyroxine-binding pre-albumin (TBPA), and thyroxine-binding globulin (TBG), a glycoprotein. T_4 is 99.9 per cent protein bound, predominantly to TBG (67 per cent) and TBPA (20 per cent), and has a half-life of 7 days. T_3 is 99.7 per cent protein bound, mainly to albumin (53 per cent) and TBG (46 per cent), and has a half-life of 24 h.

METABOLISM OF THYROID HORMONES

T_4 and T_3 are de-iodinated in the liver, kidney, skeletal muscle and other tissues. Some 45 per cent T_4 is converted to T_3 by 5'-deiodinase and 55 per cent is converted to reverse T_3 (rT_3) by 5-deiodinase.

REGULATION OF THYROID HORMONE SECRETION

Specific feedback mechanisms involving the hypothalamic–hypophysial–thyroid axis control the rate of thyroid secretion. Thyrotropin (TSH) is a glycoprotein hormone secreted by the anterior pituitary gland. The secretion of TSH is controlled by two factors: (1) thyrotropin-releasing hormone (TRH) and (2) the blood concentration of unbound T_4 and T_3.

TRH is a tripeptide secreted by the median eminence of the hypothalamus, and is transported to the anterior pituitary by the hypothalamic–hypophyseal portal system. TRH binds to receptors in the pituitary cell membrane and activates the phospholipase system to release phospholipase C, thereby increasing DAG and calcium ions and releasing TSH.

TSH increases all the secretory processes of the thyroid cells by increasing intracellular cAMP levels in the thyroid. The specific functions of TSH are: (1) increased iodine trapping, (2) increased iodination and coupling reactions to form T_4 and T_3, and (3) increased proteolysis of thyroglobulin, increasing the release of T_4 and T_3 into the circulation.

PHYSIOLOGICAL EFFECTS OF THYROID HORMONES

The thyroid hormones increase gene transcription in virtually all cells, increasing cellular enzymes, transport proteins and structural proteins.

T_4 is first deiodinated to T_3, which binds to intracellular receptors. This initiates gene transcription in the nucleus, producing mRNA. The mRNA is then translated in the cytoplasmic ribosomes, resulting in increased protein synthesis. The mitochondria in most cells also increase in size and numbers, together with an increase in cellular activity.

Excessive amounts of T_3 and T_4 can increase the basal metabolic rate (BMR) by 60–100 per cent. T_3 is three to five times more active than T_4. It is thought that this increase in BMR may be due to stimulation of the Na^+/K^+-ATPase enzyme by thyroid hormones. As a result of enhanced Na^+/K^+-ATPase activity, transport of Na^+ and K^+ ions through cell membranes is increased.

Following the overall increase in cellular enzyme activities caused by thyroid hormones, carbohydrate metabolism is enhanced by T_4 and T_3. There is rapid cellular uptake of glucose, increased glycolysis and gluconeogenesis and enhanced absorption from the gastrointestinal tract.

In the case of fat metabolism, T_4 increases lipid mobilization from adipose tissue, causing an increase in free fatty acids but a decrease in plasma cholesterol, phospholipids and triglycerides. In physiological amounts, T_4 and T_3 have a protein anabolic effect, but in large doses they have a catabolic effect.

Growth in childhood is dependent on the presence of thyroid hormones, which promote brain development in the intrauterine and neonatal periods. Thyroid deficiency may lead to mental retardation. In growing children, T_4 is important in the regulation of bone growth.

SYSTEMIC EFFECTS OF THYROID HORMONES

Thyroid hormones are essential for the normal function of the central and peripheral nervous systems. Hypothyroidism can lead to depression and psychosis and delayed peripheral reflexes.

Thyroid hormones have a direct chronotropic and inotropic effect on the heart. As a result of increased activity of all tissues, there is an increased systemic blood flow with a concomitant rise in cardiac output. Tachycardia is common in hyperthyroid states, but the mean arterial pressure is unchanged. As a result of vasodilatation, the diastolic pressure falls and pulse pressure is increased.

The rate and depth of respiration may increase with hyperthyroidism in response to increased O_2 consumption and CO_2 production. Thyroid hormones can increase appetite and food intake as well as gastrointestinal secretions and motility.

Thyroid hormones have mixed effects on muscles. Excessive amounts of T_4 can lead to muscle wasting due to enhanced protein catabolism, whilst hypothyroidism leads to sluggish muscle activity. A fine muscle tremor may be present in hyperthyroidism, and this may be due to increased excitability in the synapses in the spinal cord, controlling muscle tone.

Sexual dysfunction may be a symptom of thyroid dysfunction. In men, hypothyroidism may cause loss of libido, whilst hyperthyroidism may cause impotence. In women, hypothyroidism may lead to menorrhagia and polymenorrhoea, whilst hyperthyroidism causes oligomenorrhoea. These varied effects of thyroid dysfunction on sexual function are from direct effects on the gonads and both positive and negative feedback mechanisms on the anterior pituitary controlling sexual functions.

CALCIUM METABOLISM

The adult human body contains 25 mol of calcium, 90 per cent of which is in the skeletal system as hydroxyapatite, phosphates and carbonates. The daily turnover of calcium is 15 mol, mainly from the remodelling of bone.

Functions of calcium

Calcium has a number of biochemical functions:

- *Membrane excitation*. Excitable membranes of nerves and muscle contain specific calcium ionic channels. Calcium ions therefore control membrane excitability. An influx of calcium ions occurs during the excitation of nerves and muscles.
- *Haemostasis*. Calcium ions are necessary for activation of clotting factors in the plasma.
- *Muscle contraction*. Calcium ions are also essential for the excitation–contraction coupling of muscles.
- *Excitation–secretion processes*. An influx of calcium ions is required for the secretion of both endocrine and exocrine organs. Calcium is also necessary for the release of neurotransmitters.
- *Structural support*. Calcium is bound to cell surfaces and is important for membrane stability and intercellular adhesion. It is also an important component of bone, which acts as a store for both calcium and phosphate.

Distribution of calcium (Table 11.1)

Total body calcium may be regarded as being distributed in two major pools: (1) a readily exchangeable pool comprising 1 per cent of total body calcium which is in physiochemical equilibrium with the extracellular fluid compartment; this consists of calcium phosphate salts and acts as

Table 11.1   Distribution of calcium

Type of calcium	Percentage
Total diffusible	55
Ionized Ca^{++}	45
Complexed with HCO_3^-, citrate	10
Total non-diffusible	45
Bound to albumin	36
Bound to globulin	9

an immediate reserve for sudden changes in plasma calcium; and (2) a not readily exchangeable pool comprising 99 per cent of total body calcium and consisting of bone, which is not available for rapid mobilization. The plasma concentration of total calcium (ionized and non-ionized) is 2.5 mmol/L.

Plasma calcium is present as: (1) free or ionized calcium (45 per cent), (2) complexed with citrate, carbonate or hydrogen phosphate (10 per cent), and (3) protein (albumin)-bound calcium (45 per cent), which forms the non-diffusible fraction.

The ionized and complexed calcium constitute the diffusible fraction (55 per cent) of plasma calcium. In addition, the bony skeleton contains 1000 g of calcium, and this may also serve as a storage depot of phosphorus (80 per cent). By virtue of its carbonate, bicarbonate and phosphate content, bone can serve as a third-line defence in acid–base regulation.

Calcium metabolism: An overview

Calcium and phosphate metabolism is regulated by three hormones (parathyroid hormone (PTH), calcitonin and vitamin D) acting on bone (osteoblasts, osteoclasts and osteocytes), kidney and intestine.

In the gastrointestinal tract, calcium is absorbed by both passive (the amount absorbed depends on the plasma calcium concentration) and active mechanisms stimulated by 1:25-dihydroxycalciferol (Fig. 11.19).

In the kidney, 98 per cent of calcium filtered at the glomerulus is reabsorbed at the proximal tubule (60 per cent), the loop and distal tubules (40 per cent) under the influence of parathyroid hormone.

Bone formation is due to osteoblasts that lay down a collagen matrix with glycosaminoglycans mineralized by calcium and phosphate deposition, forming hydroxyapatite crystals. Osteoclasts reabsorb bone, releasing calcium, phosphate and matrix fragments. Local and systemic factors also regulate osteocyte numbers and activity.

Parathyroid hormone

Parathyroid hormone (PTH) is produced by four small glands located at the superior and inferior

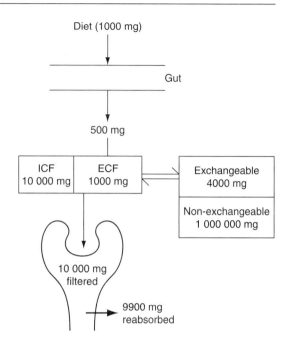

Figure 11.19 *Calcium metabolism. ECF, extracellular fluid; ICF, intracellular fluid.*

poles of both lobes of the thyroid gland. The parathyroid gland contains two cell types, namely, the chief cells that secrete PTH, and the oxyphil cells of unknown function.

PTH is a polypeptide containing 84 amino acid residues. It is formed in the ribosome as a preprohormone (110 amino acids) which is split to the prohormone (90 amino acids) and then to the hormone in the endoplasmic reticulum and the Golgi apparatus. Amino acid residues 1 and 2 of PTH are essential for the activation of the receptor, whilst residues 3 to 34 bind to the receptor.

REGULATION OF PTH

The rate of synthesis of PTH is controlled by extracellular ionized calcium; decreased extracellular ionized calcium increases PTH synthesis. The secretion of PTH is due to exocytosis of PTH granules. A decrease in extracellular calcium and β-adrenergic stimulation increase PTH secretion. The chief cells of the parathyroid possess a Ca^{++}-sensing receptor on their plasma membrane. When plasma Ca^{++} is increased, Ca^{++} binds to

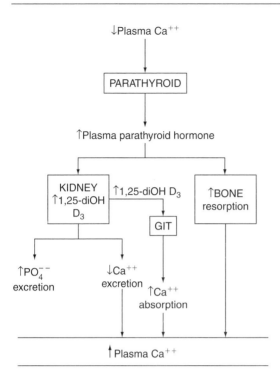

Figure 11.20   *Effects of parathyroid hormone.*

this receptor and activates G_q and reduces PTH secretion.

ACTIONS OF PTH

PTH increases ionized plasma calcium and lowers plasma phosphate concentration. It acts on bone, the kidney and, indirectly, on the gastrointestinal tract. Its actions on bone and kidney are mediated by cAMP, resulting in hypercalcaemia and hypophosphataemia.

PTH increases the rate of bone resorption by stimulating the activity of osteocytes and osteoclasts and this raises serum calcium concentration (Fig. 11.20).

Within a lag time of 2–3 h PTH increases the permeability of osteocytes so that their uptake of Ca^{++} from the bone interstitial fluid is enhanced with subsequent release of Ca^{++} into the capillaries.

A second slow phase is due to activation of osteoclastic activity mediated by the induction of the RANK (receptor activator of nuclear factor κβ) membrane receptors on osteoclasts. The

osteoclasts digest bone, releasing calcium and phosphate into the extracellular fluid.

PTH decreases the renal proximal tubular reabsorption of phosphate and increases active calcium reabsorption in the late distal tubules, the collecting tubules and the early collecting ducts. The enhanced reabsorption of calcium through the gastrointestinal tract is caused by increased formation of 1,25-dihydroxycalciferol due to PTH.

Vitamin D

Vitamin D is a steroid compound derived from cholecalciferol (D_3). 25-Hydroxycalciferol is the predominant circulating form, but this is inactive. 1,25-Dihydroxycholecalciferol is the active form.

SYNTHESIS

Cholecalciferol (vitamin D_3) is produced in the skin from 7-dehydrocholesterol by ultraviolet light. In the liver, cholecalciferol is hydroxylated to 25-hydroxycholecalciferol. In the kidney, the proximal nephrons convert 25-hydroxycholecalciferol to 1,25-dihydroxycholecalciferol by the action of renal 1-hydroxylase which is stimulated by PTH.

ACTIONS OF VITAMIN D

1,25-Dihydroxycholecalciferol raises plasma calcium concentration. It acts on the small intestine to promote the absorption of calcium and phosphate which is necessary for bone formation, and facilitates bone mineralization by increasing the extracellular fluid concentration of calcium and phosphate. In the intestine 1,25-dihydroxycholecalciferol increases the synthesis of a calcium-binding protein which promotes calcium absorption in the small intestine.

1,25-Dihydroxycholecalciferol, together with PTH, mobilizes calcium and phosphate from bone, thus raising plasma calcium and phosphate. This action is important for bone remodelling. Calciferol has receptors in the nucleus of osteoblasts. The calcitriol-receptor complex induces gene expression in bone and this leads to the synthesis of

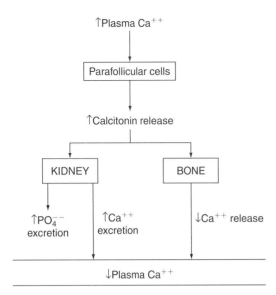

Figure 11.21   *Effects of calcitonin.*

matrix proteins (osteocalcin and osteopontin) in the osteoblasts.

Calcitonin

Calcitonin is secreted by the parafollicular (C) cells in the thyroid gland. It is a single-chain polypeptide, comprising 32 amino acids.

CONTROL OF CALCITONIN SECRETION

The secretion of calcitonin is increased when extracellular calcium rises to 2.4 mmol/L, and also by gastrin. The parafollicular C cells have calcium-sensing receptors. When Ca^{++} binds to these receptors, it stimulates calcitonin secretion by the C cells.

ACTIONS

Calcitonin decreases plasma calcium and phosphate levels principally by direct inhibition of osteoblasts (Fig. 11.21).

Calcitonin increases the renal excretion of phosphate and calcium by decrease their reabsorption in the kidney. Renal 1-α-hydroxylase activity is inhibited by calciferol, resulting in a decrease in the synthesis of 1,25-dihydroxycholecalciferol.

ADRENAL CORTEX

The adrenal cortex consists of three distinct zones of cells: (1) the zona glomerulosa, the outermost layer, which is the site of aldosterone, the principal mineralocorticoid, and corticosterone synthesis; (2) the zona fasciculata, the wider middle layer; and (3) the zona reticularis, the innermost layer, which functions as a unit synthesizing cortisol and some corticosterone, and the androgen dehydroepiandrosterone (DHEA) sulphate.

Chemistry of adrenocortical hormones

The adrenal cortex secretes two steroids: C_{19} steroids (androgenic); and C_{21} steroids (mineralocorticoid and glucocorticoid activity). The steroids are formed from cholesterol. Low-density lipoproteins in the blood are absorbed directly by endocytosis. Cholesterol is converted by 20, 22 desmolase (a rate-limiting enzyme) to pregnenolone, which is then hydroxylated to 17-OH-pregnenolone and then dehydrogenated to progesterone by 3-β-hydroxysteroid hydrogenase and isomerase in the endoplasmic reticulum.

The hydroxylation reactions following the production of pregnenolone and progesterone require NADPH and O_2. Hydroxylation at C_{21}, C_{11} and C_{18} of progesterone forms corticosterone and aldosterone, whilst hydroxylation at C_{17}, C_{21} and C_{11} forms cortisol (Fig. 11.22). As the zona glomerulosa lacks 17-α-hydroxylase, it cannot synthesize cortisol. The cells of the adrenal cortex produce and secrete the glucocorticoids, androgens and aldosterone on demand, rather than storing them.

The glucocorticoid activity of cortisol is due principally to the presence of keto-oxygen at C_3 and hydroxylation of C_{11} and C_{21}. The mineralocorticoid activity of aldosterone is due principally to the oxygen atom bound to C_{18}.

Glucocorticoids

SECRETION AND TRANSPORT

The adrenal cortex secretes two glucocorticoids, cortisol and corticosterone. About 10 times more

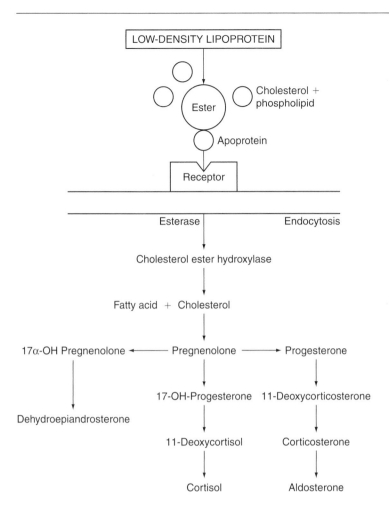

Figure 11.22 *Synthesis of steroids.*

cortisol compared to corticosterone is secreted during an average day. Cortisol is functionally the most important glucocorticoid.

Under physiological conditions, 75 per cent of plasma cortisol is bound to cortisol-binding globulin and 15 per cent to plasma albumin. The 10 per cent unbound is the physiologically active steroid. The plasma half-life of cortisol is 60–90 min, whereas that of corticosterone is about 50 min.

CONTROL OF ADRENOCORTICAL FUNCTION

There is a diurnal variation in ACTH levels, which are higher in the morning than in the evening. This accounts for the diurnal changes in plasma cortisol concentration. Physiological control of cortisol occurs via negative feedback loops. ACTH stimulates the secretion of cortisol by the adrenal cortex. This is brought about by the activation of adenyl cyclase by ACTH, resulting in the formation of cAMP. cAMP induces protein kinases which promote the conversion of cholesterol to pregnenolone, thus promoting the synthesis of glucocorticoids. The negative feedback of cortisol is exerted at the pituitary gland and the ventral diencephalon. Plasma cortisol concentrations exert control over ACTH secretion by the pituitary gland. The hypothalamus controls the anterior pituitary via the hypophyseal portal system by the release of adrenocorticotrophin releasing factor from the tubero-infundibular neurons. Acetylcholine and serotonin stimulate the release of corticotrophic-releasing hormone whereas epinephrine and γ-aminobutyric acid inhibit its release.

The normal hypothalamic–hypophyseal adrenocortical control may be stimulated by trauma, hyperpyrexia, hypoglycaemia, exercise and other stresses.

ACTIONS

The steroids bind to specific receptors in the nucleus, thereby promoting gene transcription, leading to increased mRNA and ribosomal translation and increased synthesis of some proteins and inhibition of others. Some of the effects of the steroid–receptor complex are due to the interaction with activator protein (AP-1), a transcription factor activator protein.

The glucocorticoids have numerous metabolic effects. With regard to carbohydrate metabolism, cortisol exerts an anti-insulin and a carbohydrate-sparing effect, leading to hyperglycaemia. The hyperglycaemia is due principally to enhanced gluconeogenesis by: (1) activation of DNA transcription in the liver cell nucleus with increased mRNA and thus increased synthesis of enzymes required for gluconeogenesis such as glucose-6-phosphatase, fructose-1-6-phosphatase, pyruvate carboxylase and phosphoenol carboxylase; and (2) increased protein catabolism (especially in the muscle), thereby making more amino acids available to the liver for gluconeogenesis. Cortisol also reduces glucose utilization by all cells (except brain cells) by decreased NADH oxidation. Glucose uptake is also reduced.

The most important effect of cortisol on protein is a catabolic effect. It increases the release of amino acids from proteins in muscle and other extrahepatic sites including the bone matrix, thus leading to diminished tissue stores and a negative nitrogen balance. In the liver, cortisol enhances amino acid transport into hepatocytes. Glucogenic amino acids, (especially alanine) are utilized for gluconeogenesis.

Cortisol is a direct lipolytic and this enhances cellular fatty acid oxidation. As a result of increased fatty acid oxidation, glucocorticoids may predispose to ketosis.

It also has an indirect effect, reducing glucose transport into fat cells and potentiating the lipolytic actions of other hormones such as GH, catecholamines, glucagon and thyroid hormone. In the liver, fatty acid synthesis is inhibited. The overall effect of cortisol is a redistribution of fat with a characteristic centripetal distribution.

Cortisol has a variety of effects on the haematological system. A mild increase in the number of red blood cells may result, together with increased platelets and neutrophils. However, cortisol may cause a decrease in lymphocytes and eosinophils.

In the stomach, cortisol causes increased acid and pepsin secretion. There is an increased tendency for peptic ulceration, which may be due to decreased prostaglandin synthesis required to maintain the mucosal barrier to acid and pepsin.

Cortisol increases the reactivity of peripheral blood vessels to catecholamines. This permissive effect leads to a positive inotropic effect.

The glucocorticoids also cause osteoporosis by diminished collagen synthesis by osteoblasts and increased collagen breakdown by collagenase, probably by modifying gene transcription.

Stress induces secretion of corticotrophin (ACTH), resulting in an increase in cortisol, and this mobilizes the labile proteins available for energy and synthetic process in damaged tissues.

The anti-inflammatory effects of cortisol are brought about in two ways: (1) by stabilizing lysosomal membranes and therefore reducing proteolytic enzymes; and (2) by decreasing capillary permeability, thus reducing capillary leakage and leukocyte diapedesis. This is associated with decreased bradykinin and histamine release.

The immune system is suppressed by cortisol. Some of the immunosuppressive actions may be due to inhibition of gene transcription for phospholipase A_2, which is involved in the formation of inflammatory mediators, e.g. leukotrienes, platelet-activating factor. Cytokines and complement are decreased due to inhibition of gene transcription. This is associated with decreased circulating lymphocytes, monocytes, basophils, eosinophils, T cells and antibodies. The release of interleukin-1 from white cells is also inhibited.

METABOLISM

The liver is the main site of corticosteroid metabolism. Cortisol is converted to cortisone by 11-OH steroid dehydrogenase and excreted as tetrahydrocortisone glucuronide. The keto substitution at C_{20} may be hydroxylated, conjugated and excreted as glucuronides.

Sex steroids

The adrenal cortex secretes four androgenic hormones: androstenedione, testosterone, DHEA and DHEA sulphate. DHEA and DHEA sulphate are converted to testosterone in the peripheral tissues. The adrenal cortex produces minute amounts of oestrogens, but its major role is the supply of androstenedione and DHEA as substrates which may be converted to oestrogen by fat, mammary gland and other tissues.

Aldosterone

Aldosterone is a C_{21} corticosteroid and is the major mineralocorticoid produced by the zona glomerulosa in humans. Aldosterone has a half-life of 20 min, and is transported mainly by plasma albumin and by corticosteroid-binding globulin (10 per cent).

The biosynthesis of aldosterone is enhanced by ACTH, angiotensin II (via the phosphatidylinositol system) and increased plasma K^+ (Fig. 11.23). About 90 per cent of aldosterone is inactivated by a single passage through the liver by the reduction of the double bond of the A ring of the steroid structure of aldosterone. Tetrahydroaldosterone is formed and conjugated with glucuronide, and excreted readily by the kidney.

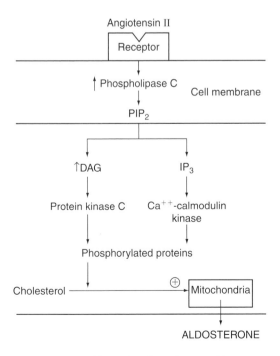

Figure 11.23 *Effects of angiotensin II on aldosterone synthesis.*

PHYSIOLOGICAL EFFECTS OF ALDOSTERONE

Aldosterone binds to intracellular receptors that induce DNA transcription, producing proteins that increase Na^+/K^+-ATPase in the basement membrane of cells in the kidney, colon and bladder.

In the kidney, aldosterone enhances Na^+ reabsorption in the collecting segment of the distal tubule and the cortical and medullary collecting tubules in exchange for K^+ and H^+ which are excreted. Some 1–2 per cent of the Na^+ filtered through the glomerulus is actively reabsorbed by the aldosterone-dependent mechanism in the distal nephron. Aldosterone combines with cytoplasmic receptors to form a protein complex. The aldosterone–protein complex increases protein synthesis, which include the Na^+/K^+-ATPase pump at the basolateral membrane and sodium and potassium channels in the apical membrane of the tubular cells. Aldosterone also increases the number of K^+ channels in the apical membrane of the distal tubular cells of the kidney, and thus enhances K^+ secretion. More than 75 per cent of K^+ excreted in the urine is due to distal K^+ secretion. Aldosterone also promotes the renal excretion of H^+ and NH_4^+ in the distal tubule. Chloride reabsorption is increased, leading to hyperchloraemic alkalosis. Thus, aldosterone act on the principal cells of the collecting duct to increase sodium reabsorption and potassium secretion. Aldosterone also promotes Na^+ reabsorption in the sweat glands, salivary glands and gastrointestinal mucosa of the distal colon.

Together with increased Na^+ reabsorption, aldosterone produces increased H_2O reabsorption, leading to an increased extracellular fluid volume (ECF). However, the increase in ECF volume is limited to about 5–15 per cent. An 'escape' phenomenon, secondary to a release of atrial natriuretic factor caused by the initial expansion of the ECF, causes a decrease in Na^+ reabsorption in the proximal tubule.

CONTROL OF ALDOSTERONE SECRETION

Aldosterone secretion is controlled by three well-defined mechanisms: ACTH, serum K^+, and the renin–angiotensin system. ACTH enhances aldosterone synthesis by catalysing the conversion of cholesterol to pregnenolone.

A 1 per cent increase in plasma K^+ can promote the release of aldosterone from the zona glomerulosa. This is probably due to the depolarization of the zona glomerulosa cell membrane by raised plasma K^+.

A 10 per cent decrease in plasma Na^+ can also stimulate the release of aldosterone from the zona glomerulosa. However, this stimulus may be over-ridden by changes in ECF volume. Aldosterone secretion may be increased in a patient with hyponatraemia and hypovolaemia, but decreased in a patient with hyponatraemia and hypervolaemia.

Intrarenal control mechanisms regulate aldosterone secretion by the renin–angiotensin system which is controlled by the sympathetic nervous system. Renin, with a half-life of 40–120 min, is released by the juxtaglomerular cells in response to a decrease in effective circulating blood volume due to acute hypovolaemia or hypotension. Renin combines with angiotensinogen, an α_2 globulin, to produce angiotensin I, an inactive decapeptide. Angiotensin I is then converted to angiotensin II by angiotensin-converting enzyme. Angiotensin II, with a half-life of 1–3 min, stimulates the secretion and synthesis of aldosterone by the zona glomerulosa. Angiotensin II also is a potent vasoconstrictor.

ADRENAL MEDULLA

The adrenal medulla is derived from neural crest tissue, and functionally is an analogue of sympathetic postganglionic fibres of the autonomic nervous system. Essentially, the adrenal medulla represents an enlarged sympathetic ganglion and is innervated by long sympathetic preganglionic cholinergic neurons that synapse with the cells of the adrenal medulla.

The adrenal medulla contains two types of modified ganglion cells called chromaffin cells. These contain granules of either norepinephrine or epinephrine associated with soluble acidic proteins called chromagranin, lipids and ATP. Some 80 per cent of the chromaffin cells have large and less dense granules containing epinephrine, and 20 per cent have small and dense granules containing norepinephrine. The preganglionic sympathetic fibres that innervate the adrenal medulla are derived from T5–T9.

Adrenomedullary hormones

The adrenal medulla synthesizes and secretes bioamines, namely the dihydroxyphenolic amines, epinephrine and norepinephrine, and metenkephalin. Epinephrine is mainly produced in the adrenal medulla, with small amounts being synthesized in the brain. Norepinephrine is produced to a lesser extent in the adrenal medulla, and is widely produced and distributed in all neural tissues. Most of the circulating norepinephrine is derived from peripheral nerve terminals. In early fetal life, the adrenal medulla contains predominantly norepinephrine.

CATECHOLAMINE SYNTHESIS

The catecholamines are synthesized from L-tyrosine, which may be present in the diet or derived from the hydroxylation of L-phenylalanine in the liver. Tyrosine is hydroxylated in the cytoplasm by tyrosine hydroxylase to L-dopa (3,4-dihydroxyphenylalanine). Dopa decarboxylase coverts dopa to dopamine in the cytoplasm. The dopamine is taken up by the chromaffin granules, where it is converted to norepinephrine by dopamine β-hydroxylase. Some 20 per cent of the chromaffin cells of the adrenal medulla contain norepinephrine; in the other 80 per cent of cells, norepinephrine diffuses into the cytoplasm where it is N-methylated by phenylethanolamine-N-methyl transferase (PNMT). S-adenosyl methionine is the methyl donor, and epinephrine is synthesized. The catecholamines are bound to ATP and chromogranin (a soluble protein) and stored as specific granules (chromaffin granules).

CONTROL OF CATECHOLAMINE SECRETION

Catecholamine release is stimulated by the release of acetylcholine from preganglionic sympathetic nerve fibres in the greater splanchnic nerve. Following depolarization of the chromaffin cells, calcium influx occurs resulting in exocytosis and release of catecholamines.

Stress factors leading to the 'fight or flight' reaction activate the sympathetic system. These include trauma, pain, haemorrhage, exercise, hypoglycaemia and anxiety. During hypoglycaemia, the adrenal medulla is activated selectively. Angiotensin II also potentiates the release of catecholamines.

INACTIVATION OF CIRCULATING CATECHOLAMINES

The plasma half-lives of epinephrine and norepinephrine are 10–15 and 20–30 s, respectively. The rapid termination of the actions of catecholamines is brought about by non-enzymatic and enzymatic mechanisms.

The non-enzymatic mechanisms include: (1) extraneuronal uptake (tissue uptake) of the catecholamines in the lungs, liver, kidney and gut; and (2) neuronal uptake by sympathetic nerve endings that are able actively to take up amines from the circulation, resulting in intraneuronal storage. A proportion of the catecholamines taken up by the nerve endings are also inactivated by monoamine oxidase (MAO), a non-specific deaminase found in the mitochondria of the liver, kidney, stomach and small intestine, and in the cytoplasm of the nerve endings. MAO catalyses the oxidative deamination of circulating and intraneuronal catecholamines. 3,4-Dihydroxymandelic acid is formed from epinephrine and norepinephrine; the methylated metabolites of epinephrine and norepinephrine produce 3-methoxy-4-hydroxymandelic acid by oxidative deamination by the action of MAO.

Catechol-O-methyl transferase (COMT) is an extraneuronal enzyme found mainly in the liver and kidney, but also in postsynaptic membranes. S-adenosylmethionine is required as a methyl donor, producing normetaphrine from norepinephrine and metaphrine from epinephrine, and 3-methoxy-4-hydroxymandelic acid (also known as vanillylmandelic acid or VMA) from 3,4-dihydroxymandelic acid.

Only 2–3 per cent of circulating catecholamines are conjugated with sulphuric or glucuronic acid and excreted directly into the urine. Most of the catecholamines are excreted as their deaminated metabolites, VMA and 3-methoxy-4-hydroxy phenyl glycol (MOPG). As a majority of the urinary metabolites are derived from norepinephrine, urinary VMA and MOPG reflect the activity of the sympathetic nervous system rather than that of the adrenal medulla. Plasma or urinary concentrations of epinephrine are better indices of adrenal medullary function.

ADRENORECEPTORS

In 1913, Dale showed that epinephrine had two distinct effects, namely, vasoconstriction in certain vascular beds (causing a rise in blood pressure) and vasodilatation in others. In 1948, Ahlquist postulated that the catecholamines acted on two receptors, termed alpha (α) and beta (β). α Receptors mediated vasoconstriction and had the order of agonist potency norepinephrine > epinephrine > isoprenaline, while β receptors mediated vasodilatation and showed agonist potency of isoprenaline > epinephrine > norepinephrine. The use of selective antagonists confirmed Ahlquist's original classification. Subsequent studies with agonists and antagonists confirmed the existence of subtypes of both α and β receptors.

Cloning studies have demonstrated that all adrenoreceptors are G protein-coupled receptors with seven transmembrane α-helical segments, with three extracellular and three intracellular loops. The receptor is linked to a guanine nucleotide binding protein. Each of these pharmacological groups is generally associated with a specific messenger system.

The α_1 receptors are coupled to phospholipase C and produce their effects by releasing intracellular calcium. Molecular cloning has confirmed the existence of at least four subtypes of α_1 receptors (α_{1A}, α_{1B}, α_{1C}, α_{1D}). α_2 Receptors inhibit adenylate cyclase and reduce cAMP; molecular cloning has identified at least three subtypes of α_2 receptors.

Molecular cloning has demonstrated at least three subtypes of β receptors: β_1, β_2, β_3. All three types of β receptors stimulate adenylate cyclase, leading to an increase in cAMP. Activation of adenylate cyclase is mediated by the stimulatory coupling G protein.

The main effects of receptor activation may be summarized as follows:

- α_1 *Receptors*. Vasoconstriction, relaxation of gastrointestinal smooth muscle, contraction of the genitourinary smooth muscles.
- α_2 *Receptors*. Inhibition of norepinephrine release from nerve endings, platelet aggregation, vascular smooth muscle contraction.
- β_1 *Receptors*. Increase in heart rate and myocardial contractility, relaxation of gastrointestinal smooth muscle.

- β_2 *Receptors.* Bronchodilatation, vasodilatation, relaxation of visceral smooth muscle, hepatic glycogenolysis and muscle tremors.
- β_3 *Receptors.* Lipolysis in fat cells, especially brown fat tissues.

BIOCHEMICAL EFFECTS OF CATECHOLAMINES

The main biochemical effect on carbohydrate metabolism is increased glycogenolysis. In the liver, glycogenolysis is increased by epinephrine (β_2 effect) as it enhances glycogen phosphorylase and inhibits glycogen synthetase. Liver glycogenolysis is enhanced as epinephrine suppresses insulin secretion and stimulates glucagon secretion.

Glycogenolysis in the muscle is also increased by a β_1 adrenergic mechanism which stimulates adenyl cyclase and thus enhances cAMP stimulation of glycogen phosphorylase. Glucose-6-phosphate is produced, but as muscle lacks glucose-6-phosphatase, this does not directly increase blood glucose. The glucose-6-phosphate is converted to lactate or pyruvate, which becomes a major precursor for hepatic gluconeogenesis.

In physiological concentrations, epinephrine does not have a direct glycogenolytic action. However, epinephrine in high concentrations causes hyperglycaemia by several mechanisms such as increased glycogenolysis in the liver and increased glucagon secretion. The increased glucagon causes increased hepatic gluconeogenesis from increased lactate and pyruvate produced by muscle glycogenolysis, and inhibition of glucose uptake by suppression of glucose transporter proteins in the cell membranes of skeletal and cardiac muscle and fat cells.

With regard to fat metabolism, epinephrine activates intracellular lipase via cAMP (β adrenergic effect) and hence stimulates lipolysis. Free fatty acids are mobilized from fat tissues and form a substrate for ketogenesis in the liver.

MAJOR PHYSIOLOGICAL EFFECTS

The different adrenoreceptors mediate a large number of physiological effects. The predominant action of the catecholamines on the heart is mediated by β_1 receptors, resulting in an increase in the heart rate (chronotropic effect) and the force of contraction (inotropic effect) giving rise to an increased cardiac output and myocardial oxygen consumption. Electrical conduction is enhanced, and this can result in cardiac arrhythmias (dromotropic effect).

Vascular smooth muscle contraction caused by α_1-receptor stimulation results mainly from the release of intracellular calcium via IP_3. The skin and splanchnic vascular beds are markedly constricted. Large arteries, arterioles and veins are also constricted, resulting in decreased vascular compliance, increased peripheral vascular resistance and increased central venous pressure. The cerebral, coronary and pulmonary vasculature are relatively unaffected. Overall, these vasoconstrictor effects lead to an increase in systolic and diastolic arterial pressure, activating baroreceptor reflexes and reflex bradycardia.

β_2 stimulation increases intracellular cAMP and produces vascular smooth muscle relaxation. β_2-mediated vasodilatation is most marked in the vascular bed of the skeletal muscle, although it can be demonstrated in other vascular beds. Bronchial and uterine smooth muscles are dilated by the activation of β_2 receptors.

In the eye, the radial muscles of the iris contract as a result of α_1 stimulation, as do the smooth muscles of the vas deferens and the splenic capsule.

The marked inhibitory effects of the sympathetic nervous system on the gastrointestinal smooth muscle are mediated by both α and β receptors. Part of this inhibitory effect is mediated by activation of presynaptic α_2 receptors in the myenteric plexus. The α receptors on the smooth muscle cells cause hyperpolarization due to increased potassium permeability, and therefore this results in inhibition of action potential discharge. The sphincters of the gastrointestinal system contract via α-receptor activation.

The physiological effects of the catecholamines on skeletal muscle are mediated by β receptors. The most significant effect is glycogenolysis mediated by β_1 receptors; β_2 effects are less marked. The twitch tension of the slow red muscle fibres is reduced, whereas the twitch tension of the fast contracting white fibres is increased. Muscle tremors may also be present with β_2 activity, probably as a result of an increase in muscle spindle discharge causing instability in the reflex control of muscle length.

As described earlier, the metabolic effects of the catecholamines promote the conversion of energy stores such as glycogen and fat to glucose and free fatty acids by β activity. Lipolysis is increased by β_3 receptor activity, and the free fatty acids produced may be utilized for energy production.

Histamine release by mast cells is inhibited by the catecholamines, probably mediated by β_2 receptors. Platelet aggregation is enhanced by an α_2-mediated action of the catecholamines. Lymphocyte proliferation and lymphocyte-mediated cell killing are also inhibited by catecholamines but the clinical significance of these effects is unknown.

ERYTHROPOIETIN

Erythropoietin is a glycoprotein hormone of 165 amino acids with a molecular mass of 30 kDa. The carbohydrate moiety of the hormone prevents its breakdown by the liver.

About 75–90 per cent of circulating erythropoietin in the adult is produced by the kidneys, and the remainder by the liver. In the fetus and the neonate, the liver is the chief source of erythropoietin. Synthesis in the kidneys is localized to the peritubular interstitial cells, whilst in the liver the centrilobular hepatocytes are the main source of erythropoietin. The normal serum level of erythropoietin is 5–25 μg/mL, and the plasma half-life is estimated to be 3–8 h. Hepatic degradation is the main route of elimination, although renal excretion and catabolism may also contribute.

Factors controlling erythropoietin production

The primary physiological stimulus for erythropoietin production is hypoxia or decreased availability of oxygen in the tissues sensed by the renal peritubular interstitial cells. The production of erythropoietin increases exponentially with decreasing arterial oxygen tension, which is responsible for the increased erythropoietin levels found in natives of high-altitude regions.

A decrease in oxygen-carrying capacity associated with chronic anaemia is also associated with increased erythropoietin levels. This becomes significant when the haemoglobin concentration

falls below 10.5 g/dL. However, with anaemia associated with renal failure, decreased erythropoietin production may be due to non-functioning renal tissue.

Local mediators such as renal prostaglandins PGE_1 and PGE_2 also increase the release of erythropoietin. Thyroid hormone, GH and the androgens have been shown to increase erythropoietin levels. A circadian variation in erythropoietin production is also present, with decreased production between midnight and 4:00 AM, suggesting that the hypothalamus may influence erythropoietin production.

Mechanism of action

Erythropoietin is the main regulator of red cell production in the bone marrow. Red cell development is from bone marrow stem cells and is regulated by several growth factors. The bone marrow stem cells differentiate to form burst-forming unit-erythroid cells (BFU-E) which progressively become more responsive to erythropoietin.

Erythropoietin stimulates mitosis in the BFU-E cells, thereby increasing the number of red cell precursors. However, erythropoietin appears to exert its major effect on the next stage of red cell production by preventing DNA breakdown so that proerythroblasts are formed. Erythropoietin is also important for regulating the rate of maturation of red blood cells (Fig. 11.24). Erythropoietin shortens the time between recruitment of precursor stem cells and release of reticulocytes.

ATRIAL NATRIURETIC FACTOR

Atrial natriuretic factor (ANF) is a peptide that is produced by the atrial muscle cells and which exerts a physiological influence on the kidneys.

Chemistry

ANF is a 28-amino acid peptide that is synthesized and secreted by the atria. It is stored in the atrial myocyte as a 126-amino acid prohormone. When secreted, the prohormone is cleaved into an inactive N-terminal portion of 98 amino acids

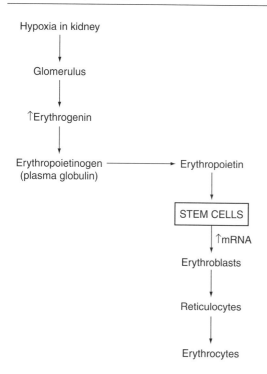

Hypoxia in kidney

↓

Glomerulus

↓

↑Erythrogenin

↓

Erythropoietinogen ————————→ Erythropoietin
(plasma globulin)

↓

STEM CELLS

↓ ↑mRNA

Erythroblasts

↓

Reticulocytes

↓

Erythrocytes

Figure 11.24 *Synthesis and effects of erythropoietin.*

(N-ANF) and the active ANF. The half-life of ANF is 2.5 min.

Control of secretion

Atrial wall stress and stretch are the predominant stimuli for ANF release, There is a linear relationship between plasma ANF levels and atrial pressure.

Physiological effects of ANF

The physiological effects of ANF involve the kidney, the adrenal cortex and the peripheral vascular system:

- *Renal effects.* The main effect of ANF is an increased glomerular filtration rate (GFR). This is brought about by constriction of the efferent arteriole and dilation of the afferent arteriole of the glomerulus, thus increasing the hydrostatic pressure in the glomerular capillaries. As a result of the increase in GFR as well as an increase in renal medullary blood flow, an increased urinary sodium excretion (natriuresis) occurs. An increase in the delivery of sodium and chloride to the macula densa, as well as an increase in the hydrostatic pressure at the juxtaglomerular apparatus secondary to the afferent arteriolar vasodilation, causes an inhibition of renin secretion.
- *Adrenal secretion of aldosterone.* ANF blocks the secretion of aldosterone that is stimulated by angiotensin II, ACTH and cyclic AMP.
- *Peripheral vasculature.* Vasorelaxation of the smooth muscles of the peripheral blood vessels leads to a decrease in mean systemic arterial blood pressure. This is partly brought about by the suppression of renin release.

Brain natriuretic peptide

Brain natriuretic peptide (BNP) is a 32-amino acid peptide that is structurally similar to ANF and shares the same guanylate cyclase receptors on the endothelial cells. BNP was first isolated in porcine brain but subsequently has been found in cardiac tissue. BNP is secreted predominantly from the ventricular muscle cells in response to ventricular dilation, although smaller amounts may be released from atrial cells.

The physiological actions of BNP are similar to those of ANF, resulting in enhanced diuresis and natriuresis, vasodilatation and a reduction in aldosterone.

SEX HORMONES

The endocrine control of the reproductive system involves the sex steroids from the gonads, the hypothalamic trophic peptides and the gonadotrophins from the anterior pituitary.

The sex steroids

OESTROGENS

The oestrogens are synthesized mainly by the ovary and the placenta during pregnancy, and in small amounts in the adrenal cortex and in the testes in males. Oestradiol is the principal and most active oestrogen. Oestrone and oestriol are weak oestrogens. FSH is the main hormone controlling oestrogen secretion. The oestrogens are important for the growth and maintenance of the fallopian tubes, uterus, vagina and external genitalia in the female, as well as the endometrial lining. Oestrogens also increase contractile proteins in the myometrium and uterine contractility, cervical mucus production and ductal development in the breast. The oestrogens promote hepatic synthesis of transport proteins and closure of the epiphyseal plate. The mechanism of action of the oestrogens is by binding of the steroid–receptor complex to DNA, the result being gene transcription.

PROGESTOGENS

Progesterone is the natural progestogen secreted in large amounts by the corpus luteum and the placenta during pregnancy, and in small amounts by the adrenal cortex and the testes in the male. Progesterone enhances the secretory function of the endometrium by proliferation of the endometrial glands and decreases the frequency and strength of myometrial contractions. It also enhances alveolar development in the breast.

ANDROGENS

Testosterone is the main androgen produced by the testes in the male. It is produced in small amounts by the adrenal cortex and by the ovary in the female. Cholesterol is the chief substrate for testosterone synthesis. The reproductive functions of the androgens are the control of spermatogenesis, the maturation of sex organs and development of secondary sex characteristics. Testosterone has anabolic effects in muscle by promoting cell division, growth and maturation, resulting in an increase in muscle strength.

Reflections

1. Hormones are released by glands and transported in blood to act on distant target organs. There are three categories of hormones: peptides (the largest group), steroids (derivatives of cholesterol), and tyrosine derivatives (catecholamines, thyroid hormones). The peptides and catecholamines are transported in free solution in the plasma, while steroid and thyroid hormones are largely bound to plasma proteins. Hormones act by binding to specific receptors in their target cells. Peptides and catecholamines bind to plasma membrane receptors whereas steroid and thyroid hormones enter cells and bind to intracellular receptors and modulate gene expression. The secretion of most hormones is under negative feedback control except the secretion of gonadotrophins which is subject to positive feedback regulation during the preovulatory phase of the female reproductive cycle.

2. The anterior pituitary gland secretes growth hormone, thyroid-stimulating hormone, adrenocorticotrophic hormone, follicle-stimulating hormone, luteinizing hormone, prolactin and a number of related peptides. The hypophyseal portal blood vessels carry hormones synthesized in the median prominence region of the hypothalamus to the anterior pituitary gland. Releasing hormones control the release of anterior pituitary hormones except prolactin which is inhibited by dopamine. The hypothalamo–hypophyseal axis is regulated by feedback mechanisms involving its target organs.

3. Growth hormone (GH) is a peptide hormone secreted by somatotrophs of the anterior pituitary gland. GH secretion is stimulated by hypothalamic GH-releasing factor and suppressed by hypothalamic somatostatin. The effects of GH-releasing factor are dominant. GH exerts a wide range of direct metabolic actions and also indirect actions mediated by somatomedins or insulin-like growth factors synthesized in the liver. GH promotes protein synthesis and has anti-insulin effects that are glucose sparing. GH also promotes lipolysis which provides a non-carbohydrate source of substrate for ATP generation. GH is essential for skeletal growth in children and maintains tissues in adults. Skeletal growth occurs in response to somatomedins

(IGF) which enhance division of cartilage cells and increase the deposition of cartilage at the epiphyses (growth plates). The most important metabolic stimulus for GH secretion is hypoglycaemia. GH output is also increased by increased plasma concentrations of amino acids and reduced concentrations of free fatty acids.

4. The posterior pituitary gland secretes two peptide hormones; vasopressin (ADH, antidiuretic hormone) and oxytocin. These hormones are synthesized in the cell bodies of the neurons in the paraventricular and supraoptic nuclei. Although vasopressin and oxytocin are structurally similar, they have very different actions. Vasopressin is released in response to an increase in the osmotic pressure of plasma or a decrease in blood volume (greater than 10 per cent). Vasopressin enhances the reabsorption of water from the medullary collecting ducts of the nephrons. It also causes vasoconstriction with higher plasma concentrations. Oxytocin is released by a neuroendocrine reflex in response to suckling. It stimulates the ejection of milk from the lactating breast and increases the contractile activity of the uterine myometrium.

5. The follicular cells of the thyroid gland secrete T_3 and T_4 (thyroxine) while the parafollicular cells secrete calcitonin. The thyroid hormones play an important role in metabolism, maturation of the skeleton, and maturation of the central nervous system. TSH secreted by the anterior pituitary gland controls all activity of the thyroid gland. Active uptake of iodide by the follicular cells concentrates iodide within the cell. Iodide is oxidized to iodine and then incorporated into thyroglobulin. Thyroid hormones are released from the gland by enzymatic hydrolysis of iodinated thyroglobulin. T_3 and T_4 are transported in blood bound to carrier proteins including thyroxine- binding globulin. In the tissues, T_4 is converted to T_3. Thyroid hormones increase oxygen consumption in most tissues (except the brain, spleen and gonads) which is important for the maintenance of body temperature. Thyroid hormones increase ventilation, cardiac output and erythropoiesis. The metabolic actions of thyroid hormones are dose dependent. Low concentrations of thyroid hormones cause hypoglycaemia while higher concentrations stimulate glycogenolysis and gluconeogenesis. Thyroid hormones cause lipolysis and stimulate the oxidation of free fatty acids. Low concentrations of thyroid hormones stimulate protein synthesis while higher concentrations are catabolic. Normal thyroid hormone levels are necessary for the normal development and maturation of the skeletal and nervous tissue.

6. Insulin is synthesized in the β cells of the pancreatic islets. It is an amino acid and binds to a plasma membrane receptor which activates tyrosine kinase. It is secreted chiefly in response to a rise in the plasma glucose and amino acids. It stimulates the uptake of glucose by most cells via the GLUT-4 transporter especially in adipose tissue and skeletal muscle. It stimulates both glycogen and protein synthesis.

7. Glucagon is secreted by the α cells of the pancreatic islets and promotes the release of glucose into the blood by stimulating glycogenolysis, lipolysis and gluconeogenesis. Hypoglycaemia is a potent stimulus for the secretion of glucagon.

8. The adrenal gland consists of an outer adrenal cortex and an inner medulla. The cortex secretes glucocorticoids, mineralocorticoids and a small amount of sex hormones. Adrenocortical hormones are transported in blood in combination with plasma proteins such as albumin and a specific protein called transcortin. Glucocorticoids have important metabolic actions and are vital for the body's response to stress. They stimulate gluconeogenesis and glycogen production. In general they have an anti-insulin effect. They also stimulates lipolysis. High levels of glucocorticoids reduce lymphoid tissue mass and suppress the immune response to infection, and have effects on the central nervous system causing depression or euphoria by an unknown mechanism. Aldosterone is the chief mineralocorticoid and acts to conserve body sodium by stimulating its absorption in exchange for potassium in the distal nephron. The adrenal medulla is composed of chromaffin cells innervated by cholinergic terminals of the

splanchnic nerve; secretes epinephrine and norepinephrine and cause increases in heart rate, contractility and cardiac output. Epinephrine causes vasodilation and reduces diastolic blood pressure, and it reduces gut motility, promotes glycogenolysis, lipolysis and increases oxygen consumption. It is also a potent bronchodilator.

9. Parathyroid hormone is a peptide hormone secreted by the parathyroid gland in response to a fall in plasma calcium. PTH increases mobilization of bone, and stimulates the reabsorption of calcium in the distal tubules of the kidney while decreasing the reabsorption of phosphate by the proximal tubules. PTH enhances the production of 1,25-dihydroxy-cholecalciferol which increases intestinal calcium absorption. Overall, PTH increases plasma calcium concentration.

10. Calcitonin is a peptide secreted by the parafollicular cells of the thyroid gland and it decreases plasma calcium levels.

11. Dihydroxycholecalciferol is an active metabolite of vitamin D. It increases the absorption of dietary calcium by the intestine and enhances the turnover of bone so that old bone is resorbed and new bone is laid down.

12. Calcium plays an important role in many aspects of cellular function and is a major structural component of the bony skeleton. Plasma concentrations of calcium and phosphate are regulated by actions of parathyroid hormone, active metabolites of vitamin D and calcitonin. These hormones act on bone, gut and kidney to regulate the movement of calcium into and out of the extracellular pool.

Metabolism, nutrition, exercise and temperature regulation

After studying this chapter the reader should be able to:

1. Explain the principles of nutrition and describe the metabolic pathways involved
2. Explain the physiological changes during starvation
3. Define basal metabolic rate and describe how it is measured and the factors that influence it
4. Describe the cellular mechanisms that produce heat and energy
5. Explain the concept of energy equivalent of oxygen for carbohydrates, fats and proteins

6. Describe the energy requirements of muscles during exercise and how there are met
7. Describe the cardiorespiratory and metabolic responses to exercise
8. Explain the importance of thermoregulation and the consequences of hypothermia and hyperthermia
9. Describe the responses of the body to cold including behavioural changes, shivering, vasoconstriction and non-shivering thermogenesis

METABOLISM

NORMAL ENERGY METABOLISM

Metabolism can be defined as the sum of the chemical changes that occur in the cell and involve the breakdown and synthesis of stored energy sources. Metabolism can be divided into two basic processes: (1) anabolism, involving the synthesis of cellular macromolecules, and (2) catabolism, the breakdown of energy stores to adenosine triphosphate (ATP) and reducing equivalents for cell function and provision of precursors for anabolism. The balance between the anabolic and catabolic processes is by the actions of hormones, which coordinate and control tissue responses. Acute control of these processes is via hormones that rapidly modify the activity of existing enzymes, whilst chronic control is effected by the amount of enzymes in the cells.

A continuous supply of energy is required for the survival of the body, and this is provided by oxidation of exogenous organic molecules. The oxidation process requires an adequate supply of oxygen. Metabolic fuels which exist as macromolecules in the diet (carbohydrates, fats and proteins) are hydrolysed during digestion so that they can be absorbed as simple units and utilized, although fats are resynthesized immediately after absorption. These simple units are then oxidized to provide energy (with nitrogen excreted as urea, carbon as CO_2 and hydrogen as water), or converted into various storage forms or synthesized into membranes, enzymes, etc. The metabolic fuels may therefore be derived either from digestion of a meal, or from the breakdown of internal stores such as storage organelles or cell constituents.

In a normal person, 85–90 per cent of energy requirements is provided by the oxidation of carbohydrate and fat (approximately in equal proportions) to CO_2 and water. Protein oxidation to CO_2, water and nitrogen containing end-products (urea, the principal end-product) provides about 10–15 per cent of the energy needs. The energy is used to supply basal requirements (maintenance of cell membrane potentials, respiration and heart beat), in thermoregulation, and in the performance of external work.

ENERGY COMPOUNDS

The basic chemical currency of energy in all living cells consists of the two high-energy phosphate bonds contained in adenosine triphosphate (ATP). To a lesser extent, other purine and pyrimidine nucleotides (guanosine triphosphate, cytosine trophosphate, inosine triphosphate) also serve as energy sources after energy from ATP is transferred to them. The brain and muscle can use the high-energy phosphate bond in creatine phosphate as a back-up store for energy.

NADH, NADPH and flavoproteins are intermediate energy storage compounds that conserve energy as ATP. They function as coenzymes and are normally reoxidized to generate ATP. One NADH or NADPH is equivalent to three ATP, and one reduced flavoprotein is equivalent to two ATP. During a catabolic reaction, a large fall in energy leads to the formation of NADH, an intermediate energy fall forms reduced flavoprotein, whilst a small energy fall produces ATP. Generally, the efficiency of energy conservation is approximately 60 per cent in a resting human. One-third of the total energy is used for ion transport and the synthesis of neurotransmitters by the brain and nervous tissue. Mechanical work such as muscular contraction and chemical processes such as synthesis of storage compounds (e.g. glycogen) or cell constituents (e.g. enzymes) also utilize ATP. Although approximately 100 mol of ATP are used each day, only 25 mmol of ATP is present at any given moment, indicating the rapid turnover.

CATABOLIC PATHWAYS

Simple units of carbohydrate (hexoses), fat (free fatty acids) and protein (amino acids) undergo three phases of catabolism. At Phase 1, these simple units are partially oxidized to three major compounds (acetyl CoA, α-ketoglutarate and oxaloacetate) and three minor compounds (pyruvate, fumarate and succinyl CoA). Only one-third of the total energy is released by Phase 1 reactions. At Phase 2, complete oxidation of the products of Phase 1 by the citric acid cycle produces CO_2, with the remaining two-thirds of the total energy being released. Energy may be conserved as NADH, reduced flavoprotein and small amounts of ATP. At Phase 3, the reduced enzymes are reoxidized and their hydrogen released as water, with energy transferred to ATP by phosphorylation of ADP (this is usually termed 'oxidative phosphorylation').

The hormonal profile controlling catabolism is that plasma insulin is decreased while the concentration of catecholamines, glucagon, growth hormone and glucocorticoids is elevated.

Phase 1 reactions

The combustion of fatty acids, the major energy component of fats, commences with their activation to CoA derivatives such as palmitoyl-CoA. Palmitoyl-CoA must be first converted to palmitoyl-carnitine by carnithine-palmitoyl transferase in the outer mitochondrial membrane before it can enter the mitochondria. At the inner mitochondrial membrane, palmitoyl-carnitine is reconverted to palmitoyl CoA and then oxidized by β-oxidation which releases two carbon compounds as acetyl CoA until the entire fatty acid molecule is broken down. β-Oxidation of free fatty acids provides a major source of acetyl CoA, an important substrate for the citric acid cycle. Free fatty acids in blood, derived from the diet or by the action of lipoprotein lipase on lipoproteins at the endothelial cell

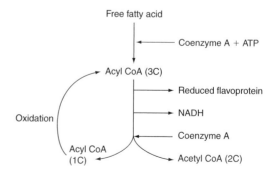

Figure 12.1 *β-oxidation of free fatty acids.*

layer of the tissue, are oxidized in the mitochondria. Growth hormone and glucocorticoid increase the mobilization of fat stores by increasing the amount of triglyceride lipase. Initially, free fatty acid is converted to acyl CoA utilizing one ATP. Acyl CoA is oxidized to acetyl CoA, and the residual carbon atoms re-enter the cycle to produce more acetyl CoA (Fig. 12.1). This partial oxidation of free fatty acids produces hydrogen ions that are removed as NADH and reduced flavoproteins.

The partial oxidation of carbohydrates (also known as glycolysis) in the form of glucose is the second major source of acetyl CoA. Glycogen stores in the liver and muscle are mobilized by the activity of glycogen phosphorylase to produce glucose. This mobilization occurs as a result of raised epinephrine or glucagon concentrations, or decreased insulin concentration. The other sources of glucose are from absorption in the gastrointestinal tract or gluconeogenesis.

Glycolysis is the pathway by which glucose units are broken down to pyruvate in all tissues to provide energy. Glucose is converted to fructose diphosphate, utilizing one ATP. The 6-carbon units are then cleaved into two triose phosphate units, which are then partially oxidized to pyruvate (Fig. 12.2). The pyruvate enters the mitochondrion, where it is further oxidized to acetyl CoA. Some energy is conserved as NADH and ATP. Oxidation of pyruvate to acetyl CoA is irreversible, as animal cells cannot synthesize glucose from acetyl CoA. In the liver, glycolysis is inhibited by raised plasma glucagon concentration, as a result of activation of protein kinase.

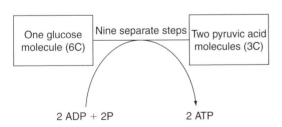

Figure 12.2 *Simplified outline of glycolysis.*

Glycolysis serves as a sole source of energy only briefly because the body stores of glucose are limited and pyruvate that is accumulated is reduced to lactate.

The control of protein breakdown in the liver is by raised plasma glucagon concentration, and in muscle and liver by elevated glucocorticoid concentration. The combustion of proteins first requires hydrolysis to its component amino acids. Each amino acid undergoes degradation by individual pathways which ultimately produces intermediate compounds of the citric acid cycle and then to actyl CoA and CO_2.

The catabolism of amino acids involves oxidative deamination. The first step involves transamination where the amino group is removed, leaving a carbon moiety. The amino groups pass through several amino acids and finally form glutamate and aspartate. The glutamate is converted to ammonia by glutamate dehydrogenase. Ammonia, together with aspartate and CO_2, enters the urea cycle to form urea, utilizing two ATPs (Fig. 12.3).

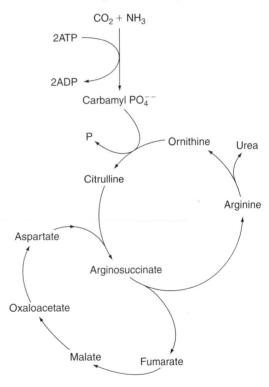

Figure 12.3 *The urea cycle.*

As muscle tissue cannot synthesize urea, transaminated amino acids from the muscle must be transported to the liver by the glucose–alanine cycle (Fig. 12.4). The carbon residues of the amino acids are partially oxidized to form intermediates of the citric acid cycle (such as α-ketoglutarate, oxaloacetate, fumarate, succinyl CoA) or glycolysis (pyruvate, acetyl CoA).

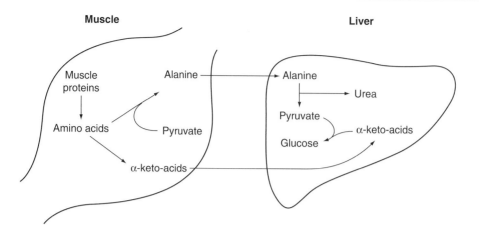

Figure 12.4 *The glucose–alanine cycle.*

Phase 2 reactions: The citric acid cycle

The citric acid cycle (Fig. 12.5) is an important aerobic metabolic reaction within the mitochondria as it provides reduced coenzymes required for the final phase, oxidative phosphorylation. Simple catabolic units of fat, carbohydrate and protein are channelled through this cycle so that complete oxidation of the compounds occurs, resulting in carbon dioxide formation; in addition, hydrogen

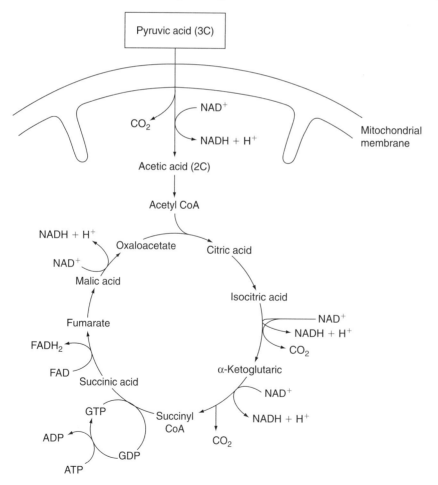

Figure 12.5 *The citric acid cycle.*

is removed as NADH and reduced flavoproteins. The chief precursor entering into the citric acid cycle is acetyl CoA, which is formed from free fatty acids and carbohydrates. Overall, the cycle results in the oxidation of one acetate to two CO_2 molecules. Products of the catabolism of amino acids can enter the cycle via α-ketoglutarate, fumarate, succinyl CoA or oxaloacetate. In mammalian tissues, the main limitation of this pathway is the redox state especially the $NADH/NAD^+$ ratio. NADH inhibits the dehydrogenases of the cycle.

Phase 3: oxidative phosphorylation

Oxidative phosphorylation consists of a sequence of reactions within the mitochondrion that results in the re-oxidation of the reduced coenzymes to produce NAD^+ and flavoprotein (Fig. 12.6). The oxidation of reducing equivalents by the respiratory chain is coupled to the formation of ATP by the flow of protons extruded across the inner mitochondrial membrane and then re-entering the matrix. NADH is oxidized to NAD^+, and three ATP are formed by the phosphorylation of ADP. As the reduced flavoproteins are reoxidized at a later stage, only two ATPs are formed. Oxygen is involved in the last reaction, where it is reduced to H_2O. The oxidation rate is controlled by the $NADH/NAD^+$ ratio in the mitochondrion. Some 95 per cent of the ATP produced by the body is derived from oxidative phosphorylation, the remaining 5 per cent being derived from the phosphorylation of glycolytic substrates (phosphoenolpyruvate, diphosphoglycerate) and the conversion of succinyl CoA to succinate in the citric acid cycle.

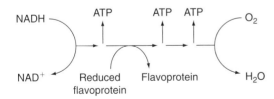

Figure 12.6 *Outline of oxidative phosphorylation.*

Anaerobic glycolysis

As glycolysis is the only sequence of metabolic reactions that does not require oxygen, it is able to operate anaerobically to produce pyruvate and release two ATP (Fig. 12.7). In the absence of oxygen, oxidative phosphorylation ceases so that NADH is not re-oxidized and therefore accumulates. The citric acid cycle also ceases, the acetyl CoA is not utilized, and as a result NADH is used to convert pyruvate to lactate. The NADH is re-oxidized in this process and lactate accumulates. ATP is produced by substrate phosphorylation during glycolysis, which is the only pathway able to operate under anaerobic conditions to produce energy. Substrate phosphorylation does not occur in the catabolism of amino acids. The lactate produced is utilized by the liver to resynthesize glucose.

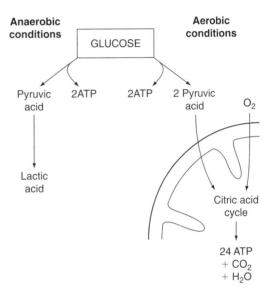

Figure 12.7 *Energy yield in anaerobic and aerobic conditions.*

Gluconeogenesis

Gluconeogenesis is an important metabolic process by which glucose is synthesized from non-carbohydrate precursors derived from fat and protein metabolism, for example lactate, pyruvate, glycerol and amino acids. The major site for this pathway, is in the cytoplasm of the liver, although it can occur in the kidney to a limited extent (Fig. 12.8).

The distinct reactions that occur in gluconeogenesis include the hydrolytic reactions converting glucose-6-phosphate to glucose (catalysed by glucose-6-phosphatase), fructose-1-6-diphosphate

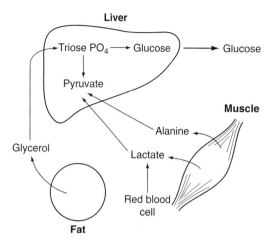

Figure 12.8   *Tissues involved in gluconeogenesis.*

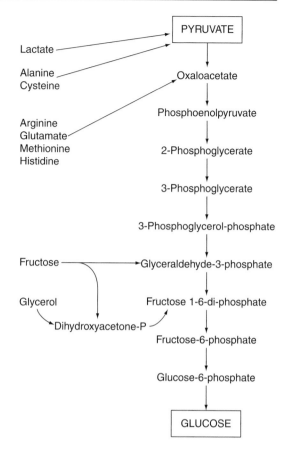

Figure 12.9   *Gluconeogenesis: entry points.*

to fructose-6-phosphate, and the conversion of pyruvate to phosphoenolpyruvate (Fig. 12.9). The conversion of pyruvate to phosphoenolpyruvate occurs in two separate reactions that require the enzymes pyruvate carboxylase and phosphoenolpyruvate carboxykinase. Pyruvate carboxylase enhances the conversion of pyruvate to oxaloacetate, which is then converted to phosphoenolpyruvate by phosphoenolpyruvate carboxykinase. The gluconeogenetic enzymes are present in the cytoplasm of the cells, except pyruvate carboxylase, which is present in the mitochondria.

The gluconeogenic pathway may be considered to start with the generation of alanine from muscle, lactate from muscle and other tissues and glycerol from fat tissues. Lactate is formed mainly in the muscles during exercise, whilst glycerol is released into the blood as a result of the hydrolysis of triacylglycerol from fat tissues and phosphorylated to glycerol-3-phosphate which is converted to dihydroxyacetone phosphate. Amino acids also provide a source of precursors for gluconeogenesis. In the liver, alanine and lactate derived from muscle are converted to pyruvate and therefore are important for gluconeogenesis. During starvation, alanine and glutamine derived from the breakdown of skeletal muscle are important precursors for gluconeogenesis.

The regulatory hormones for gluconeogenesis are glucagon and glucocorticoids, which have a promotional effect, and insulin, which has an inhibitory effect.

Ketone body formation and utilization

Ketone bodies (primarily acetoacetate and β-hydroxybutyrate) are formed in the liver, especially when large amounts of acetyl CoA are formed by β-oxidation (Fig. 12.10). The production of ketoacids results from an imbalance between the flow of fatty acids into mitochondria of hepatocytes and the capacity of the citric acid cycle to remove acetyl CoA. Two acetyl CoA units condense to form acetoacetate, which may be reduced with NADH to produce β-hydroxybutyrate. The liver is the only organ with the enzymes required for ketone body formation, but it cannot utilize ketone bodies as it lacks the oxo-acid-CoA transferase enzyme that catalyses the transfer of CoA from succinyl CoA to acetoacetic acid. Acetoacetate and β-hydroxybutyrate are released by the liver and are then utilized by skeletal and cardiac muscle and the kidney in preference to glucose. Glucagon and

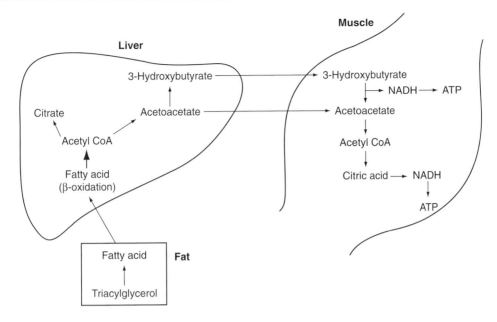

Figure 12.10   *Ketone acid synthesis.*

epinephrine stimulate lipolysis and increase ketone body formation. Nervous tissues can also utilize ketone bodies, especially during fasting. β-Hydroxybutyrate is converted into acetoacetate, which enters the citric acid cycle via acetyl CoA.

ANABOLIC PATHWAYS

Anabolism is promoted when there is an elevated plasma concentration of insulin relative to plasma concentrations of glucagon, catecholamines and glucocorticoids. Insulin is the main hormone responsible for the synthesis of energy stores (glycogen, triglycerides) and protein synthesis.

Glycogen synthesis

Glycogen is synthesized from glucose (especially in the liver and muscle) when excess glucose is available, and this requires the enzyme glycogen synthase. A high blood sugar concentration increases the influx of glucose into the liver, although insulin is required to promote the entry of glucose in the muscle. In addition, insulin stimulates glycogen synthase so that glycogen formation is enhanced. In a normal adult, the liver stores 70–100 g of glycogen, whilst muscle stores approximately 400 g. The amount of

glycogen stored in the brain is relatively small, and can provide anaerobic glycolysis within the brain for approximately 4 min.

Glycogen breakdown (glycogenolysis) to glucose is controlled by epinephrine and glucagon. Both hormones activate glycogen phosphorylase, which converts glycogen to glucose. Glucose released from the liver is utilized by other tissues. Muscle glycogen can be used only by muscle because muscle lacks the enzyme glucose-6-phosphatase which is required for the release of glucose into the blood.

Fat synthesis

A major function of fat synthesis is to store the chemical energy of foods as triglycerides when there is an excess of chemical energy above the immediate requirements of the body. Further fat synthesis is important for the formation of phospholipids, which are essential components of cell membranes.

Free fatty acid synthesis occurs mainly in the liver and adipose tissue. In the liver, the main precursor for fatty acid synthesis is endogenous glucose derived from glycogen, lactate and blood glucose. Pyruvate is the main source of acetyl CoA, and this process is enhanced by a raised plasma insulin concentration and lowered glucagon concentration. Acetyl CoA is an important substrate for the synthesis of free fatty acids under the control of acetyl-CoA-carboxylase. The acetyl-CoA is converted first to malonyl CoA and then to fatty acid. Citrate formed in the citric acid cycle diffuses out of the mitochondrion and splits into acetyl CoA and oxaloacetate in the cytoplasm (Fig. 12.11). The NADPH required for free fatty acid synthesis is supplied by the hexose monophosphate shunt and by the conversion of citrate to pyruvate in the cytoplasm. The hexose monophosphate shunt is highly active in the cytoplasm of the liver and adipose tissue.

FAT STORAGE

A major storage form of excess energy is as fats, or triacylglycerols. The synthesis of fat stores occurs in both the cytoplasm and the endoplasmic reticulum of the fat cells, utilizing acyl CoA and glycerol phosphate derived from glycolysis. The liver and adipose tissue are the main sites of fat synthesis, although it can also occur in the intestine during the absorption of fat.

The breakdown of fat in the liver and adipose tissue produces free fatty acids and glycerol. In the liver, the glycerol can be converted to glycerol

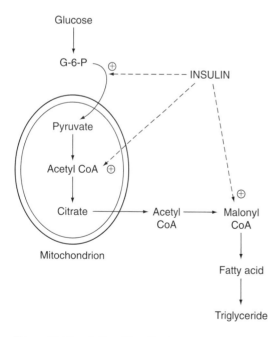

Figure 12.11 *Fatty acid synthesis.*

phosphate, but this does not occur in adipose tissue, which releases glycerol with free fatty acids.

Protein synthesis

Proteins are synthesized from amino acids. However, as essential amino acids cannot be synthesized within the body, protein synthesis may be limited by the availability of one or more of these. Proteins synthesized form muscle protein, serum proteins, connective tissue and enzymes. In fasting states, gluconeogenesis utilizes amino acids from the breakdown of muscle and liver proteins.

CONTROL OF METABOLIC PATHWAYS

Energy demands within the cell control the energy production. High ATP concentrations produced by the citric acid cycle and oxidative phosphorylation inhibit oxidative phyosphorylation, leading to a build-up of NADH. This will inhibit dehydrogenase enzymes and hence retard the citric acid cycle. When there is a deficiency of ATP, high concentrations of ADP can stimulate oxidative phosphorylation.

The concentration and activity of enzymes are important rate-limiting factors of metabolism. Hormones are important in regulating the activity and the rate of synthesis of enzymes. Glycolysis is controlled by hexokinase, phosphofructokinase and pyruvate dehydrogenase, and the activity of these enzymes is increased by insulin. Glycogen synthesis is enhanced by glycogen synthetase, promoted by insulin and inhibited by glucagon. Gluconeogenesis is controlled by fructose diphosphatase and pyruvate carboxylase, enhanced by glucagon. The citric acid cycle is controlled by dehydrogenases and citrate synthetase, which are both inhibited by NADH.

There are major differences in the fuels metabolized by different tissues. The liver is the only organ capable of taking up most compounds and performing most conversions between fuels. Skeletal muscle oxidizes free fatty acids and ketone bodies for energy production. Glucose becomes an important fuel for skeletal muscle only during hyperglycaemia or local anaerobic conditions. Skeletal muscle releases amino acids (mainly alanine and glutamine) from protein during starvation. In contrast, the main fuels for cardiac muscle are free fatty acids, lactate and ketone bodies. Brain and nervous tissue use glucose as a chief source of energy normally, but during fasting, acetoacetate and β-hydroxybutyrate may become important sources of energy. Adipose tissues store energy as fat and can convert glucose to triacylglycerols. Adipose tissues also accept triacylglycerols for storage. The breakdown of fat by adipose tissues to free fatty acids and free glycerol releases energy.

NUTRITION

NUTRIENTS

Nutrients are the chemical constituents of the diet, and consist of carbohydrates, fats, protein, electrolytes, minerals and vitamins. The kilocalorie is the unit of energy used in metabolic studies of humans. The standard unit of heat energy is the calorie, defined as the amount of heat energy necessary to raise the temperature of 1 g of water from 15 to 16°C. One kilocalorie (kcal, Cal) is equal to 1000 calories (4.18 kJ). The caloric value of food is defined as the overall energy released from its oxidation inside the body.

Carbohydrates

In affluent societies carbohydrates provide approximately 50 per cent of the total calories contained in the diet but this proportion may rise to 85 per cent in poor communities. Starch is the chief dietary carbohydrate in plant food sources, but sugar in the form of sucrose and lactose may also be present in food. A small quantity of glycogen in meat contributes a trivial amount of carbohydrate. Nearly all the dietary carbohydrate is oxidized to carbon dioxide and water. A normal diet may contain 300–500 g of carbohydrate; therefore, since the caloric value of carbohydrate is 4 kcal/g, this will provide a total of 1200–2000 kcal.

Fats

Fats form the second largest source of energy, providing approximately 40 per cent of our total calories. A normal daily diet usually contains 140 g of fat, providing 1260 kcal/day. Fat has a high caloric value (9 kcal/g), and thus provides a more concentrated form of energy. The most important source of energy from fats is the triglycerides. The greater part of dietary fat is oxidized in the body and excreted as CO_2. The urine contains traces of acetoacetic acid and β-hydroxybutyrate acid, intermediary products of fat metabolism which increase in starvation states and diabetes mellitus.

Proteins

Proteins are required in the diet to replace the protein lost continuously by catabolism – approximately 200–400 g protein per day. Although amino acids formed by protein breakdown are re-utilized, an additional protein intake of 20–40 g/day is required to maintain protein balance. Amino acids such as isoleucine, leucine, valine, lysine, methionine, threonine, phenylalanine and tyrosine are called essential amino acids because the body cannot synthesize them from other amino acids by transamination. The sulphur-containing amino acids, cysteine and methionine, provide the sulphur contained in proteins and other biologically important compounds and are the source of sulphates in the urine. The caloric value of protein is 4 kcal/g. It is recommended that the daily diet should contain 1 g protein per kg body weight to allow for the great variation in the minimal needs of individuals. Newborns require approximately five times as much protein as an adult for growth.

Vitamins

Vitamins are organic compounds that the body is unable to synthesize and which are components of enzyme systems. Vitamins may be classified into fat-soluble and water-soluble groups.

FAT-SOLUBLE VITAMINS

Fat-soluble vitamins are stored in large amounts in the liver, and excessive intake can lead to toxicity. The fat-soluble vitamins include vitamins A, D, E and K. Vitamin A can be formed in the gut from β-carotene, which is present in fruits and vegetables; it is essential for the retinal pigment (rhodopsin), normal bone formation, and epithelial tissue repair. Vitamin D is important for bone formation as it promotes calcium absorption from the gut and facilitates the deposition of bone by the formation of insoluble calcium salts. Vitamin E is a mixture of tocopherols, which are antioxidants. Vitamin K is necessary for prothrombin and other clotting factors in the liver.

WATER-SOLUBLE VITAMINS

Water-soluble vitamins circulate freely in the body, but are not stored in large amounts as they are readily excreted in the urine.

Vitamin C (ascorbic acid) is an important water-soluble vitamin that prevents scurvy (haemorrhage in the skin or internal organs). Vitamin B_1 (thiamine) is essential for pyruvate dehydrogenase activity and deficiency may lead to peripheral neuropathy or cardiac failure. Nicotinic acid is essential for NAD or NADP synthesis, and deficiency may lead to dermatitis and diarrhoea. Riboflavin is essential in the synthesis of flavoproteins, which are important for oxidative phosphorylation.

Vitamin B_{12} (cyanocobalamin) consists of a porphyrin ring containing cobalt, and absorption in the ileum requires intrinsic factor, a mucoprotein secreted by the stomach. The daily requirement for B_{12} is less than $1\,\mu g$, and its biological half-life is approximately 1 year. Vitamin B_{12} is required for nucleic acid synthesis and is concerned with the *de-novo* synthesis of the labile methyl group from 1-carbon precursors (e.g. glycine). Folic acid is usually associated with vitamin B_{12} activity, and is concerned with the movement of methyl groups from one acceptor to another. Vitamin B_{12} and folic acid are required for the maturation of red cells. In addition, vitamin B_{12} is essential for the integrity of myelin, and a deficiency may lead to peripheral neuritis and patchy degeneration of the dorsal and lateral columns of the spinal cord.

INTER-RELATION BETWEEN FAT AND CARBOHYDRATE METABOLISM

Fatty acids and carbohydrates are broken down to acetyl CoA, and hence can supply energy through the citric acid cycle and the respiratory chain. The liver will take up most compounds and can perform most conversions between fuels. Normally, the brain and the central nervous system use glucose as the sole source of energy, but during prolonged fasting, the brain utilizes ketone bodies as energy sources. Skeletal muscles generally oxidize free fatty acids and ketone bodies to produce mechanical work. Glucose is an important fuel for skeletal muscle only when the muscle becomes anaerobic. The main fuels for cardiac muscle are free fatty acids, lactate and ketone bodies.

The proportions of fuels utilized by a tissue may be altered by food intake and exercise, the mechanism for the switch-over being hormonal in nature. Following the ingestion of a high-glucose meal, insulin is secreted and the glucose is partly converted in the liver and adipose tissue to glycogen, the remainder being broken down to triose phosphate by the glycolytic and the hexose monophosphate pathways. α-Glycerophosphate is produced especially in adipose cells, and esterifies the released free fatty acids to triglycerides. This increases glucose utilization for oxidation, which in turn produces pyruvate and acetyl CoA, finally yielding CO_2 and $NADH_2$. Excess acetyl CoA is converted to fatty acids in the liver and adipose cells. Therefore, excess dietary carbohydrate is stored as glycogen or converted to fat when the glycogen storage capacity is reached.

During a short burst of muscle activity, glycogenolysis within the muscle cells provides the immediate source of energy. However, after a few minutes the muscles extract plasma free fatty acids to provide another source of energy.

During fasting, liver glycogen is depleted and a metabolic shift occurs to fat – the main reserve fuel – as a source of energy. Less glucose enters adipose tissue, and free fatty acids are released into the blood due to catecholamine release and consequent activation of lipase. The free fatty acids are oxidized, producing acetyl CoA and $NADH_2$. The acetyl CoA enters the citric acid cycle, whilst the $NADH_2$ is utilized for oxidative phosphorylation.

DIETARY ENERGY SOURCES

Although fats, carbohydrates and proteins are exchangeable sources of energy, all diets must contain proteins to provide essential amino acids and nitrogen. In affluent societies, the provision of energy is approximately 48 per cent by carbohydrate, 40 per cent by fat and 12 per cent by protein.

Metabolic heat

The chemical energy of food is transformed to heat, even when no external work is done. When work is performed, not more than 20 per cent of the chemical energy of food is converted to external work.

The energetic equivalent of O_2 is the amount of energy released by food for each mole of O_2 consumed and this varies for different substrates. The ratio of the volume of CO_2 produced to the volume of O_2 used when different nutrients are oxidized is known as respiratory quotient (RQ). An RQ of 0.7 denotes all energy is derived from fat; an RQ of 1, all energy is derived from carbohydrate; and an RQ of 0.85, energy is derived from both fat and carbohydrate (50:50 ratio). A RQ greater than 1 denotes fat synthesis from an excess of carbohydrate. Protein oxidation can be assessed from urinary nitrogen excretion; 1 g of nitrogen denotes the oxidation of 6.3 g of protein, oxygen consumption of 5.9 L of oxygen, and the production of 4.8 L of CO_2.

The caloric or energertic equivalent of O_2 and RQ for different food substrates are summarized in Table 12.1.

Respiratory quotient reflects cellular activity and in practice cannot be measured. Measurements of the volume of CO_2 expired and volume of O_2 consumed via the respiratory tract is termed, more correctly, as respiratory exchange ratio (RER). The 'body stores' of CO_2 are extremely large compared with body stores of oxygen because of its higher solubility. Therefore any factor that alters CO_2 'production' such as ventilation, temperature or acid–base disturbances, will affect RER measurement.

In the body, the volume of CO_2 expired and volume of O_2 consumed vary with metabolic, as well as non-metabolic, factors. During exercise, the RER approaches a value of 2 because a greater volume of CO_2 is expired due to hyperventilation. In metabolic acidosis, the RER also increases because the respiratory compensation for acidosis causes the amount of CO_2 expired to increase, while, in contrast, the RQ decreases in metabolic alkalosis as the hypoventilatory compensatory response reduces the amount of CO_2 expired.

BASAL METABOLIC RATE

Basal metabolism or energy expenditure is the energy required to sustain life, and is the largest component of 24-h energy expenditure. The resting energy output is referred as the basal metabolic rate (BMR), defined as the energy output or heat production in a subject in a state of mental and physical rest in a comfortable environment, 12 h after a meal. BMR is expressed as watts (1 W = 1 J/s) or W/m^2 body surface area. The BMR of a 70-kg man is 100 W, or 58 W/m^2 (approximately 1.43 kcal/min, or 2000 kcal/day).

Various factors influence BMR, with body size as reflected by surface area being important. Females have a lower BMR due to their higher proportion of body fat. When lean body mass is used as an index of comparison, there is no difference in metabolic rate between sexes.

Age also influences BMR. For example, a newborn has an O_2 consumption of approximately 7.0 mL/kg per min, about twice that of an adult on a weight basis, the difference being due to the increased needs for growth of the infant. As growth declines with increasing age, BMR decreases; typically BMR declines at a rate of 2 per cent per decade throughout adult life largely because of differences in the proportions of metabolically more active lean tissue, and fat. Body fat increases and lean tissue decreases with age.

After a meal, the BMR rises for 4–6 h by about 10–15 per cent, an effect known as the specific dynamic action (SDA) of food. Most of this SDA is due to the oxidative deamination of food in the liver. Dietary-induced thermogenesis is energy expended during the digestion and assimilation of food and is greater for protein (30 per cent) compared with carbohydrate (4–5 per cent) or fat (1–2 per cent). Starvation decreases BMR because of a decreased cell mass and reduced tissue metabolism.

Table 12.1 *Energetic equivalents of O_2 and respiratory quotient for various food substrates*

Nutrient	Respiratory quotient	Calorie equivalent of oxygen (kcal/L O_2)
Glucose	1	5.01
Fat	0.7	4.7
Protein	0.8	4.6
Ethyl alcohol	0.66	4.86

Climatic factors also influence BMR; for example, individuals living in the tropics have a BMR 10 per cent less than those living in temperate climates.

Hormones such as thyroxine and epinephrine also influence BMR. Thyroxine stimulates oxidation and increases heat production within cells. BMR is increased progressively to about 20 per cent above normal throughout pregnancy, especially during the second and third trimesters. These increased requirements are for the metabolism of the fetus and placenta, increased cardiac and respiratory work, and metabolism of additional uterine and breast tissue. Lactation also increases BMR as the mother is required to produce and secrete milk.

Human calorimetry

In 1900, Atwater and Rosa developed a human calorimeter to measure directly the total heat output of a human. This consisted of a chamber where a person would live and work for several days, and at the same time the net energy intake could be related to the total output of heat. Direct calorimetry is difficult to measure, however, and as heat output is related quantitatively to oxygen utilization, the measurement of O_2 consumption is normally used in indirect calorimetry. This is based on the assumption that 1 L of O_2 used is equivalent to the production of 4.8 kcal of energy.

There are three main methods used in indirect calorimetry:

- The Benedict–Roth spirometer is a simple closed-circuit breathing system that is filled with 6 L of oxygen and held in a drum, floating on a water seal. The subject breathes in from this drum through an inspiratory valve, and expired air is passed back to the drum through an expiratory valve and a soda-lime canister which removes the CO_2 produced. As oxygen is consumed, the volume of the drum decreases, and this is recorded. The rate of oxygen consumption is determined and the metabolic rate calculated.
- In the Douglas bag technique, all expired air is collected using a mouthpiece with inspiratory and expiratory valves. The expired air collected in the Douglas bag is analysed for the content of oxygen and carbon dioxide so that oxygen utilization and carbon dioxide production can be calculated.
- The Max Planck respirometer is based on the Douglas bag method, and the volume of expired gas is measured directly in a dry gas meter. A device within the spirometer diverts an adjustable volume of expired gas into a breathing bag, from which the expired gas may be sampled and analysed. This type of respirometer is used for measuring very high rates of oxygen consumption, and for prolonged periods.

STARVATION

Starvation is a state of relative or absolute inadequate energy supply. The body has to obtain its energy supply from endogenous reserves. There are no reserves of protein as it forms the functional compartment of body mass. The responses to a lack of nutrient intake are designed to conserve energy, minimize protein losses, and to maintain a supply of glucose to tissues (brain and nerves) that cannot derive energy from any other source. Red blood cells and the renal medulla utilize glucose derived from glucose via glycolysis with the production of lactate. There are three phases in the adaptation to starvation: the glycogenolytic, the gluconeogeneic and the ketogenic phases.

Glycogen stores in the liver (70–100 g) and the muscle (400 g) are rapidly exhausted within 24 h and thereafter glucose is obtained by gluconeogenesis. The next major source of energy is free fatty acids released by adipose tissues. During the first 24 h, glucose is produced predominantly by liver glycogenolysis because of the lower concentrations of insulin. Small amounts of acetoacetate and β-hydroxybutyrate are produced by the liver from free fatty acids. Only a small quantity of glucose is produced by gluconeogenesis from lactate and glycerol by the liver predominantly, and to a minor extent in the kidneys (Fig. 12.12).

After approximately 24 h of fasting, glucose is produced almost entirely by gluconeogenesis via amino acids, glycerol (from adipose tissue) and lactate from erythrocytes. The increase in gluconeogenesis coincides with the increase in plasma glucagon concentration over the first 24–48 h and a decrease in plasma insulin concentration. Alanine is the most important amino acid for gluconeogenesis by the alanine–glucose cycle, and is formed in muscle mainly by the transamination of pyruvate derived from the oxidation of isoleucine, leucine and valine. Glutamine is the major precursor for renal gluconeogenesis.

Plasma cortisol and epinephrine concentrations rise, and these mobilize fat stores, resulting in an increase in plasma free fatty acids and glycerol. Oxidation of the free fatty acids increases the plasma concentration and urinary excretion of acetoacetate and β-hydroxybutyrate. The increased plasma concentration of cortisol may also reduce protein synthesis, especially in skeletal muscle. Growth hormone concentrations also rise over the first 24–48 h and then decrease. Lactate and pyruvate may be formed by glycolysis in the renal medulla and erythrocyte, and transferred to the liver, where they are used to resynthesize glucose by the Cori cycle (Fig. 12.13). The changes described occur between 2 and 4 days of fasting, whilst plasma glucagon concentration reaches a peak at about 4 days.

After 3–4 days gluconeogenesis declines as the body adjusts to get an energy supply from fat and urinary nitrogen falls. The decrease in plasma insulin, the consequent rise in glucagon, and the decline in tri-iodothyronine stimulates lipolysis. Glycerol is synthesized into glucose by the liver and the kidney, whereas the fatty acids provide energy for gluconeogenesis. Ketone bodies gradually replace glucose as the fuel for brain and nervous tissue (Fig. 12.14). In a fully adapted state, the brain obtains about 50 per cent of its energy needs from ketone oxidation. Ketone body formation by the liver is maintained at a high rate, but other tissues revert to using free fatty acids as a source of energy. Both cardiac and skeletal muscles obtain their energy from fatty acid oxidation. The rate of gluconeogenesis is reduced as a protein-sparing mechanism. Plasma glucagon concentration is

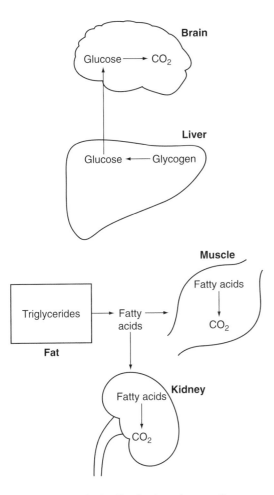

Figure 12.12 *Fuel utilization in early starvation.*

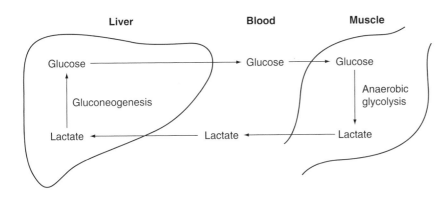

Figure 12.13 *The Cori cycle.*

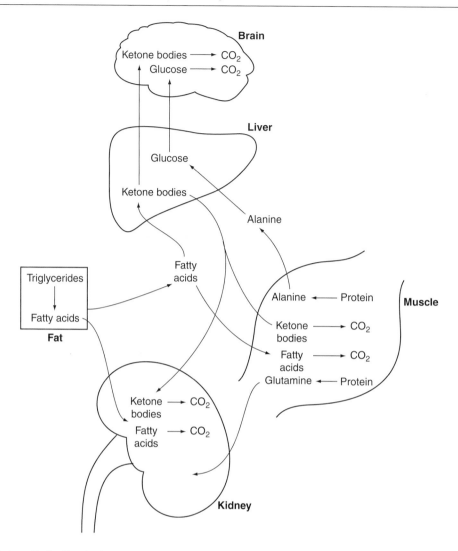

Figure 12.14 *Fuel utilization in intermediate starvation.*

reduced to its prefasting levels at 10 days and this is responsible for the decreased rate of gluconeogenesis. During this time (fasting over 2 weeks) plasma insulin concentration remains low while plasma cortisol and adrenaline concentrations are increased. The body becomes fully adapted to fat metabolism in the second week. Glucose is highly conserved and any lactate derived from red blood cells or the renal medulla is recycled for synthesis into glucose. During the first week of starvation, protein breakdown is approximately 75 g per day during the first few days but decreases to 20 g per day by the third week due to ketone body formation. During prolonged fasting, there is a decrease in

resting metabolism rate of about 30 per cent, mainly due to a decrease in the mass of active tissues such as the liver, kidney and the gastrointestinal tract (Fig. 12.15).

EXERCISE

ENERGY DEMANDS

Exercise involves a complex physiological response to the voluntary activation of skeletal muscles. There is a coordinated response to increased

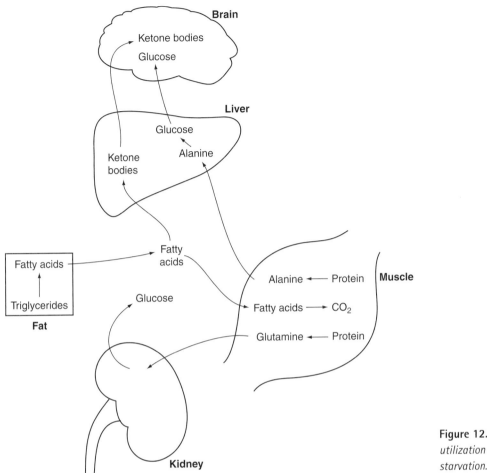

Figure 12.15 *Fuel utilization in prolonged starvation.*

muscle energy needs that involves almost every organ. In a trained athlete, physiological adaptations modify these responses.

Muscle cells convert chemical energy into mechanical energy. ATP is the energy source used during muscle contraction. Myosin ATPase located on the globular head hydrolyses ATP to release energy to the myosin head and ADP is formed. The phosphagen system consists of cellular ATP and phosphocreatine. The ATP pool in skeletal muscle is small and can only support only a few contractions. However, it is continually replenished during contraction. Muscle creatine phosphate is used to convert ADP into ATP so that the ATP store is replenished during muscle contraction. The creatine phosphate store is only five times the size of the ATP store and therefore can sustain maximal muscle contraction for only 8–10 s.

The muscle cell replenishes the creatine phosphate pool during recovery by utilizing ATP derived from oxidative phosphorylation.

During muscle contraction that lasts from a few seconds to a minute, energy is obtained from glycolysis or anaerobic metabolism. Glycogen stored in muscle rapidly splits into glucose that is used for energy. Muscle cells can also take up glucose from blood, a process that is stimulated by insulin. During glycolysis glycogen is converted to pyruvate. However, the muscle also releases lactate because glycolysis forms pyruvate faster than the muscle can oxidize it. Glycogen in this process is mostly converted to lactate and supplies four ATP molecules for each molecule of glucose. The glycogen–lactic acid system forms ATP two and a half times as fast as oxidative phosphorylation in the mitochondria but is self limiting. This process supplies enough energy for maximal muscle contraction for 1.3–1.6 min.

During exercise or when caloric intake is insufficient pyruvate can be converted into alanine, a non-essential amino acid. Alanine produced by skeletal muscle can be converted to glucose in the liver. This is called the alanine cycle. For longer periods of exercise, energy for muscle contraction is supplied by the aerobic system in which glucose, fatty acids and amino acids are oxidized in the mitochondria to form ATP. During the steady state of exercise, the oxidation of glucose from muscle glycogen and blood glucose accounts for 40–50 per cent of the energy released in the active muscle with fatty acid oxidation accounting for the remainder.

When O_2 is limited in muscle, muscle glycogen is converted to lactate. The lactate then enters the blood and is carried to the liver where it is converted to glucose (i.e. gluconeogenesis). This hepatic glucose enters the blood and reaches the muscle to be used for muscle contraction. This is called the Cori cycle, whereby liver converts the anaerobic metabolic product (lactate) to a fuel (glucose) that can be used anaerobically.

Fatty acids are an important source of energy for muscle cells in prolonged exercise. β-Oxidation of the fatty acids within the mitochondria produces acetyl CoA which enters the citric acid cycle and produces ATP.

An increasing difficulty to sustain a given level of exercise initiates a sensation called 'fatigue'. Fatigue occurring within the first few seconds of very intense exercise is due to depletion of ATP and creatine phosphate stores in the active muscles. When fatigue occurs with less intense exercise, it is associated with the accumulation of lactic acid in the muscles. In marathon runners, depletion of muscle glycogen stores also causes fatigue.

The workload above which the blood concentration of lactic acid rises is called the 'anaerobic threshold'. Oxygen consumption during exercise reaches a steady state after several minutes, when no net energy is provided by non-oxidative sources (Fig. 12.16). During the first few minutes of exercise, oxygen uptake by the exercising muscle is less than the steady-state maximal oxygen consumption. This inadequacy of oxygen supply relative to demand during exercise is called the 'oxygen deficit'. The oxygen deficit is accounted for by the oxygen released by myoglobin in active muscles, increased oxygen extraction (hence lower venous O_2) by the active muscles and depletion of ATP and creatine phosphate in the active muscles.

The resting metabolic rate of skeletal muscle is 1.5–2 mL O_2 per minute per kg. During maximal muscle exercise, the skeletal muscle metabolic rate can exceed 150 mL O_2 per min per kg. The maximal oxygen uptake of exercise reaches a ceiling approximately 20 times the basal consumption, but it can be increased by endurance training. The steady-state maximal oxygen consumption is a predictor of the ability to perform prolonged dynamic external work and represents the physiological limit to oxygen transport and utilization. It is decreased by age, bed rest and increased body fat.

When exercise is stopped, there is an apparent excess of oxygen consumption during the first few minutes of recovery, and this is the 'oxygen debt'. The oxygen debt represents the repayment of an oxygen deficit. ATP and creatine phosphate stores, and the oxygen content of myoglobin, return to baseline values within a minute. The increased oxygen consumption can last for minutes to hours after exercise ceases, this being attributed to the energy needed to dissipate heat and restore intracellular electrolytes to normal, and to the high metabolic rate due to increased body temperature, circulating levels of catecholamines and thyroxine. The metabolism of accumulated lactate, which occurs within 30 min following exercise, does not contribute significantly to the oxygen debt.

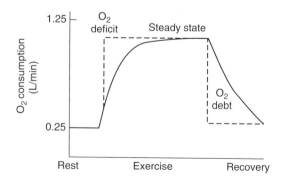

Figure 12.16 *Oxygen consumption during exercise.*

CARDIOVASCULAR RESPONSES TO EXERCISE

Exercise involves the cardiovascular centres in the brainstem and the autonomic nervous system controlling the heart and resistance vessels acting in association with local control mechanisms. Initially, neural activity of the motor cortex communicates with the cardiovascular centres to reduce vagal tone to the heart and reset the arterial baroreceptors to respond to a higher range of blood pressures.

The increased energy requirements during exercise are met by enhanced energy delivery. Dynamic exercise invokes an increase in perfusion of active muscles to a maximum of 20 times resting blood flow, brought about by vasodilatation. This vasodilatation is initially due to increased cholinergic sympathetic neural activity to the vasculature in skeletal muscles. The vasodilatation, however, is predominantly due to local factors such as decreased Pa_{O_2} and pH, increased Pa_{CO_2}, temperature and K^+, and the release of local metabolites such as adenosine. These factors dilate the arterioles, resulting in a large increase in the number of perfused capillaries, and in turn allowing enhanced oxygen extraction from haemoglobin by a shortened oxygen diffusion distance between capillaries to the mitochondria in muscle tissue. The oxygen dissociation curve is shifted to the right by the Bohr effect, resulting in 90 per cent oxygen extraction from blood perfusing muscles at maximal exercise.

Changes in regional blood flow occur during exercise. Coronary blood flow increases to meet the increased myocardial oxygen demands of increased cardiac output. Skeletal muscle blood flow increases, predominantly due to local factors, as described earlier. Skin blood flow increases to dissipate heat, but may decrease at maximal exercise as the cardiac output cannot meet the demands of exercise. Cerebral blood flow is maintained at a constant level, whatever the level of exercise. Splanchnic and renal blood flow are reduced by sympathetic activity.

The heart rate increases linearly with exercise to a maximum determined by a person's age and training. Stroke volume increases in light or moderate exercise by an increase in end-diastolic volume. At higher levels of exercise, a decrease in end-systolic volume due to enhanced myocardial contractility

contributes to a further increase in stroke volume. Endurance training causes an increased volume of the ventricles, resulting in an increased stroke volume.

Overall, the cardiac output can increase five-fold during severe exercise. This is by increased venous return due to venoconstriction and muscle pump (repeated compression of veins by active muscles), increased myocardial contractility, and decreased systemic vascular resistance (by about one-third of the resting systemic vascular resistance). Blood pressure is maintained as the increased sympathetic activity limits vasodilatation in active muscles in very severe exercise.

The time course of the increase in cardiac output during exercise is influenced by a variety of factors (Fig. 12.17). There is a sudden increase in cardiac output at the start of exercise brought about by the stimulation of the cardiovascular system by neural activity and the increased venous return due to the muscle pump. This is followed by a gradual rise to the steady state due to vasodilatation in active muscles and stimulation of the cardiovascular system.

Isometric or static exercise also increases muscle blood flow and cardiac output. However, these increases in flow are limited by the higher mean intramuscular pressure. As a result of the limited increase in muscle blood flow, anaerobic metabolism occurs more readily, with the production of lactic acid and increased ADP/ATP ratio. Different cardiovascular responses occur during isometric exercise. The blood pressure is increased to a greater extent compared with isotonic exercise, but there is

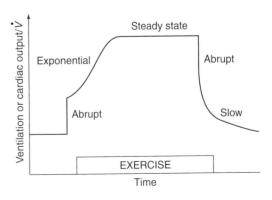

Figure 12.17 *Time course of changes in ventilation and cardiac output during exercise.*

a lesser increase in cardiac output because of a relatively greater increase in afterload.

Prolonged training produces adaptatory changes in the heart, depending on the nature of exercise. With chronic dynamic exercise, left ventricular volume increases, leading to a large resting and exercise stroke volume, an increased vagal tone and decreased β-adrenergic sensitivity with resting bradycardia. With isometric exercise, the heart adapts to the high afterload (systemic pressure) by left ventricular hypertrophy. Coronary blood flow also adapts to the effects of endurance training. The peak coronary blood flow is increased by training due to increased responsiveness to adenosine, endothelium-mediated regulation and control of intracellular free calcium. Endurance-trained muscles have increased concentration of oxidative enzymes, myoglobin, mitochondria and more capillaries.

RESPIRATORY RESPONSES TO EXERCISE

Minute ventilation increases with exercise in a linear fashion until the anaerobic threshold is reached, when it increases more steeply due to the liberation of lactic acid. The increased minute ventilation is brought about by increases in tidal volume and respiratory rate.

The changes in minute ventilation show a rapid response followed by a gradual response. The rapid responses at the start and end of exercise result from neural inputs to the respiratory centre from the motor cortex and proprioceptive receptors in the exercising muscles.

In moderate steady exercise, minute ventilation is proportional to oxygen uptake, CO_2 output and to the metabolic rate. Several mechanisms have been implicated to explain the increase of ventilation during exercise.

Venous P_{CO_2} is increased during exercise. Although the mean arterial P_{CO_2} is unchanged, fluctuations of arterial P_{CO_2} occur, and it is suggested that the oscillatory discharge of central chemoreceptors provides a potent respiratory stimulus in exercise. Decreased arterial P_{O_2} and increased H^+ ions stimulating respiration via the peripheral chemoreceptors have also been implicated. Increased temperature of the blood from exercising muscles may be partially responsible by an action on the respiratory centres. Activation of mechanoreceptors in muscles and joints also contributes to respiratory changes during exercise. It would appear that multiple factors are responsible for the regulation of ventilation in exercise, as no single mechanism can explain the respiratory changes seen.

At higher levels of exercise, above the anaerobic threshold, there is a disproportionate increase in minute ventilation and CO_2 production compared with oxygen consumption. The release of lactic acid by the muscles results in metabolic acidosis, and this stimulates the peripheral chemoreceptors.

MUSCLE AND BONE RESPONSES TO EXERCISE

Several factors limit exercise bringing about fatigue. H^+ ions and ADP accumulation in the muscle have been implicated. Decreased activity of muscles leads to atrophy. Increased muscle activity with low loads results in enhanced oxidative metabolic capacity without hypertrophy, whilst increased activity with increased (high) loads produces hypertrophy mediated by increased mRNA.

Bone remodelling and bone density is also related to muscle strength and activity. With prolonged immobilization there is a loss of bone density.

GASTROINTESTINAL AND ENDOCRINE EFFECTS

Chronic physical activity increases gastric emptying and small bowel motility, probably as a result of adaptation to increased energy requirements. Blood flow to the gut decreases in proportion to the intensity of exercise.

Exercise suppresses insulin secretion by increasing sympathetic activity to the islet cells. However, glucose uptake by muscles is enhanced as exercise appears to recruit glucose transporters from their intracellular sites to the plasma membranes of active skeletal muscle cells. This leads to increased insulin sensitivity.

TEMPERATURE REGULATION

Temperature is a measure of the average kinetic energy of a substance per degree of freedom of its constituent molecules. Mammals maintain a constant body temperature, and enzyme systems of the body function optimally within a narrow range of between 35 and 41°C. Above 45°C, enzymes become denatured. A convenient expression of the relationship between the rate of a reaction and temperature is the Q_{10} value. This is the ratio of the velocity of a reaction at $T+10$°C to its velocity at T°C. The enzyme reactions in the body increase 2- to 2.5-fold for each 10°C rise in temperature (i.e. $Q_{10} = 2$–2.5). The specific heat of water is 4.2 kJ/kg per °C (1 kcal/kg per °C), whereas the specific heat of body tissue is about 85 per cent that of water, i.e. 3.6 kJ/kg per °C. Body temperature is regulated by a balance between heat loss and heat production. Heat is lost from the body via radiation, convection, evaporation and conduction from the skin, as well as evaporation from the respiratory tract and via urination and defaecation. Radiation, or emitting electromagnetic energy, contributes to about 40–50 per cent of the heat loss of the body. About 15 per cent of heat loss is through conduction and convection, and 30 per cent via evaporation (the latent heat of vaporization of water = 2.4 MJ/kg or 580 cal/g at 37°C). Heat loss via respiration is approximately 5 per cent.

Over a wide range of ambient temperatures, the core temperature is kept constant within 0.4°C of its set point (37°C), with the main internal organs being kept almost at the same temperature. The deep body temperature of the main internal organs is referred to as the core temperature and is usually measured as the rectal temperature. The rectal temperature, which is approximately 0.5°C higher than the axillary temperature, shows a diurnal variation, being at its highest in the evening (37.3°C) and lowest in the early morning (35.8°C). In females, the core temperature is 0.5°C higher in the latter half of the menstrual cycle. The peripheral temperature varies widely and is less than the core temperature.

The principal site of temperature regulation is the hypothalamus, which initiates negative feedback mechanisms to keep variations from normal values to a minimum. This is achieved by integrating signals from the brain, spinal cord and skin.

AFFERENT TEMPERATURE SENSORS

In the skin, distinct warm and cold receptors sense the ambient temperature. Cold receptors (bulbs of Krause) located in the dermis transmit impulses to the hypothalamus by Aδ fibres. The cold receptors exhibit static (regular and periodic) discharges at constant temperature and these increase below 24–25°C. The face is especially sensitive to thermal stimulation.

Warm receptors in the skin are elliptical structures called the bulbs of Ruffini, and have static discharges between 30 and 40°C. Maximal discharge rates occur at about 44°C but cease at 46°C. Information from the warm thermal receptors are transmitted to the hypothalamus by unmyelinated C fibres.

There is evidence that temperature sensors are also present in the spinal cord and in the intestinal walls.

The ascending thermal information travels along the lateral spinothalamic tracts in the anterior spinal cord, synapses at the reticular system of the medulla and reaches the posterolateral and ventromedial nuclei of the medulla. The skin and deep-body thermal receptors are connected to the anterior and posterior hypothalamus.

CENTRAL REGULATION

The hypothalamus is the main body temperature regulatory centre and it regulates body temperature by generating an optimal set temperature, integrating afferent thermal inputs and initiating physiological and behavioural responses to minimize the difference between the set and actual temperatures.

The anterior hypothalamus is sensitive to local warming of blood which increases the firing rate, producing sweating and vasodilatation. The posterior hypothalamus responds to cold afferent impulses from the peripheral temperature receptors, and causes increased shivering thermogenesis. The posterior hypothalamus is also responsible for establishing the reference or set point around which body temperature is maintained. This set point may be determined by the ratio of sodium and calcium ions in the posterior hypothalamus.

Various neurotransmitters in the hypothalamus are involved in temperature regulation. Norepinephrine, 5-hydroxytryptamine, dopamine and prostaglandins appear to be mediators in the anterior hypothalamus. Acetylcholine is the main neurotransmitter in the posterior hypothalamus.

Control of the autonomic responses is determined predominantly by the input from core structures. The range of core temperature over which no autonomic thermoregulatory responses occur is called the 'interthreshold range'. The interthreshold range is approximately 0.2°C at normal temperature (37°C) in a non-anaesthetized state. The threshold temperatures are influenced by circadian rhythms, food intake, thyroid function, drugs and thermal adaptation to warm or cold ambient temperatures. Impaired central regulation is present in the elderly and the critically ill patient. At the upper end of the threshold, sweating begins, whilst vasoconstriction commences at the lower end. These thresholds may be 0.3–0.5°C higher in women. The slope of the intensity of response (i.e. sweating or vasoconstriction) to the difference between the thermal input and threshold temperature is called the 'gain' of the response (Fig. 12.18). In contrast, afferent input from cutaneous receptors appears more important in controlling behavioural responses to temperature.

EFFERENT RESPONSES

Thermoregulatory centres in the hypothalamus sense the core temperature and receive inputs from peripheral thermal receptors. Thermal comfort is a mental phenomenon from an awareness of the thermal state and emotional factors. It is influenced by ambient temperature, air movement, humidity and radiation intensity, so that the skin temperature is maintained at about 33°C. Below the critical temperature (27°C), the metabolic rate rises linearly as the temperature decreases. As the ambient temperature rises above the critical temperature, the metabolic rate remains constant until an upper limit is reached and the metabolic rate rises again. This range of environmental temperature in which the metabolic rate (and oxygen consumption) is minimal and steady is called the 'thermoneutral zone' (Fig. 12.19).

Deviations from threshold temperature evoke thermoregulatory responses that increase metabolic

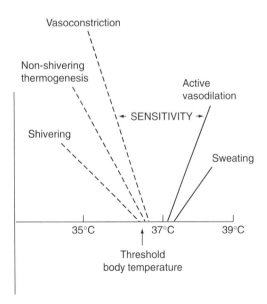

Figure 12.18 *Thermoregulatory thresholds.*

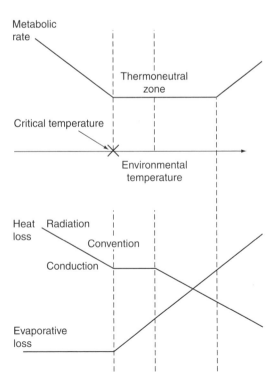

Figure 12.19 *Effects of environmental temperature.*

heat production (by shivering and non-shivering thermogenesis), decrease heat production (by active vasoconstriction and behavioural changes), or increase heat loss (active vasodilatation, sweating and behavioural changes).

CUTANEOUS RESPONSES TO HEAT

The control of cutaneous blood flow is important in regulating heat loss by convection, radiation and conduction. The skin circulation, consisting of capillaries and arteriovenous shunts, receives 8 per cent of the cardiac output. The superficial cutaneous network of arterioles, capillaries and venules and the deep venous plexus are innervated by α-adrenergic sympathetic fibres. The deep venous plexus is a capacitance system and can contain up to 1.5 L of blood.

The arteriovenous shunts of the skin circulation are mainly located in the skin of the hands, feet, ears, nose and lips. These shunt vessels are $50–100\,\mu m$ in diameter, and their smooth vascular muscles are innervated by α-adrenergic receptors regulated by the sympathetic fibres. When body temperature equals or exceeds 30°C, sympathetic activity to the skin decreases under the control of the anterior hypothalamus. This results in vasodilatation, especially of the arteriovenous anastomoses, and is enhanced by bradykinin released by sweat glands activated by cholinergic sympathetic fibres. The fall in total vascular resistance triggers an increase in cardiac output, and a massive increase in skin blood flow accommodated in the dilated arteriovenous shunts and venous plexus. This provides a large area for heat transfer between the skin and the environment. The cutaneous blood flow can increase 30-fold in heat stress and decrease 10-fold in cold stress.

Cutaneous veins provide an important countercurrent system for heat conservation. The calibre of the cutaneous veins is under noradrenergic control. In cold conditions, blood returning from the deep veins of the limbs (vena communicantes) acquires heat from the arteries and thus prevents heat loss from body surfaces.

Shivering thermogenesis is an involuntary contraction of muscles, controlled by the hypothalamus, which can increase the metabolic rate by 100 per cent in adults. The muscle contractions consist of rapid tremors up to a frequency of 250 Hz with unsynchronized muscle activity, suggesting a peripheral mechanism. Superimposed on the fast tremors are slow (4–8 cycles/min) synchronous waves that wax and wane, suggesting central control. Cyclical muscle rigors at 10–20 Hz appear as shivering becomes more intense.

Non-shivering thermogenesis (Fig. 12.20) increases metabolic heat production without mechanical work. Heat production is activated by β_3-sympathetic activity, which uncouples oxidative phosphorylation in brown fat and skeletal muscle. In adults, lipolysis of adipose tissue releases glycerol and free fatty acids, and the free fatty acids are energy sources for skeletal muscle and myocardium. In the newborn, brown fat is found in the interscapular areas, the perinephric fat and around the intra-abdominal vessels. Brown fat is highly vascular with abundant mitochondria and is richly innervated by adrenergic fibres. In the neonate, non-shivering thermogenesis involving brown fat can increase the metabolic rate two-fold above the resting value. In the newborn, triglycerides are hydrolysed to glycerol and free fatty acids, the latter then being re-esterified with the formation of fatty acid acyl CoA. The CoA moiety is replaced by glycerol derived from glucose, whilst the acyl CoA is broken down with the liberation of heat. As each molecule of fatty acid recycles, one ATP is converted to heat. Thus, a small fraction of fatty acid is oxidized rather than re-esterified, and this is reflected by the raised oxygen consumption.

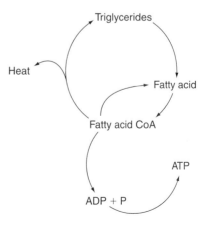

Figure 12.20 *Non-shivering thermogenesis in brown fat.*

EFFECTS OF ANAESTHESIA ON THERMOREGULATION

General anaesthesia increases the interthreshold range by decreasing the thermoregulatory threshold to cold by approximately 2.5°C and increasing the threshold temperature by approximately 1.3°C. Within this expanded interthreshold range, the patients are poikilothermic as active thermoregulatory responses are absent so that body temperature changes passively in proportion to the difference between metabolic heat production and heat lost to the environment. The only thermoregulatory responses available to anaesthetized, paralysed and hypothermic patients are vasoconstriction and non-shivering thermogenesis. The core temperature at which thermoregulatory threshold triggers peripheral vasoconstriction is agent- and dose-dependent. Non-shivering thermogenesis appears to increase metabolic heat production significantly in hypothermic infants.

PHYSIOLOGY OF ALTERED TEMPERATURE

THERMONEUTRAL ZONE

Humans are homeothermic animals and maintain their core temperature within a narrow range of 36–38°C, despite large fluctuations in their environmental temperature and metabolic activity. For this it is necessary to balance the heat gained by the body with the heat lost to the environment. Heat gained by the body is usually derived from metabolism, but, if the ambient temperature exceeds skin temperature, heat is then gained from the environment. Heat losses occur by radiation, convection, conduction and evaporation from the skin to air.

The thermoneutral zone is the range of environmental temperatures over which metabolic heat production is minimal and thermoregulation is maintained by vasomotor activity. In a 70-kg naked man, the thermoneutral zone is 27–31°C. The lower limit of the thermoneutral zone is called the 'critical temperature' (Fig. 12.19). The critical temperature of a clothed individual is much lower

than 27°C, depending on the amount and type of clothing used.

Skin blood flow varies from 1 to 150 mL/min per 100 g skin, and therefore is extremely effective for thermoregulation within the thermoneutral zone. Peripheral constriction reduces the amount of blood flowing from the warm deep areas to the skin, and less heat is lost to the environment. Peripheral vasodilatation has the opposite effect. There are two types of peripheral temperature-sensitive receptors: cold receptors (Aδ fibres) stimulated by a lower range of temperatures, and warm receptors (C fibres), which are stimulated by a higher range of temperatures. The static discharge of both cold and warm receptors varies with skin temperature. The response curves of the cold and warm receptors overlap, and the mid-point of the overlap of the two curves represents the preferred mean skin temperature of a naked individual. The discharge of the cold receptors increases as skin temperature falls, whereas the discharge of the warm receptors increases as skin temperature rises. In the hypothalamus and the spinal cord, cold and warm central temperature receptors respond to changes in core temperature. The preoptic region and the anterior hypothalamus increase heat loss via responses to warm temperatures, whereas the posterior hypothalamus is concerned with heat production and conservation. The hypothalamus regulates peripheral blood flow by modulating sympathetic control of the cutaneous arterioles.

Thyroid hormones act upon mitochondrial and nuclear receptors in most tissues and regulate the activity of membrane-bound Na^+/K^+-ATPase, which balances metabolic heat production with heat loss within the thermoneutral zone.

RESPONSES TO HYPOTHERMIA

The clinical definition of hypothermia in humans is a core temperature below 35°C. Behavioural responses to hypothermia are as important as the physiological responses that increase heat production and reduce heat loss.

Voluntary and involuntary muscle activity increases heat production. Increased voluntary muscle activity enhances heat production by ATP hydrolysis associated with muscle fibre contraction. Shivering is an involuntary contraction of muscle fibres occurring in an uncoordinated pattern in which the fibres contract and relax out of phase with one another and, again, increased ATP hydrolysis increases heat production. Shivering is more pronounced in the extensor and proximal muscles of the upper limb and trunk, and in severe cases, the jaw muscles. Shivering can increase heat production five-fold, but cannot be sustained for long periods as energy reserves are exhausted. Skin receptors provide the main stimulus for shivering; it does not readily occur with a moderate fall in core temperature. The shivering response is poorly developed in neonates.

Non-shivering thermogenesis is a heat-producing mechanism by which the basal metabolic rate can be increased two- to three-fold. In neonates, it occurs in brown fat, the thermogenic capacity of which is about 150 times that of skeletal muscle. However, the contribution of skeletal muscle to thermogenesis is still significant because it forms about 40 per cent of total body mass. Two mechanisms have been suggested to explain non-shivering thermogenesis by brown fat. One theory is that the presence of large amounts of non-esterified fatty acids uncouples mitochondrial phosphorylation from respiration, and energy is lost directly as heat. The other theory suggests that recycling of triglycerides and fatty acids leads to increased ATP turnover. Lipolysis and re-esterification result in ATP hydrolysis and heat production. Non-shivering thermogenesis is stimulated by catecholamines. Thyroid hormones have a permissive effect in thermogenesis as they increase the response of brown fat to catecholamines.

There is some evidence that humans can acclimatize to cold environments, and most homeothermic animals increase non-shivering thermogenesis in response to repeated exposure to cold temperatures. Studies in adult humans who spent at least 6 months in Antarctica show that the core temperature increased when exposed to cold, suggesting an increased metabolic response to cold by increased non-shivering thermogenesis. As there is no increase in brown fat mass, the enhanced non-shivering thermogenesis is likely to involve the liver and skeletal muscles, with an increased sensitivity to the calorigenic effects of norepinephrine in humans following cold acclimatization. Eskimos and the Alakaluf Indians at the southern tip of South America do not shiver in ambient temperatures of 2–5°C, and have a basal metabolic rate 30–40 per cent higher than other populations. Cold environments increase the appetite, and increased food intake increases metabolic heat production.

Decreased heat loss is accomplished by behavioural changes, increases in insulation and cutaneous vasoconstriction. Humans may curl up to reduce the surface area available for heat loss. Goose-pimples brought about by contraction of the piloerector muscles of the hairs present increase insulation by trapping air next to skin, thus reducing convectional heat loss. Subcutaneous fat can provide some insulation, but the changes in fat with cold temperature are too small to be significant to influence overall thermoregulation.

Cutaneous vasoconstriction is the primary response to a reduced environmental temperature, and is mediated by increased sympathetic activity. Reduced cutaneous blood flow decreases heat loss from the skin. However, prolonged cooling can induce a paradoxical vasodilatation by a direct cold-induced paralysis of peripheral blood vessels that become unresponsive to catecholamines. Thereafter, vasoconstriction alternates with vasodilatation ('hunting reaction') which serves to prevent tissue damage such as frostbite. There is some evidence of vascular adaptation to cold. In people who are frequently exposed to cold the vasoconstrictor effect is less severe, and the onset of vasodilatation is more rapid.

When the core temperature falls below 35°C there is muscle weakness, resulting in decreased mobility and decreased shivering. At temperatures below 34°C, mental confusion occurs and consciousness is lost between 32 and 30°C. Hypothermia also decreases the heart rate by slowing the rate of discharge of the sinoatrial node. At core temperatures below 28°C cardiac arrhythmias are frequent and ventricular fibrillation may occur.

Frostbite is the most severe form of cold injury and is due to freezing of the peripheral tissues. Damage to tissues is by cell dehydration and the mechanical effects of ice crystals, associated with an increase in permeability of blood vessels. In mild forms only the skin freezes, but the muscle and tendons may also freeze in severe cases. There is loss of fluid from the circulation on thawing, and the increased haematocrit in the blood vessels of affected tissues can reduce blood flow and cause gangrene.

Prolonged cooling rather than freezing can cause neuromuscular damage. Sensory and motor paralysis in cold immersion injury are due to the direct effects of cold on nerves and muscles.

RESPONSES TO HIGH TEMPERATURES

Responses to high temperatures are aimed at decreasing heat gain and increasing heat loss. Decreased heat gain is achieved by behavioural changes (reduced activity, reduced feeding and reduced heat gain from the environment using appropriate clothing and housing). Sweating is the main method of increasing heat loss as the environmental temperature increases. There are about two million sweat glands in the body, but these are not evenly distributed. About 50 per cent of the sweat production occurs in the chest and back, and women have lower rates of sweat production compared with men. Although the maximum rate of producing sweat is 3 L/h this cannot be sustained, and the maximum amount of sweat produced is 12 L/day. The proximal region of the sweat gland produces a hypotonic solution that is modified by solute reabsorption as the fluid moves along the duct towards the skin surface. At low rates of secretion, the sodium content of sweat is low (5 mmol/L), but at high secretion rates it can reach 10 times the basal value because there is less time for ductal reabsorption. Evaporation of sweat is important for heat loss; the latent heat of evaporation of water at 37°C is 2.4 kJ/mL, and the majority of the heat for this comes from the body.

As the environmental temperature increases, cutaneous vasodilatation occurs rapidly, promoting heat transfer from the deep to the superficial tissues and then to the environment. If heat loss is less than heat production, then the core temperature rises and induces other cardiovascular responses such as tachycardia and increased cardiac output. Blood pressure does not increase, as the vasodilatation described earlier reduces peripheral vascular resistance.

Acclimatization to heat is achieved primarily by increased production of sweat, which starts at a lower threshold core temperature, and by a diminished sweat sodium concentration.

Under conditions of extreme heat the thermoregulatory mechanisms may fail and cause either heat stroke, heat exhaustion or heat collapse. Heat stroke is characterized by a loss of energy and irritability progressing to neurological disturbances caused by a complete loss of thermoregulation. Cessation of sweating appears to be the primary cause of loss of thermoregulation. The individual becomes unconscious as the core temperature rises to 42°C. Cellular damage and coagulation of proteins with high core temperatures lead to death. Heat exhaustion can result from excessive water or salt loss associated with high core temperatures. Water-deficiency heat loss occurs as a result of inadequate water replacement of fluid losses. Dehydration and a gradual decrease in plasma volume occur. With a loss of 5 per cent body fluids there is fatigue and dizziness, and physical and mental deterioration with a loss of over 10 per cent. Salt-deficiency heat exhaustion occurs when the salt losses in sweat are not replaced adequately, with cramps in the legs, arms or back and fatigue and dizziness. The extracellular fluid compartments contract as tissue osmolality decreases.

Heat collapse or fainting is characterized by dizziness and a temporary loss of consciousness in the heat; it is caused by pooling of blood in the dilated vessels of the skeletal muscles and skin in the lower limbs, leading to diminished cerebral blood flow. It usually occurs in unacclimatized individuals in hot climates.

Reflections

1. Energy is derived from the combustion of carbohydrates, fats and proteins. Energy expenditure is necessary for basal thermogenesis and for sedentary activity, voluntary exercise and heavy work. Fatty acids are a major fuel in most tissues except for the central nervous system and red blood cells where glucose is the only substrate for energy. Energy is mainly stored as triglycerides in fat tissue. Proteins are less readily available energy stores. Carbohydrate stores

are limited and therefore efficient glucose production by gluconeogenesis by the liver is necessary to maintain a supply of glucose for the brain.

2. Protein metabolism requires intake of essential amino acids such as leucine. Essential amino acids are irreversibly degraded. Non-essential amino acids such as alanine are degraded and re-synthesized daily. Fat metabolism involves lipoproteins that transfer triglycerides and cholesterol derived from food or from hepatic synthesis to peripheral tissues and the liver.

3. Metabolism is due to the chemical processes of the body that give rise to heat and provides energy in the form of ATP. Energy is measured in kilocalories or joules where 1 kcal is equal to 4.187 kJ. The rate at which chemical energy is expended is called the metabolic rate. Basal metabolic rate (BMR) refers to the energy production under conditions of complete mental and physical rest and a comfortable ambient temperature, and 12 h after a meal. BMR varies with age, sex and body build. The body metabolizes variable quantities of fats, carbohydrates, and sometimes proteins to produce energy. Metabolic rate can be calculated from oxygen consumption and the respiratory quotient which is related to the energy equivalent of oxygen for foods being metabolized. Metabolic rate is increased by exercise, food ingestion and by fever. It is reduced by malnutrition and during sleep. Hormones can modify the metabolic rate; catecholamines and thyroid hormones are potent stimulators of metabolism whilst growth hormone and sex steroids exert a mild stimulatory effect.

4. The energy for muscle contraction during exercise is provided by the breakdown of ATP. ATP levels in exercising muscles are maintained by the transfer of a high-energy phosphate group from creatine phosphate. Phosphocreatine is synthesized when extra amounts of ATP are available, and when ATP utilization increases then the energy in phosphocreatine is transferred back to ATP. Phosphocreatine serves to act as an 'ATP buffer' to maintain ATP levels constant so long as phosphocreatine is available. Skeletal muscle contains large amounts of glycogen which is broken down to glucose during exercise. The glucose is then metabolized by either anaerobic pathways via glycolysis to lactate or aerobically via the tricarboxylic acid cycle to generate ATP. Aerobic metabolism is more efficient in generating ATP than anaerobic metabolism. At the onset of aerobic exercise oxygen consumption rises exponentially to its steady state. During maintained exercise oxygen consumption is directly proportional to the work rate. At the end of exercise oxygen consumption falls rapidly but does not return to resting values for some time. Oxygen debt is the extra oxygen consumption after completion of strenuous muscle activity. This excess oxygen is used to (1) convert accumulated lactic acid back to glucose; (2) reconvert adenosine monophosphate and diphosphate to ATP; and (3) re-establish phosphocreatine levels.

5. In exercise the cardiac output increases as a result of an increase in heart rate and an increase in stroke volume. Blood flow is redistributed from the splanchnic circulation to the exercising muscle. Systolic blood pressure rises but diastolic pressure is stable, and may even fall due to a reduction in peripheral resistance as a result of vasodilatation of skeletal muscle blood vessels. In mild and moderate exercise pulmonary ventilation increases in direct proportion to the work done. At workloads below the anaerobic threshold the Pa_{O_2} and Pa_{CO_2} of arterial blood do not change significantly. In the venous blood there is a fall in P_{O_2} and a rise in P_{CO_2}. The oxygen requirements of exercising muscles are met by an increase in cardiac output, increased blood flow in the skeletal muscles brought about by arteriolar vasodilation, and by increased extraction of oxygen by the muscles.

6. Both cardiac output and pulmonary ventilation are precisely adjusted to meet the metabolic demands during exercise. The cardiovascular response is initiated by signals from the brain which inhibit parasympathetic activity and increase sympathetic activity. Consequently, the increased cardiac output and blood flow is preferentially distributed to the exercising muscles. Afferent signals arising from the joints and muscles activate cardiovascular reflexes that act to maintain a cardiovascular response

at a level appropriate to the intensity of exercise. The associated rise in body temperature initiates a reflex vasodilatation of the skin blood vessels which enhance heat loss.

7. The ventilatory response to exercise is initiated by signals from the brain that are supported by signals from the muscle spindles and mechanoreceptors in the muscles and joints. The arterial partial pressures of the respiratory gases and pH are maintained at a steady level except in severe exercise.

8. During fasting, endocrine mechanisms provide energy for basic cellular functions. Blood glucose levels are maintained by an increase in glucagon, growth hormone, cortisol and catecholamine concentrations and a decrease in insulin concentration. Growth hormone enhances lipolysis and decreases peripheral utilization of glucose. Although cortisol is weakly lipolytic, it exerts a permissive effect on the mobilization of fatty acids by growth hormone and epinephrine during fasting. During long-term fasting, gluconeogenesis from amino acids, glycerol, and lactate is required to sustain the central nervous system and other systems that are dependent on glucose. Increased use of fatty acids during fasting increases the production of ketoacids such as acetoacetate and β hydroxybutyrate.

9. Humans maintain a constant core body temperature between 36 and 38°C to maintain optimal conditions for enzyme activity. Heat is produced by metabolic reactions and is lost from the surface of the body by radiation, convection, conduction and by evaporation. Heat loss is balanced by heat gain to achieve effective thermoregulation. The cutaneous circulation plays an important role in thermoregulation. Vasodilatation of cutaneous blood vessels increases heat loss, whereas vasoconstriction of the skin blood vessels reduces heat loss. The hypothalamus receives input from temperature receptors in the skin and the body core, and acts as a thermostat to initiate appropriate mechanisms to conserve or lose heat so that the core temperature is kept at a set point around 37°C. Physiological responses that conserve heat during exposure to cold include cutaneous vasoconstriction, shivering, and non-shivering thermogenesis. Physiological responses to high body temperatures include vasodilatation and sweating, which increases heat loss by evaporation. Hypothermia occurs when the core temperature falls below 35°C and this activates heat-conserving mechanisms and may cause mental confusion, cardiovascular complications, followed by coma. Newborn infants and the elderly patients are at a high risk of hypothermia. Hyperthermia occurs when heat loss mechanisms fail, and may lead to cerebral oedema and later irreversible neuronal damage.

Physiology of pain

PHILIP J. SIDDALL AND MICHAEL J. COUSINS

LEARNING OBJECTIVES

After studying this chapter the reader should be able to:

1. Define pain and explain its biopsychological aspects
2. Describe nociceptors and explain how they transform the various stimuli to a nerve signal
3. Describe the chemical mediators of pain and the mode of action of these mediators
4. Describe the characteristics of the primary afferent neuron and the nerve fibres involved in pain
5. Explain how pain perception can be changed by modulation at all levels of the nervous system

6. Describe modulation of pain at the spinal level, including the phenomenon of 'wind-up' and the 'gate control theory'
7. Explain the transmission of nerve impulses from the dorsal horn to the thalamus
8. Describe the endogenous ligands involved in the modulation of pain via spinal mechanisms or descending tracts

INTRODUCTION

It should be stated at the outset that the biological response to a noxious stimulus is not pain. The International Association for the Study of Pain has defined pain in the following way:

Pain is an unpleasant sensory and emotional experience associated with actual or potential tissue damage, or described in terms of such damage.

It must always be remembered that the perception of pain is a complex interaction that involves sensory, emotional and behavioural factors.

A person's emotional and behavioural responses must always be considered as an important component in the perception and expression of pain. The person in pain must always be seen in the context of the interactions between biological and psychosocial processes. Any attempts to manage pain that fail to take these interactions into account will, inevitably, lead to frustration and failure.

The biological processes involved in our perception of pain are no longer viewed as a simple 'hard-wired' system with a pure 'stimulus–response' relationship. The more recent conceptualization of pain seeks to take into account the changes which occur within the nervous system following any prolonged, noxious stimulus. Trauma to any part of the body – and nerve damage in particular – can lead to changes within other regions of the nervous system which influence subsequent responses to sensory input. Long-term changes occur within the peripheral and central nervous system following noxious input. This 'plasticity' of the nervous system then alters the body's response to further peripheral sensory input.

Pain be divided into two entities: 'physiological' and 'pathophysiological' (or 'clinical'). Physiological pain describes the situation in which a noxious stimulus activates peripheral nociceptors which then transmit sensory information through several relays to the brain and is recognized as a potentially harmful stimulus. More commonly, the insult to the body which produces pain also causes inflammation and tissue or nerve injury. The pathophysiological processes which occur following injury result in a stimulus–response pattern that is different from that seen following physiological pain and has therefore been termed 'pathophysiological' or 'clinical' pain.

Nociception describes the somatosensory response of the nervous system to a potentially harmful stimulus and serves to avoid tissue damage. Various stimuli that initiate nociceptive responses can cause tissue damage at different levels: mechanical, electrical, thermal and chemical stimuli can cause superficial damage at the skin surface, ischaemia, distention or stretch, and inflammation in deeper tissues; ischaemia and chemical stimuli can damage nerves. The psychological responses associated with pain include emotional, cognitive, and behavioural changes.

PERIPHERAL MECHANISMS OF PAIN

Primary afferent nociceptors

The primary afferent nociceptor is generally the initial structure involved in nociceptive processes. Nociceptors are widespread in skin, muscle, connective tissues, blood vessels and thoracic and abdominal viscera. They are responsive to a variety of mechanical, thermal and chemical stimuli. The main molecular events that control the excitability of a primary neuron include the opening and closing of voltage-gated sodium or potassium channels. Depending on the response characteristics of the nociceptor, stimulation results in propagation of impulses along the afferent fibre toward the spinal cord.

Various molecular mechanisms allow different stimuli to be transduced to neural signals transmitted to the brain for pain perception. Activation of cationic ion channels are responsible for the generation of nociceptive signals. The main channels responsible for inward membrane currents in nociception are voltage-activated sodium and calcium channels whilst outward current is mediated by potassium ions. In addition, activation of non-selective cation channels is also responsible for the excitation of sensory neurons. Sodium (Na^+) channels open rapidly and transiently when the membrane is depolarized beyond -60 to $-40\,mV$, causing rapid membrane depolarization and action potential generation. Ectopic discharges of electrical activity arise from injured sensory nerves and these are mediated by Na^+ channels. The Na^+ channels may be sensitive or resistant to tetrodotoxin. These ectopic discharges can blocked by tetrodotoxin. Tetrodotoxin-resistant channels are down-regulated in damaged sensory neurons after spinal nerve injury but are up-regulated in undamaged but sensitized neurons. These changes indicate that some types of Na^+ channels are involved in the generation and maintenance of neuropathic pain. Voltage-dependent calcium channels (VDCC) exert their functions in sensory transduction by increasing intracellular Ca^{++} in response to depolarization. On activation of VDCC, substance P and calcitonin gene-related peptide (CGRP) are released. K^+ channels are the main channels that stabilize the membrane potential by producing hyperpolarizing outward currents. A variety of voltage-gated K^+ channels are found in sensory neurons. An increase in the excitability of a sensory neuron may be mediated by a decrease in the expression of K^+ channels. The transient receptor potential (TRP) ion channels are a family of non-selective cation channels with a variable permeability to Ca^{++} ions. The vanilloid receptor-related TRP channels (TRPV) are activated by a variety of sensory stimuli. TRPV1 is expressed in small sensory neurons and is activated by capsaicin, heat ($>43°C$), acid, inflammation, ischaemia, and endogenous lipids (anandamide, polyunsaturated fatty acids). The acid-sensing ion channel (ASIC) is activated by extracellular acid (during inflammation and ischaemia of tissues) and is widely distributed in sensory neurons. A purinergic receptor (subtype P2X3) is an ATP-gated receptor that is expressed in nociceptors and is responsible for hyperalgesia in neuropathic pain. Serotonin is an endogenous pain-producing mediator released by platelets and enterochromaffin cells. Serotonin has been shown to depolarize C and A nerve fibres via a ionotropic ($5-HT_3R$) receptor.

Primary afferent nociceptors are pseudounipolar, with the cell body located in the dorsal root ganglion (DRG). The peripheral processes of these neurons innervate a variety of tissues where they they lose their perineural sheath. Nociceptors may be free nerve endings or have specialized terminal structures (e.g. Pacinian corpuscles). The nerve terminals do not only transduce mechanical, thermal or chemical stimuli into a series of action potentials relayed to the spinal cord; they also releases peptides (e.g. substance P, calcitonin gene-related peptide (CGRP), neurokinin A) that mediate inflammation.

There are two main categories of cutaneous receptors associated with noxious stimulation: Aδ-fibre-mechanothermal and C-fibre polymodal nociceptors. Approximately 10 per cent of cutaneous myelinated fibres and 90 per cent of unmyelinated fibres are nociceptive. The smallest myelinated fibres (Aδ fibres) involved in nociception have a diameter of 2–5 μm and conduction velocity of 6–30 m/s. Unmyelinated C fibres are less than 2 μm in diameter and have a conduction velocity of 0.5–2 m/s. The C fibres may be peptidergic (as they synthesize and release peptides such as substance P, calcitonin gene-related peptide, somatostatin, galanin, and vasoactive intestinal peptide) or non-peptidergic-expressing purinoreceptors and lectin.

The C-fibre polymodal nociceptors respond to noxious thermal (>45°C), noxious mechanical, and noxious chemical stimuli. However, there are a number of other receptors capable of transmitting noxious information from skin, muscle, joints and viscera. Activation of faster-conducting Aδ fibres generally results in short-lasting, pricking-type pain. Activation of slower-conducting C fibres generally results in dull, poorly localized, burning-type pain.

Trigeminal system

The face, head and parts of the mouth are innervated by the ophthalmic, mandibular and maxillary divisions of the trigeminal nerve. The cell bodies of these nerves are located in the trigeminal or Gasserian ganglion. Central terminals of these nerves enter the trigeminal sensory nucleus, which is the equivalent of the spinal dorsal horn.

Neurons responsive to noxious stimuli are located in the nucleus caudalis of the trigeminal complex. Second-order neurons in the trigeminal nucleus project to the contralateral ventroposterior region of the thalamus.

Trigeminal neurons also appear to receive inputs from spinal afferent collateral fibres. Stimulation from upper cervical segments can result in excitation of neurons that receive trigeminal sensory input. This means that pain which is perceived in the distribution of the trigeminal nerve, e.g. atypical facial pain, may be due at least in part to pathology that provides sensory input at upper cervical levels.

Visceral afferents

Nociception from primary afferent fibres in deeper tissues is less well understood. Pain from deep structures such as muscles, joints, bone, and viscera is more diffuse and difficult to localize. Sensation is often associated with autonomic effects such as sweating, increased blood pressure and increased respiratory rate mediated by the C fibres.

Visceral afferents transmit noxious information from visceral organs. The cell bodies are located in the dorsal root ganglia, and fibres travel with sympathetic and parasympathetic axons. The number of afferent fibres is low compared with the surface that is innervated, so that pain is poorly localized. Visceral afferents converge onto second-order dorsal horn cells which also receive cutaneous input. Convergence gives rise to the phenomenon of referred pain in dermatomal segments corresponding to their cutaneous innervation.

Silent nociceptors

Silent nociceptors are a class of unmyelinated primary afferent neurons that do not respond to excessive mechanical or thermal stimuli under normal circumstances, but in the presence of inflammation and chemical sensitization they become responsive and discharge vigorously, even during ordinary movement, and display changes in receptive fields. Their activation increases the transmission of nociceptive impulses reaching the dorsal horn of the spinal cord when the tissues are inflamed and may be important in the development of sensitization following inflammation.

Inflammation

Many forms of pain arise from direct activation or sensitization of primary afferent neurons, especially C-fibre polymodal nociceptors. However, the process of nociceptor activation sets in train other processes which contribute to and modify responses to further stimuli. For example, a relatively benign noxious stimulus such as a scratch to the skin initiates an inflammatory process in the periphery which then changes the response properties to subsequent sensory stimuli. Under normal conditions, thermal, mechanical and chemical stimuli activate high-threshold nociceptors which signal this information to the first relay in the spinal cord. However, under clinical conditions, the application of a noxious stimulus is usually prolonged, traumatic and associated with tissue damage. Tissue damage results in inflammation, which directly affects the response of the nociceptor to further stimulation.

PERIPHERAL SENSITIZATION

Part of the inflammatory response is the release of intracellular contents from damaged cells and inflammatory cells such as macrophages, lymphocytes and mast cells. Nociceptive stimulation also results in a neurogenic inflammatory response with the release of substance P, neurokinin A and calcitonin gene-related peptide (CGRP) from the peripheral terminals of nociceptive afferent fibres. Release of these peptides results in a changed excitability of sensory and sympathetic nerve fibres, vasodilatation, extravasation of plasma proteins as well as action on inflammatory cells to release chemical mediators. These interactions result in the release of a 'soup' of inflammatory mediators such as potassium, serotonin, bradykinin, substance P, histamine, cytokines, nitric oxide and products from the cyclooxygenase and lipoxygenase pathways of arachidonic acid metabolism (Fig. 13.1). These chemicals then act to sensitize high-threshold nociceptors, which results in the phenomenon of peripheral sensitization.

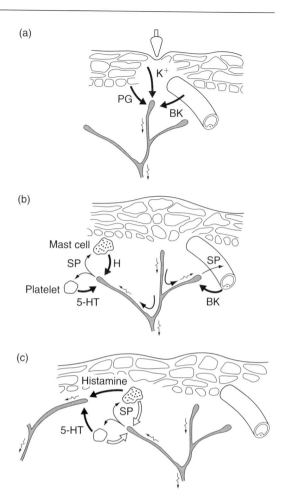

Figure 13.1 *Events leading to activation, sensitization and spread of sensitization of primary afferent nociceptor terminals. (a) Direct activation by intense pressure and consequent cell damage. Cell damage leads to release of potassium (K$^+$) and to synthesis of prostaglandins (PGs) and bradykinin (BK). Prostaglandins increase the sensitivity of the terminal to bradykinin and other pain-producing substances. (b) Secondary activation. Impulses generated in the stimulated terminal propagate not only to the spinal cord but also into other terminal branches, where they induce the release of peptides including substance P (SP). Substance P causes vasodilation and neurogenic oedema with further accumulation of bradykinin. Substance P also causes the release of histamine (H) from mast cells and serotonin (5-HT) from platelets. (c) Histamine and serotonin levels rise in the extracellular space, secondarily sensitizing nearby nociceptors. This leads to a gradual spread of hyperalgesia and/or tenderness (Reproduced, with permission, from Pain, H. Fields, 1987, McGraw Hill, New York.)*

Following sensitization, low-intensity mechanical stimuli which would not normally cause pain are now perceived as painful. There is also an increased responsiveness to thermal stimuli at the site of injury. This zone of 'primary hyperalgesia' surrounding the site of injury is due to peripheral changes and is a feature commonly observed following surgery and other forms of trauma.

Peripheral action of opioids

Opioids have traditionally been viewed as centrally acting drugs. However, there is now evidence for the action of endogenous opioids on peripheral sites following tissue damage. Opioid receptors are manufactured in the cell body (dorsal root ganglion) and transported toward the central terminal in the dorsal horn and toward the periphery (Fig. 13.2). These peripheral receptors then become active following local tissue damage. This occurs with unmasking of opioid receptors and the arrival of immunocompetent cells that possess opioid receptors and have the ability to synthesize opioid peptides.

Peripheral nerve injury

Nociceptors do not act simply as inert conductors of sensory information. Section of, or damage to, a peripheral nerve results in a number of biochemical, physiological and morphological changes (Fig. 13.3). Damage to peripheral nerves will result in ectopic discharges near the site of damage and adjacent to the dorsal root ganglion (Fig. 13.4). Nerve damage also results in an increased production of peptides, such as nerve growth factor (NGF), which normally regulate neuronal growth. Nerve growth factor is a neurotropic peptide which activates the tyrosine kinase (Trk) receptor. Inflammation is associated with increased NGF expression and synthesis in peripheral tissues. Nerve growth factor is important for the development of peripheral sensitization mediated by direct and indirect actions of inflammatory mediators on nociceptive afferents, mast cells and post-ganglionic efferents. Axonal transport of NGF has tropic effects within the spinal cord dorsal horn resulting in central sensitisation.

Growth factor changes may be responsible for the neuroplastic structural changes aimed at

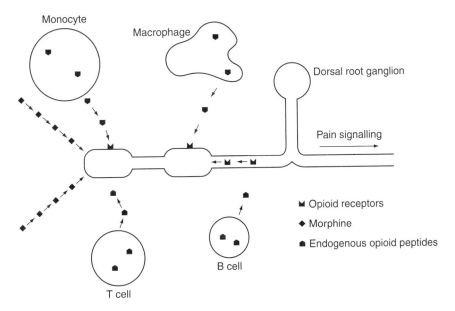

Figure 13.2 *Peripheral inflammation results in production of opioid receptors by the dorsal root ganglion (DRG) and transport of opioid receptors toward the peripheral terminal. Peripheral opioid receptors are activated by exogenous application of morphine and endogenous opioid peptides released by monocytes, T cells, B cells and macrophages. (Reproduced, with permission, from Stein C. Morphine – a local 'analgesic'.* Pain: Clinical Updates *3: 1–4, 1995.)*

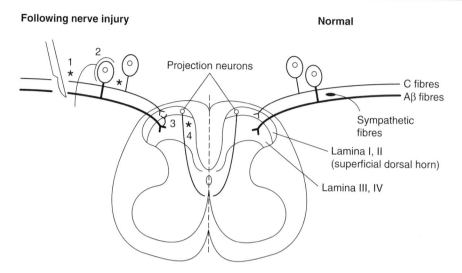

Figure 13.3 *Summary of events following peripheral nerve injury that may lead to the perception of pain.*
(1) Ectopic firing of damaged nerves either at the site of injury or close to the dorsal root ganglion. (2) Sprouting
of sympathetic nerve fibres around the dorsal root ganglion. (3) Sprouting of large-diameter afferent fibres into
the superficial dorsal horn. (4) Ectopic firing of cell bodies in the dorsal horn that have lost their normal afferent
input.

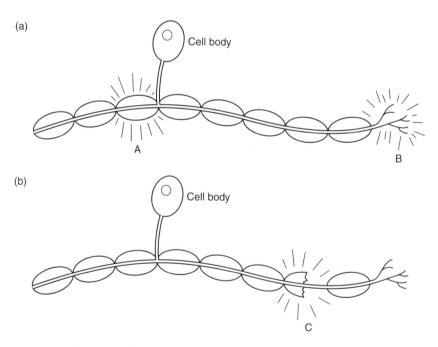

Figure 13.4 *Sites of ectopic discharge in damaged primary afferent nociceptors. (a) A transected nerve begins to*
regenerate, sending out sprouts (B) that are mechanically sensitive, sensitive to α-adrenergic agonists, and
spontaneously active. In addition, a secondary site of hyperactivity (A) develops near the cell body in the dorsal root
ganglion. (b) Ectopic impulses may arise from a short patch of demyelination on a primary afferent (C). (Reproduced, with
permission, from Pain, H. Fields, 1987, McGraw Hill, New York.)

maintaining sensory input that occur at peripheral and/or spinal levels (Fig. 13.5). The damaged end of the nerve fibre sprouts and may produce a

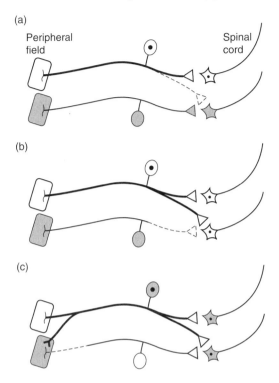

(a)

Peripheral field

Spinal cord

(b)

(c)

Figure 13.5 *Neuroplasticity following damage to primary afferent fibres. (a) Normal connectivity of primary afferent. Primary afferents innervate a defined peripheral region and activate a specific population of spinal cord neurons. In addition, the primary afferent has central connections that are normally ineffective (dashed line). In the normal situation, each spinal cord cell responds only to stimulation of its own peripheral field. (b) When the central process of a primary afferent that innervates an adjacent peripheral field (tinted) is interrupted (dotted line), the formerly ineffective central connection of the intact primary afferent (heavy line) becomes effective. Both spinal cord cells now respond only to stimulation of the innervated peripheral field (not tinted). (c) When the peripheral process of the adjacent primary afferent is cut (dotted line), changes occur in the spinal cord that are similar to those produced by cutting the central process. In addition, the peripheral process of the intact primary afferent (heavy line) sprouts and grows into the denervated peripheral region (tinted). In this case, both spinal cord cells respond to stimulation of both peripheral fields. (Reproduced, with permission, from* Pain, *H. Fields, 1987, McGraw Hill, New York.)*

spontaneously firing neuroma. It may also demonstrate changed properties in response to various stimuli. These properties include sensitivity to mechanical stimuli, and sensitivity to noradrenaline (norepinephrine). Similar changes occur within the cell body of the afferent nociceptor, the dorsal root ganglion. Reduction in the blood supply to myelinated fibres results in demyelination and the production of ectopic impulses. These impulses may give rise to the perception of sharp, shooting or burning pain in conditions such as diabetic neuropathy. Other clinical conditions that may have a peripheral neuropathic component include post-herpetic neuralgia and post-amputation pain; however, almost invariably there is a central component (see below).

Sympathetic nervous system

The sympathetic nervous system also has an important role in the generation and maintenance of chronic pain states. Nerve damage and even minor trauma can lead to a disturbance in sympathetic activity (Fig. 13.6) which then leads to a sustained condition termed a 'complex regional pain syndrome', which now replaces the previously used term 'reflex sympathetic dystrophy'. Complex regional pain syndromes are associated with features of sympathetic dysfunction including vasomotor and sudomotor changes, abnormalities of hair and nail growth, and osteoporosis as well as sensory symptoms of spontaneous burning pain, hyperalgesia and allodynia, and often disturbance of motor function.

Basic studies demonstrated that several changes involving the sympathetic nervous system may be responsible for the development of these features. Inflammation can result in the sensitization of primary nociceptive afferent fibres by prostanoids that are released from sympathetic fibres. Following nerve injury, sympathetic nerve stimulation or administration of norepinephrine can excite primary afferent fibres via an action at α adrenoceptors. There is also innervation of the DRG by sympathetic terminals. This means that activity in sympathetic efferent fibres can lead to abnormal activity or responsiveness of the primary afferent fibre.

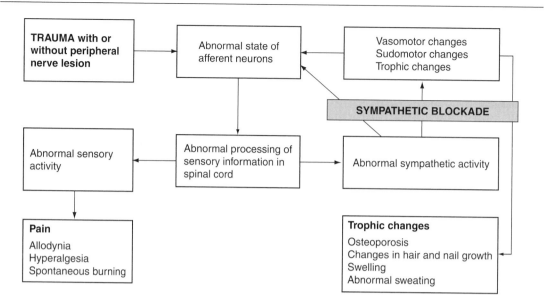

Figure 13.6 *General hypothesis about the neural mechanisms underlying the generation of pain and other symptoms associated with complex regional pain syndromes (types I and II) following peripheral trauma with and without nerve lesions.*

DORSAL HORN MECHANISMS

Termination sites of primary afferents

The dorsal horn is the site of termination of primary afferents. Small myelinated and unmyelinated fibres tend to aggregate in the lateral aspect of the dorsal root and enter the dorsal horn laterally, whilst larger fibres tend to travel medially. Whilst the principal route of entry for primary afferents is through the dorsal root, a significant number of primarily unmyelinated afferent neurons enter via the ventral root. Small myelinated fibres terminate principally in the superficial dorsal horn (lamina I) and deeper in lamina V. Unmyelinated fibres terminate principally in lamina II. Some small-diameter fibres also ascend and descend several segments in Lissauer's tract before terminating on neurons that project to higher centres.

Within the dorsal horn, there is a complex interaction among afferent fibres, local intrinsic spinal neurons and the endings of descending fibres from the brain. There are two main classes of second-order dorsal horn neurons associated with sensory processing. The first class of neurons is termed 'nociceptive-specific' or 'high-threshold'. The second class of neurons is termed 'wide dynamic range'

(WDR) or 'convergent'. Nociceptive-specific neurons are located within the superficial laminae of the dorsal horn and respond selectively to noxious stimuli. WDR neurons are generally located in deeper laminae and respond to both noxious and non-noxious input. WDR neurons increase their firing rate in proportion to the intensity of the stimulus.

WDR neurons normally do not signal pain in response to a tactile stimulus at a non-noxious level. However, if they become sensitized and hyper-responsive, they may discharge at a high rate following a tactile stimulus. If the activity of the WDR neuron exceeds a threshold level following this stimulus, then the non-noxious tactile stimulus will be perceived as painful and give rise to the phenomenon of allodynia.

Endogenous ligands

Pharmacological studies have identified the many neurotransmitters and neuromodulators that are involved in pain processes in the dorsal horn. The excitatory amino acid glutamate has a major role in nociceptive transmission in the dorsal horn. Glutamate acts at N-methyl-D-aspartate (NMDA) receptors, non-NMDA receptors such as AMPA (α-amino-3-hydroxy-5-methyl-4-isoxazolepropionic acid), kainate and metabotropic glutamate receptors (Fig. 13.7).

A number of peptides released by primary afferents have a role in nociception. and these include substance P, neurokinin A and CGRP. Substance P and neurokinin A act on neurokinin receptors. Other receptors such as opioid (μ, κ and δ), α-adrenergic, γ-aminobutyric acid (GABA), serotonin (5-HT) and adenosine receptors are also involved in nociceptive transmission or modulation. New drugs which act at the nicotinic acetylcholine receptor appear to provide strong analgesia without the withdrawal symptoms associated with opioids.

Traditional approaches in pain management have focused on classical ligand–receptor blockade as a means to reduce nociceptive or neuropathic input. Our understanding of the molecular and genetic mechanisms involved in nociception provides a new and potentially useful approach to pain management. It may be possible to develop drugs that regulate gene expression and selectively modify the expression of specific receptors that are involved in the transmission of nociceptive and neuropathic messages.

NMDA receptor

Non-NMDA receptors such as the AMPA receptor may mediate responses in the 'physiological' processing of sensory information. With prolonged release of glutamate or activation of neurokinin receptors, a secondary process occurs which appears to be crucial in the development of abnormal responses to further sensory stimuli. This sustained activation of non-NMDA or neurokinin receptors 'primes' the NMDA receptor so that it is in a state ready for activation.

NMDA receptors are involved in a number of phenomena that may contribute to the medium- or long-term changes observed in chronic pain states. These phenomena include the development of 'wind-up', facilitation, central sensitization, changes in peripheral receptive fields, induction of

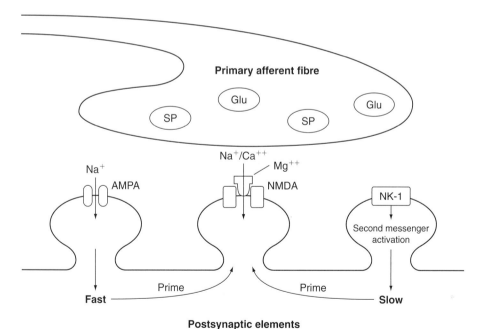

Figure 13.7 *Release of glutamate (Glu) and substance P (SP) from the terminals of nociceptive primary afferents activates AMPA and neurokinin-1 (NK-1) receptors, respectively, on the postsynaptic membrane. Activation of these receptors results in sodium (Na^+) influx at the AMPA receptor and activation of second messengers. These processes then act to 'prime' the NMDA receptor with removal of the magnesium (Mg^{++}) 'plug' and sodium and calcium (Ca^{++}) influx. AMPA, α-amino-3-hydroxy-5-methyl-4-isoxazolepropionic acid; NMDA, N-methyl-d-aspartate.*

immediate early genes and long-term potentiation. Long-term potentiation, in particular, relates to the changes in synaptic efficacy that occur as part of the process of memory, and may play a role in the development of a cellular 'memory' for pain or enhanced responsiveness to noxious inputs. NMDA antagonists can attenuate these responses, indicating a role for such compounds in the prevention of chronic pain states.

INTRACELLULAR EVENTS

Activation of NMDA receptors activates a cascade of secondary events in the cell which lead to changes within the cell to increase the responsiveness of the nociceptive system. The NMDA receptor channel in its resting state is 'blocked' by a magnesium 'plug'. Priming of the NMDA receptor by co-release of glutamate and the peptides acting on the neurokinin receptors removes the magnesium plug and results in the subsequent calcium influx into the cell, leading to secondary events such as immediate early gene induction, production of

nitric oxide (NO), and activation or production of a number of second messengers including phospholipases, polyphosphoinosites (IP_3, DAG), cGMP, eicosanoids and protein kinase C (Fig. 13.8). These second messengers then directly change the excitability of the cell or induce the production of oncogenes which may result in long-term alterations in the responsivity of the cell. Prolonged stimulation, through sustained and excitotoxic release of glutamate, may result in cell death. It has been demonstrated experimentally that interference with the function of second messenger systems such as protein kinase C results in normal responses to acute painful stimuli ('good pain') and attenuation of the development of neuropathic pain states ('bad pain').

The exact role of NO in nociceptive processing is unclear, and it does not appear to be important in acute nociception. However, NO is implicated in the induction and maintenance of chronic pain states, and may contribute to cell death which has been demonstrated to occur under these

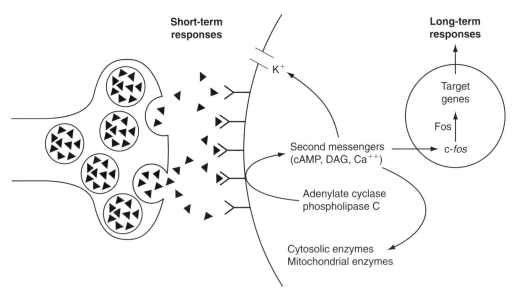

Figure 13.8 *Neurotransmitter release from the central terminal of peripheral afferents results in activation of receptor sites on the postsynaptic membrane. Activation of phospholipase C and adenylate cyclase leads to the production of the second messengers cyclic AMP and diacylglycerol (DAG). Mobilization of these second messengers may result in a decrease in K^+ efflux and elevation of intracellular calcium. The increase in intracellular calcium results in the induction of the proto-oncogene c-fos, production of Fos protein and a presumed action on target genes to alter long-term responses of the cell to further stimuli. (Reproduced, with permission, from* Neural Blockade in Clinical Anesthesia and the Management of Pain, *M. Cousins and P. Bridenbaugh, 1998, Lippincott-Raven, Philadelphia.)*

conditions (Fig. 13.9). It has been suggested that NO acts as a positive-feedback mechanism in the maintenance of pain. Blockade of NO in neuropathic animal pain models results in a decrease in the behavioural correlates of pain.

The production of arachidonic acid metabolites as part of the cascade that occurs following NMDA receptor activation also raises an interesting potential avenue of intervention that is already being explored. Although the peripheral effects of NSAIDs have been emphasized in the past, it appears that there may be a role for the spinal administration of NSAIDs. Spinal NSAIDs either act directly on receptors, such as the NMDA, neurokinin and strychnine-sensitive glycine receptors, or influence the production of metabolites within the cell.

Central sensitization

Changes in the periphery following trauma can lead to the phenomenon of peripheral sensitization and primary hyperalgesia. The sensitization that occurs, however, can only be partly explained by the changes in the periphery. Following injury, there is an increased responsiveness to normally innocuous mechanical stimuli (allodynia) in a zone of 'secondary hyperalgesia' in uninjured tissue surrounding the site of injury. In contrast to the zone of primary hyperalgesia, there is no change in the threshold to thermal stimuli. These result from processes that occur in the dorsal horn of the spinal cord following injury. This is the phenomenon of central sensitization.

A barrage of nociceptive input, such as occurs with surgery, results in changes to the response properties of dorsal horn neurons. A painful stimulus, which is at a level sufficient to activate C fibres, not only activates dorsal horn neurons but neuronal activity also progressively increases throughout the duration of the stimulus. Therefore, with clinical pain associated with nociceptive input, there is not a simple stimulus–response relationship, but a 'wind-up' of spinal cord neuronal activity. Wind-up refers to the phenomenon when a repeated stimulus (with no

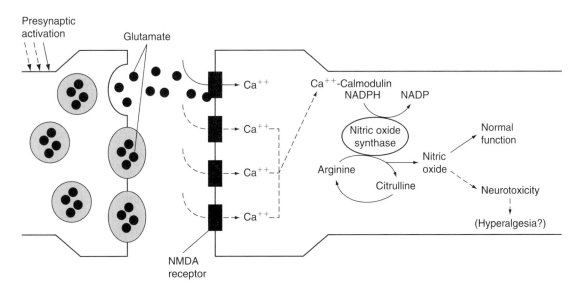

Figure 13.9 *Diagram illustrating postsynaptic events following release of glutamate from central terminals of primary afferents in the spinal cord. Following priming of the NMDA receptor, complex subsequent glutamate release results in NMDA receptor activation with subsequent calcium influx. Intracellular calcium then acts on a calmodulin-sensitive site to activate the enzyme nitric oxide synthase (NOS). In the presence of a co-factor NADPH, NOS uses arginine as a substrate to produce nitric oxide and citrulline. Nitric oxide has a role in normal cellular function, but increased production may be involved in hyperalgesia and may lead to neurotoxicity. (Reproduced with permission, from* Neural Blockade in Clinical Anesthesia and the Management of Pain, *M. Cousins and P. Bridenbaugh, 1998, Lippincott-Raven, Philadelphia.)*

change in strength) causes an increase in response from dorsal horn neurons mediated by the release of excitatory neuromediators. Wind-up is dependent on activation of the NMDA receptor. This 'wind-up' may make these neurons more sensitive to other input and is a component of central sensitization.

Several other changes occur in the dorsal horn with central sensitization. First, there is an expansion in receptive field size so that a spinal neuron will respond to stimuli which would normally be outside the region that responds to nociceptive stimuli. Second, there is an increase in the magnitude and duration of the response to stimuli which are above threshold in strength. Third, there is a reduction in threshold so that stimuli which are not normally noxious activate neurons that normally transmit nociceptive information. These changes may be important both in acute pain states such as postoperative pain and in the development of chronic pain.

The demonstration of the phenomenon of 'wind-up' has led to a surge of interest in approaches such as pre-emptive analgesia. The rationale behind pre-emptive analgesia lies in an attempt to reduce the development of 'sub-acute' or chronic pain by abolishing or reducing acute pain and thus preventing the changes associated with wind-up.

However, the development of chronic pain may have more to do with the phenomenon of long-term potentiation (LTP) than it does with wind-up. LTP is the strengthening of the efficacy of synaptic transmission that occurs following activity across that synapse, and shares many of the physiological and biochemical features that are associated with the development of chronic pain.

Nerve damage enhances calcium flux, NO production and protein kinase C (PKC) generation. PKC increases activity at the NMDA receptor, thus causing a vicious circle. Intrathecal, or systemic, administration of morphine for neuropathic pain may also increase NMDA activity, whilst at the same time decreasing the efficacy of morphine at the μ opioid receptor. Thus, unwittingly, morphine administration in neuropathic pain may progressively contribute to increasing the pain.

Morphological changes occur within the dorsal horn following peripheral nerve injury. Peripheral nerve injury results in a redistribution of central terminals of myelinated afferents, with sprouting of these terminals from lamina IV to lamina II. If functional contact is made between these terminals which normally transmit non-noxious information and neurons that normally receive nociceptive input, this may provide a framework for the pain and hypersensitivity to light touch (allodynia) that is seen following nerve injury.

Modulation at a spinal level

Transmission of nociceptive information is modulated at several levels of the neuraxis, including the dorsal horn. Afferent impulses arriving in the dorsal horn initiate inhibitory mechanisms which limit the effect of subsequent impulses. Inhibition occurs through the effect of local inhibitory interneurons and descending pathways from the brain (Fig. 13.10). Melzack and Wall proposed the 'gate control theory of pain' in 1965 to explain how pain can be modulated in the spinal cord. They proposed that transmission or T cells located in the dorsal horn project information to the brain. The output from the transmission cells is regulated by inhibitory interneurons at the synapse between the Aδ or C fibres and the T cells in the substantia gelatinosa. Synaptic transmission between primary and secondary nociceptive afferent neurons can be 'gated' by interneurons in the substantia gelatinosa. Non-noxious (touch, pressure, temperature) sensory information carried by large-diameter Aβ fibres activate the inhibitory interneurons and inhibit the transmission cells and suppress the flow of pain information towards the brain. Noxious or pain input along the small-diameter Aδ or C afferent fibres inhibit the inhibitory interneurons and therefore increase output from the transmission cells. Descending pathways from the brain can also inhibit transmission of information by transmission cells. In summary, activity in the large-diameter fibres tends to close the gate whilst activity in the smaller pain fibres tends to open the gate and facilitate transmission.

In the dorsal horn, incoming nociceptive messages are modulated by endogenous and exogenous agents which act on opioid, α adrenoreceptors, GABA and glycine receptors located at both pre- and postsynaptic sites.

OPIOID RECEPTORS

Opioid receptors are found both pre- and postsynaptically in the dorsal horn, although the majority

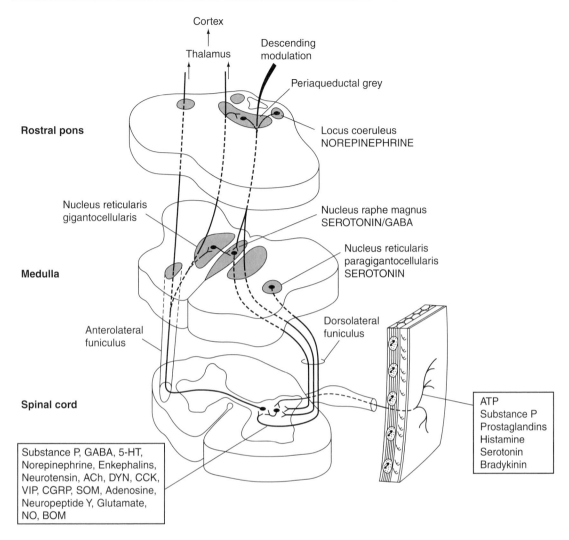

Cortex

Thalamus

Descending
modulation

Periaqueductal grey

Rostral pons

Locus coeruleus
NOREPINEPHRINE

Nucleus reticularis
gigantocellularis

Nucleus raphe magnus
SEROTONIN/GABA

Nucleus reticularis
paragigantocellularis
SEROTONIN

Medulla

Anterolateral
funiculus

Dorsolateral
funiculus

Spinal cord

ATP
Substance P
Prostaglandins
Histamine
Serotonin
Bradykinin

Substance P, GABA, 5-HT,
Norepinephrine, Enkephalins,
Neurotensin, ACh, DYN, CCK,
VIP, CGRP, SOM, Adenosine,
Neuropeptide Y, Glutamate,
NO, BOM

Figure 13.10   *Simplified schema of afferent sensory pathways and descending modulatory pathways arising from the midbrain and medulla. Note the various sites for enhancement or reduction in pain signalling. Release of chemicals from peripheral terminals of primary afferents results in peripheral sensitization. The main ascending nociceptive pathway travels primarily via the anterolateral funiculus and terminates in the thalamus (spinothalamic) and reticular formation (spinoreticular). Incoming signals from the periphery are then modulated at the spinal dorsal horn by intrinsic interneurons and descending influences from the brainstem. These descending influences travel primarily via the dorsolateral funiculus and arise from several regions including the periaqueductal grey matter, locus coeruleus and nucleus raphe magnus. Neurotransmitters released from the terminals of these descending pathways then act to inhibit incoming peripheral nociceptive input. (Reproduced, with permission, from* Neural Blockade in Clinical Anesthesia and the Management of Pain, *M. Cousins and P. Bridenbaugh, 1998, Lippincott-Raven, Philadelphia.)*

(about 75 per cent) are located presynaptically. Activation of presynaptic opioid receptors results in a reduction in the release of neurotransmitters from the nociceptive primary afferent. However, the changes that occur with inflammation and neu-ropathy can produce significant changes in opioid sensitivity via a number of mechanisms. These include an interference with opioid analgesia by cholecystokinin (CCK); loss of presynaptic opioid receptors; and the formation of the morphine metabolite, morphine-3-glucuronide, which may antagonize the analgesic action normally produced by opioid receptor activation.

NMDA receptor is also involved in the development of tolerance to opioids. Animal studies indicate that administration of an NMDA antagonist reduces the development of tolerance to morphine and prevents the withdrawal syndrome in morphine-tolerant rats.

ALPHA-ADRENOCEPTORS

Activation of α adrenoceptors in the spinal cord has an analgesic effect either by endogenous release of norepinephrine by descending pathways from the brainstem; or by exogenous spinal administration of agents such as clonidine. Furthermore, α-adrenoceptor agonists appear to have a synergistic effect with opioid agonists. There are a number of α-adrenoceptor subtypes, and the development of selective α-adrenoceptor-subtype agonists has the potential to provide effective new analgesic agents with reduced side effects.

GABA AND GLYCINE

Both GABA and glycine are involved in tonic inhibition of nociceptive input, and loss of their inhibitory action can result in features of neuropathic pain such as allodynia. Both $GABA_A$ and $GABA_B$ receptors have been implicated at both pre- and postsynaptic sites. However $GABA_A$-receptor-mediated inhibition occurs through largely postsynaptic mechanisms. In contrast, $GABA_B$ mechanisms may be preferentially involved in presynaptic inhibition via suppression of excitatory amino acid release from primary afferent terminals.

ASCENDING TRACTS

Spinal structures

Second-order projection neurons in the dorsal horn, as well as some in the ventral horn and central canal region, project to supraspinal structures. Fibres may ascend one or two segments from their point of origin before crossing in the dorsal commissure. They then ascend predominantly in the contralateral anterolateral quadrant (ventrolateral funiculus) of the spinal cord, but a significant proportion travel ipsilaterally.

There appears to be a somatotopic organization within the ventrolateral funiculus. Fibres arising from more caudal segments tend to be located laterally, whilst those entering from more rostral segments tend to be located medially and ventrally. Fibres form several tracts within the ventrolateral funiculus, including the spinothalamic, spinoreticular and spinomesencephalic tracts.

The distribution of fibres associated with pain transmission within the anterolateral quadrant would suggest that section of these tracts using an anterolateral cordotomy should be a useful procedure in abolishing or relieving pain. However, results are variable and often transient. There is a latent ipsilateral pathway which progressively takes over from the contralateral spinothalamic pathway following cordotomy, and this may account for eventual failure and, sometimes, also the lack of analgesia.

The dorsal columns mainly contain large-diameter fibres associated with the transmission of information related to light touch and vibration. Stimulation of the dorsal columns produces a sensation of vibration rather than pain. However, there are some fibres which have their origin in lamina V and respond to noxious thermal and mechanical stimuli.

Supraspinal structures

Second-order neurons ascend the spinal cord to terminate in many supraspinal structures throughout the brainstem, thalamus and cortex.

Axons from the spinoreticular tract terminate in several brainstem nuclei such as nucleus reticularis gigantocellularis, nucleus reticularis paragigantocellularis lateralis, nucleus subcoeruleus, nucleus reticularis pontis caudalis and medullary raphe nuclei (magnus and pallidus). These projections may activate the descending modulatory pathways, generalized arousal mechanisms, and motor and autonomic reflexes.

Axons from the spinomesencephalic tract terminate in the superior colliculus, nucleus cuneiformis and periaqueductal grey matter. Like the projections to the lower brainstem, midbrain projections may be involved in autonomic reflexes and activation of modulatory pathways. They may also be responsible for the activation of more integrated and coordinated affective motor responses to pain such as confrontation, escape or quiescence.

In the thalamus, axons within the spinothalamic pathway have been divided into two main groups. One pathway terminates more laterally in the ventral posterior nuclei of the thalamus and is involved in the sensory discriminative component of pain. Another pathway terminates more medially in the intralaminar nuclei, including the central lateral nucleus. This pathway is believed to be involved in the affective–motivational aspects of pain. Animal studies indicate that there is a large spinothalamic projection with terminations within the ventral posterior thalamic nucleus, and neurons within this region respond preferentially to noxious stimuli. Nuclei within this region act as a relay for the transmission of nociceptive information. Stimulation of the ventrocaudal nucleus (analogous to the ventral posterior nucleus in animals and supposedly part of the 'pain' pathway) in awake humans rarely results in pain, except in those people who have central deafferentation pain.

Neurons that project to lateral and medial areas of the thalamus also display different properties. Neurons that project to the lateral thalamus have smaller receptive fields and require more intense stimulation for activation. Neurons that project to the medial thalamus have larger receptive fields, often including the whole body.

Positron emission studies identified a number of subcortical structures that are involved in nociceptive transmission and pain perception. These include the thalamus, putamen, caudate nucleus, hypothalamus, amygdala, periaqueductal grey, hippocampus and cerebellum. Also of interest is the finding from positron emission tomography (PET) studies that an acute experimental painful stimulus results in an increase of activity in the thalamus, whilst those with chronic pain due to cancer and chronic neuropathic pain demonstrate a decrease in activity in the thalamus.

Cortical structures

The higher neural centres involved in pain processing can be divded into those which are involved in the sensory–discriminative component of pain perception (somatosensory cortex) and the affective component of pain perception (cingulate cortex). However, this may be an oversimplification, and the role of the cortex in pain perception remains unclear.

The effect of cortical stimulation and lesions on pain perception is confusing and intriguing. Patients who have had a complete hemispherectomy can have almost normal pain sensation. In the awake human, stimulation of primary somatosensory cortex typically evokes non-painful sensations. Neurosurgical lesions of cortical regions produce varying effects depending on the region ablated. Lesions of the frontal lobe and cingulate cortex result in a condition in which pain perception remains, but the suffering component of pain appears to be reduced – the person only reports pain when queried, and spontaneous requests for analgesia are reduced. Following lesions of medial thalamus and hypothalamus there is pain relief, but without demonstrable analgesia to peripheral stimuli which would normally be perceived as painful.

Both PET and functional magnetic resonance imaging (fMRI) have been helpful in elucidating supraspinal mechanisms of pain processing. Using both techniques, painful stimuli result in activation of sensory, motor, premotor, parietal, frontal, occipital, insular and anterior cingulate regions of the cortex. It has been suggested on the basis of PET findings that the parietal regions of the cortex are responsible for evaluation of the temporal and spatial features of pain, and the frontal cortex, including anterior cingulate, is responsible for the emotional response to pain.

DESCENDING MODULATION

Since the turn of the 20th century, considerable interest has focused on the presence of descending influences which modulate sensory input. This concept was further developed by Melzack and Wall with the proposal of the 'gate theory'. It is now known that there are powerful inhibitory (and facilitatory) influences on nociceptive transmission acting at many levels of the neuraxis. Descending inhibition may be activated by external factors such as stress, acupuncture and spinal cord stimulation.

Descending influences arise from a number of supraspinal structures including hypothalamus, periaqueductal grey (PAG) matter, locus coeruleus, nucleus raphe magnus and nucleus paragigantocellularis lateralis (Fig. 13.10). They descend in the spinal cord in the dorsolateral funiculus.

Descending modulation involves a number of neurotransmitters. Serotonin and norepinephrine are released from descending fibres in the dorsal horn and appear to have an important role in descending modulation. Other transmitters which appear to be important are substance P, cholecystokinin, GABA, thyrotrophin-releasing hormone (TRH), somatostatin and enkephalin.

There is also evidence for the modulation of 'higher' structures. For example, stimulation in the PAG matter can produce inhibition of the responses of neurons in the medial thalamus. Although this inhibition may occur through the activation of descending pathways, it indicates that there are multiple interactions at many levels of the nervous system.

Reflections

1. Pain is an unpleasant sensory and emotional experience associated with actual or potential tissue damage, or described in terms of such damage. Physiological pain occurs when a noxious stimulus activates peripheral nociceptors and is recognized as a potentially harmful stimulus. The pathophysiological processes, e.g. inflammation or nerve damage following injury, result in an altered stimulus–response pattern and the response has therefore been termed 'pathophysiological' or 'clinical' pain.

2. Nociceptors are widespread in skin, muscle, connective tissues, blood vessels and thoracic and abdominal viscera. They respond to mechanical, thermal and chemical stimuli. The main molecular events that control the excitability of a primary neuron include the opening and closing of voltage-gated sodium or potassium channels.

3. The main channels responsible for inward membrane currents in nociception are voltage activated sodium and calcium channels whilst outward current is mediated by potassium ions. The transient receptor potential (TRP) ion channels are a family of non-selective cation channels with a variable permeability to Ca^{++} ions. The vanilloid receptor-related TRP channels (TRPV) are activated by a variety of sensory stimuli: capsaicin, heat ($>43°C$), acid, inflammation, ischaemia, and endogenous lipids. The acid-sensing ion channel (ASIC) is activated by extracellular acid (during inflammation and ischaemia of tissues). The purinergic receptor (P2X3) is an ATP-gated nociceptor responsible for hyperalgesia in neuropathic pain. Serotonin is an endogenous pain-producing mediator released by platelets and enterochromaffin cells.

4. Primary afferent nociceptors are pseudounipolar, with the cell body located in the dorsal root ganglion (DRG). The two main cutaneous receptors associated with noxious stimulation are the Aδ-fibre mechanothermal and C-fibre polymodal nociceptors. The C-fibre polymodal nociceptors respond to noxious thermal ($>45°C$), noxious mechanical, and noxious chemical stimuli. Activation of faster-conducting Aδ fibres generally results in short-lasting, pricking-type pain. Activation of slower-conducting C fibres generally results in dull, poorly localized, burning-type pain.

5. Nociception from primary afferent fibres in deeper tissues such as muscles, joints, bone, and viscera is more diffuse and difficult to localize and is often associated with autonomic effects such as sweating, increased blood pressure and increased respiratory rate mediated by the C fibres.

6. Silent nociceptors are unmyelinated primary afferent neurons that do not respond to excessive mechanical or thermal stimuli under normal circumstances, but become responsive in the presence of inflammation

and chemical sensitization. Various interactions result in the release of a 'soup' of inflammatory mediators such as potassium, serotonin, bradykinin, substance P, histamine, cytokines, nitric oxide and products from the cyclooxygenase and lipoxygenase pathways of arachidonic acid metabolism. Nociceptive stimulation also causes release of substance P, neurokinin A and calcitonin gene-related peptide (CGRP) from the peripheral terminals of nociceptive afferent fibres. These chemicals then sensitize high-threshold nociceptors.

7. Peripheral nerve damage can cause ectopic discharges near the site of damage and adjacent to the dorsal root ganglion. Nerve damage also increases production of peptides, such as nerve growth factor (NGF), which normally regulate neuronal growth. Nerve growth factor is important for the development of peripheral sensitization mediated by direct and indirect actions of inflammatory mediators on nociceptive afferents, mast cells and postganglionic efferents. Axonal transport of NGF has trophic effects within the spinal cord dorsal horn resulting in central sensitization. It may be responsible for the neuroplastic structural changes. The damaged end of the nerve fibre sprouts and may produce a spontaneously firing neuroma and show increased sensitivity to mechanical stimuli, and sensitivity to noradrenaline (norepinephrine). Similar changes occur within the cell body of the afferent nociceptor, the dorsal root ganglion.

8. Nerve damage and even minor trauma can lead to increased sympathetic activity. Complex regional pain syndromes are associated with features of sympathetic dysfunction including vasomotor and sudomotor changes, abnormalities of hair and nail growth, and osteoporosis as well as sensory symptoms of spontaneous burning pain, hyperalgesia and allodynia, and often disturbance of motor function.

9. There are two main classes of second-order dorsal horn neurons: the 'nociceptive-specific' or 'high-threshold' neurons, and the 'wide dynamic range' (WDR) or 'convergent' neurons. Nociceptive-specific neurons, located within the superficial laminae of the dorsal horn, respond selectively to noxious stimuli. WDR neurons are generally located in deeper laminae and normally do not signal pain in response to a tactile stimulus at a non-noxious level. However, if they become sensitized and hyper-responsive, they may discharge at a high rate following a tactile stimulus and then the non-noxious tactile stimulus will be perceived as painful and give rise to 'allodynia'.

10. The excitatory amino acid glutamate has a major role in nociceptive transmission in the dorsal horn. Glutamate acts at N-methyl-D-aspartate (NMDA) receptors, non-NMDA receptors such as AMPA (α-amino-3-hydroxy-5-methyl-4-isoxazolepropionic acid), kainate and metabotropic glutamate receptors. Peptides released by primary afferents that have a role in nociception include substance P, neurokinin A and CGRP. α-Adrenergic, β-aminobutyric acid (GABA), serotonin (5-HT) and adenosine receptors are also involved in nociceptive modulation.

11. NMDA receptors may contribute to the medium- or long-term changes such as 'wind-up', facilitation, central sensitization, changes in peripheral receptive fields, induction of immediate early genes and long-term potentiation. Wind-up refers to the phenomenon when a repeated stimulus (with no change in strength) causes an increase in response from dorsal horn neurons and is a component of central sensitization. There is an expansion in receptive field size so that a spinal neuron will respond to stimuli which would normally be outside the region that responds to nociceptive stimuli. These changes may be important both in acute pain states and in the development of chronic pain.

12. Transmission of nociceptive information is modulated at several levels of the neuraxis, including the dorsal horn. Inhibition occurs through the effect of local inhibitory interneurons and descending pathways from the brain. Melzack and Wall proposed that output from the transmission cells in the spinal cord is regulated by inhibitory interneurons at the

synapse between the Aδ or C fibres and the T cells in the substantia gelatinosa (gate control theory). Non-noxious (touch, pressure, temperature) sensory information carried by large-diameter Aβ fibres activates the inhibitory interneurons and inhibits the transmission cells and suppresses the flow of pain information towards the brain. Noxious or pain input along the small-diameter Aδ or C afferent fibres inhibits the inhibitory interneurons and therefore increases output from the transmission cell. Therefore activity in the large-diameter fibres tends to close the gate whilst activity in the smaller pain fibres tends to open the gate and facilitate transmission.

13. Activation of α-adrenoceptors in the spinal cord has an analgesic effect brought about by the endogenous release of norepinephrine by descending pathways from the brainstem. Both GABA and glycine mediate tonic inhibition of nociceptive input, and loss of their inhibitory action can result in features of neuropathic pain such as allodynia.

14. Second-order projection neurons in the dorsal horn, as well as some in the ventral horn and central canal region, project to supraspinal structures. The fibres associated with pain transmission lie within the antero-lateral quadrant The dorsal columns mainly contain large-diameter fibres associated with the transmission of information related to light touch and vibration. Second-order neurons ascend the spinal cord to terminate in many supraspinal structures throughout the brainstem, thalamus and cortex. The higher neural centres involved in pain processing include the somatosensory cortex (responsible for sensory–discriminative component of pain perception) and the cingulate cortex (the affective component of pain).

15. Descending influences arising from supraspinal structures (hypothalamus, periaqueductal grey matter, locus coeruleus, nucleus raphe magnus and nucleus paragigantocellularis lateralis descend in the spinal cord in the dorsolateral funiculus. Serotonin, substance P, cholecystokinin, GABA, thyrotrophin-releasing hormone (TRH), somatostatin, enkephalin and norepinephrine released from descending fibres in the dorsal horn have an important role in descending modulation of pain.

Maternal and neonatal physiology

After studying this chapter the reader should be able to:

1. Describe the physiological changes in the mother and explain the role of hormones in the maintenance of pregnancy, breast development, milk production and lactation
2. Describe the functions of the placenta
3. Explain the mechanisms involved in the transfer of gases and nutrients between the mother and the fetus
4. Describe the differences in the organization of the fetal and adult circulations and explain why these are essential for the survival of the fetus

5. Describe the carriage of oxygen in fetal blood
6. Describe the physiological changes that take place following birth: the cardiovascular changes associated with the first breath, the factors responsible for closure of foramen ovale, ductus arteriosus, and ductus venosus, and the role of surfactant in lung inflation
7. Describe the differences in the respiration of a neonate and that of an adult
8. Describe the mechanisms underlying temperature regulation in the neonate

MATERNAL PHYSIOLOGY

Physiological changes in every organ system enable the mother to provide for the nutritional and metabolic demands of the fetus and the newborn and to meet the physiological stresses of labour. Progressive anatomical, physiological and biochemical alterations occur throughout pregnancy and postpartum. This chapter reviews the physiological changes that occur during pregnancy, and details the basis of these changes.

THE DEMANDS OF PREGNANCY

There is a large increase in fetal growth during the last trimester of pregnancy, and this causes a metabolic demand on the mother. Fat stores (approximately 3 kg) are laid down in the mother – primarily in the first half of pregnancy – to provide a metabolic store for the third trimester when fetal growth predominates. Placental growth occurs steadily, and this is important for fetal growth because net nutrient transfer from mother to fetus is directly proportional to the placental surface area. The major components of maternal weight gain are increases in uterine and breast tissue, extracellular fluid (ECF) and fat. The large increase in uterine size is due mainly to stretching and hypertrophy of existing muscle cells by the stimulatory effects of oestrogens and progesterone. The ECF volume increases by about 3 L at term.

Labour is a stressful period for the mother, with marked increases in cardiac output and changes in intravascular volume. The energy demands of labour are met by the breakdown of carbohydrates, and this causes a significant increase in maternal blood lactate concentrations.

Endocrine changes in pregnancy

The physiological changes in the mother are caused by or associated with changes in the endocrine system.

The placenta acts as an endocrine organ as it produces both peptide (human chorionic gonadotrophin, human placental lactogen) and steroid hormones (oestrogen and progesterone). Human chorionic gonadotrophin (HCG) is produced by the trophoblast cells from 8–9 days after fertilization, and maintains corpus luteal oestrogen and progesterone production during the first trimester to maintain pregnancy until the placenta takes over. The plasma concentration of HCG peaks at 10–12 weeks of pregnancy, and then declines to term. The plasma concentration of human placental lactogen (HPL), produced by the placenta, rises throughout pregnancy and peaks near term (Fig. 14.1). The actions of HPL are to mobilize free fatty acids, antagonize the actions of insulin, and retain potassium and nitrogen.

Throughout pregnancy the placenta synthesizes progesterone and oestrogen from precursors derived from the fetal adrenal cortex. These steroid hormones are important for the maternal physiological changes observed. The high concentrations of oestrogens during pregnancy are important for the growth and enlargement of the uterus, and the development of the mother's breast with the growth of the ductal structures. Progesterone is required for pregnancy for the following: to promote of storage of nutrients in the endometrial cells and transform them into decidual cells; to reduce uterine smooth muscle contractions; to promote the development of the alveoli of the breasts; and to promote the secretion of nutrients from the epithelium of the fallopian tubes to sustain the zygote before implantation.

The pituitary gland increases the secretion of prolactin, adrenocorticotrophin (ACTH), and melanocyte-stimulating hormone. The production of growth hormone is reduced, possibly by HPL. Placental steroids reduce the pituitary gonadotrophin production, whilst adrenal hormone

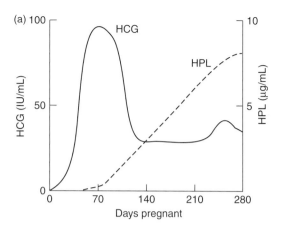

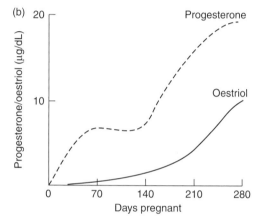

Figure 14.1   *Hormone levels during pregnancy: (a) peptide hormones; (b) steroid hormones.*

production is increased. Plasma concentrations of both free and total cortisol rise, aldosterone secretion increases because of the natriuretic effect of progesterone, and plasma renin and angiotensin concentrations rise. The thyroid gland increases thyroxin and tri-iodothyronine synthesis during pregnancy. As there is a higher production of thyroid-binding globulin, the free plasma concentrations of these thyroid hormones are unchanged. Plasma ionized calcium concentration decreases during pregnancy because of increased utilization by the fetus, and this increases parathyroid hormone secretion. Absorption of calcium by the gut is also enhanced. Plasma concentrations of prostaglandins increase during pregnancy; prostaglandin A levels increase three-fold during the first trimester and cause systemic vasodilatation, but prostaglandin E only increases significantly in the third trimester.

Metabolic changes in pregnancy

The basal metabolic rate increases to 20 per cent above non-pregnant levels at 36 weeks of pregnancy and then falls slightly to 15 per cent above baseline levels at term. The increased metabolic demand is caused by the increased demands of the fetus, the hypertrophy of maternal tissues and the increased respiratory work and heart rate. In total, the oxygen consumption is increased by 20 per cent.

Carbohydrate metabolism changes as a result of the increased plasma concentrations of oestrogen, HPL, free cortisol and progesterone. Insulin secretion increases from the end of the first trimester to 32 weeks and then declines to non-pregnant levels at term because of increased glucose utilization by the fetus and glycosuria. At the same time, tissue sensitivity to insulin diminishes, leading to a progressive reduction in glucose tolerance. The anti-insulin effect of HPL may contribute to this impaired glucose tolerance.

Fat metabolism is characterized by storage in the first half of pregnancy and mobilization during the second half. Thus, the plasma concentrations of free fatty acids (FFAs) and glycerol decrease from early to mid-pregnancy and then rise towards term. The increase in plasma FFAs enhances lipid transfer across the placenta and provides substrate to the fetal liver for fat synthesis. The increased maternal cholesterol and phospholipid plasma concentrations also enhance placental transfer of FFAs to the fetus.

Plasma amino acids fall with their utilization for gluconeogenesis, transplacental transfer and loss in the urine. The fetus uses amino acids for protein synthesis and as an energy substrate.

Cardiovascular changes in pregnancy

Significant cardiovascular changes occur within the first 8 weeks of pregnancy. Heart rate (HR) may increase as early as 4 weeks after conception, and there is also a decrease in the mean arterial blood pressure (MAP). HR increases by 17 per cent by the end of the first trimester and to 25 per cent at the middle of the third trimester, after which no further rise is observed (Fig. 14.2(a)). Central venous pressure and pulmonary capillary wedge

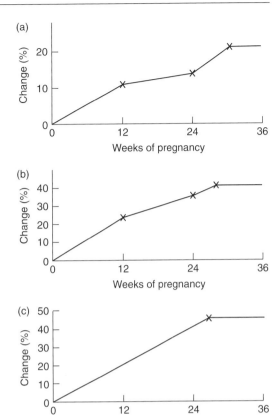

Figure 14.2   *Changes in (a) maternal heart rate; (b) stroke volume and (c) cardiac output (in lateral position) during pregnancy.*

pressure remain stable throughout pregnancy. Colloid oncotic pressure falls by 14 per cent, and this may predispose to oedema.

Stroke volume increases by 20–30 per cent, predominantly in the first trimester. Total peripheral vascular resistance decreases by 30 per cent at the 12th week, 35 per cent by the 20th week and then remains at 30 per cent below the non-pregnant values (Fig. 14.2(b)). The decrease in vascular resistance is due to vasodilatation mediated by progesterone, prostaglandins and down-regulation of α receptors. Systolic and diastolic arterial blood pressures decrease slightly (about 10 per cent) and reach a nadir at 20 weeks of pregnancy.

The cardiac output increases progressively throughout pregnancy to approximately 40–45 per cent above non pregnant values at the 12th to the 28th week, reaches a peak of 50 per cent during the 32nd to the 36th week, and then decreases

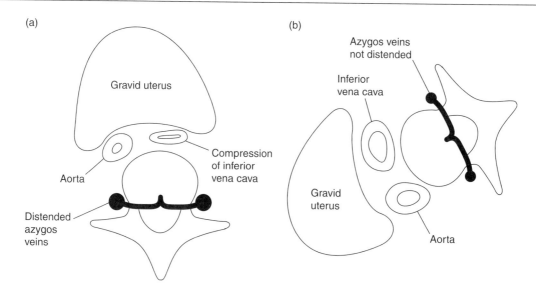

Figure 14.3 *Cross-sectional views of the aorta and inferior vena cava in the pregnant woman in the supine (a) and lateral (b) positions.*

slightly (to 47 per cent above non-pregnancy values) after that (Fig. 14.2(c)). The cardiac output increase is produced by an increased venous return due to venodilatation and an increased vascular volume caused by oestrogens. A large proportion of the cardiac output is directed to the uteroplacental circulation that increases its blood flow 10-fold to about 750 mL/min at term. Renal blood flow increases by 80 per cent in the first trimester, but may fall slightly towards term. There is also increased blood flow to the breasts, gastrointestinal tract and skin.

About 15 per cent of pregnant women, when near term, develop hypotension, pallor, nausea and vomiting when they are supine. This is known as the supine hypotension or aortocaval compression syndrome. The ill effects of the supine hypotension syndrome may be seen as early as the 20th week of gestation. Compression of the inferior vena cava by the gravid uterus decreases the venous return and reduces the cardiac output (Fig. 14.3). Blood returns to the heart via the paravertebral epidural veins draining into the azygos vein. Uterine perfusion is diminished because of increased uterine venous pressure. Compression of the aorta may also be present and may be associated with uterine arterial hypotension and reduced uteroplacental perfusion. The supine hypotension syndrome can be prevented by positioning the mother on her left side.

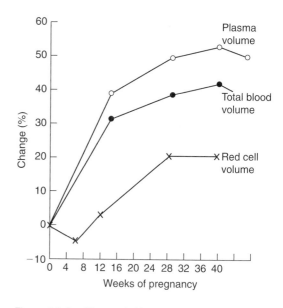

Figure 14.4 *Changes in blood volume, plasma volume and red blood cell volume during pregnancy.*

Maternal blood volume begins to rise in the first trimester. Near term, the maternal blood volume is increased by 35–40 per cent, approximately 1000–1500 mL, compared with non-pregnant values (Fig. 14.4). The plasma volume increases by 45 per cent as a result of sodium and water retention by oestrogen stimulation of the renin–angiotensin system. The red blood cell volume increases by 20 per cent due to increased renal erythropoietin synthesis. The slower rate of rise in red cell mass compared with that of the plasma volume accounts for the fall of the maternal haematocrit to 33 per cent.

During labour, each uterine contraction squeezes about 300 mL blood from the uterus into the central maternal circulation. The cardiac output increases by about 15 per cent during the latent phase of labour, by 30 per cent during the active phase, and by 45 per cent during the expulsive stage (Fig. 14.5). Immediately after delivery the cardiac output is about 60–80 per cent above prelabour values as a consequence of autotransfusion and increased venous return associated with uterine involution. Maternal systolic and diastolic arterial blood pressures increase by 10–20 mmHg during uterine contraction. The cardiac output and arterial blood pressures return to non-pregnant values by 2 weeks after delivery.

Respiratory changes

There are marked anatomical changes in the respiratory system that alter lung volumes. The diaphragm is displaced upwards by about 4 cm, but its contraction is not markedly restricted. The anteroposterior and transverse diameters of the thoracic cage increase by 2–3 cm, because the lower ribs flare out and the subcostal angle increases from 68° to 103° at term. The circumference of the thoracic cage is increased by 5–7 cm. These changes are produced by relaxin (secreted by the corpus luteum) that relaxes the ligamentous attachments of the ribs. There is also capillary engorgement throughout the respiratory tract, so that the vocal cords may be swollen or oedematous. The large airways are dilated, decreasing airway resistance by 35 per cent.

Although respiratory changes begin early in pregnancy, significant changes in lung volumes are only detected from the 20th week onwards. The expiratory reserve volume (ERV) and residual volume (RV) gradually decrease as pregnancy progresses. At term, the ERV and RV, and consequently the functional residual capacity (FRC), are 20 per cent less than the non-pregnant values. These changes are caused by the progressive elevation of the diaphragm and, to a lesser extent, by an increase in pulmonary blood volume (Fig. 14.6). In the

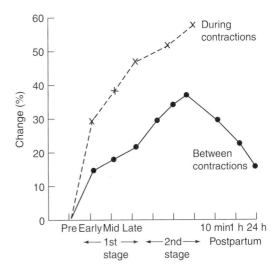

Figure 14.5 *Changes in cardiac output during labour with mother in the lateral position.*

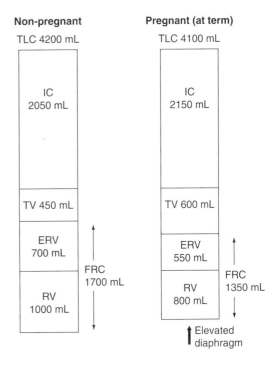

Figure 14.6 *Lung volumes in non-pregnant and pregnant (at term) women.*

supine position the FRC falls to 70 per cent of its value measured with the patient sitting. Tidal volume (TV) begins to increase in the first trimester, rising to 28 per cent above non-pregnant values at term. Inspiratory capacity (IC) increases 10 per cent at term, whilst expiratory capacity decreases 20 per cent. Total lung capacity decreases by 5 per cent and vital capacity remains unchanged.

Studies carried out on pregnant women (sitting) found no changes in airway closure, closing capacity or flow–volume curves during pregnancy. Although lung compliance remains unchanged in pregnancy, chest wall compliance, and thus total respiratory compliance, decreases by 20 per cent, by elevation of the diaphragm.

Recent studies have shown that anatomical dead space increases by 45 per cent due to the larger conducting airways, but the dead space/tidal volume ratio remains unchanged.

Minute ventilation increases in the early weeks of pregnancy and reaches 50 per cent above non-pregnant values at term (Fig. 14.7). This is produced by a 40 per cent increase in tidal volume and a 10 per cent increase in respiratory rate. Recent studies have indicated that maximal hyperventilation occurs as early as the 8th to 10th week of pregnancy. Progesterone stimulates the respiratory centres and shifts the ventilation/carbon dioxide response curves to the left. As a result, the arterial carbon dioxide tension is reduced to about 26–32 mmHg at the end of the first trimester. The respiratory alkalosis of pregnancy is compensated

for by renal excretion of bicarbonate, a decrease in plasma bicarbonate (18–21 mmol/L) and a base deficit of -2 to -3 (Fig. 14.8).

During labour, minute ventilation increases further due to pain (70 per cent), and the uterine contractions increase oxygen consumption (60 per cent). After the painful contractions and at the beginning of uterine relaxation, there is a hypocapnia-induced transient hypoventilatory period that produces brief desaturation of oxygen. After delivery of the baby the FRC and RV return to normal within 48 h, and the tidal volume declines within 5 days. The respiratory centre's sensitivity to carbon dioxide decreases rapidly after delivery. The respiratory changes during pregnancy have

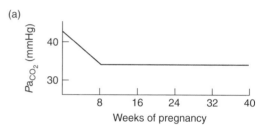

(a)

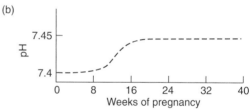

(b)

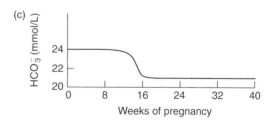

(c)

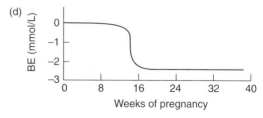

(d)

Figure 14.8   *Changes in acid–base balance during pregnancy: (a) arterial* P_{CO_2}; *(b) pH; (c) plasma bicarbonate; and (d) base excess (BE).*

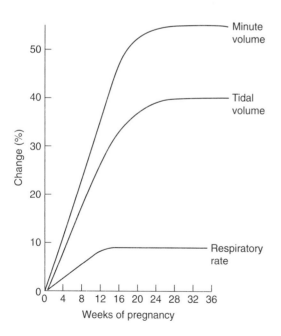

Figure 14.7   *Ventilatory changes during pregnancy.*

important anaesthetic implications. The decreased FRC and the higher oxygen consumption reduce the oxygen reserve of the mother. Anatomical changes in the upper airway may make endotracheal intubation difficult.

Haematological changes in pregnancy

Because of the relative greater increase in plasma volume, red cell count, haemoglobin and haematocrit values decrease to $3.7–3.8 \times 10^6/mm^3$, 12–13 g/dL, and 33–35 per cent, respectively, during pregnancy. However, the red cell mass increases by 18 per cent. The white cell count is 8000–9000/mm^3 due to an increase in neutrophils and monocytes (Fig. 14.9).

During pregnancy there is a significant increase in the concentrations of factors VII, VIII, IX, X and fibrinogen. The platelet count remains unaltered, or decreases slightly due to haemodilution. The coagulation system seems to prepare for blood loss at delivery by increasing the concentration of factors throughout pregnancy. Recent studies have shown an increase in fibrinolysis and fibrin formation in late pregnancy. Total circulating proteins increase during pregnancy, but the concentrations of total proteins and albumin decrease as a result of haemodilution. There is an increase in total globulins, especially α-globulin and some β-globulins, but a slight decrease in γ-globulin. Fibrinogen increases from 300 to 450 mg/dL at term. Serum pseudocholinesterase activity is reduced by 20–30 per cent at the end

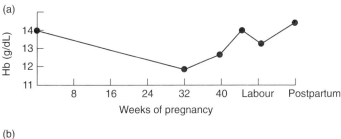

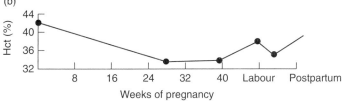

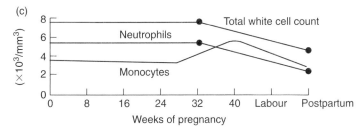

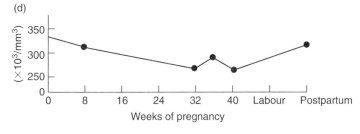

Figure 14.9 *Haematological changes during pregnancy: (a) haemoglobin; (b) haematocrit; (c) white cell count; and (d) platelet count.*

of the first trimester and remains at that level until term.

Gastrointestinal changes in pregnancy

During pregnancy the stomach and intestines are displaced cephalad by the gravid uterus. Progesterone relaxes smooth muscles and inhibits the contractile response of the gastrointestinal tract to acetylcholine and gastrin. These factors are important for changes in the gastrointestinal tract during pregnancy. The lower oesophageal sphincter (LOS) tone progressively decreases, and, with changes in the angle of the gastro-oesophageal junction, the LOS becomes incompetent and results in gastric reflux. This makes pregnant women more prone to pulmonary aspiration during general anaesthesia. Gastric motility is reduced, and there is delayed gastric emptying at 12–14 weeks of gestation. Further prolongation of gastric emptying occurs during labour as a result of anxiety and pain. Gastrin production progressively increases during pregnancy, as it is produced by the placenta. Gastrin stimulates the secretion of water and enzymes from the gastrointestinal tract. Gastric acid production is increased during the third trimester. Administration of opioids during the post-partum period delays gastric emptying. The motility of the small and large intestines is reduced due to a reduced plasma concentration of motilin. In the second and third trimesters the contractility of the gallbladder is reduced as a result of diminished release of cholecystokinin from intestinal mucosa caused by progesterone.

During pregnancy, liver blood flow remains unaltered. Histological changes in the liver consist of mild fatty changes, mild glycogen depletion and lymphocytic infiltration. The smooth endoplasmic reticulum proliferates, suggesting an increase in hepatic microsomal activity. There is also an increase in serum alkaline phosphatase and serum cholesterol levels.

Renal changes in pregnancy

Progressive dilatation of the renal pelvis, calyces and ureters begins from the second or third month of pregnancy, primarily due to obstruction of urine flow by the gravid uterus or dilated ovarian plexuses. Glomerular filtration rate (GFR) and effective renal plasma flow increase by 50 per cent during the first trimester. These increases reflect the change in cardiac output. Consequently, the plasma concentrations of urea and creatinine fall in the first two trimesters.

Although the tubular function of the nephrons is not altered in pregnancy, glycosuria is common, most likely due to an increase in the GFR with a slightly reduced proximal tubular reabsorption. Excretion of most amino acids also increases, but the cause is unknown. Proteinuria is present in 20 per cent of normal pregnant women, and this may be related to increased renal venous pressure.

Central nervous system changes in pregnancy

The placenta produces endorphins and enkephalins that may be analgesic during pregnancy. Endorphin production increases significantly in proportion to the frequency and duration of uterine contractions during labour and delivery, but their role in pregnancy is not completely understood. Progesterone has sedative actions, and levels increase 10- to 20-fold in the third trimester. The minimum alveolar concentration of volatile agents is reduced by 30–40 per cent during pregnancy, partly because of endorphins and progesterone. The epidural veins are engorged and the epidural pressures are higher ($+1\,cmH_2O$) than in non-pregnant women ($-1\,cmH_2O$) as a result of the increased intra-abdominal pressure. As labour progresses, the epidural pressures increase to 4–$10\,cmH_2O$. At the second stage of labour the epidural pressure can increase to $60\,cmH_2O$ when the patient is bearing down. Resting cerebrospinal fluid pressure is not altered in pregnancy but can rise to $70\,cmH_2O$ during bearing-down efforts.

PHYSIOLOGY OF THE PLACENTA

The placenta is a unique, disc-shaped organ that acts as an interface between the mother and fetus. Its functions are to act as:

- An endocrine organ of pregnancy, as described earlier

- An immunological barrier to protect the fetus from the maternal immune system
- An interface between maternal and fetal plasma for the transfer of nutrients and waste products

Anatomy

The basic structural unit of the human placenta is the chorionic villus. The villi are vascular projections of fetal tissue surrounded by chorion, the outermost layers of fetal tissue. The chorion consists of two layers, the syncytiotrophoblast that is in direct contact with maternal blood within the intervillous space, and the cytotrophoblast (Fig. 14.10). Substances in the maternal blood are carried into the intervillous space and pass through the two layers of trophoblast, fetal connective tissue, and the endothelium of fetal capillaries into fetal blood. During pregnancy, the placenta grows to provide an ever-larger surface area for maternal–fetal exchange.

The blood supply to the uterus is by the uterine and ovarian arteries that form the arcuate arteries from which radial arteries arise and penetrate the myometrium. The radial arteries divide into spiral arteries that supply the intervillous space, and basal arteries that supply the myometrium and decidua (Fig. 14.11). The maternal blood in the intervillous space bathes the chorionic villi. Continuous inflow of blood into the intervillous space pushes blood into venous openings that drain into uterine veins. Uteroplacental blood flow at term is approximately 600 mL/min. Blood flow in the uteroplacental circulation depends on maternal arterial blood pressure, but increased intrauterine pressure during uterine contractions can reduce placental blood supply. As the uteroplacental arteries have α-adrenergic receptors, sympathetic stimulation leads to uterine artery vasoconstriction.

Maternal blood from the spiral arteries is ejected into the intervillous space and passes haphazardly over the villous surface. Blood enters the fetal side of the placenta from two umbilical ateries and returns to the fetus via a single umbilical vein. Although the fetal and intervillous blood flows should effectively be a counter-current system, the human uteroplacental blood flow is no more efficient than a concurrent system (where flow on both

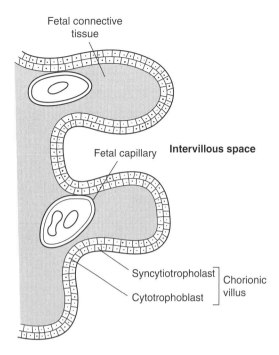

Figure 14.10 *Structure of the chorionic villus in the placenta.*

sides runs in the same direction) because of considerable shunting. However, the maternal placental blood flow is almost double the umbilical blood flow, and this improves the efficiency of transfer of substances across the placental barriers.

Synthetic and metabolic functions of the placenta

The placenta contains enzymes that synthesize hormones such as oestrogen, progesterone, chorionic gonadotrophin and placental lactogen. It also contains pseudocholinesterase, alkaline phosphatase, monoamine oxidase and catechol-*O*-methyl transferase.

Immunological functions of the placenta

The placenta is a selective immunological barrier as it permits transport of maternal IgG antibodies to provide passive immunity to the fetus. The

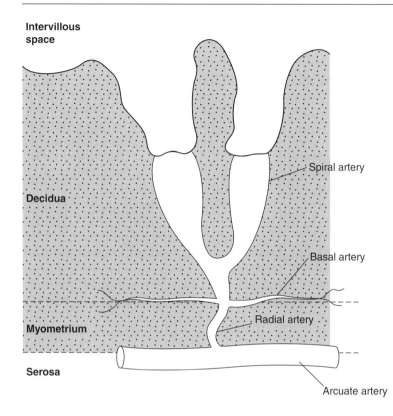

Figure 14.11 *Maternal blood supply to the placenta.*

syncytiotrophoblast possesses receptors for the Fc fragments of IgG, the bound IgG is endocytosed into a vesicle, and the IgG is then released from the syncytiotrophoblast by exocytosis into the fetal blood. In Rh isoimmunization, maternal antibodies against fetal red blood cells cross the placenta and cause fetal haemolysis. Autoimmune antibodies that cause maternal autoimmune disorders (thyrotoxicosis, myasthenia gravis and idiopathic thrombocytopenia) can cross the placenta and affect fetal tissues.

Placental exchange or transport

A variety of nutrients, waste products and toxins cross the placental barrier by simple diffusion, facilitated transport, active transport, endocytosis and bulk flow.

Most drugs and respiratory gases cross by simple diffusion. The rates of transfer of these substances follow Fick's law. Substances such as glucose cross through the placenta more rapidly than predicted by Fick's law, because of facilitated diffusion. Amino acids, calcium, iron and vitamins A and C are transported by active transport of substances against a concentration gradient. Bulk flow of water by osmotic and hydrostatic forces may transport small molecules, whilst large molecules such as IgG cross the placenta by endocytosis.

Respiratory exchange

Oxygen and carbon dioxide are small hydrophobic molecules that have a high permeability and therefore are transferred by flow-limited passive diffusion. The rates of blood flow in the maternal and fetal sides of the placenta have major importance for the maintenance of fetal oxygenation. Animal studies have suggested that fetal oxygen uptake starts to fall when uterine blood flow decreases by 50 per cent. Other important determinants of fetal oxygenation include maternal arterial oxygen tension, the high oxygen affinity and the oxygen capacity of fetal blood.

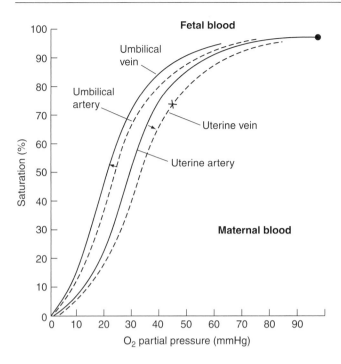

Figure 14.12 *Transport of oxygen from the mother to the fetus: the double Bohr effect.*

Oxygen transfer from the mother to the fetus depends mainly on the difference between the oxygen tension of maternal blood in the intervillous space and the fetal blood in the umbilical artery. The oxygen tension of maternal blood in the intervillous space is about 50 mmHg (6.7 kPa), but it varies widely in different areas. The oxygen tension of fetal blood in the umbilical artery flowing into the placenta is about 20 mmHg (2.7 kPa), resulting in an oxygen partial pressure gradient of 30 mmHg (4 kPa) which is the driving force for diffusion of oxygen from the maternal to fetal blood. The oxygen partial pressure of the blood returning to the fetus in the umbilical vein is about 30 mmHg (4 kPa). Oxygen transfer to the fetus is enhanced by the Bohr effect. As it releases its carbon dioxide to the maternal blood, fetal blood develops a greater affinity for oxygen because of a shift to the left in the oxygen dissociation curve. At the same time, carbon dioxide reaches the maternal blood and consequently shifts the maternal blood oxygen dissociation curve to the right, enhancing oxygen release from the maternal haemoglobin. The shifts of the oxygen dissociation curves in both fetal and maternal blood promote the transfer of oxygen from

mother to fetus and are referred to as the 'double Bohr effect' (Fig. 14.12). The high haemoglobin content of fetal blood also increases the oxygen-carrying capacity of fetal blood. Fetal haemoglobin has a greater baffinity for oxygen than does adult haemoglobin. At the partial pressures of oxygen in the placenta, fetal haemoglobin can carry 20–50 per cent more oxygen than maternal blood.

Carbon dioxide can easily move across the layers of the placenta from the fetus to the mother because it is extremely soluble in biological membranes. The carbon dioxide tension of blood in the fetal umbilical artery is about 50 mmHg (6.7 kPa), and averages about 37 mmHg (4.9 kPa) in the blood in the intervillous space. The high haemoglobin concentration in fetal blood increases its capacity for carriage of carbon dioxide as carbaminohaemoglobin. The Haldane effect facilitates the transfer of carbon dioxide from the fetus to the mother. As maternal blood releases oxygen, it is able to carry more carbon dioxide as carbaminohaemoglobin, without any increase in carbon dioxide tension. At the same time as the fetal blood takes up oxygen, it releases carbon dioxide that is combined with fetal haemoglobin. The combination of these two

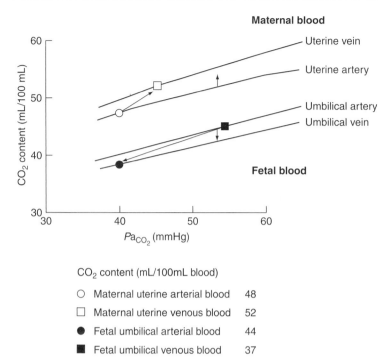

CO$_2$ content (mL/100mL blood)

○   Maternal uterine arterial blood   48
□   Maternal uterine venous blood   52
●   Fetal umbilical arterial blood   44
■   Fetal umbilical venous blood   37

Figure 14.13   *Transport of carbon dioxide from the fetus to the mother: the double Haldane effect.*

events is referred to as the 'double Haldane effect' (Fig. 14.13).

Transfer of electrolytes, glucose and amino acids

The transfer of sodium and chloride ions across the human placenta is mainly by passive diffusion. However, carrier-mediated transport may have a role in the transfer of both ions. Na^+–H^+ exchangers and Na^+–amino acid co-transporters are present on the plasma membranes of the syncytiotrophoblast facing the maternal blood. An anion exchanger is found on the plasma membrane of the syncytiotrophoblast. The transfer of calcium ions is by active carrier-mediated transport. The transfer of glucose across the placenta is by facilitated diffusion. However, the rate of transfer of glucose to the fetus is dependent on the maternal–fetal concentration gradient. Amino acids are actively transported between the mother and the fetus. Several transporter proteins specific for anionic, cationic and neutral amino acids are present. Small neutral amino acids (e.g. alanine) are transported by a Na^+-dependent carrier system.

PERINATAL PHYSIOLOGY

Before birth, the fetus relies on the mother for oxygen, nutrition, excretion, temperature regulation and homeostasis, and most of these functions are performed by the placenta. After birth, these placental functions are taken over by the baby's organs and physiological adaptations occur in all systems. The changes that occur in the neonatal period (the first 28 days of life) and infancy (1–12 months) are continuous, and the rate of change is dependent on the gestational (i.e. postconceptional (weeks after conception)) age. At the end of the neonatal period most physiological systems have matured adequately in a healthy baby born at term, but those of low postconceptional age may take a longer time to mature.

The neonate differs in many ways from an adult, and one important difference is that the surface area/body weight ratio of the neonate is 2–2.5 times greater than adults. This greater surface area results in increased heat loss, causing impaired thermoregulation in neonates. Resting oxygen consumption is 6–8 mL/kg per min in the neonate and 5–6 mL/kg per min in infants, compared with

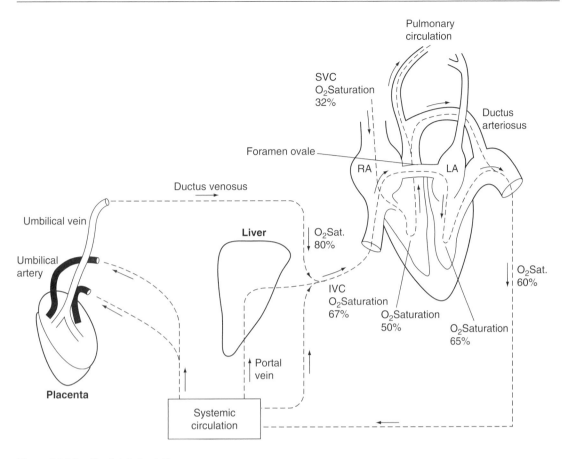

Figure 14.14   *The fetal circulation.*

3–4 mL/kg per min in adults. The higher oxygen consumption in the neonate requires an increased pulmonary ventilation and cardiac output to enhance oxygen and carbon dioxide transport.

FETAL CIRCULATION

In the fetus, oxygenated blood, with a P_{O_2} of about 30 mmHg (4 kPa) and an oxygen saturation of 80 per cent, returns from the placenta through the umbilical vein into the left branch of the hepatic portal vein (Fig. 14.14). About 60 per cent of the blood from the umbilical vein bypasses the liver through a shunt between the portal vein and the inferior vena cava called the ductus venosus. The remaining 40 per cent is mixed with blood from the gastrointestinal tract and perfuses the liver, especially the left lobe. Blood in the inferior vena cava (IVC) (oxygen saturation about 67 per cent)

enters the right atrium (RA). About 60 per cent of this blood is directed across the foramen ovale to the left atrium (LA) by the crista terminalis, a muscular ridge in the right atrium. As a result, this highly oxygenated blood (Pa_{O_2} of 25–28 mmHg (3.3–3.7 kPa) and oxygen saturation 65 per cent) flows into the left ventricle and is ejected into the aorta, thus supplying the coronary arteries and the brain with the most oxygenated blood. The left ventricular afterload consists of the high resistance of the cerebral circulation and the circulation of the upper body.

The remainder of the blood from the inferior vena cava mixes with deoxygenated blood (oxygen saturation of 40 per cent) flowing into the right atrium from the head and neck via the superior vena cava (SVC), and passes into the right ventricle and then into the pulmonary artery (with an oxygen saturation of 55 per cent). The lungs receive about 10 per cent of the right ventricular output because they are collapsed, and pulmonary vascular resistance is high. The remainder of the right ventricular output flows through the ductus arteriosus, a wide muscular arterial channel between the pulmonary artery and the aorta, into the descending aorta. The shunted blood in the descending aorta (Pa_{O_2} of 19–20 mmHg (2.5–2.7 kPa) and oxygen saturation of 60 per cent) perfuses the lower half of the body or returns to the placenta. Some 70 per cent of the venous return to the heart is via the inferior vena cava, and 20 per cent via the superior vena cava; the remaining 10 per cent is from the lungs and coronary sinus. The right ventricle afterload consists of the low resistance of the ductus arteriosus and the placenta, and the high resistance of the pulmonary vasculature and the circulation of the lower body.

As a result of the presence of shunts (foramen ovale and ductus arteriosus), the right and left ventricles work in parallel. Echocardiographic evidence indicates that the right and left ventricles are of equal size and wall thickness. During the intrauterine period the cardiac output in the fetus is a function of heart rate, which is normally 120–140 beats/min. Heart rates below 100 and above 180 beats/min indicate fetal distress. Heart rate changes are pronounced during labour, with pressure on the fetal skull causing fetal tachycardia. Bradycardia late in uterine contraction suggests fetal hypoxaemia.

Transitional circulation at birth (Fig. 14.15)

At birth, the circulation changes from a parallel system to a system in series as a result of resistances changing throughout the circulation of the neonate. The low-resistance placenta is excluded, as the umbilical cord vessels are clamped and closed, with an increase in systemic vascular resistance and left ventricular end-diastolic pressure and a fall in right atrial pressure due to reduced inferior vena cava flow. The lungs expand, reducing pulmonary vascular resistance and right ventricular end-diastolic pressure. Pulmonary vascular resistance gradually falls further under the influence of the increasing arterial oxygen tension and pH, and a decreasing arterial carbon dioxide tension. In a normal neonate, the pulmonary artery (PA) pressure decreases to adult values in about 2 weeks, with most of the change occurring in the first 3 days.

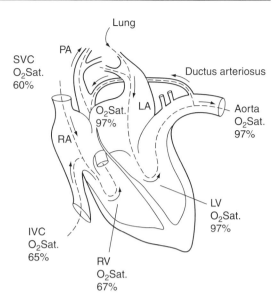

Figure 14.15 *Transitional circulation.*

The left atrial pressure (LAP) rises because of the increased blood flow through the lungs and the increased left ventricular end-diastolic pressure (Fig. 14.16). The foramen ovale closes when the LAP exceeds the right atrial pressure: permanent closure by the fusion of the septum secundum with the edges of the foramen ovale takes 4–6 weeks.

The ductus arteriosus, with dense spirally arranged smooth muscles in its media, constricts in response to the increasing Pa_{O_2} after the first breath and the closure of the foramen ovale, and to the decreasing concentrations of circulating and locally produced prostaglandins E_1 and E_2. This physiological closure occurs within 10–15 h, and the permanent closure takes place in 2–3 weeks by thrombosis and fibrosis.

The ductus venosus closes a few hours after birth, but the exact mechanism is unknown. With the closure of the shunts, the right (RV) and left ventricles (LV) are in series, and the adult configuration of the circulation is complete.

Neonatal cardiovascular function

In the first week of life after birth the cardiac output ranges from 280 to 430 mL/min per kg body weight and decreases to about 150 mL/min per kg

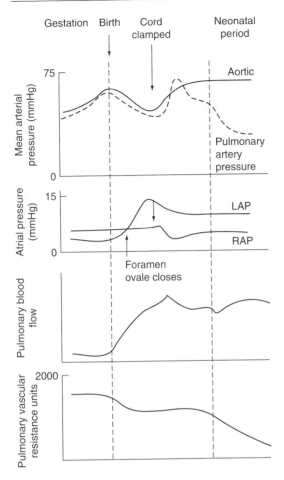

Figure 14.16 *Haemodynamic changes at birth. LAP, left atrial pressure; RAP, right atrial pressure.*

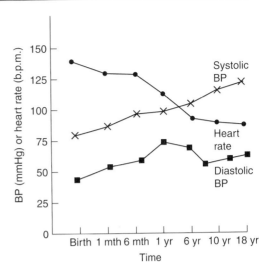

Figure 14.17 *Changes in heart rate and blood pressure in children.*

at 8 weeks. As there is a higher proportion of non-contractile proteins in the immature myocardial cells the resting tension of the neonatal myocardium is higher and less tension per unit area is developed at any preload. The newborn's ventricle is therefore less compliant or 'stiffer'. Consequently, neonates and infants have a relatively fixed stroke volume and their cardiac output is dependent on heart rate. The left ventricle grows rapidly early, its wall thickness increasing by 50 per cent in the first 6 months of life.

At birth, the heart rate is 120–160 beats/min and then gradually falls to about 100 beats/min by 5 years of age. Normal systolic blood pressure at birth is between 70 and 90 mm Hg and then gradually increases to 100 mmHg by 1 year, remaining constant until about 6 years of age. A gradual increase then occurs to 120 mmHg at the age of 18 years (Fig. 14.17). The aortic chemoreceptors are important for cardiovascular control in the newborn; hypoxia causes hypotension, vasoconstriction and variable heart rate changes. In the newborn, the physiological right to left shunt is normally about 20 per cent compared with 7 per cent in the adult.

The neonatal sympathetic system is only partially developed at birth, and significant increases of myocardial content of norepinephrine occur with age. The bradycardia and hypotension in response to hypoxia, and the less efficient response to postural changes in the neonate reflect the functional immaturity of the myocardial sympathetic innervation. The circulation of the neonate is labile and can revert to the pattern of fetal circulation if pulmonary vasoconstriction occurs. The pulmonary vasculature is also labile and constricts in response to hypoxaemia, hypercapnia and acidaemia through an α-adrenergic mechanism.

FETAL RESPIRATORY SYSTEM

The bronchial tree of the fetal lungs is fully developed by 16 weeks of gestation. By 28 weeks, the pre-acinar pattern of airways, arteries and veins are formed with capillaries in the alveolar walls. Type II pneumocytes of the alveolar epithelium are seen by 24 weeks of gestation, and surfactant can be detected in lung extracts from 23 weeks onwards and in fetal tracheal fluid by 28 weeks. Surfactant production in the fetal lung is increased by the administration of cortisol and thyroxine to the mother. Surfactant production is associated with an increase with lecithin in the amniotic fluid. The concentration of sphingomyelin is constant throughout pregnancy and a lecithin/sphingomyelin ratio of greater than 2 is normally present by 36 weeks of gestation.

Fetal breathing movements can be detected by ultrasonic methods, and initially are very irregular. As pregnancy progresses they become more regular and rapid (about 60 breaths/min), usually associated with rapid eye movements shown on fetal electroencephalogram.

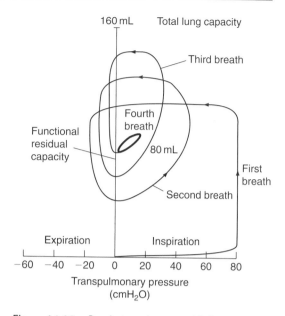

Figure 14.18 *Respiratory changes at birth.*

Respiratory changes at birth

During vaginal delivery, about 35 mL of fluid from the lungs is squeezed out by compression of the fetal thorax, and is reabsorbed into the pulmonary capillaries and lymphatics as the pulmonary vascular pressure decreases. Factors that stimulate the newborn to take its first breath include environmental stimuli such as sound, touch, temperature and gravity. The increase in sensory activity arising at the moment of birth activates the reticular system and increases the sensitivity of the respiratory centres. At delivery, the arterial oxygen tension falls from 30 mmHg (4 kPa) to about 15 mmHg (2 kPa) and CO_2 tensions rise to 55–60 mmHg (7.3–8 kPa) with acidaemia. Both the central and peripheral chemoreceptors become more responsive and exert more control over respiration, and may be due to an increased blood flow through them. The first breath generates a high negative inspiratory pressure of 70–100 cmH$_2$O (Fig. 14.18). Within a few minutes the functional residual capacity increases to 20 mL/kg body weight, and rises to 30–35 mL/kg within 1 h of birth. Tidal volume is about 20 mL (5–6 mL/kg) and the respiratory rate is about 30 breaths per minute. The physiological dead space is about 1.5–2 mL/kg body weight. The rapid rise in arterial oxygen tension following the onset of respiration leads to a fall in pulmonary vascular resistance, and an uptake of 100 mL of blood into the pulmonary circulation.

Respiratory system in the neonate

Anatomical differences in the airway include a longer 'U'-shaped epiglottis, and a larynx situated at a cephalic level opposite the third and fourth cervical vertebrae, descending to the fifth cervical vertebra during the first 3 years and then at puberty to the final position at the sixth vertebra. The narrowest part of the larynx is the cricoid ring. After puberty, the cricoid enlarges and the narrowest part of the larynx is the vocal cords. The length of the trachea varies from 3.2 to 7 cm depending on the size of the baby. The angle at which the bronchi branch is similar to that of adults, 30° on the right and 47° on the left. The tongue is relatively large, and the angle of the mandible is 140° compared with 120° in an adult. The shape of the chest wall in neonates influences the mechanics of breathing. The anteroposterior expansion is limited because the ribs are more horizontal, while the transverse expansion is reduced due to the lack of the buckle handle mechanism of the ribs. There are also fewer type I muscle fibres (slow contracting and highly oxidative fibres used for sustained contractions) in the diaphragm and intercostal muscles, and hence these respiratory muscles fatigue easily. The intercostal muscles comprise 20, 45 and 65 per cent type I muscle fibres in the premature, neonate at term and at full maturity, respectively. The diaphragm comprises 10, 25 and 55 per cent type I muscle fibres in the premature, neonate and at 9 months of age, respectively.

The respiratory rate of a newborn infant is 30–40 breaths/min, and this gradually falls to 15 breaths/min by late childhood. The high respiratory rate is the optimal frequency for the minimal work of breathing to overcome the compliance of the respiratory system. The respiration is irregular and mainly diaphragmatic. It is generally thought that neonates are obligate nasal breathers. The cephalad position of the larynx and the large tongue make mouth breathing more difficult. The minute ventilation of 220 mL/kg and alveolar ventilation of 140 mL/min are about twice that of the adult.

In neonates, the outward recoil of the chest wall is very low because the rib cage is cartilaginous and the respiratory muscles are not well developed. The inward recoil of the lungs is slightly lower than that of adults. As a result, during general anaesthesia, the functional residual capacity (FRC) decreases to very low values because of a decrease in intercostal muscle tone. However the FRC in spontaneously breathing infants is maintained at 40 per cent of total lung capacity (TLC). As the elastic recoil of the lungs is low, small airway closure occurs. Therefore, the closing capacity as a per centage of the TLC is relatively high in infants and exceeds FRC in neonates. Neonates have a lower oxygen reserve and will develop hypoxaemia more rapidly as they have, in addition, an increased oxygen consumption.

Lung compliance increases during the first few hours after birth from 1.5 to 6 mL/cmH₂O (Fig. 14.19). Specific compliance is similar in the neonate, infant and adult. The chest wall is very compliant because of the soft rib cage of the infant. Airway obstruction results in sternal retraction, and any restriction of diaphragmatic movements can precipitate respiratory failure.

Airway resistance decreases from about 90 cmH₂O/L per second in the first minute to about 25 cmH₂O/L per minute at the end of the first day of life. The resistance of the nasal passages in the neonate is about 50 per cent of the total airway resistance. Significant ventilation–perfusion mismatch occurs in the newborn, with a ventilation perfusion ratio of 0.4 due to small airway closure. This results in a lower normal arterial oxygen

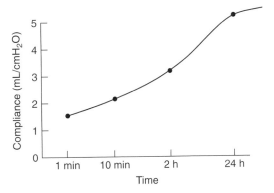

Figure 14.19 *Lung compliance changes during the first 24h after birth.*

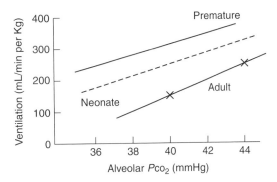

Figure 14.20 *Carbon dioxide response curves.*

tension of 50–70 mmHg (6.7–9.3 kPa) in the neonate. The central and peripheral chemoreceptors are well developed in the neonate. The ventilatory response to carbon dioxide is mature, with the CO₂ response curve shifted to the left compared with the adult, so that ventilatory increases take place at a lower level of CO₂ tension (Fig. 14.20). The increase in ventilation is mainly achieved by an increased tidal volume. Hypoxaemia causes a transient increase in ventilation for about 2 min in the immediate postnatal period, but a sustained response is seen by the 10th day after birth. The respiratory centre of the neonate is depressed by hypothermia.

The chest wall muscle spindles are important mechanoreceptors that detect forces applied to the chest wall, and workload. The large airways have receptors that sense lung inflation and deflation and changes in the interstitial lung fluid. The Hering–Breuer reflex is evoked by gradual inflation of the lungs, and results in a transient apnoea following inflation. Apnoeic spells lasting 5 s normally occur five to six times an hour, and this risk decreases at 52–60 weeks post-gestation. Apnoeic spells lasting longer than 15 s and associated with bradycardia and cyanosis are regarded as significant.

HAEMATOLOGY

In the fetus, haemopoiesis occurs in the yolk sac at 14 days' gestation. The liver then becomes the primary organ for blood formation until the first week after birth. Haemopoiesis begins in the bone marrow in the fifth month of gestation. The mean haemoglobin concentration of the newborn is about 17–18 g/dL, and this may rise by 1–2 g/dL in the first days of life as a result of excretion of fluids. A week after birth the haemoglobin concentration returns to 18 g/dL, and then decreases steadily to about 11–12 g/dL at 4–8 weeks due to a decrease in red cell mass (Fig. 14.21). Red cell survival is approximately 60–70 days at term and 30–40 days in the premature baby. The haemoglobin concentration remains low in childhood, but increases to adult levels by puberty.

Before birth, fetal haemoglobin (HbF) accounts for 90 per cent of all Hb production, but production declines after 35 weeks of gestation. At birth, HbF forms 75–80 per cent of the total haemoglobin, but this gradually decreases so that at 6 months after birth it is replaced by adult haemoglobin (HbA). HbF has two α and two γ chains in its molecule and

has an increased affinity for oxygen because of reduced binding of 2,3-diphosphoglycerate (2,3-DPG). The haemoglobin oxygen dissociation curve of HbF is shifted to the left, with a P_{50} of about 19 mmHg (2.5 kPa). Neonatal blood has an oxygen-carrying capacity about 1.25 times that of adult blood. As concentrations of HbA and 2,3-DPG increase, the oxygen dissociation curve gradually shifts to the right.

Platelet numbers are in the same range as adults, but they have a transient mild defect in function. Neonatal platelets have lower levels of serotonin and adenine nucleotides. There is a deficiency of vitamin K-dependent factors (II, VII, IX and X) as synthesis by the liver is suboptimal. The levels of these factors may be as low as 5–20 per cent of adult levels on the second or third day after birth. Vitamin K stores are deficient at birth, but the administration of vitamin K does not fully correct the coagulation deficiencies because the liver is immature.

The blood volume of a neonate is estimated to be 80–85 mL/kg body weight. For the premature baby the blood volume is about 100 mL/kg. The plasma volume is approximately 5 per cent of body weight.

The total body water content of the newborn is proportionately higher than the adult because of the relative larger ECF compartment. Body water is about 80–85 per cent of body weight in the premature, and 75 per cent in the neonate at term (Fig. 14.22). At birth, ECF constitutes 40 per cent

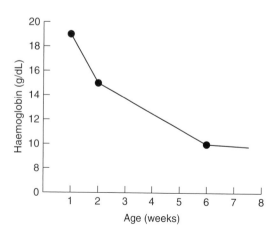

Figure 14.21   *Changes in haemoglobin concentration.*

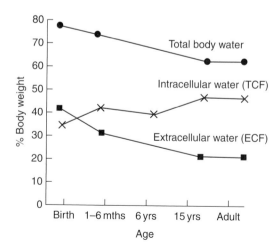

Figure 14.22   *Changes in water compartments.*

of the body weight, and intracellular fluid (TCF) 35 per cent. During the first few days after birth excess ECF is excreted. At about 4–6 months after birth the volumes of the ICF and ECF compartments are similar, after which the intracellular fluid increases and ECF decreases. Insensible water loss in a premature newborn is between 2.5 and 3 mL/kg per hour and in neonates at term 0.7–1 mL/kg per hour.

ACID–BASE STATUS

Blood gases measured from umbilical cord blood show a combined respiratory and metabolic acidosis; pH 7.26, Pa_{CO_2} 55 mmHg (7.3 kPa), and Pa_{O_2} 20 mmHg (2.7 kPa). After the cord is clamped, the Pa_{CO_2} falls to 32–36 mmHg (4.3–4.8 kPa), and rises to 40 mmHg (5.3 kPa) by 2 weeks of age. The pH increases rapidly to 7.34–7.36 and stays there for 2–3 weeks before rising again to 7.38–7.4.

RENAL FUNCTION

At birth, fluid and electrolyte homeostasis is taken over by the kidneys, which gradually improve their function over the first few days. Renal blood flow in the newborn is low because of incomplete glomerular development and arteriolar constriction. In the first 12 h after birth renal blood flow is about 150 mL/min, increasing to 250–300 mL/min during the first week. The proportion of cardiac output supplying the kidneys is about 5 per cent during the first 12 h, 10 per cent in the first week, and this increases to 25 per cent at maturity. In the newborn a relatively greater proportion of the renal blood flow perfuses the medulla, but cortical blood flow increases with an increased ability to excrete a sodium load. Over the 24–48 h after birth the GFR falls and recovers after the first week. GFR reaches adult values at about 2 years of life. The ability of the neonatal kidney to concentrate urine is less than that of an adult. The neonate cannot concentrate urine above 600 mOsm/L during the first week of life, but this increases during the first month. The neonate has no diuretic response to a water load in the first 48 h after birth, but by the end of the first week dilute urine can be produced. Tubular function of the kidney develops at different rates, the distal tubule maturing earlier than the proximal tubule, and the loop of Henle even later. When the cortical nephrons develop, sodium reabsorption improves. The secretion of substances by the proximal tubule is deficient in the neonate.

LIVER FUNCTION

In the neonate many liver functions are poorly developed, especially carbohydrate metabolism and detoxication. Liver enzyme systems mature rapidly after birth and function at adult levels by 3 months of age. Albumin synthesis in the liver starts at 3–4 months' gestation and increases towards term. The ability of the neonate to conjugate bilirubin and drugs with glucuronide is less because of the low activity of hepatic uridine diphosphoglucuronyl transferase. The activity of this enzyme system increases to adult levels about 70 days after birth. The glycogen reserves of the neonate are low (about 4 g/kg body weight) as rapid synthesis of glycogen by the fetal liver ends at 36 weeks of gestation. Energy is derived from the use of fat and proteins, as well as carbohydrates. Glucose is the main energy source in the first few hours after delivery. The blood sugar level in a normal neonate at term is 2.7–3.3 mmol/L, and 2.2 mmol/L in the premature. As liver and muscle glycogen stores decrease, fat metabolism becomes a more significant energy supply.

METABOLIC BALANCE

The metabolic needs of the fetus *in utero* are for growth, and fat and glycogen stores are laid down in the third trimester. Glucose is the main metabolic substrate for the fetus, and this is transferred from the mother across the placenta by facilitated diffusion, although small amounts may be produced from amino acids and fats in the fetal liver. Fetal blood glucose concentration is 70 per cent of that of the mother. In the third trimester of gestation glycogen is laid down in the liver, myocardium and skeletal muscle. There is about 9 g of stored glycogen in the fetus at 33 weeks' gestation, increasing to 34 g at 40 weeks.

At birth, there is an immediate increase in metabolic requirements and oxygen consumption rises to 7–8 mL/kg per min from the oxygen consumption *in utero* of 4–5 mL/kg per min. Glycogen stores are exhausted by 3–4 h in response to catecholamine secretion, after which the fat stores are mobilized with an increase in the plasma free fatty acid and glycerol concentrations. As premature babies have inadequate stores of glycogen, hypoglycaemia frequently occurs, resulting in apnoea, seizures and sometimes cerebral damage.

NERVOUS SYSTEM

At birth the brain is relatively large – about 10 per cent of the total body weight – and by 1 year its weight is trebled as a result of myelination and growth of dendritic processes. Newborns have a higher circulation of β endorphins than adults. The blood–brain barrier is immature and may allow the passage of drugs. The water content of the brain decreases from 92 to 82 per cent by the end of the first year.

Neuromuscular junction

The peripheral muscles are innervated by motor nerve fibres by the end of the first trimester. At 28 weeks, the motor nerve endings differentiate to form end-plates. However there is less acetylcholine available within the neuromuscular junction, and the neuromuscular junction of the neonate takes a longer time to recover from a neuromuscular block.

THERMOREGULATION

The ability to maintain a stable core temperature in the face of changes in the ambient temperature is accomplished by balancing heat production and heat loss. Heat balance in the newborn is seriously affected by the environmental temperature because:

- The surface area/weight ratio is large, approximately three times that of an adult
- The insulating capacity of subcutaneous tissue in a full neonate is half that of an adult
- The shivering mechanism is poorly developed

The temperature-regulating centre is situated in the hypothalamus and responds to changes in blood temperature of 0.1–0.2°C, as well as to information from skin temperature receptors. Heat loss is regulated by physiological changes mediated via the vasomotor centre. The hypothalamus also regulates metabolism to influence heat production.

Heat production in the neonate is by non-shivering thermogenesis, with increased metabolism of brown fat (a specialized fat tissue with a high mitochondrial content, and a rich sympathetic innervation that can be activated by stimulation of the ventromedial nucleus of the hypothalamus). Brown fat is found largely in the interscapular region, mediastinum, in perinephric tissues, in the axillae and near major blood vessels in the neck, constituting about 11 per cent of the total body fat. Cold exposure increases sympathetic activity, and norepinephrine is released at sympathetic nerve endings, binding to β_3 receptors with activation of adenyl cyclase and protein kinases. The protein kinases enhance the actions of lipase, resulting in hydrolysis of triglycerides to free fatty acids and glycerol. During cold exposure, heat is also produced by glucose metabolism and gluconeogenesis, especially in the brain and the liver. The neonate is capable of a three-fold increase in its basal metabolic rate by non-shivering thermogenesis.

Heat loss

The neonate loses heat by radiation, convection, evaporation and conduction. Various studies have estimated that radiation, convection, evaporation and conduction account for 39, 34, 24 and 3 per cent, respectively, of heat loss in newborns in incubators. Radiant heat loss decreases as the environmental temperature rises. Because the newborn has a large surface area/volume ratio, radiant heat loss is greater with smaller neonates. Convective heat loss (the transfer of heat by the movement of surrounding air) is proportional to the temperature difference between the body surface and the air and the velocity of movement of the air. Evaporative heat loss occurs from the body surface and the respiratory tract. Evaporation from the skin is related to the relative humidity and the amount of sweating. Although the newborn has six times more sweat glands per unit area than an adult, its ability to sweat is less, the peak response being about one-third that of adults. Full-term neonates sweat when the rectal temperature is between 37.5° and 37.9°C, and at ambient temperature greater than 35°C. Premature infants of less than 30 weeks' gestation do not sweat at all, as the sweat glands are immature.

Thermoneutral environment

The thermoneutral temperature is the range of environmental temperatures within which the body will maintain its temperature with minimal oxygen consumption (Fig. 14.23). The lower end of the thermoneutral temperature range is called the 'critical temperature', the point at which extra heat must be generated to prevent a fall in body temperature. The thermoneutral range for a naked newborn baby is narrow. On the first day of life, the thermoneutral temperature is 32–34°C in full-term infants and 35–36°C in low-birth-weight babies. In the thermoneutral range, body temperature is controlled by changes in skin blood flow alone, with minimal oxygen consumption. The critical temperature at which oxygen consumption begins to rise depends on maturity and body size. In a full-term baby of 3 kg, the critical temperature is 33°C at birth, and this decreases to 32°C at 2 weeks of age. In contrast, for a premature neonate of 1 kg, the critical temperature is 35.5°C. The increase in oxygen consumption below the critical temperature depends on the temperature gradient between the body and the environment and insulation of the infant.

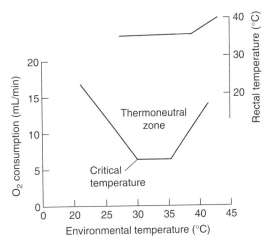

Figure 14.23 *Effect of temperature on oxygen consumption.*

The main danger of hypothermia in the neonate is an increase in oxygen consumption and mortality from hypoxaemia. It is also associated with coagulopathies, decreased surfactant synthesis and altered pharmacokinetics of drugs. When children become hypothermic during anaesthesia, respiration may be depressed, and cardiac output, blood pressure and heart rate are decreased, with delayed recovery. Because of increased metabolism, hypoglycaemia may develop.

Hyperthermia can also be harmful to neonates. An increase in the rectal temperature to above 37°C leads to a three-fold increase in evaporative water loss, although sweating is not well developed until 36–37 weeks' gestational age. Low-birth-weight neonates exposed to high environmental temperatures, or who are febrile, are more prone to apnoeic attacks.

Reflections

1. The mother is in an anabolic state early in pregnancy. The anabolic state facilitates growth of her reproductive tissues and energy stores for pregnancy. The physiological changes during pregnancy are due to steroid and peptide hormones produced by the placenta. Maternal cardiac output progressively increases by about 40 per cent at the third trimester. The substantially larger increase in plasma volume compared to the increase in red cell mass leads to a state of haemodilution. The haemocrit and viscosity of blood decreases and, combined with vasodilatation mediated by progesterone, maternal systemic vascular resistance decreases. The gravid uterus can reduce venous return, especially in the third trimester of pregnancy, leading to a marked decrease in cardiac output and blood pressure when the mother is in a supine position, a phenomenon known as 'supine hypotension syndrome'. The mother hyperventilates

as a result of the resetting of the central chemo-receptors by progesterone and produces a chronic respiratory alkalotic state with renal compensation. The diaphragm is elevated and this leads to a reduction of the FRC by about 20 per cent. There is an increase in the anterior posterior diameter of the chest and this diminishes the loss of lung volume caused by the elevation of the diaphragm. The lower oesophageal sphincter is less competent as a consequence of the raised intra-abdominal pressure and the muscle-relaxation effects of progesterone and this increases the risk of gastric aspiration. There is an increase in the functions of all the endocrine glands of the mother and this contributes to the anabolic state in early pregnancy. The later catabolic state is marked by insulin resistance and this facilitates the supply of nutrients to the growing fetus.

2. The placenta forms an interface between the maternal and fetal circulations. Following implantation, the trophoblastic tissues of the zygote invade the endometrial tissue via chorionic villi that possess fetal capillaries, resulting in the erosion of spiral arteries and the formation of intervillous spaces containing blood between adjacent chorionic villi. The arrangement of the two circulations within the placenta enables the exchange of solutes and respiratory gases across the placental barrier according to concentration gradients and other transport mechanisms. Although the lipid-rich placental barrier is relatively impermeable to polar molecules, the process is facilitated by the large surface area available as a result of the extensive branching of the chorionic villi. During fetal life the placenta performs the functions normally undertaken by the lungs, kidneys and the gastrointestinal tract in the adult. Oxygen diffuses passively from maternal blood to the fetus by means of a concentration gradient which is enhanced by the 'double Bohr' effect. Carbon dioxide diffuses in the opposite direction as a result of the 'double Haldane' effect. Essential nutrients cross the placenta via passive diffusion or carrier-mediated transport. Glucose and amino acids moves across the placenta from the maternal to fetal blood by a carrier-mediated transport mechanism whilst free fatty acids diffuse passively across the placenta. Fetal waste products such as urea diffuse from fetal to maternal blood down their concentration gradients. The placenta also secretes peptide and steroid hormones. Human chorionic gonadotrophin (HCG) and human placental lactogen (HPL) are the major peptide hormones, and progesterone and oestrogen are the major steroid hormones. HCG, a potent luteotropic hormone, prevents the regression of the corpus luteum so that continued secretion of progesterone occurs during early pregnancy. Progesterone is essential for pregnancy because it maintains the endometrium and reduces myometrial excitability and stimulates the development of breast glands (alveoli) to facilitate lactation. Oestrogens are produced and released by the syncytiotrophoblast of the placenta and are responsible for some of the physiological changes in the mother. HPL is secreted from the 10th week of gestation and exerts important metabolic effects in the mother. It stimulates an increase in maternal plasma levels of glucose, amino acids, and free fatty acids to ensure optimal transfer of these nutrients from the mother to the fetus. High concentrations of oestrogens in the mother dominate the hormonal profile in the last days of pregnancy. Oestrogens increase myometrial contractility and this may be one of the triggers for parturition.

3. The fetal circulation is arranged so that the right side and left side of the heart work in parallel, and three fetal shunts bypass blood from those organs with minimal or no function. The walls of the right and left ventricle in the fetus are equal in thickness. Fetal blood pressure is low and heart rate is high. The fetal lungs are filled with fluid and are virtually collapsed, causing a high pulmonary vascular resistance. The fetal pulmonary circulation receives only 10 to 20 per cent of the right ventricular output; the other 80 to 90 per cent passes through the ductus arteriosus into the aorta. Fetal blood carries about 16 mL O_2 per dL although its Pa_{O_2} is low. It has a high capacity for oxygen because it has a high haemoglobin concentration and fetal haemoglobin has a high affinity for oxygen. After delivery, several

factors may stimulate the baby to breathe. A most likely trigger for the first breath is hypercapnia following cord compression during delivery. Other physical factors such as temperature changes, proprioceptive and tactile stimuli. At the first breath the baby must first overcome enormous surface tension forces at the gas-liquid interface in the alveoli. Although surfactant, which reduces these surface tension forces, is produced during the last few weeks of a normal pregnancy, a massive inspiratory effort is still required by the baby to generate the large negative intrathoracic (-60 to $-70\,cmH_2O$) pressures to inflate the lungs. The peripheral chemoreceptors in the fetus also respond to reductions in oxygen tension following delivery. Neonatal respiration is different from that in an adult in several ways. The respiratory rate is higher but is irregular and diaphragmatic in nature. Airway resistance is higher and the work of breathing is greater in the neonate.

The circulatory system of the infant adapts to the pulmonary changes that occur at birth. The parallel arrangement of the right and left sides of the heart converts to a series arrangement as a result of the closure of the three fetal shunts. Closure of the shunts depends on ventilation of the lungs. As the lungs expand pulmonary vascular resistance decreases and pulmonary perfusion increases while at the same time the umbilical vessels close. This causes left atrial pressure to rise above right atrial pressure, resulting in the closure of the septa which form the foramen ovale. The ductus venosus and the ductus arteriosus vasoconstrict in reponse to a rise in arterial Pa_{O_2}.

The fetal kidneys cannot concentrate urine effectively and produce hypotonic urine from about 8 weeks' gestation. Glucose absorption is comparable to that in adults but sodium reabsorption is relatively low. From birth, glomerular filtration rate (GFR) and urine output increase gradually, as does the ability to concentrate urine. The fetal gastrointestinal system is relatively immature and the chief metabolic substrate is glucose derived solely by placental transfer from the mother. The contents of the large intestine of the fetus accumulate as meconium which may be passed into the amniotic fluid during fetal distress. After birth the baby relies on the fat and carbohydrtaes stores that are laid down during late gestation. With milk feeds the major metabolic substrate changes to fats. Digestive juice secretion and gastrointestinal motility increase.

The large surface area to volume ratio, lack of insulating fat and the relatively high cardiac output combine to cause the newborn infant to lose body heat very rapidly and become hypothermic. Babies cannot shiver and generate large quantities of heat through brown fat metabolism, which is stimulated by catecholamines released in response to cold stress. Brown fat is a well-vascularized fat tissue located around the kidneys, between the scapulae, in the axillae and at the nape of the neck.

4. Lactation is the synthesis and secretion of milk by the mammary glands. Progesterone stimulates the development of alveoli which are spherical collections of cells that produce milk. Under the influence of placental hormones (progesterone and oestrogens) the alveoli mature and the breast is able to secrete milk. During pregnancy the oestrogens and progesterone inhibit the lactogenic action of prolactin but this inhibitory action is lost after delivery and lactation commences. Colostrum is secreted in the first few days after delivery. Colostrum is rich in proteins, minerals and immunoglobulins but low in fats and sugar. The composition of milk gradually changes and by 3 weeks' postpartum mature milk that is rich in fats, proteins and sugars is produced. Lactose is the chief milk sugar while casein, lactoglobulin and α-lactalbumin are the chief milk proteins.

15

Physiology of ageing

FUNCTIONAL DECLINE WITH AGEING

Ageing implies a decreased viability or an increased vulnerability to stress with a diminished ability to maintain homeostasis. The World Health Organization regards people in the age range 45–59 years as 'middle aged', 60–74 as 'elderly', 75–89 as 'old' and over 90 as 'very old'. Anatomical and physiological changes with ageing usually begin in middle life in almost every body system (Fig. 15.1), but their magnitude and clinical significance vary considerably. There is considerable confusion about what is part of the ageing process and what is due to disease. The exact cause of ageing is unknown. It has been suggested that cellular ageing may occur as a result of some alteration in the information carried by DNA, or by programmed cell death. Other theories of ageing suggest that the changes result from wear and tear, accumulation of substances such as lipofuscin, collagen, amyloid and calcium in abnormal sites, progressive reduction of endocrine function and altered immune function.

CHANGES IN THE NERVOUS SYSTEM

There is a continual loss of neuronal tissue with advancing age. The brain weight may decrease by

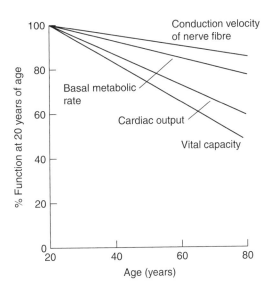

Figure 15.1 *Decline in physiological functions as a percentage of that function at 20 years of age.*

6–7 per cent between 20 and 80 years of age. About 10 000 brain cells are lost per day from the age of 20 years onwards, but the dendrites of the remaining cells become longer and form more connections to compensate for this loss. Lipofuscin accumulates in many nerve cells, but its relation to ageing is not known. The grey matter of the brain decreases from 45 to 35 per cent of the total brain

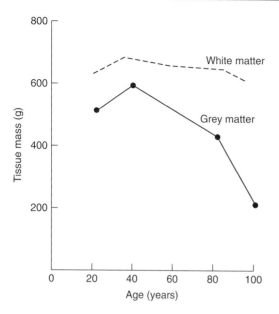

Figure 15.2   *Changes in the brain related to age.*

weight between the ages of 20 and 80 years (Fig. 15.2).

The amount of neurotransmitters in the brain is also depleted with age. For example, a decrease in dopamine in the substantia nigra can occur with Parkinsonian symptoms, whilst a decrease in the acetylcholine concentration in the hypothalamus can lead to senile dementia. Depression may result from a decrease of norepinephrine and 5-hydroxytryptamine in the hypothalamus. Mental confusion with difficulty in temporal and spatial orientation is common in the elderly. The short-term memory is impaired with ageing, but the long-term memory may be preserved. Conceptual skills and reasoning capabilities decline early, but verbal skills are usually well preserved. Sleep patterns are also altered and the amount of rapid eye movement sleep progressively decreases with age.

Regulatory functions are impaired in later life. Postural control is impaired, and may be associated with slow reflexes. Autonomic regulation of cardiovascular functions and maintenance of body temperature may also be impaired. There is a loss of myelin in peripheral nerves, a reduction of axons and synapses, and a reduced number of motor neurons in the spinal cord. However, the peripheral nervous system is less affected than the central nervous system.

CHANGES IN THE CARDIOVASCULAR SYSTEM

The changes in the cardiovascular system in the elderly may be due to the ageing process, prolonged deconditioning, and age-related disease. Although the resting heart rate does not alter with age, the maximum heart rate that can be achieved decreases from about 200 to 160 beats/min. The intrinsic heart rate (i.e. without autonomic influence) is reduced. There is fibrous infiltration of the sinoatrial node with a loss of pacemaker cells leading to an increased susceptibility to supraventricular arrhythmia and ventricular ectopic beats. The atrioventricular node and bundle of His are usually unchanged histologically, but there may be some loss of Purkinje fibres in the left ventricle.

There is an increase in connective tissue in the heart with age due to the replacement of fragmented elastin by collagen. Scattered deposits of amyloid and lipofuscin are present within and between the myocardial cells. Myocardial wall thickness may increase as a result of an increase in myocyte size in response to increases in impedance to left ventricular output. Increasing fibrosis of the endocardium leads to a decreased compliance of the heart. Calcification in the heart valves can distort the valve cusps and produce valvular incompetence. Although cardiac output is thought to decrease with age by about 1 per cent per year beyond 30 years of age, recent studies suggest that there is no significant decline in cardiac output at rest or during exercise in healthy subjects between the ages of 25 and 79 years. The maximum stroke volume that can be achieved is reduced. Traditional studies have demonstrated that maximum exercise performance (measured by heart rate, stroke volume and cardiac output) is reduced with age. It is suggested that the decline in cardiac output in elderly people may be due to reduced preconditioning (i.e. sedentary lifestyle) or age-related disease. Healthy elderly people can increase their cardiac output during exercise by increased reliance on the Frank–Starling mechanism, primarily by increasing stroke volume in response to an increase in left ventricular end-diastolic volume and pressure.

Large artery elasticity is decreased with age, resulting in stiffening of the arterial vasculature, and as a result the mean and systolic arterial blood

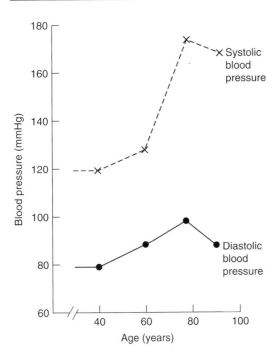

Figure 15.3   *Changes in blood pressure with ageing.*

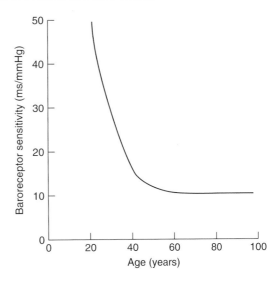

Figure 15.4   *Changes in baroreceptor activity with age.*

pressures increase with age (Fig. 15.3). Diastolic blood pressure may increase because of increased peripheral resistance. However, studies have shown there is a decrease in diastolic blood pressure in people over 75–80 years of age because of the rapid run-off of blood in the stiff large arteries. In healthy elderly people, pulmonary artery systolic pressure increases from the value of 20 mmHg in young adults to 26 mmHg, pulmonary artery diastolic pressure rises from 9 to 11 mmHg, and pulmonary vascular resistance rises from 70 to 120 dyn · s/cm^5.

Baroreceptor mechanisms are impaired in the elderly, and this may cause postural hypotension (Fig. 15.4). There is a reduced responsiveness of the cardiac β-adrenergic agonists due to either reduced receptor numbers or affinity, or diminished generation of cyclic adenosine monophosphate after β-receptor activation.

CHANGES IN THE RESPIRATORY SYSTEM

Ageing is associated with decreased lung volumes and reduced efficiency of gas exchange. From the ages of 20 to 70 years, total lung capacity is decreased by 10 per cent because of increased thoracic cage rigidity that restricts chest expansion. Progressive kyphosis causes an upward and anterior rotation of the ribs and further restriction of chest expansion. These anatomical changes contribute to the reductions in vital capacity and maximum breathing capacity seen in the elderly. At the age of 20 years, the maximum voluntary ventilation is 100 L/min, 12–15 times that required for basal metabolism. In the elderly, this is reduced to 30–40 L/min (approximately seven times the basal requirements) because of a loss of elasticity of the thoracic cage and lung parenchyma.

Lung parenchymal changes are similar to those seen in emphysema. Alveolar septa are lost, and the alveolar surface area is reduced. The elastic recoil of the lung is reduced by the loss of functional alveoli with age. These changes increase both the ratio of residual volume to total lung capacity (TLC), and the ratio of functional residual capacity (FRC) to TLC in the elderly (Fig. 15.5).

The closing volume of the lungs increases as small airways collapse at larger lung volumes because of the diminished radial traction of the terminal bronchioles due to the reduction in alveolar septa in the elderly (Fig. 15.6). As the closing volume increases with ageing, a greater proportion of the tidal volume will occur at lung volumes below closing volume, resulting in increased

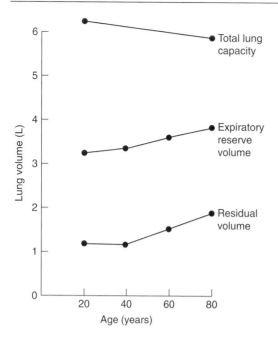

Figure 15.5 *Changes in lung volumes with ageing.*

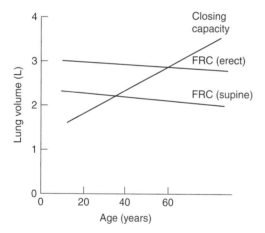

Figure 15.6 *Changes in functional residual capacity (FRC) and closing volume with ageing.*

ventilation–perfusion (V/Q) inequalities. The resting arterial oxygen tension decreases with age at a rate described by the following equation:

$$P_{aO_2} = 100 - (0.33 \times \text{age [years]}) \text{ mmHg}$$
$$= 13.6 - 0.044 \times \text{age (years) kPa}$$

The efficiency of gaseous exchange in the elderly declines as a result of increased closing volume, reduced alveolar surface area, increased V/Q

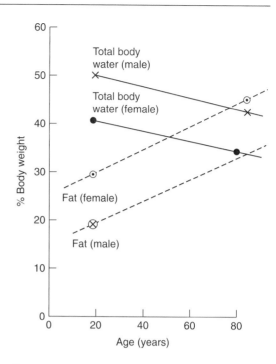

Figure 15.7 *Changes in body composition with ageing.*

inequalities and increased alveolar capillary membrane thickness.

CHANGES IN BODY COMPARTMENTS

There is a 10 per cent decrease in lean body mass (skeletal muscle mass) with ageing, with an average loss of 6 kg of muscle mass at the age of 80 years. This is associated with an increase in the percentage of body fat and a decrease in intracellular water content. These changes are more marked in women (Fig. 15.7). Total body water is reduced not only by a reduction in intracellular water but also by a reduction in blood volume. It is estimated that the blood volume is reduced by 20–30 per cent by the age of 75 years. Plasma albumin concentration gradually decreases from 4 g/dL at 40 years of age to 3.6 g/dL in patients at 80 years.

CHANGES IN RENAL FUNCTION

The number of functional renal glomeruli decreases with age. Glomerular filtration decreases by about 1–1.5 per cent per year from the age of 30 years, and there is a parallel decrease in tubular excretion. The decrease in GFR is by a reduced renal plasma flow due to a reduction of the renal vascular bed and decreased cardiac output. A disproportionately large loss of cortical glomeruli occurs with ageing.

CHANGES IN LIVER FUNCTION

A progressive decline in the hepatic clearance of a variety of substances occurs in elderly people due primarily to reduced liver size. Hepatic blood flow decreases, but hepatic enzyme function does not change with ageing.

ENDOCRINE CHANGES

Pancreatic function declines with age, and it is thought that this explains the increased incidence of glucose intolerance and diabetes mellitus in people over 70 years of age. Insulin secretion in response to hyperglycaemia is slower, and there is also insulin resistance at peripheral sites. Plasma renin concentration decreases by about 30 per cent in the elderly, and this may lead to a reduction in plasma aldosterone concentration.

THERMOREGULATION

Elderly patients have a reduced ability to maintain body temperature, due to decreased heat production, increased heat loss and less efficient thermoregulation. Basal metabolic rate decreases by about 1 per cent per year beyond the age of 30 years. The reduced autonomic control of the peripheral vasculature of the elderly leads to a diminished ability to vasoconstrict upon exposure to a cold environment.

Reflections

1. Anatomical and physiological changes with ageing usually begin in middle life in almost every body system. Cellular ageing may occur as a result of programmed cell death. Other theories of ageing suggest that the changes result from wear and tear, accumulation of substances such as lipofuscin, collagen, amyloid and calcium in abnormal sites, progressive reduction of endocrine function and altered immune function. There is a continual loss of neuronal tissue with advancing age. There is a loss of myelin in peripheral nerves, a reduction of axons and synapses, and a reduced number of motor neurons in the spinal cord. The amount of neurotransmitters in the brain is also depleted with age. A decrease in dopamine in the substantia nigra can lead to Parkinson's disease, whilst a decrease in the acetylcholine concentration in the hypothalamus can lead to senile dementia. Depression may result from a decrease of norepinephrine and 5-hydroxytryptamine in the hypothalamus. Autonomic regulation of cardiovascular functions and maintenance of body temperature may also be impaired. The changes in the cardiovascular system in the elderly may be due to the ageing process, prolonged deconditioning, and age-related disease. There is an increased susceptibility to supraventricular arrhythmia and ventricular ectopic beats. Increasing fibrosis of the endocardium leads to a decreased compliance of the heart. Calcification in the heart valves can distort the valve cusps and produce valvular incompetence. The maximum cardiac stroke volume that can be achieved is reduced. Maximum exercise performance (measured by heart rate, stroke volume and cardiac output) is reduced with age. Healthy elderly people increase their cardiac output during exercise by increased reliance on the Frank–Starling mechanism. Large artery elasticity is decreased with age, resulting in stiffening of the arterial vasculature, and diastolic blood pressure may increase because of increased peripheral resistance. However, a decrease in diastolic blood pressure in people over 75–80 years of age occurs because of the rapid run-off of blood in the stiff large arteries. Baroreceptor mechanisms are impaired in the elderly, and this may cause postural hypotension. A reduced

responsiveness of the cardiac β-adrenergic agonists occurs as a result of either reduced receptor numbers or affinity.

2. Ageing is associated with decreased lung volumes and reduced efficiency of gas exchange. Anatomical changes contribute to the reductions in vital capacity and maximum breathing capacity seen in the elderly. The elastic recoil of the lung is reduced by the loss of functional alveoli with age. These changes increase both the ratio of residual volume to total lung capacity (TLC), and the ratio of functional residual capacity (FRC) to TLC in the elderly . The resting arterial oxygen tension decreases with age. The efficiency of gaseous exchange in the elderly declines as a result of increased closing volume, reduced alveolar surface area, increased V/Q inequalities and increased alveolar capillary membrane thickness.

3. The 10 per cent decrease in lean body mass (skeletal muscle mass) with ageing is associated with an increase in the percentage of body fat and a decrease in intracellular water content. The blood volume is reduced by 20–30 per cent by the age of 75 years. Plasma albumin concentration gradually decreases.

4. Glomerular filtration decreases by about 1–1.5 per cent per year from the age of 30 years, and there is a parallel decrease in tubular excretion. This is due to a reduction of the renal vascular bed and decreased cardiac output. A disproportionately large loss of cortical glomeruli occurs with ageing. Hepatic blood flow decreases, but hepatic enzyme function does not change with ageing. Pancreatic function declines with age, and it is thought that this explains the increased incidence of glucose intolerance and diabetes mellitus in elderly people. There is also insulin resistance at peripheral sites. Plasma renin concentration decreases by about 30 per cent in the elderly, and this may lead to a reduction in plasma aldosterone concentration.

5. Elderly patients have impaired thermoregulation due to decreased heat production, increased heat loss and less efficient thermoregulation. Basal metabolic rate decreases by about 1 per cent per year beyond the age of 30 years.

Special environments

After reading this chapter the reader should be able to:

1. Describe the diving response and explain the physiological consequences of diving
2. Explain the physiological changes that occur following ascent to high altitude
3. Explain the physiological problems associated with space travel and zero gravity

PHYSIOLOGY OF DIVING

Life underwater exposes the diver to a major rise in the ambient pressure of the environment, with physiological changes and problems from the direct effects of pressure on the body. In addition, the diver must be supplied with a mixture of gases to breathe at a pressure equal to ambient pressure, and this can also give rise to other problems.

PHYSICAL LAWS

As the SI unit for pressure (N/m^2 (pascals)) is small, the 'bar' has been used for ambient pressures. One bar is equal to $10^5 N/m^2$, 750 mmHg and approximately 1 atmosphere. However, in the diving industry, pressure is described in terms of depth in sea water, metres of sea water (msw). For every 10 m under sea water (density 1.025), there is an increase in ambient pressure of about 1 bar. Thus, assuming that normal atmospheric pressure is 1 bar, at a depth of 50 msw the ambient pressure is 6 bar.

Three physical principles must be used to understand the physiological as well as the pathophysiological effects of life underwater. Gases are compressible and follow Boyle's law. During a breath-hold dive, as the volume of a gas is inversely related to its pressure for a given mass of gas at constant temperature, the volume of air in the lungs decreases with increasing depth and ambient pressure.

As the pressure of the gas increases, its density (mass per unit volume) increases. According to Dalton's law, an increase in the total gas pressure is associated with an increase in partial pressures of the constituent gases by the same proportion. This results in an increase in the amount of gas dissolved in body fluids, according to Henry's law.

DIRECT EFFECTS OF INCREASED PRESSURE

Several cardiovascular changes are associated with diving. On first immersion and exposure to water, especially when it is colder than 15°C, there is a dramatic decrease in heart rate, apnoea and selective vasoconstriction (especially cutaneous) and this is called the 'diving reflex'. The vasoconstriction occurs in those organs that can utilize anaerobic metabolism such as the skin, muscle, kidneys and the gastrointestinal tract. The brain requires a constant supply of oxygen as it relies on oxidative metabolism. During the diving reflex the oxygen supply to the brain is maintained by the redistribution of the cardiac output to the cerebral circulation. The initial stimulus for the diving reflex appears to be the immersion of the skin of the face to cold water. The afferent fibres responsible for this reflex are in the trigeminal nerve. Immediately following face immersion, the heart rate decreases by about 50 per cent and this is mediated by increased vagal activity. This is usually associated with an increase in arterial pressure that occurs because of profound peripheral vasoconstriction due to increased sympathetic activity. The apnoea is induced partly by voluntary control and partly by reflex inhibition of respiration mediated by stimulation of the trigeminal receptors. Prolonged breath-holding leads to severe hypoxaemia and hypercapnia, which stimulate the carotid body chemoreceptors and initially cause bradycardia and peripheral vasoconstriction. Under these conditions the chemoreceptor stimulus is strongly inhibited by the activation of the trigeminal receptors. Eventually the hypercapnia is so severe that the trigeminal inhibition is overcome and the desire to breathe is compelling. This is called the break point, which is determined largely by the arterial P_{CO_2}.

When the body is immersed in water, the water exerts an external hydrostatic pressure gradient down the body, which opposes the internal hydrostatic pressure gradient due to gravitational effects of gravity along the longitudinal axis of the body. As a result of this, venous pooling in the dependent parts of the body does not occur, and about 500 mL of blood moves from the dependent or lower half of the body to the thorax. This increase in the central blood volume raises the right atrial pressure and increases the stroke volume and cardiac output. Consequently, pulmonary blood flow rises with an increased pulmonary diffusing capacity and improved ventilation/perfusion ratio.

The increased central blood volume stretches the atrial receptors and decreases vasopressin secretion and increases atrial natriuretic factor, resulting in diuresis.

The increased ambient pressure during diving affects the mechanics of respiration. The external hydrostatic gradient down the thoracic cage exaggerates the normal intrapleural pressure gradient that is present from the apex to the base of the lungs. If a person is immersed in water with his or her head out (head-out immersion), the alveolar pressure is equal to the atmospheric pressure and the increased pressure on the chest wall opposes the outward elastic recoil. Consequently, the residual volume and the functional residual capacity are reduced and the work of breathing rises by about 60 per cent. If the diver is supplied with air to breathe at a pressure equal to the ambient pressure acting at the level of the chest, then these changes do not occur as the transthoracic pressures are normal.

The pathophysiological effects of a hyperbaric environment are called 'barotrauma'. When a diver dives deeper, the volume of air in the lungs decreases according to Boyle's law. However, at great depths the volume of gases in the lungs may be compressed below residual volume, producing a negative intra-alveolar pressure with pulmonary oedema or haemorrhage.

During ascent from a breath-hold dive, gases in the lungs re-expand to their original volume, causing no problems. However, if the diver has been breathing compressed gases, pulmonary barotrauma can occur during ascent if he or she fails to exhale, or if a cyst or bulla is present in the lung. Alveolar rupture occurs when the intra-alveolar pressure exceeds the ambient pressure by 80 mmHg (10.7 kPa), and may occur during ascent from a dive to 2 m. Upon alveolar rupture, the gas may track via the perivascular sheaths to the hilum into the mediastinum or into the pleural cavity. In addition, the gas can escape into the pulmonary circulation and reach the systemic circulation via the left heart, with the risk of cerebral arterial gas embolism.

Gas-containing cavities in the paranasal passages are bounded by rigid structures and cannot undergo any volume changes. As the pressure in such cavities cannot equilibrate with the ambient pressure during descent in a dive, the transmural pressure across the capillaries may produce exudation of fluid and capillary rupture with haemorrhage. The middle ear is protected from barotrauma by equilibration with the ambient pressure through the Eustachian tube. As descent continues, a significant pressure gradient can develop across the tympanic membrane, and an inward bulging of the tympanic membrane can be painful. If the pressure gradient exceeds 100 mmHg (13.3 kPa) there may be haemorrhage or perforation of the tympanic membrane. Gases in the gastrointestinal tract are compressed during a dive, and re-expansion on ascent can produce abdominal discomfort and flatulence.

The high-pressure neurological syndrome of tremor, dizziness, nausea, and a loss of dexterity and attentiveness can occur at 200 msw. At such high pressures lipids are compressed (they are more compressible than water), and in neurons this changes membrane permeability and ionic transport properties to produce the symptoms of high-pressure neurological syndrome.

EFFECTS OF BREATHING HYPERBARIC GASES

When a diver uses a snorkel, a depth greater than 0.5 msw is not safe as the increased ambient pressure decreases the volume of air in the lungs and the inspiratory muscles would have to generate a pressure of 90 mmHg (12 kPa) to prevent this diminution of lung volume. In practice, the maximum safe depth for snorkelling is less than 1 m, because the alveolar pressure is less than the ambient pressure and pulmonary exudation and haemorrhage can occur. Also a snorkel longer than 0.5 m increases dead space, causing inadequate alveolar ventilation.

In dives deeper than 0.5 m, the diver must be supplied with a mixture of gases at a pressure equal to the ambient pressure, and this is normally from compressed air in SCUBA (self-contained underwater breathing apparatus) diving. In deeper diving, a demand valve ensures that air is supplied at ambient pressure. However, breathing under pressure leads to problems from the increased density of gases and increased partial pressure of the inspired gases.

Because of the increased density of inspired gases, the work of breathing rises and there is a decrease in the maximal voluntary ventilation proportional to the reciprocal of the gas density. Airflow in the respiratory tract becomes more turbulent, and this further increases airway resistance. Intra-alveolar diffusion of gases is slowed. The

reduction in ventilatory performance can be prevented by replacing nitrogen with the lower-density gas, helium. The other advantage of helium is that its decreased solubility (about 40 per cent that of nitrogen) reduces the likelihood of decompression sickness.

Nitrogen narcosis can occur because the partial pressure of nitrogen increases when compressed air is used for breathing. At high pressures nitrogen may be soluble in lipids, leading to membrane expansion and interference with neural transmission by modulating ion channels. At a depth of 30 msw divers develop euphoria, and reduced dexterity and mental agility. At 50 msw there may be loss of concentration and decreased neuromuscular coordination. Beyond 50 msw the symptoms become severe, and divers may become unconscious at depths beyond 90 msw.

Oxygen toxicity can occur if the inspired oxygen tension exceeds 1350 mmHg (180 kPa or 1.8 bar), as at a depth of 8 msw when 100 per cent oxygen is inspired. Symptoms of oxygen toxicity include vertigo, paraesthesia of the arms and legs and muscle twitching around the mouth and eyes, progressing to convulsions. Pulmonary oxygen toxicity can develop at a depth of 16 msw when compressed air is inspired, with an inspired oxygen concentration of 375 mmHg (50 kPa or 0.5 bar). The first sign of pulmonary oxygen toxicity is dyspnoea, and pulmonary oedema and intra-alveolar haemorrhage are seen later. These symptoms usually occur after 30 h of exposure to high inspired oxygen tensions. The latency of onset decreases with higher inspired oxygen tensions so that symptoms occur after 5 h if the inspired oxygen tension is 1500 mmHg (200 kPa or 2 bar).

Decompression sickness occurs on ascent from a dive. The inspired gas pressures decrease as the diver ascends, and a partial pressure gradient for gases develops between the tissues and the alveoli. If the rate of ascent is rapid, the gases come out of solution in the tissues and form bubbles. The signs and symptoms of decompression sickness usually occur within 6 h of decompression. Joint pain in the limbs is due to bubble formation in the ligaments, tendons and joints. Large intravascular bubbles trapped in the pulmonary circulation cause dyspnoea and cough ('chokes'). Bubbles may form in the spinal cord, leading to motor and sensory deficits. Bubbles in the vestibular apparatus produce vertigo ('staggers'). Avascular necrosis of the head and neck of the humerus, femur and upper tibia is a long-term adverse effect of decompression sickness. The process of ascending in several stages, with stops at depths where the ambient pressure is half that at the depth of the previous stop, can prevent decompression sickness. Breathing helium–oxygen mixtures rather than compressed air also reduces this risk.

The diver is weightless whilst underwater because of the buoyancy of the body in water. Hearing is impaired as sound is attenuated. Heat is rapidly conducted from the body, and hypothermia can result.

PHYSIOLOGICAL EFFECTS OF ALTITUDE

ENVIRONMENTAL CHANGES WITH ALTITUDE

With increasing altitude environmental changes include:

- A colder temperature than at sea level at any given latitude, with a 1°C fall in ambient temperature for every 150 m above sea level.
- A decrease in relative humidity that results in increased insensible water loss by evaporation from the skin and respiration.
- Increased solar radiation as a result of reduced cloud cover.
- The saturated vapour pressure of water at body temperature remains constant. This is important, as at 19 000 m the barometric pressure equals water vapour pressure, so that alveolar P_{O_2} and P_{CO_2} become zero.
- The barometric pressure and the inspired oxygen partial tension fall with increasing altitude, although the fractional concentration of oxygen in air (0.21) remains constant.

HYPOBARIC ENVIRONMENTS

The hypobaric environment is the major problem associated with life at high altitudes, as the other problems can be avoided by behavioural changes (e.g. cold and the use of appropriate clothing). The barometric pressure decreases exponentially as altitude increases and is approximately halved for every 5500 m above sea level. As the concentration of oxygen in air remains constant at 20.93 per cent, there is a parallel decrease in the ambient oxygen tension (P_{O_2}), and a fall of alveolar and arterial P_{O_2}. The decreases in alveolar and arterial P_{O_2} are greater than the decrease in ambient P_{O_2} because inspired air is always saturated with water vapour (exerting a pressure of 47 mmHg, or 6.3 kPa), and this results in a form of hypoxic hypoxia called hypobaric hypoxia. There are a number of adverse physiological effects during a rapid exposure to altitude, and compensatory changes allow humans to acclimatize to the hypoxic environment.

EFFECTS OF RAPID ASCENT TO ALTITUDE

Healthy individuals do not demonstrate any adverse effects below an altitude of 2500 m. This is why aircraft cabins are usually maintained at a pressure equivalent to the ambient pressure at an altitude of about 2000–2500 m. Rapid exposure to altitudes in the range of 3000–6000 m results in acute mountain sickness. The symptoms usually appear in the first 24 h, and include headache, somnolence, nausea, vomiting, insomnia and muscle fatigue. These symptoms usually decrease after 3–4 days. At altitudes above 4000 m, cerebral hypoxia occurs with psychomotor impairment (diminished sensory acuity, manual skills, judgement and response times). Consciousness is lost within minutes above 6000 m, or within seconds at higher altitudes.

Respiratory changes

The minute volume of ventilation increases at altitudes in proportion to the reduction in barometric pressure that decreases the alveolar and, hence, arterial, P_{O_2}. The reduced arterial P_{O_2} stimulates the peripheral chemoreceptors of the carotid and aortic bodies. Hyperventilation commences at altitudes above 3000 m and increases progressively with increasing altitude to reach a maximum at 6000 m, where minute ventilation is approximately 160 per cent of that at sea level. This leads to respiratory alkalosis and alkalaemia (a rise in arterial pH). The decrease in arterial P_{CO_2} also reduces cerebrospinal fluid (CSF) P_{CO_2} and produces a rise in CSF pH. These changes act to oppose the increased respiratory drive by altitude. However, if the individual remains at altitude, the minute ventilation continues to rise. The time taken for this secondary hyperventilation to reach a maximum increases with increasing altitude. At 3000–4000 m, it takes about 1–2 weeks, persists as long as the individual remains at altitude, and continues for some time after returning to sea level.

The mechanisms responsible for the continued increase in respiratory drive are:

- The pH of the CSF bathing the central chemo-receptors is restored to normal by choroid plexus active transport of bicarbonate ions out of the CSF to the blood.
- Renal excretion of the excess bicarbonate in blood to restore arterial pH toward normal.
- The respiratory centre is reset to function at a lower arterial P_{CO_2}, so that the carbon dioxide response curve is shifted to the left with an increased slope, and the apnoeic threshold is decreased.

The increased ventilation seen at altitude increases alveolar P_{O_2} to a limited extent only, and there is no evidence to suggest significant changes in pulmonary diffusing capacity in an individual ascending rapidly to an altitude. This leads to an increased alveolar–arterial oxygen tension gradient and a fall in arterial P_{O_2}, especially during exercise.

Haematological changes

Polycythaemia occurs in a person exposed to altitude. There is a linear increase in the polycythaemic response with increasing altitude up to 3700 m, and after that the rise is more rapid. Erythropoietin production is increased because the decreased arterial P_{O_2} is sensed by the macula densa of the kidneys. Plasma erythropoietin increases within the first 2 h of exposure to altitude and reaches a maximum within a day. There is a lag between erythropoietin secretion and red cell production so that the red cell count rises after 3–5 days of exposure to altitude. The haemoglobin concentration rises and increases the oxygen-carrying capacity of blood. As the haematocrit increases, so does blood viscosity, and this tends to

decrease blood flow and increase cardiac work. The oxyhaemoglobin dissociation curve is shifted to the right within 12 h of altitude exposure by an increased concentration of erythrocyte 2,3-diphosphoglycerate that is induced by respiratory alkalosis.

Cardiovascular changes

Within the first few hours of altitude exposure there is an initial increase in heart rate as a result of a generalized increase in sympathetic activity. Stroke volume may be unchanged or slightly decreased initially. After a few days, stroke volume is decreased and the heart rate remains elevated so that the cardiac output is equal to that at sea level. However, the maximal cardiac output that can be achieved during exercise at altitudes is reduced, mainly because of a decreased maximal stroke volume. An increase in coronary blood flow occurs initially, due to the vasodilator action of hypoxia at altitude. There is also a redistribution of cardiac output associated with a decrease in cutaneous and splanchnic blood flow, and an increased perfusion of the vital organs.

Endocrine changes

Within a week, the hypoxia associated with altitude induces the secretion of catecholamines, glucocorticoids, thyroid hormones and antidiuretic hormone. Fluid retention occurs initially, but with acclimatization after about a week the increased red cell mass and blood volume decrease secretion of aldosterone and antidiuretic hormone.

Acute mountain sickness

The symptoms associated with a rapid ascent to altitudes above 2500 m are referred to as 'acute mountain sickness'. The first symptom of acute mountain sickness in a mountain climber is exertional dyspnoea at about 2000 m above sea level. Above this height, headache, nausea, insomnia and muscle fatigue occur. Cheyne–Stokes breathing and other types of sleep apnoea occur at 4000 m above sea level, with dizziness, amnesia and feelings of unreality developing at about 5000 m.

These symptoms usually develop within the first 8–24 h at altitude, but may be delayed for 3–7 days. The pathogenesis of acute mountain sickness is not clearly known, but the primary stimulus appears to be hypoxaemia. Increased secretion of adrenal glucocorticoids and antidiuretic hormone leads to fluid retention and accumulation in the lungs, splanchnic bed and brain. Acute mountain sickness can be prevented by limiting the rate of ascent to 300 m per day at altitudes between 3000 and 4200 m, and to 150 m per day above that. If symptoms occur, the first measures to be taken are descent to a lower altitude and oxygen therapy. Acetazolamide, a carbonic anhydrase inhibitor, can be used to prevent acute mountain sickness by correcting respiratory alkalosis, and it enhances acclimatization more rapidly. Dexamethasone may be used to reduce cerebral oedema.

Malignant forms of acute mountain sickness present as high-altitude pulmonary or cerebral oedema. The symptoms of high-altitude pulmonary oedema are marked dyspnoea, dry cough, followed by productive cough with pink foamy sputum. The pathogenesis of this syndrome is unclear, but it is probably multifactorial. Hypoxic pulmonary vasoconstriction of the pulmonary arterioles and arteries leading to pulmonary hypertension may be an important factor. There is increased filtration of fluid across the pulmonary capillaries, aggravated by increased pulmonary capillary permeability and inflammatory changes. The treatment of this condition is oxygen therapy and descent to a lower altitude.

Cerebral oedema presents initially as ataxia, irritability and irrational behaviour, and progresses to drowsiness and coma. Papilloedema may be seen, and cerebral oedema with intracranial haemorrhage, and thrombosis may be found at post-mortem studies.

HIGH–ALTITUDE RESIDENTS

A person who is permanently resident at high altitude may show a number of physiological adaptations. The high-altitude resident (HAR) does hyperventilate, but at a rate 20 per cent less than an altitude-acclimatized person. This lower rate of hyperventilation is an adaptive response that reduces energy consumption. The HAR has a blunted response to hypoxia, with little change in ventilation over a range of low inspired oxygen concentrations, although a very low alveolar P_{O_2} still induces hyperventilation. In the HAR, the blunted response to hypoxia is associated with hypertrophy and biochemical changes in the carotid bodies. The pulmonary diffusing capacity at rest in a HAR is 20–30 per cent greater than a normal person, largely because of an increased alveolar surface area and pulmonary blood volume. The HAR has a resting cardiac output equal to that of a person at sea level, and there is an increased vascularization of the myocardium. The main cardiovascular difference in a HAR is that the maximal cardiac output during exercise is similar to that of individuals of similar physical ability at sea level, unlike an altitude-acclimatized 'normal' individual. The P_{50} of the oxyhaemoglobin curve of a HAR is about 30.7 mmHg (4.1 kPa), indicating a rightward shift that enhances tissue oxygen delivery. This is caused by an increase in the 2,3-diphosphoglycerate concentration within erythrocytes. The maximal rate of oxygen consumption in a HAR is increased with increased work capacity. HAR may have pulmonary hypertension which becomes more marked during exercise, and is probably due to hypoxic pulmonary vasoconstriction. The HAR also has a larger than normal lung volume and vital capacity due to the stimulation of lung growth by hypoxia. The plasma concentrations of glucocorticoids and thyroid hormones are increased in the HAR.

Chronic mountain sickness

A small number of HARs, especially middle-aged men, develop chronic mountain sickness characterized by an impaired response to hypoxia, cyanosis, pulmonary hypertension, dyspnoea and lethargy, and they have a low arterial P_{O_2} and an elevated P_{CO_2}.

PHYSIOLOGY OF SPACE TRAVEL

GRAVITATIONAL FORCES

Changes in gravitational forces occur when the motion of an individual is being accelerated or retarded. From Newton's laws of motion, any movement causes a force to be exerted upon the body. These gravitational forces (G-forces) can cause pooling of blood in dependent areas, and various corrective reflexes are stimulated. When considering alterations in blood flow associated with changes in G-forces, it must be noted that in a column of fluid there is a difference in pressure between two points, given by the equation:

$$P = h \times d \times G$$

where P is the pressure in cmH_2O, h is the distance (cm) measured in the direction of the acceleration force, d is the density of the fluid (g/cm^3) and G is the acceleration force expressed as a multiple of the normal gravitational force.

It is evident from this equation that the pressure gradient varies in magnitude and direction depending on the orientation of the body in relation to the direction of acceleration, and the size of the acceleration force acting on the body.

INCREASED G-FORCES

When a spaceman is launched into space, the Earth's gravitational force acts vertically, and the direction of the imposed centripetal force depends on the spaceman's posture.

Effect on the central nervous system

When a subject is accelerated in a headwards direction with a force of 4 G, impaired visual perception of detail and colour (known as 'grey-out') occurs. At an acceleration of 4.5–5 G, vision is lost ('blackout'). These symptoms are caused by inadequate blood flow to the eyes, as a decrease in the intraluminal pressure and increase in vessel resistance diminish retinal blood supply. At an acceleration above 5 G, unconsciousness occurs because of reduced blood flow to the brain. Cerebral function is less susceptible than sight to the effects of acceleration, as cerebral perfusion is better preserved because of autoregulation. The pressure in the cerebrospinal fluid changes by an equal amount to that in the cerebral blood vessels, so there is little change in transmural pressure and hence vessel calibre. After about 5–10 s of acceleration, baroreceptor

reflexes mediated by the carotid sinuses restore blood flow to the brain and the eyes. Indeed, if the acceleration forces are increased gradually, these reflexes can maintain cerebral and retinal blood flows up to an acceleration of 6 G.

Effect on the cardiovascular system

As a result of the G-forces there is an increased pooling of blood in the dependent parts of the body, a decrease in venous return and, consequently, cardiac output falls. Arterial blood pressure falls and baroreceptor reflexes are induced.

Effect on the respiratory system

With headwards acceleration, the diaphragm is pulled downward; this facilitates inspiration, but hinders expiration. The gradients of blood flow and ventilation, which increase from the apex to the base in the lungs, are accentuated. The apical alveoli are fully distended, poorly ventilated, and perfusion reaches zero with abnormally high ventilation/perfusion ratios. In contrast, the basal alveoli receive a large blood flow, but as alveolar ventilation increases to a lesser amount there is a lowered ventilation/perfusion ratio. These changes increase the alveolar dead space in the apex of the lungs and increase venous shunts, resulting in hypoxaemia.

Preventive measures

The pressure changes described above in the body can be minimized if the subject lies supine. Active tensing of the leg and thigh muscles can increase venous return. Anti-gravity suits can be useful, as they prevent venous pooling and limit the descent of the diaphragm. Goggles can be used to promote retinal blood by reducing extraocular pressure, and these diminish the eye effects of G-forces by maintaining transmural pressure.

PHYSIOLOGICAL EFFECTS OF WEIGHTLESSNESS

Various studies have revealed a number of physiological changes caused by zero gravity, or weightlessness. There is a redistribution of about 2 L of body fluids from the dependent parts of the body to the thorax, neck and head. This is because, under zero gravity, there is no hydrostatic effect on the veins in the dependent limbs, and pooling of blood decreases. The neck veins are engorged and the head may be bloated. The cardiac output increases at the onset of weightlessness. Antidiuretic hormone production is decreased, and atrial natriuretic peptide increased by the increased venous pressure stretching the atrial and venular volume receptors. Consequently, diuresis occurs and this reduces the increased intrathoracic blood volume. The cardiac output and heart rate are reduced by reflex decreases in sympathetic activity and increased parasympathetic tone. Myocardial workload decreases, and some ventricular atrophy occurs as a result of this 'deconditioning'. For an unknown reason bone marrow activity ceases and red and white cell counts fall.

Patients who are bed-ridden for extended periods of time experience some of the problems of weightlessness. Even though they are subjected to gravity, it acts across the transverse width of the body and no longer along the length of the body. The bed-ridden patients are therefore not subjected to the normal venous pooling in dependent limbs that occurs in healthy individuals when standing up.

During the first few days in space, cosmonauts experience giddiness, disorientation, nausea and vomiting, probably due to an imbalance of neural inputs to the brain. There are considerable alterations in inputs from skin, muscle and joint receptors that signal postural forces acting on the body and from the otolith receptors that sense the orientation of the head in the gravitational field.

The tonic activity of the antigravity extensor muscles is reduced, and this predisposes to muscle atrophy. The individual develops fatigue, muscle weakness and increased urinary urea excretion. There is also demineralization of bones, especially those that bear the weight of the body, increasing the danger of fractures when the cosmonaut returns to Earth. When the cosmonaut does return to Earth, blood is pooled in the dependent parts of the body. As there has been an approximately 10 per cent decrease in the blood volume and reduced tissue tone, the cosmonaut will experience difficulties standing. In response, the cosmonaut drinks more, his or her kidneys conserve sodium, and reticulocytosis occurs. The cardiac and skeletal muscles recover more slowly over a period of 5 months, but recalcification of bone occurs slowly and osteoporosis can persist forever.

Preventive measures

The deconditioning and unpleasant side effects of life in space can be reduced by several measures. Restricting head movement using head caps can reduce nausea. Negative pressure applied to the legs by spacesuits increases venous pooling in space. Isometric muscle contractions and specialized exercise programmes and equipment may reduce muscle atrophy. Just before return to Earth, extra measures such as deliberate body fluid volume expansion by drinking saline can minimize the problems experienced upon return from space.

Reflections

1. The diver is exposed to a major rise in the ambient pressure of the environment, with physiological changes and problems from the direct effects of pressure on the body. As the diver descends into the water the pressure increases and gases are compressed to smaller volumes. At elevated ambient pressures the quantity of dissolved gases is directly proportional to the pressure. Rapid decompression can lead to bubble formation and consequent tissue damage. If the diver is breathing air, the quantity of nitrogen dissolved in the tissues increases and nitrogen narcosis can occur. Early symptoms of nitrogen narcosis occur at about 50 m below sea level and beyond 100 m the diver may be become unconscious. In addition a large

amount of the total oxygen is dissolved rather than being bound to haemoglobin. Breathing 100 per cent O_2 at pressures greater than 1 atmosphere leads to acute oxygen toxicity. Exposure to oxygen at 4 atmospheres causes seizures followed by coma in most people after 30 min, caused by large amounts of oxygen free radicals which damage lipid membranes and some cellular enzymes. During breath-holding diving, a diving reflex is initiated and this results in reflex slowing of the heart, profound peripheral vasoconstriction, and apnoea. The duration of the dive is limited by the breakpoint which is determined largely by the arterial P_{CO_2}.

2. A decrease in barometric pressure is the basic cause of high-altitude hypoxia. The fall in inspired P_{O_2} leads to acute hypoxia which leads to an increase in ventilation and an increase in cardiac output. The increased ventilation leads to respiratory alkalosis. These physiological changes are associated with symptoms of mountain sickness. Following exposure to chronic hypoxia, a person becomes acclimatized to the low Pa_{O_2}. Acclimatization makes it possible for the person to work without hypoxic effects or ascend to still higher altitudes. There is a maintained increase in ventilation, an increased vascularization of the tissues, and an increased oxygen carrying capacity of blood. An acute exposure to a hypoxic environment increases alveolar ventilation by about 65 per cent as the hypocapnia and respiratory alkalosis inhibit the respiratory centre and oppose the effects of low Pa_{O_2} to stimulate the peripheral respiratory chemoreceptors. This acute inhibition diminishes within 2–3 days allowing the respiratory centre to respond fully to hypoxia, enabling the respiratory centre to respond fully, and ventilation increases by about five-fold. The decreased inhibition results mainly from a reduction of bicarbonate ion concentration in the CSF and brain tissues which therefore decreases the pH in the interstitial fluid around the central chemoreceptor, thereby increasing the activity of the respiratory centre. A person who remains at a high altitude for a very long period can develop chronic mountain sickness. The features of chronic mountain sickness include an increase haematocrit, high pulmonary arterial pressure, and biventricular cardiac failure and death may occur.

3. The physiological problems associated with weightlessness are related to the translocation of fluids in the body (because of the absence of gravity to exert hydrostatic pressures) and diminished physical activity. The physiological consequences of prolonged periods in space include decreased blood volume, decreased red cell mass, decreased muscle strength and work capacity, decreased maximum cardiac output and the loss of calcium and phosphate from bones and loss of bone mass.

Key equations and tables

THE NERNST EQUATION

This calculates the potential difference that any ion would produce if the membrane was permeable to it. The actual potential will only be similar to the calculated Nernst potential if the membrane is permeable to that ion. At rest, the calculated Nernst potentials for potassium and chloride are similar to the real potential, as these ions diffuse across the membrane with ease. This is not true for sodium as the resting membrane is relatively impermeable to this ion.

$$E = \frac{RT}{zF} \ln \frac{[ion]_o}{[ion]_i}$$

E = the equilibrium potential for a specific ion (inside of the cell with respect to the outside)

R = gas constant ($8.314\,J\,deg^{-1}\,mol^{-1}$)

T = absolute temperature (degrees Kelvin = 273 + degrees centigrade)

F = Faraday's constant (96 500 coulombs mol^{-1})

z = ionic valency (+1 for K^+, Na^+; −1 for Cl^-)

ln = logarithm to base e

o = ionic concentration outside the cell

i = ionic concentration inside the cell

The equation can be simplified:

$$E = 58 \log_{10} \frac{[ion]_o}{[ion]_i}\,mV$$

(e.g. for potassium

$E_K = 58 \log_{10} \frac{[5]}{[150]}\,mV$ [approximate concentrations]

$= 58 \times -1.48$

$= -86\,mV$

The Nernst equation

The Nernst potential of potassium, chloride and sodium ions

Ion	Ionic concentration (mM)		≈Nernst potential (mV)
	Intracellular	Extracellular	
K^+	150	5	−90
Cl^-	10	125	−70
Na^+	15	150	+60

MEMBRANE PERMEABILITY

$$\text{Membrane permeability } (E_m) = \\ 58 \log_{10} \frac{P_k[K^+]_o + P_{Na}[Na^+]_o + P_{Cl}[Cl^-]_i}{P_K[K^+]_i + P_{Na}[Na^+]_i + P_{Cl}[Cl^-]_o} \, mV$$

The Goldman–Hodgkin–Katz form of the Nernst equation

The membrane permeability is the central factor influencing the membrane potential and the relative membrane permeability to different ions is important. To take account of this, the Nernst equation was expanded to the Goldman–Hodgkin–Katz form. P is the permeability to each ion; if this changes, then the membrane permeability changes.

COMPOUND ACTION POTENTIALS

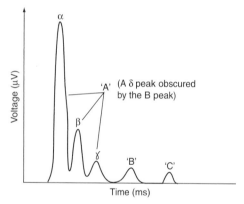

Graphical and tabular representation of compound action potential:

Fibre	Function	Diameter (μm)	Conduction velocity (m/s)
A			
α	Skeletal motor, joint position	10–20	60–120
β	Touch, pressure	5–10	40–70
γ	Muscle spindle motor	3–6	15–30
δ	Pain, temperature touch	2–5	10–30
B	Preganglionic autonomic	1–3	3–15
C	Pain	0.5–1	0.5–2

Alternative classification for sensory fibres in the nerves of mammalian muscles

Group	Sensory ending	Diameter (μm)	Conduction velocity (m/s)
Ia	Muscle spindle, primary ending	12–20	72–120
Ib	Golgi tendon organ	12–20	72–120
II	Muscle spindle, secondary ending	4–12	24–72
III	Pressure/pain receptors	1–4	6–24
IV	Pain	0.5–1	0.5–2

Peripheral nerves contain a mixture of fibres, and these have been classified according to function, diameter and conduction velocity. Monophasic extracellular recording reveals a compound potential composed of A, B and C peaks. The compound action potential is not 'all or none'; the various components have different threshold intensities because they represent simultaneous activity in fibres of different diameter and conduction velocity.

An alternative classification exists for sensory fibres in the nerves of mammalian muscles.

CALCIUM CHANNELS

Calcium channels: blocker sensitivity and threshold potentials

	T	N	L
Threshold potential (mV)	−70	−10	−10
Cadmium block	+	+++	+++
Conus toxin block	+	+++	+++
Dihydropyridine sensitivity	−	−	+++

In calcium channels T (transient channels), N (inactivated at very negative potentials, neuronal) and L (long-lasting currents) types of calcium channels differ in their sensitivity to blockers and threshold potentials.

CATECHOLAMINES

Ahlquist classified adrenergic receptors by comparing the tissue effects of isoprenaline with those

Catecholamines: agonists and antagonists

Type	Agonist	Antagonist
α_1	Phenylephrine	Prazosin
α_2	Clonidine	Yohimbin
β_1	Dobutamine	Practolol
β_2	Salbutamol	Butoxamine

of norepinephrine and epinephrine, describing β and α receptors. β and α receptors have been further classified by their response to agonists and antagonists.

MUSCLE SPINDLES, GOLGI TENDON ORGANS AND SPINAL REFLEXES

Response of muscle spindle and tendon organ afferents to active skeletal muscle contraction or passive stretching

	Spindle afferents	Tendon organ	
α motor activity	↓	↑	– shortens spindle
γ motor activity	↑	–	– reflex ↑ α activates motor activity
Passive muscle stretching	↑	↑	– 'knee jerk' reflex

Muscle spindles detect muscle length and movement, whilst tendon organs sense tension. Muscle spindles are capsules of specialized fibres arranged in parallel with the muscle. Tendon organs lie in series with the muscle. Muscle spindle and tendon organ afferents respond differently to active skeletal muscle contraction or passive muscle stretching, permitting close monitoring of movement.

CLASSIFICATION OF SENSORY RECEPTORS

Receptors can be classified by which stimulus they react to, and whether this originates from inside or outside the body.

Classification of sensory receptors

	Stimulus origin		
	Inside	Outside	
		Contact	Distance
Mechanoreceptors	Muscle length, tension Joint movement Arterial blood pressure	Touch	Hearing
Photoreceptors	–	–	Sight
Chemoreceptors	[H$^+$] body fluids	Taste	Smell
Thermoreceptors	Hypothalamic temperature receptors	Cutaneous temperature receptors	–

ARTERIAL AND VENOUS BLOOD OXYGEN AND CARBON DIOXIDE

Partial pressures and oxygen and carbon dioxide content in arterial and venous blood

	P_{O_2} (mmHg /kPa)	O_2 content (mL/100 mL blood)	P_{CO_2} (mmHg/ kPa)	CO_2 content
Arterial blood	100/13.3	20	40/5.3	48
Mixed venous blood	40/5.3	15	46/6.1	52

The partial pressures and contents of oxygen and carbon dioxide in arterial and venous blood are indicated in this table. The normal oxygen consumption per minute is 250 mL; the total amount of oxygen in the body is only approximately 1.5 L, and less than half is immediately available for use. Carbon dioxide production is 200 mL/min, and the total body content amounts to 120 L.

OXYGEN–CARRYING CAPACITY OF THE BLOOD AND OXYGEN DELIVERY

The oxygen-carrying capacity of blood is dependent upon the P_{O_2} and the haemoglobin concentration

Oxygen delivery to tissues

$$= \left[\begin{pmatrix} Hb\,(g/100\,mL) \times 1.306 \times \%Sa_{O_2} \end{pmatrix} \\ + \begin{pmatrix} 0.003 \times Pa_{O_2}\,mmHg \end{pmatrix} \times 10 \right]$$

$\times$ cardiac output (L/min)

= 1000 mL/min (under normal conditions)

Calculating oxygen delivery

of blood. The oxygen delivered to the tissues per minute by the cardiovascular system is the product of the oxygen content in arterial blood and the cardiac output, and is normally 1000 mL oxygen per minute.

THE LAW OF LAPLACE

Bubble

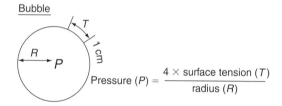

Pressure $(P) = \dfrac{4 \times \text{surface tension } (T)}{\text{radius } (R)}$

Fluid-lined alveolus

$$P = \frac{2T}{R}$$

The law of Laplace.

Surface tension develops at air–water interfaces where the forces of attraction between water molecules are much greater than those between water and gas molecules. The surface tension is the force acting along a 1-cm line in the surface of the liquid. Pressure is produced inside a bubble because of the surface tension. The pressure is determined by the surface tension and the radius of the bubble, according to the law of Laplace.

THE POISEUILLE EQUATION

The pressure required to produce laminar flow is related linearly to the flow rate, and can be calculated from the Poiseuille equation. With laminar flow, the resistance to the flow is inversely related to the fourth power of the radius.

$$\text{Flow rate} = \frac{P\pi r^4}{8\,\eta L}$$

$P = \Delta P$ (driving pressure)
r = radius of tube
η = viscosity
L = length of tube

The Poiseuille equation

$$R = \frac{8\eta L}{\pi r^4}$$

Calculation of the resistance to laminar flow

REYNOLDS NUMBER

$$Re = \frac{\rho D \upsilon}{\eta}$$

D = diameter of tube
υ = velocity of flow
ρ = density of gas
Re = Reynolds number
η = viscosity of gas

Calculation of the Reynolds number

The factors determining whether flow will be laminar or turbulent include the velocity of flow, the radius of the tube, and the density and viscosity of the gas. In smooth tubes, flow is likely to be turbulent when the Reynolds number exceeds 2000.

FICK'S LAW OF DIFFUSION

$$\text{Flow of gas} \propto \frac{A}{T} \cdot D\left(P_1 - P_2\right)$$

where:
A = lung area (50–100 m^2)
D = diffusion constant of the gas
T = lung thickness (0.3 μm)
$(P_1 - P_2)$ = partial pressure gradient
across the membrane
and:

$$D_{gas} \propto \frac{\text{solubility}}{\sqrt{\text{mol. wt.}}}$$

(D_{CO_2} is approximately equal to 20 × D_{O_2})

The flow of oxygen and carbon dioxide across the blood–gas barrier of the alveolar wall, interstitial fluid and pulmonary capillary endothelium is governed by Fick's law of diffusion. This relates the flow of gas across a membrane to the area and thickness of the membrane, the partial pressure difference of the gas across the membrane, and the diffusion constant (D) of the individual gases.

THE BOHR EQUATION FOR PHYSIOLOGICAL DEAD SPACE

Volume of CO_2 eliminated from 'ideal alveolar' gas = volume of CO_2 in mixed expired gas

= %CO_2 alveolar gas × alveolar ventilation
= %CO_2 mixed expired gas × minute volume of ventilation

For one breath:
%CO_2 ideal alveolar gas × $(V_T - V_D)$
= %CO_2 mixed expired gas × V_T

Now:
V_D = physiological dead space
V_T = tidal volume
%CO_2 mixed expired gas ($P\bar{E}_{CO_2}$)
%CO_2 ideal alveolar gas ≡ Pa_{CO_2} (arterial P_{CO_2})

Therefore:

$$\frac{V_D}{V_T} = \frac{\left(Pa_{CO_2} - P\bar{E}_{CO_2}\right)}{Pa_{CO_2}}$$

(the ratio of physiological dead space to the tidal volume can be estimated by measurement of the P_{CO_2} in arterial blood and mixed expired gas)

The basis of the Bohr equation is that only ideal alveolar gas contains carbon dioxide (i.e. the alveolar and anatomical dead spaces do not contain carbon dioxide), and that the carbon dioxide content of mixed expired gas must be equal to the product of alveolar ventilation and the concentration of carbon dioxide in ideal alveolar gas.

THE CARBON DIOXIDE CONTENT OF ALVEOLAR GAS

$$\% \text{ Alveolar } CO_2 = \frac{CO_2 \text{ output}}{\text{alveolar ventilation}}$$
$$= \frac{200\,\text{mL/min}}{4000\,\text{mL/min}} = 5\%$$

The alveolar carbon dioxide concentration is therefore determined by the ratio of the rate of carbon dioxide output divided by the rate of alveolar ventilation.

Alveolar P_{CO_2} $\left(PA_{CO_2}\right)$
= barometric pressure (dry) (mmHg)
$$\times \left[FI_{CO_2} + \frac{CO_2 \text{ output}}{\text{alveolar ventilation}}\right]$$
(normally ≃ 0)

$= PA_{CO_2} \simeq 35$–$40\,\text{mmHg}\,(4.7$–$5.3\,\text{kPa})$

The partial pressure of carbon dioxide in alveolar gas (PA_{CO_2})can be calculated if the atmospheric pressure, carbon dioxide output, alveolar ventilation and inspired carbon dioxide concentration are known.

THE OXYGEN CONTENT OF ALVEOLAR GAS AND THE ALVEOLAR AIR EQUATION

Alveolar P_{O_2}(mmHg) ≃
$$\text{inspired } P_{O_2} - \frac{\text{arterial } P_{CO_2}}{\text{respiratory quotient}}$$

Alveolar air equation for PA_{O_2} (simple form)

The partial pressure of oxygen of ideal alveolar gas must be calculated by the alveolar air equation. The basis for this equation is that the effects of shunt and ventilation-perfusion mismatch do not produce significant differences between alveolar and arterial P_{CO_2}. In addition, the respiratory quotient must be taken into account. A useful simple form of the alveolar air equation only requires knowledge of the inspired partial pressures of oxygen and carbon dioxide and the respiratory quotient (assumed to be 0.8).

THE SHUNT EQUATION

$$\begin{pmatrix} \text{Cardiac output} \\ O_2 \text{ content} \end{pmatrix} = \begin{pmatrix} \text{shunt} \\ O_2 \text{ content} \end{pmatrix} +$$

$$\begin{pmatrix} \text{pulmonary blood flow} \\ O_2 \text{ content} \end{pmatrix}$$

Now:

Cardiac output = total blood flow = $\dot{Q}t$
Shunt flow = $\dot{Q}s$
Pulmonary flow = $\dot{Q}t - \dot{Q}s$

and:

$\left. \begin{array}{l} Ca_{O_2} = \text{arterial } O_2 \text{ content} \\ C\overline{v}_{O_2} = \text{mixed venous } O_2 \\ \qquad \text{content} \end{array} \right\}$ measured from blood samples

Cc'_{O_2} = pulmonary end-capillary O_2 content (obtained from the ideal alveolar P_{O_2}, from the aveolar gas equation and oxygen–haemoglobin dissociation curve)

then:

$$\dot{Q}t \times Ca_{O_2} = (\dot{Q}s \times C\overline{v}_{O_2}) + (\dot{Q}t - \dot{Q}s) Cc'_{O_2}$$
$$\Rightarrow \frac{\dot{Q}s}{\dot{Q}t} = \frac{Cc'_{O_2} - Ca_{O_2}}{Cc'_{O_2} - C\overline{v}_{O_2}}$$

The ratio of the shunt flow to the total blood flow from the left ventricle (the cardiac output) can be calculated by considering the different oxygen contents of shunted, pulmonary end-capillary, and arterial blood. The basis for the shunt equation is that the oxygen carried by the cardiac output (arterial blood) must be the sum of the oxygen carried by the pulmonary end-capillary blood flow and the oxygen in the shunt blood flow. The assumption made is that the shunted blood has the same oxygen content as mixed venous blood.

CARDIOVASCULAR PHYSIOLOGY

THE STARLING FORCES

The capillary wall is a semipermeable membrane as it is permeable to water and solutes, but impermeable to large proteins (including albumin). A plasma ultrafiltrate is filtered by bulk flow through the capillary wall by the action of opposing hydrostatic and oncotic forces. Four 'Starling forces' are involved in this filtration process (see illustration on p. 445). The net filtration pressure (NFP) is the balance of the opposing capillary and hydrostatic pressures minus the balance of the opposing plasma and interstitial oncotic pressures.

LIVER PHYSIOLOGY

MEASUREMENT OF HEPATIC BLOOD FLOW USING CLEARANCE TECHNIQUES

Single bolus technique

A single bolus of ICG (0.5 mg/kg) is injected intravenously, and venous blood samples are collected every 2 min for 14 min. The concentration–time delay curves are analysed by non-linear regression analysis. Clearance is calculated from the formula:

$$\text{Clearance} = \frac{\text{dose}}{\text{area under concentration versus time curve}}$$

As the extraction ratio of ICG is 0.74, hepatic blood flow is calculated using the formula:

$$\text{Hepatic blood flow} = \frac{\text{clearance}}{\text{extraction ratio}}$$

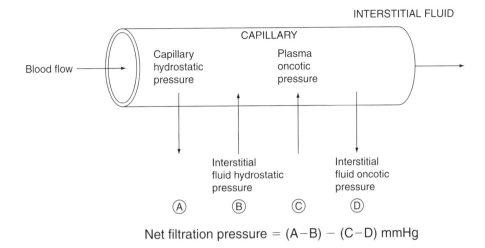

Net filtration pressure = (A−B) − (C−D) mmHg

The Starling forces.

This technique assumes that there is adequate mixing of the dye in blood, and exclusive hepatic extraction.

Continuous infusion technique

After a loading dose of ICG (0.5 mg/kg body weight of subject), a constant infusion of indocyanine green (ICG) is administered for 20 min to achieve equilibration. Samples are taken simultaneously from a peripheral artery and the hepatic vein. Hepatic blood flow is calculated using the formula:

$$\text{Hepatic blood flow} = \frac{\text{clearance}}{\text{extraction ratio}}$$

where

$$\text{Clearance} = \frac{\text{infusion rate}}{\text{art. conc. ICG}}$$

and

$$\text{Extraction ratio} = \frac{(\text{art. conc. ICG} - \text{venous conc. ICG})}{\text{arterial conc. ICG}}$$

where conc. ICG is the concentration of ICG.

Liver blood flow can be estimated by the clearance of markers such as ICG, which is eliminated by the liver without any circulation. Clearance can be estimated by either a single bolus or a continuous infusion technique. Both these clearance techniques assume that ICG is extracted exclusively by the liver and that hepatic venous samples reflect liver efflux.

RENAL PHYSIOLOGY

THE NET FILTRATION PRESSURE

NFP (mmHg) =
  − glomerular capillary hydrostatic pressure
  − Bowman's capsule hydrostatic pressure
    glomerular capillary oncotic pressure

The net filtration pressure (NFP) is a balance of the capillary hydrostatic pressure moving fluid out of the capillary, the plasma oncotic pressure in the capillary, which tends to retain fluid in the vessel, and the hydrostatic pressure in the Bowman's capsule, which tends to oppose fluid movement out of the capillary. Filtration pressure drops from the beginning to the end of the glomerular capillary as the hydrostatic pressure falls due to vessel resistance, and oncotic presure rises as protein-free fluid is filtered off into the Bowman's capsule. The average NFP acting across the surface of the glomerular capillaries is thought to be 17 mmHg.

RENAL CLEARANCE

Renal clearance =
  amount of substance in urine per unit time
  ──
  plasma concentration of substance (P)

The amount of a substance in urine over a given time is the volume of urine produced in that time (V) multiplied by the urinary concentration of the substance (U). Thus, for any substance:

$$\text{Renal clearance} = \frac{U \times V}{P}$$
(plasma volume/unit time; mL/min, L/day)

For the substance, the renal clearance is the volume of plasma completely cleared of the substance by the kidneys per unit time. The units of renal clearance are therefore volume of plasma over time.

PARA-AMINOHIPPURIC ACID (PAH)

Effective renal blood flow =

  effective renal plasma flow
  ──────────────────────────
  1 − blood haematocrit

para-aminohippuric acid (PAH) clearance can be used to calculate effective renal plasma flow, from which the effective renal blood flow can be calculated.

THE HENDERSON–HASSELBALCH EQUATION

$$\begin{aligned}
pH &= pK_a + \log\frac{[HCO_3^-]}{[CO_2]} \\
&= 6.1 + \log\frac{24}{1.2} \\
&= 7.4
\end{aligned}$$

In aqueous solution, carbon dioxide behaves as an acid, and reacts with water to release hydrogen ions. It also releases bicarbonate ions, the corresponding buffer base. The Henderson–Hasselbalch equation can be used to calculate the normal plasma pH from the plasma carbon dioxide concentration, the bicarbonate concentration and the pK_a of the carbon dioxide–bicarbonate buffer system.

DEFINITIONS OF AN ACID AND A BASE

$$HA \underset{k_2}{\overset{k_1}{\rightleftharpoons}} H^+ + A^-$$

An *acid* is a substance that donates a proton (a hydrogen ion without its orbital electron), and a *base* is a substance that accepts protons in solution.

In solution, an acid will dissociate to a hydrogen ion and a base as shown in the equation above. The proportion of the relative reactions are determined by the dissociation constants k_1 an k_2.

$$[H^+] = K\frac{[HA]}{A^-} \quad \text{where } K = \frac{k_1}{k_2}$$

Henderson applied the law of mass action and described the equation given above. Hasselbalch modified the Henderson equation using logarithmic transformation, which resulted in the equation:

$$pH = pK_a + \log\frac{[A^-]}{[HA]}$$

THE pH SYSTEM

Hydrogen ion concentration my be measured directly (nmol/L) or indirectly as pH. pH is defined as the negative logarithm (to the base 10) of the concentration of hydrogen ions. The pH is related to the concentration of hydrogen ions as follows:

(a) $pH = \log_{10}\dfrac{1}{[H^+]}$

(b) $pH = -\log_{10}[H^+]$

(c) $H^+ = 10^{-pH}$

(d) $pH = pK + \log\dfrac{\text{base}}{\text{acid}}$

BUFFER SYSTEMS: BICARBONATE–CARBONIC ACID

The Hasselbalch equation describes acid–base relationships as:

$$pH = pK + \log\left(\frac{\text{kidneys}}{\text{lungs}}\right)$$

or

$$pH = pK + \log\frac{[HCO_3^-]}{[sP_{CO_2}]}$$

For the bicarbonate–carbonic acid system, pK is the dissociation constant of carbonic acid, s is the solubility coefficient of CO_2 in plasma (0.03 mmol/L per mmHg at 37°C). Therefore, pH = 6.1 + log (24/1.2) = 7.4.

PHYSIOLOGY OF AGEING

CHANGES IN THE RESTING ARTERIAL OXYGEN TENSION WITH AGE

$$
\begin{aligned}
Pa_{O_2} &= 100 - (0.33 \times \text{age [years]}) \text{ mmHg} \\
&= 13.6 - 0.044 \times \text{age (years) kPa}
\end{aligned}
$$

The resting arterial oxygen tension decreases with age at a rate described by this equation above.

SPECIAL ENVIRONMENTS

GRAVITATIONAL FORCES

$$P = h \times d \times G$$

Changes in gravitational forces occur when the motion of an individual is being accelerated or retarded. When considering alterations in blood flow associated with changes in gravitational forces, there is a difference in pressure between two points, given by the equation above, where P is the pressure (in cm water), h is the distance measure in the direction of the acceleration force, d is the density of the fluid, and G is the acceleration force expressed as a multiple of the normal gravitational force.

Further reading

GENERAL REFERENCES

Basic

Vander, A.J., Sherman, J.H., Luciano, D.S. 1977: *Human Physiology: The Mechanisms of Body Function*, 7th edn. New York: McGraw-Hill.

Intermediate

Ganong, W.F. 1999: *Review of Medical Physiology*, 19th edn. Stamford, Connecticut: Appleton & Lange.

Guyton, A.C., Hall, J.E. 2000: *Textbook of Medical Physiology*, 10th edn. Philadelphia: W.B. Saunders.

Advanced

Gregor, R., Windhorst, V. 1996: *Comprehensive Human Physiology: From Cellular Mechanisms to Integration*. Berlin: Springer-Verlag.

CHAPTER 1: PHYSIOLOGY OF EXCITABLE CELLS

Aidley, D.J. 1998: *The Physiology of Excitable Cells*, 4th edn. Cambridge: Cambridge University Press.

CHAPTER 2: PHYSIOLOGY OF THE NERVOUS SYSTEM

Cottrell, J.E., Smith, D.S. 2000: *Anesthesia and Neurosurgery*, 4th edn. St Louis: Mosby.

Kandel, E.R., Schwartz, J.H., Jessel, T.M. 2000: *Principles of Neural Science*, 4th edn. New York: MacGraw-Hill.

CHAPTER 3: RESPIRATORY PHYSIOLOGY

Lumb, A.B. 2000: *Nunn's Applied Respiratory Physiology*, 5th edn. Oxford: Butterworth-Heinemann.

West, J.B. 1999: *Respiratory Physiology: The Essentials*, 6th edn. Baltimore: William & Wilkins.

CHAPTER 4: CARDIOVASCULAR PHYSIOLOGY

Berne, R.M., Levy, M.N. 1997: *Cardiovascular Physiology*, 7th edn. St Louis: Mosby.

Dampney, R.M. 1994: Functional organization of central pathways regulating the cardiovascular system. *Physiological Reviews* 71(2), 323–64.

Foex, P., Leone, B.J. 1994: Pressure–volume loops: a dynamic approach to the assessment of ventricular function. *Journal of Cardiothoracic and Vascular Anesthesia* 8, 84–96.

Levick, J.R. 2000: *An Introduction to Cardiovascular Physiology*, 3rd edn. Oxford: Butterworth-Heinemann.

Thompson, I.R. 1984: Cardiovascular physiology: venous return. *Canadian Anesthetic Society Journal* 31(3), S31–7.

CHAPTER 5: GASTROINTESTINAL PHYSIOLOGY

Johnson, L.R., Christensen, J., Jacobson, G. (eds) 1994: *Physiology of the Gastrointestinal Tract*, 3rd edn. New York: Raven.

CHAPTER 6: LIVER PHYSIOLOGY

Sear, J.W. 1990: Hepatic physiology. *Current Anaesthesia and Critical Care*, 1, 196–203.

Sear, J.W. 1992: Anatomy and physiology of the liver. In: *Baillière's Clinical Anaesthesiology* Vol. 6, No. 4, *The Liver and Anaesthesia*. London: Baillière Tindall.

CHAPTER 7: RENAL PHYSIOLOGY

Lote, C.J. 1994: *Principles of Renal Physiology*, 3rd edn. London: Chapman & Hall.
Vander, A. 1995: *Renal Physiology*, 5th edn. New York: McGraw-Hill.

CHAPTER 8: ACID–BASE PHYSIOLOGY

Holmes, O. 1993: *Human Acid–Base Physiology. A Student Text*. London: Chapman & Hall Medical.
Kokko, J.P., Tannen, R.C. 1996: *Fluids and Electrolytes*, 3rd edn. New York: W.B. Saunders.

CHAPTER 9: PHYSIOLOGY OF BLOOD

Hoffman. R., Benz, E.J., Shattil, S.J., Furie, B., Cohen, H.J. 1999: *Hematology: Basic Principles and Practice*, 3rd edn. New York: Churchill Livingstone.

CHAPTER 10: PHYSIOLOGY OF THE IMMUNE SYSTEM

Stites, D.P., Terr, A.I. 1997: *Basic and Clinical Immunology*, 9th edn. Stamford, Connecticut: Appleton & Lange.

CHAPTER 11: ENDOCRINE PHYSIOLOGY

Goodman, H.M. 1994: *Basic Medical Endocrinology*, 2nd edn. New York: Raven.
Greenspan, F.S., Strewler, G.J. (eds) 1997: *Basic and Clinical Endocrinology*, 5th edn. Stamford, Connecticut: Appleton & Lange.
Hedge, G.A., Colby, H.D., Goodman, R.L. 1987: *Clinical Endocrine Physiology*. Philadelphia: W.B. Saunders.
Wilson, J.D., Foster, D.W. (eds) 1992: *Williams Textbook of Endocrinology*, 8th edn. Philadelphia: W.B. Saunders.

CHAPTER 12: METABOLISM, NUTRITION, EXERCISE AND TEMPERATURE REGULATION

Pellett, P.L. 1990: Food and energy requirements in humans. *American Journal of Clinical Nutrition* 51, 711–22.
Salway, J.G. 1994: *Metabolism at a Glance*. Oxford: Blackwell Science.
Schwartz, M.W., Seeley, R.J. 1997: Neuroendocrine responses to starvation and weight loss. *New England Journal of Medicine* 336, 1802–11.
Sessler, D.I. 1994: Temperature monitoring. In: Miller, R.D. (ed), *Anesthesia*, 4th edn. New York: Churchill Livingstone, 1363–82.
Wasserman, K. 1994: Coupling of external to cellular respiration during exercise: the wisdom of the body revisited. *American Journal of Physiology* 226, E19–39.

CHAPTER 13: PHYSIOLOGY OF PAIN

Siddall, P.J., Cousins, M.J. 1998: Introduction to pain mechanisms: implications for neural blockade. In: Cousins, M.J., Bridenbaugh, P.O. (eds), *Neural Blockade in Clinical Anesthesia and Management of Pain*, 3rd edn. Philadelphia: Lippincott-Raven.
Wall, P.D., Melzack, R. (eds) 1999: *Textbook of Pain*, 4th edn. New York: Churchill-Livingstone.
Yaksh, T.L. 1998: Physiological and pharmacological substrates on nociception and nerve injury. In: Cousins, M.J., Bridenbauch, P.O. (eds), *Neural Blockade in Clinical Anesthesia and Management of Pain*. Philadelphia: Lippincott-Raven Publishers.

CHAPTER 14: MATERNAL AND NEONATAL PHYSIOLOGY

Chestnut, D.H. 1994: *Obstetric Anesthesia: Principles and Practice*. Mosby Year Book.
Gregory, G.A. (ed.) 1994: *Pediatric Anesthesia*, 3rd edn. New York: Churchill Livingstone.
Norris, M.C. (ed.) 1999: *Obstetric Anesthesia*, 2nd edn. Philadelphia: Lippincott Williams & Wilkins.

CHAPTER 15: PHYSIOLOGY OF AGEING

Muravchick, S. 1997: Anesthesia for the geriatric patient. In: Barash, P.G., Cullen, B.F., Stoetling, R.K. (eds), *Clinical Anesthesia*, 3rd edn. Philadelphia: Lippincott-Raven.

CHAPTER 16: SPECIAL ENVIRONMENTS

Edmonds, C., Lowry, C., Pennefather, J. 1994: *Diving and Subaquatic Medicine*, 3rd edn. Oxford: Butterworth-Heinemenn.

Ward, M.P., Milledge, J.S., West, J.B. 1995: *High Altitude Medicine and Physiology*. London: Chapman Hall.

Various authors. 1987: Physiological adaptation of man in space. In: *Aviation, Space and Environmental Medicine* 58(Supplement), A1–276.

Index

Note: page numbers in *italic* refer to figures and those in **bold** to tables.